Handbook of Aquaculture Engineering

Handbook of Aquaculture Engineering

Edited by Olando Martin

SYRAWOOD
PUBLISHING HOUSE

New York

Published by Syrawood Publishing House,
750 Third Avenue, 9th Floor,
New York, NY 10017, USA
www.syrawoodpublishinghouse.com

Handbook of Aquaculture Engineering
Edited by Olando Martin

International Standard Book Number: 978-1-68286-670-2 (Hardback)

Cataloging-in-Publication Data

Handbook of aquaculture engineering / edited by Olando Martin.
 p. cm.
Includes bibliographical references and index.
ISBN 978-1-68286-670-2
1. Aquacultural engineering. 2. Agricultural engineering. I. Martin, Olando.
SH137 .H36 2019
639.8--dc23

TABLE OF CONTENTS

Permissions

List of Contributors

Index

PREFACE

In my initial years as a student, I used to run to the library at every possible instance to grab a book and learn something new. Books were my primary source of knowledge and I would not have come such a long way without all that I learnt from them. Thus, when I was approached to edit this book; I became understandably nostalgic. It was an absolute honor to be considered worthy of guiding the current generation as well as those to come. I put all my knowledge and hard work into making this book most beneficial for its readers.

Aquaculture engineering is a branch of engineering that aims to solve the challenges faced in aquaculture systems. It includes the study of sustainable farming of aquatic vertebrates, invertebrates and algae. This field is significant to the growth and expansion of aquaculture industry. It employs knowledge of mechanical, environmental and biological systems in a multidisciplinary manner. Some significant aspects of aquaculture engineering include aquaponics, wastewater treatment, recirculating aquaculture system, etc. This book contains some path-breaking studies in the field of aquaculture engineering. It also discusses the modern methodologies and their practical applications. It will help new researchers by foregrounding their knowledge in this subject. Scientists and students actively engaged in this area will find this book full of crucial and unexplored concepts.

I wish to thank my publisher for supporting me at every step. I would also like to thank all the authors who have contributed their researches in this book. I hope this book will be a valuable contribution to the progress of the field.

Editor

Assessment of Physico-chemical Water Quality of Bira Dam, Bati Wereda, Amhara Region, Ethiopia

Tessema A[1]*, Mohammed A[1], Birhanu T[2] and Negu T[3]

[1]Department of Biology, Wollo University, Dessie, Ethiopia
[2]Tehulederie Wereda Office of Water Resource Development, Hayq, Ethiopia
[3]Kemissie Zonal Agriculture Office, Ehiopia

Abstract

Bira dam was constructed in Bati district in 1986 through the aid of International Red Cross Association for food security purpose. The study was conducted from January to September 2013. The objective of the study was to assess physico-chemical parameters of Bira dam. Current total area, average depth of the reservoir were measured using GPS and rope respectively, physico-chemical parameters were taken monthly from January to September 2013 from three sites. Digital Multimetres were used to measure pH, Temperature, conductivity and Turbidity value. SPSS Version 16 was used to analyze the collected data. Univarate test was used to test the physico-chemical parameters difference among sites and months. The mean value of pH, Temperature, Turbidity and conductivity 7.02, 24.11°C, 24.60 NTU and 399.00 µS/cm respectively. There was no significant difference in all physico-chemical parameters among sites ($P>0.05$). There was significant difference in water Temperature, Turbidity and conductivity by month ($P<0.05$). The current total area of the dam is 18 hectare which was 42 hectare when the dam was constructed; the depth also reduces from 20 to 4.33 m. Since the watershed of the dam is highly degraded, the dam will be totally dried if the situation continues. The turbidity value of Bira dam was higher than most studied dams in Ethiopia, therefore watershed of the dam should be properly managed though full participation of dam users.

Keywords: Millimeters; Turbidity; Red cross; Physico-chemical parameters; Conductivity; Association; Temperature

Introduction

Ethiopia is uniquely rich in water resources. It has numerous water bodies including ponds, lakes, rivers, reservoirs and wetlands. Based on the estimation of FAO [1,2] the surface area of major lakes and reservoirs is 7,334 Km^2 and the length of rivers is 7,185 km.

Ethiopia could be called a water tower of Eastern Africa in a continent where its most part is arid. The inland water body of Ethiopia is estimated at about 7,400 km^2 of lake area and about 7,000 km total length of [3]. These water bodies contain large population of commercially important fish species. However, the territory of Ethiopia seems to be among regions of the African continent which are least explored in ichtyofauna perspectives [4].

The development of aquatic life (flora and fauna) in surface waters is influenced by a variety of environmental conditions that determine the species as well as the physiological performance of individual organisms. The flora and fauna present in specific aquatic systems are a function of the combined effects of various hydrological, physical and chemical factors [5]. Aquatic ecosystems are dynamic and their tropic state is controlled by physical and chemical conditions. Thus, monitoring and evaluating the tropic state of lakes have become an essential prerequisite to develop control mechanisms.

Expanding human population brought about by the opportunities of good water supply, irrigation, fish production recreation and navigation offered by reservoirs has put enormous pressure and stress on the quality of water impounded by the reservoir. The impact of human activities in and around the reservoir is felt on the unique physical and chemical properties of water on which the sustenance of fish that inhabit the reservoir is built as well as to the functions of the reservoir. Water quality is determined by the physical and chemical limnology of a reservoir [6] and includes all physical, chemical and biological factors of water that influence the beneficial use of the water. Water quality is important in drinking water supply, irrigation, fish production, recreation and other purposes to which the water must have been impounded.

Water quality deterioration in reservoirs usually comes from excessive nutrient inputs, eutrophication, acidification, heavy metal contamination, organic pollution and obnoxious fishing practices. The effects of these "imports" into the reservoir do not only affect the socio-economic functions of the reservoir negatively, but also bring loss of structural biodiversity of the reservoir [7,8] have used the physico-chemical properties of water to assess the water quality of a reservoir. The use of the physico-chemical properties of water to assess water quality gives a good impression of the status, productivity and sustainability of such water body. The changes in physical characteristics like temperature, transparency and chemical elements of water such as dissolved oxygen, chemical oxygen demand, nitrate and phosphate provide valuable information on the quality of the water, the source(s) of the variations and their impacts on the functions and biodiversity of the reservoir.

The quality of surface water has deteriorated in many countries in the past few decades. As a result of the growing population, increasing industry, agriculture, and urbanization, the inland water bodies are confronted with the increasing water demand, as facing with extensive anthropogenic inputs of nutrients and sediments, especially the lakes and reservoirs [9]. To handle this problem, it is necessary to carry out water quality assessment, planning, and management, in which water

*Corresponding author: Tessema A, Wollo University, Department of Biology, Dessie, Ethiopia, E-mail: atecklie@yahoo.com

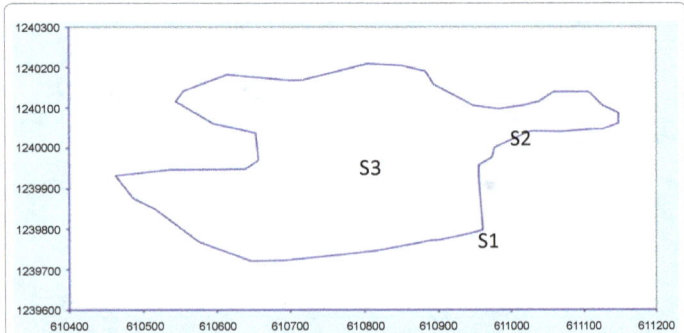

Figure 1: Sketch of Bira dam drawn using the collected X-coordinate and Y-coordinate, Longitudes belongs to X-axis and Latitudes belongs to Y-axis. Site1 (S1) is the littoral area without vegetation, Site2 (S2), Littoral area with Typha latifola species coverage and Site3 (S3), the open water, pelagic.

			Sum of Squares	Df	Mean Square	F	Sig.
pH * Month	Between Groups	(Combined)	1.346	5	.269	.328	.886
	Within Groups		9.846	12	.821		
	Total		11.192	17			
Temp * Month	Between Groups	(Combined)	62.518	5	12.504	22.574	.000
	Within Groups		6.647	12	.554		
	Total		69.165	17			
Turbidity * Month	Between Groups	(Combined)	1542.401	5	308.480	4.338	.017
	Within Groups		853.413	12	71.118		
	Total		2395.814	17			
Con * Month	Between Groups	(Combined)	12778.000	5	2555.600	48.989	.000
	Within Groups		626.000	12	52.167		
	Total		13404.000	17			

Table 1: Physico-chemical parameters variation among the different sampling months.

quality monitoring plays an important role [10]. This study aimed at assessing the water quality of Bira dam used for irrigation, livestock watering and fish production using some selected physico-chemical parameters. The results will form the baseline for monitoring and tracking changes in the water quality as a result of the dam's natural dynamics over time and impact of main activities on the dam and its watershed.

Objectives

General objective

The main objective of the study was to assess physico-chemical parameters of Bira dam to check the dam suitability for fish stocking.

Specific objective:

- To assess physico-chemical parameter of Bira dam.

Material and Methods

Study area

Bira kebele is one of the kebeles of Bati Woreda where Bira dam is found that was constructed for irrigation purpose by Red Cross. The dam at the beginning when it was constructed had a depth of 15 to 20m and a total area of 42 hectare, but recently its area reduced to 18.33 hectares due to siltation (Figure 1). Bati is one of the districts in Oromia zone that has different culture attracting tourists especially on market day, Monday. The economy is based on crop production (sorghum, teff

and maize) and livestock rearing. Livestock production is constrained by lack of grazing and access to fodder. Local agricultural labor, migration labor and firewood sale are important income generating activities particularly for poorer households (Figure 1) [11].

Methods

Physico-chemical parameters: Physico-chemical parameters conductivity in μs/cm, pH, Turbidity in NTU and temperature in °C were measured using digital multimetres in three sites, S1, S2 and S3 from January to June 2013.

Data analysis: Descriptive Statistics (mean, graphs) and inferential statistics (Univarate analysis) were used through SPSS Version 16 application.

Result and Discussion

Physico-chemical parameters

Most of the values of Physico-chemical water quality parameters during sampling months were in the optimum condition for fish production except for higher Turbidity value [12]. As stated below in Table 1, except pH, temperature, conductivity and Turbidity showed significant difference among sampling months (P<0.05). There were no significant difference in water quality parameters among the three sites (P>0.05) (Table 2).

pH

The pH is an important variable in water quality assessment as it influences many biological and chemical processes within a water body and all processes associated with water supply and treatment [13]. In unpolluted waters, pH is principally controlled by the balance between the carbon dioxide, carbonate and bicarbonate ions as well as other natural compounds such as humic and fluvic acids. Changes in pH can indicate the presence of certain effluents, particularly when continuously measured and recorded, together with the conductivity of a water body. Dial variations in pH can be caused by the photosynthesis and respiration cycles of algae in eutrophic waters. The pH of most natural waters is between 6.0 and 8.5, although lower values can occur in dilute waters high in organic content, and higher values in eutrophic waters, Groundwater brines and salt lakes [13]. The desirable pH range for fish is between 6.5- 9. Long term exposure to pH values beyond these limits slows fish growth and reduces health. Exceedingly alkaline water (greater than pH 9) is dangerous as ammonia toxicity increases rapidly. At higher temperatures fish are more sensitive to pH changes. The mean pH value of Bira dam ranged from 6.6 -7.3 almost similar with Tekeze dam and lower than Hashenge (8.4). The Bira dam pH value is suitable for fish production and its variation among sites described in Figure 2.

Water temperature

Fish are exothermic, their body temperature is about that of the surrounding environment; and affects all metabolic processes. Cold

	Site	pH	Temperature	Conductivity	Turbidity
Site1	Mean	6.9400	23.967	393.83	21.265
	Std. Deviation	.40714	2.4977	30.413	4.7235
Site2	Mean	6.6600	24.350	401.17	25.268
	Std. Deviation	.95714	2.0047	28.646	18.3661
Site3	Mean	7.4667	24.033	402.00	27.258
	Std. Deviation	.86832	1.8640	29.779	9.8578

Table 2: Physico-chemical parameters variation among sampling sites during sampling months.

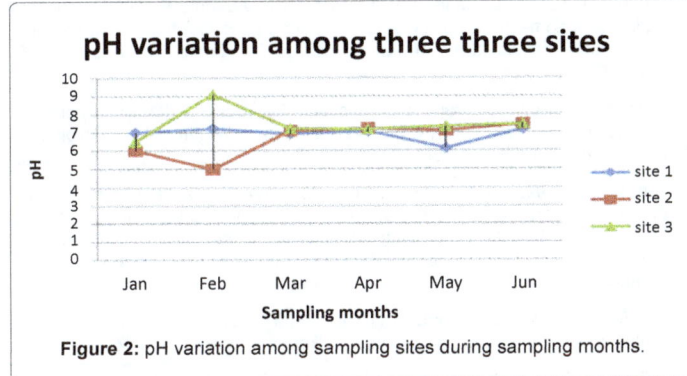

Figure 2: pH variation among sampling sites during sampling months.

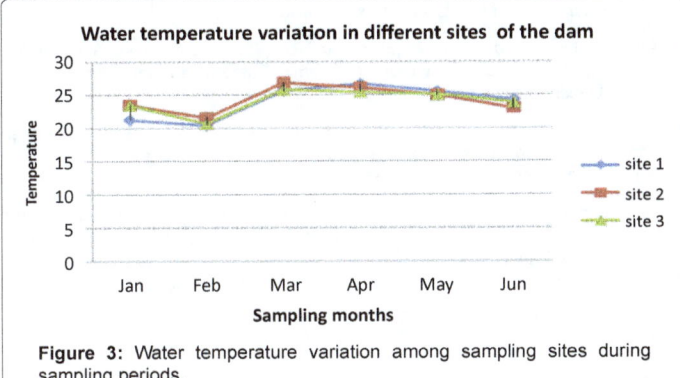

Figure 3: Water temperature variation among sampling sites during sampling periods.

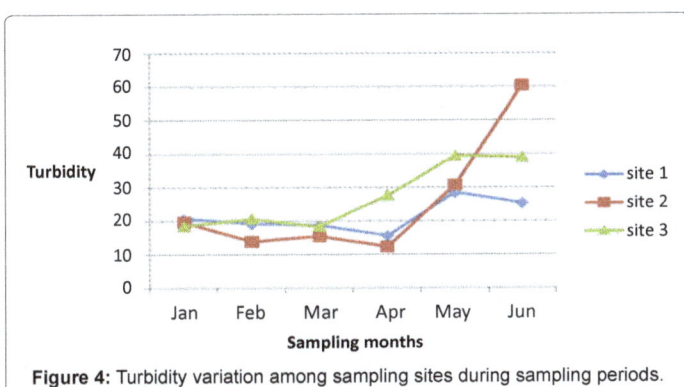

Figure 4: Turbidity variation among sampling sites during sampling periods.

water slows metabolism and warm water increases metabolic rate. Fish have adapted to a wide range of temperatures. Some cold water species can tolerate temperatures below 32°F; while desert killifish can live in pools in Death Valley at temperatures in excess of 110°F. Native warm-water fish have a temperature tolerance range of about 34- 104°F although many species will become stressed near either of these extremes. There below 55°F activity and feeding slow. Above 95°F many warm-water fish begin to reach upper lethal temperature tolerance limits. Tropical fish such as the tilapia, cannot tolerate cold water. They become stressed when water reaches 60°F and die at water temperatures below 50°F. Trout and other coldwater fish will die when water temperature exceeds 70°F. Their optimum temperature is about 55-65°F and they are active down to 40°F. Fish must adjust to temperature changes gradually. A warm-water fish may survive in 100°F water if slowly acclimated to it; however, a sudden change from a water temperature of 65°F to 75°F may shock and kill the fish. The mean water temperature of Bira dam ranged from 23.97-24.35°C suitable for

common carp fish production. Temperature variation among sampling sites is described in Figure 3.

Turbidity

Water turbidity refers to the quantity of suspended material, which interferes with light penetration in the water column. In water bodies, water turbidity can result from planktonic organisms or from suspended clay particles. Turbidity limits light penetration, thereby limiting photosynthesis in the bottom layer. Higher turbidity can cause temperature and DO stratification in water bodies. Planktonic organisms are desirable when not excessive, but suspended clay particles are undesirable. It can cause clogging of gills or direct injury to tissues of aquatic organisms. Erosion or the water itself can be the source of small (1-100 nm) colloidal particles responsible for the unwanted turbidity. The particles repel each other due to negative-charges: this can be neutralized by electrolytes resulting in coagulation. It is reported that alum and ferric sulfate are more effective than hydrated lime and gypsum in removing clay turbidity. Both alum and gypsum have acid reactions and can depress pH and total alkalinity, so the simultaneous application of lime is recommended to maintain the suitable range of pH. Treatment rates depend on the type of soil. The turbidity value measured in NTU was higher (21.26-27.27) in Bira dam than Tekeze dam (8-11).the bigger difference might be duet to highly degraded watershed of Bira dam resulted higher siltation. Turbidity variation among sites during sampling periods is described in Figure 4.

Conductivity

The values of Electrical conductivity (EC) ranged from 260 to 300 μS cm^{-1} in Tekeze . Total dissolved Conductivity is related to the concentrations of total dissolved solids and major ions. The conductivity of most freshwaters ranges from 10 to 1000 μS cm^{-1}, but may exceed 1000 μS cm^{-1}, especially in polluted waters, or those receiving large quantities of land run-off [13]. The conductivity value measured in μS cm^{-1} was higher (393.83-402.00) than Tekeze dam(260-300) and lower(569) Tendaho reservoir. The higher conductivity in Bira dam and Tendaho reservoir might be their geological characteristics containing many cations. The conductivity variation among sites during sampling duration showed in Figure 5.

Conclusion and Recommendation

Bira dam is used as source of irrigation water, livestock watering and water for washing clothes and basing. The excessive water extraction day and night without regulation, degraded watershed and absence of buffer zone resulted in siltation are major problems affecting water quality and quantity of the dam. The average depth of the dam

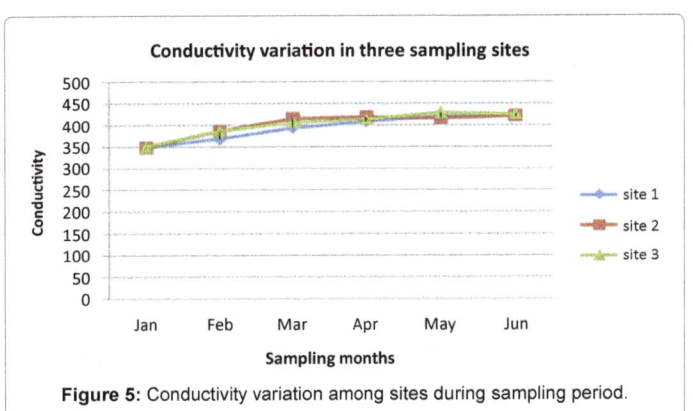

Figure 5: Conductivity variation among sites during sampling period.

reduced from 20 metre to 4.33 m and its total area from 42 hectares to 18 hectares. The dam littoral area mainly is devoid of vegetation that may support fauna and flora including fish species. As a side line activity during data collection Bira dam user association has been established and training on sustainable water utilization and watershed management was delivered to these individuals. All members of the association should actively involve in watershed Bira dam management for sustainable utilization of the resource.

Acknowledgment

The researchers would like to thank Organizations, Kemissie and Bati Agriculture and Rural Development Offices for positive response during request for transport facilities. We are also delighted to thank Mr.Fisha Woldemariam for his unreserved effort during data collection.

References

1. FAO (2001) FAO fisheries and aquaculture Ethiopia, Ethiopia.

2. Henderson HF, Welcomme RL (1974) The relationship of yield to Morpho-Edaphic-Index and number of fishermen in African inland fisheries. CIFA Occs Pap 1: 19.

3. Imevbore AMA (1970) The chemistry of the River Niger in Kainji Reservoir Area. Arch Hydrobiol 67: 412-431.

4. Golubstov AS, Darkov AA, Dgebuadze YY, Mina MV (1995) An Artificial key to fish species of the Gambela region (The White Nile basin in the limits of Ethiopia). Joint Ethio-Russian Biological expedition, Addis Ababa, Ethiopia, 84.

5. UNESCO (1996) Water Quality Assessments - A Guide to Use Of Biota, Sediments and Water in Environmental Monitoring – 2nd edn Deborah Chapman, UK.

6. Sidnei MT, Fakio ALT, Maria CR, Francises AE, Adaunto F (1992) Seasonal variation of some limnological factors of Lagoa does Guarana, a Varzea lake of the Rio Paranana State of Mato Groso do Sul, Brazil. Rev Hydrobiol 25: 269-276.

7. Djukic N, Maletin S, Pujin V, Ivanc A, Milajonovic B (1994) Ecological assessment of water quality of Tisze by physico-chemical and biological parameters. Tisca Szeged 28: 37-40.

8. Dumont HJ (1999) The species richness of reservoir plankton and the effect of reservoirs on plankton dispersal (with particular emphasis on rotifers and cladocerans). In: J.G. Tundisi and M. Straskraba edn. Theoretical Reservoir Ecology and its Applications. IIE, Backhuys Publishers, Brazilian Academy of Science: 477-491.

9. Kondratyev K, Pozdnyakov D V (1998) Water quality remote sensing in the visible spectrum. International Journal of Remote Sensing 19: 957-979.

10. Seker D Z, Goksel C, Kabdasli S, Musaoglu N, Kaya S (2003) Investigation of coastal morphological changes due to river basin characteristics by means of remote sensing and GIS techniques. Water Sci Technol 48: 135-142.

11. Amhara Livelihood Zone Reports (ALZR) (2007) Bati Woreda, Oromiya Administrative Zone. South Wollo & Oromia Eastern Lowland Sorghum and Cattle Livelihood Zone.

12. Tepe Y, Turkmen A, Mutlu E, Ates A (2005) some physico-chemical characteristics of Yarselli Lake, Turkey. Turkish Journal of Fisheries and Aquatic Sciences 5: 35-42.

13. American Public Health Association (APHA) (1995) Standard methods for the examination of water and wastewater, (19thedn). American Public Health Association, Washington DC, USA.

2

Effect of Temperatures on the Embryonic Development, Morphometrics and Survival of Macrobrachium Idella Idella (Hilgendorf, 1898)

Soundarapandian P*, Dinakaran GK and Varadharajan D

Centre of Advanced Study in Marine Biology, Faculty of Marine Sciences, Annamalai University, Parangipettai–608 502, Tamil Nadu, India

Abstract

The development of *M. idella* idella eggs incubated at four different temperatures (26, 30, 33 and 36°C). An increase in major axis length was evident at 26, 30, 33 and 36°C. However at 36°C, size variation increased with developmental stages, indicating abnormalities until 192 h, after which total mortality was observed. A distinct change in morphometric parameters (major half axis, minor half axis, area and perimeter) was demonstrated at higher temperatures, irrespective of the developmental duration of eggs. Length of hatched embryos increased with increasing incubation temperatures. Therefore a rapid rate of increase of major half axis and early hatching was observed at 33°C. The larva hatched first in 33°C (241 hrs) then followed by 30°C (265 hrs) and 26°C (302 hrs). In 36°C (182 h) there was total mortality during embryonic development so there was no hatching.

Keywords: *M. idella* Idella; Embryonic development; Temperatures; Effect; Survival

Introduction

The emission of greenhouse gases and carbon dioxide are expected to increase global mean temperature by 1.5-4.5°C over the next half-century [1]. The impact of such a large temperature will affect the biological functions of freshwater and marine fishes and shellfishes, as most of the species are poikilothermic in nature. Thermal tolerance of aquatic animals is also dependent on acclimation temperature and the duration of acclimation [2]. Over the years, attention has focused on the thermal tolerance of embryos and larvae [3] that are more sensitive to temperature changes than adult fishes [4]. The earlier investigation on Labeo rohita revealed similar results [5]. The embryos of temperate species are more sensitive to extreme temperatures than embryos of tropical species [6]. In general, thermal limits are narrower for early stages and reduced survival of embryos and juveniles. Whereas it was wider for adults [7]. It is also reported that upper lethal temperatures of embryos, larvae [8] and adults [9] of the freshwater Mozambique tilapia (Oreochromis mossambica) varies in the range of 2°C among different life stages. In *M. rosenbergii* and *P. serratus* higher temperature seems to be shortening the incubation period [10]. Hence, the present study was undertaken to assess the effect of low and higher temperatures on the incubation period and embryonic development of edible prawn *M. idella* idella.

Materials and Methods

Experimental animals

Gravid or Berried females (80-90 mm) with opaque, greenish, round or oval in shape fertilized eggs in their brood pouch were used for the present experiment.

Experimental conditions

Twelve newly spawned brooders were stocked; one in each 120L fiber glass tank to assess the effect of incubation temperature on embryonic stages of *M. idella* idella Acclimation was carried out at one degree per day from water temperature (25°C). Since the experiment was carried out during December to January the normal water temperature was found to be 25-26°C.

Temperature maintenance

The test temperatures (26, 30, 33 and 36°C) were regulated by using automatic thermostat until hatching of the eggs. Sampling of eggs was carried out once the cleavage was completed, since this early development phase was not easily observed and was considered as the initial period (0 h). Eggs were sampled aseptically by gently removing a bunch of eggs from the brood pouch using sterilized forceps and separated with the help of needle and forceps without damaging the eggs. After each sampling, brooders were given a 1-min prophylactic fungus dip treatment in malachite green (5 mg L^{-1}) before being returned to incubation tanks.

Sampling

In four different temperatures the embryonic stages of two brooders were rarely matched. Therefore, embryos were sampled at several intervals (i.e. 0, 24, 48, 72, 96, 120, 144, 168, 192, 216, 240, 264, 288 and 312 h) from each brooder. Organogenesis, developmental changes and physiological processes were recorded under a light microscope. Eggs collected from four sampling points of brood pouch (anterior to posterior) were pooled to minimize sampling error due to position of eggs in brood pouch and assessed the percentage mortality of fully developed embryos (from an aggregate of 12 embryos/brooder) at the onset of hatching. Embryo dimensions (major axis, minor axis, area and perimeter) were measured at 48-h intervals (0, 48, 96, 144, 192h) until total mortality or hatching occurred.

Results

Organogenesis and morphophysiology of eggs

Development of *M. idella* idella eggs incubated at four different

---

*Corresponding author: Soundarapandian P, Centre of Advanced Study in Marine Biology, Faculty of Marine Sciences, Annamalai University, Parangipettai –608 502, Tamil Nadu, India, E-mail: soundsuma@gmail.com

temperatures is presented in Plate 1. At 0 h, eggs in all the treatments were in similar phase of development (morula stage) and primodial mesodermal and endodermal cells were visible. After 24 h, the blastocyst was visible in all the four-temperature treatments. Embryos were visible after 48 h in all the treatments. However, abnormal embryonic development was evident at 36°C. At 72 h, the primodial compound eye was visible. At 96 h, a considerable increase in the length of the major axis was seen at higher temperatures (30, 33 and 36°C). At 120 h, star shaped and round protoplasmic islands was visible at 33 and 36°C, respectively and heartbeat was discernible at all temperature treatments. The compound eye with a visible optic lobe was visible at higher temperatures (30, 33 and 36°C). At 144 h, star shaped protoplasmic islands appeared at lower temperatures (26 and 30°C). Rudiments of appendages started developing at all acclimation temperatures. At 168 h, paired compound eyes were visible in all treatments. At 192 h, star shaped protoplasmic islands were seen at lower temperatures (26 and 30°C) while at 36°C, protoplasmic islands appeared was degenerated. After 192 h, complete mortality was noticed at 36°C (Plate 1).

At lower temperatures, rudiments of appendages were visible. The primodial digestive canal developed in segments as a dotted line in the posterior region, and appeared to originate from primodial hepatopancreas. The primodial brain was visible at the anterior part of primodial hepatopancreas. The major half axis attained maximum length at the time of hatching. Hatching was initiated by a sudden twitching movement in the posterior region (below the compound eye) by means of the rudimentary antennule. At the time of hatching, telson (or tail) and rudiments of uropod (folded below compound eye) unfolded and the embryonic case was removed from anterior portion by straightening of abdominal segments. Embryos hatched out with a jerky movement and all the anterior appendages in the cephalothoracic region (including walking legs) were started moving vigorously. The larva hatched first in 33°C (241 hrs) then followed by 30°C (292 hrs) and 26°C (340 hrs). In 36°C (192 h) there was total mortality during embryonic development so there was no hatching. Finally, larvae appeared bilaterally symmetrical (zoea 1) (Tables 1-3 and Figure 1). The embryonic development duration decreased with increasing temperatures as indicated by the negative slope in the linear regression equation, y=14.909x + 732.15 and R2=0.9932 (Figure 2).

Embryonic morphometry

The data on embryonic morphometry of *M. idella* idella exposed to four incubation temperatures (26, 30, 33 and 36°C) are reported in Table 1. An increase in major axis length was evident at 26, 30, 33 and

Parameter	Duration of embryonic development (h)	Acclimation temperatures (°C)			
		26	30	33	36
Major half axis (µm)	0	260.15 ± 3.54	270.3 ± 3.33	275.71 ± 2.63	275.06 ± 3.21
	48	273.26 ± 2.96	276.65 ± 2.64	279.17 ± 1.36	281.61 ± 2.36
	96	279.18 ± 3.08	285.95 ± 1.86	291.11 ± 3.23	289.23 ± 2.15
	144	287.23 ± 2.67	290.60 ± 4.38	298.20 ± 2.21	303.10 ± 3.25
	192	293.12 ± 3.56	303.72 ± 3.25	308.16 ± 2.06	311.69 ± 7.31
	Mean ± SE	278.58 ± 5.72	285.44 ± 5.77	290.47 ± 5.99	292.13 ± 6.75
Minor half axis (µm)	0	224.61 ± 3.18	233.07 ± 4.37	234.06 ± 1.62	235.16 ± 3.02
	48	227.15 ± 2.96	233.5 ± 2.38	235.61 ± 1.68	232.22 ± 2.12
	96	231.38 ± 2.24	235.61 ± 2.52	237.56 ± 3.18	234.81 ± 4.16
	144	234.77 ± 2.52	236.04 ± 1.86	238.11 ± 2.20	239.16 ± 3.69
	192	236.03 ± 2.37	237.30 ± 2.45	238.65 ± 1.85	240.56 ± 3.28
	Mean ± SE	230.79 ± 2.17	235.10 ± 0.79	236.80 ± 0.85	236.38 ± 1.52
Area (x10⁵) (µm²)	0	1.83 ± 0.38	1.98 ± 0.28	2.03 ± 0.23	2.03 ± 0.36
	48	1.95 ± 0.56	2.03 ± 0.18	2.07 ± 0.32	2.05 ± 0.26
	96	2.03 ± 0.28	2.12 ± 0.42	2.18 ± 0.19	2.13 ± 0.19
	144	2.12 ± 0.26	2.16 ± 0.33	2.23 ± 0.16	2.28 ± 0.42
	192	2.17 ± 0.41	2.26 ± 0.27	2.31 ± 0.40	2.35 ± 0.32
	Mean ± SE	2.02 ± 0.07	2.11 ± 0.06	2.16 ± 0.06	2.17 ± 0.08
Perimeter (µm)	0	1522.16 ± 24.36	1580.60 ± 32.21	1600.68 ± 30.12	1602.09 ± 26.62
	48	1571.32 ± 32.41	1601.86 ± 28.06	1616.41 ± 22.02	1613.44 ± 17.24
	96	1603.19 ± 30.21	1637.73 ± 31.68	1660.04 ± 19.83	1645.47 ± 42.33
	144	1639.06 ± 28.25	1653.67 ± 22.63	1684.04 ± 34.22	1702.71 ± 36.23
	192	1661.53 ± 22.32	1698.82 ± 33.72	1717.01 ± 28.43	1734.07 ± 23.54
	Mean ± SE	1599.45 ± 22.5	1634.54 ± 18.8	1655.64 ± 19.5	1659.56 ± 24.6

Table 3: Effect of incubation temperatures (26, 30, 33, 36°C) on morphological changes during the embryonic development of *M. idella idella* eggs until hatching.

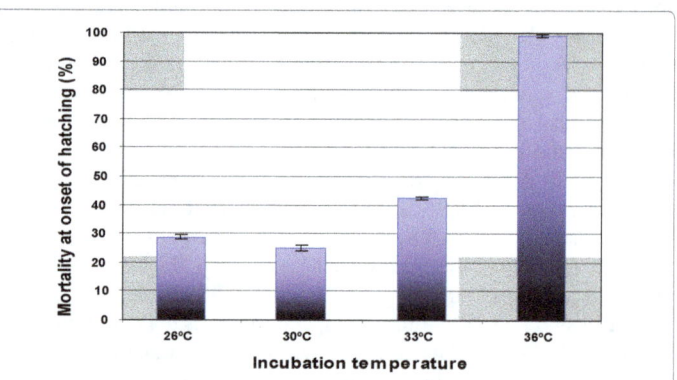

Figure 1: Percentage mortality of fully developed embryos at the onset of hatching.

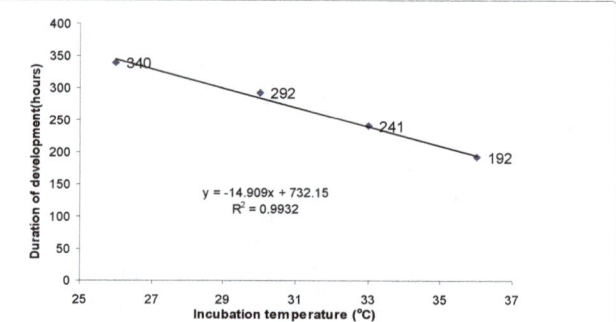

Figure 2: Duration of embryonic development with increasing incubation temperatures (morula to hatching). Data labels against each incubation temperature indicate the actual duration of development (mean of 12 values).

S. No.	Incubation temperatures (°C)	Mortality at hatching (%)
1	26	28.75 ± 0.70
2	30	25.0 ± 0.77
3	33	42.33 ± 0.66
4	36	99.00 ± 0.49

Table 1: Percentage mortality of fully developed embryos at different incubation temperatures (mean of 12 values ± SE).

S.No.	Incubation temperatures (°C)	Duration of development
1	26	340.00 ± 0.81
2	30	292.08 ± 0.79
3	33	241.00 ± 0.95
4	36	182.07 ± 0.53

Table 2: Duration of embryonic development with increasing incubation temperatures from morula to eggs hatching (mean of 12 values ± SE).

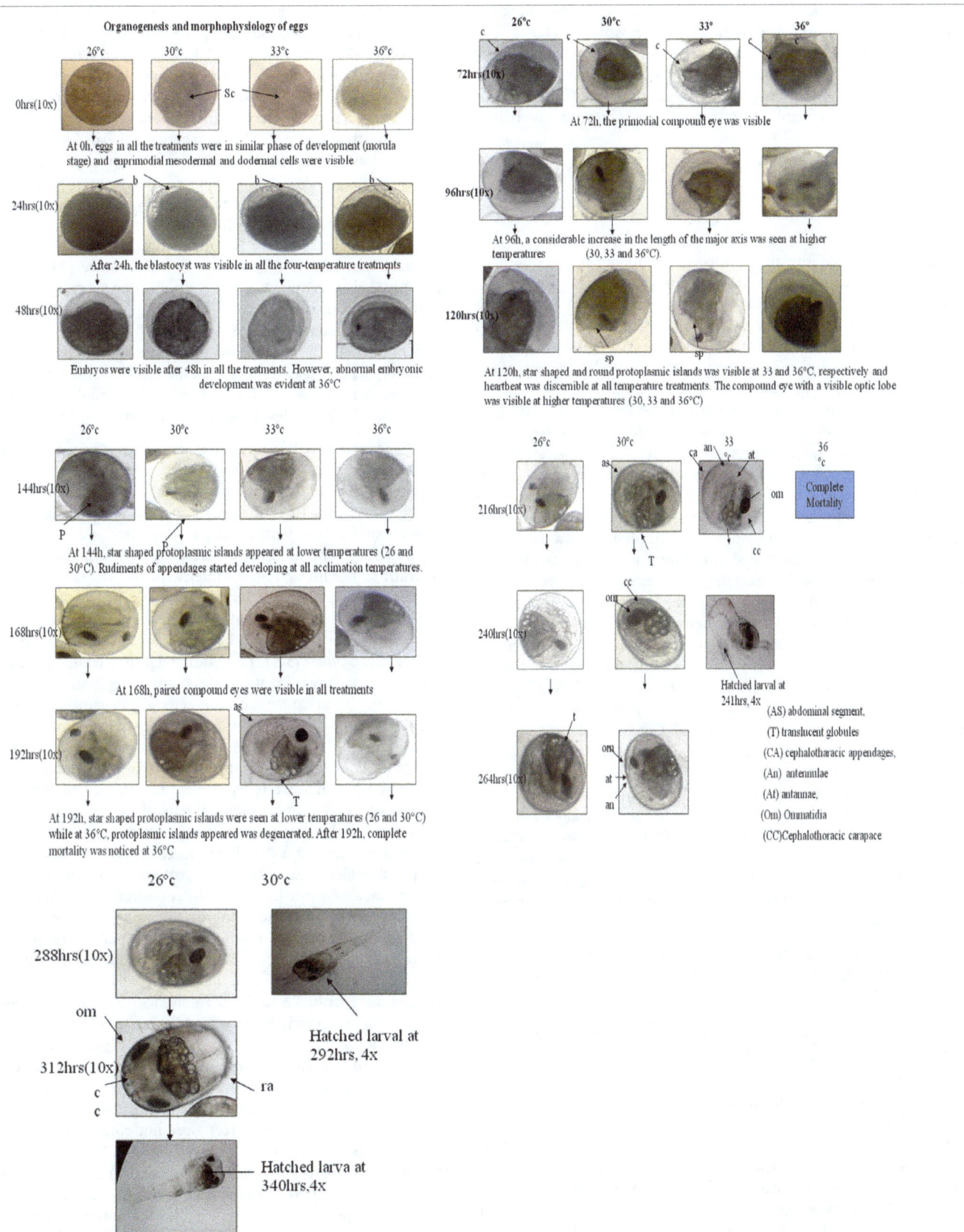

Plate 1: Development of *M. idella idella* eggs incubation at four different temperatures. sc-superfical cleavage, b-blastocyst, c-compound eye, p-protoplasmic island, sp-starshaped protoplasmic island, ra-rudimentary alimentary canal, an-antennule, at-antannae, om-Ommatidia, cc-Cephalothoracic carapace, ca-ephalotharacic appendages, as-abdominal segment.

36°C. However at 36°C, size variation increased with developmental stages, indicating abnormalities until 192 h, after which total mortality was observed. A distinct change in morphometric parameters (major half axis, minor half axis, area and perimeter) was demonstrated at higher temperatures, irrespective of the developmental duration of eggs. Length of hatched embryos increased with increasing incubation temperatures.

Therefore a rapid rate of increase of major half axis and early hatching was observed at 33°C (Plate 1).

Discussion

Prior knowledge on the effects of temperature on cultured aquatic organisms, especially during embryogenesis, is a prerequisite for successful hatchery operation and seed production. Embryonic development is a complex process in which cellular differentiation and proliferation occurs simultaneously but at different rates [11]. Both organogenesis and somatic growth are controlled by enzymatic activities. Embryonic development of ectotherms mainly depends on the differential expression of certain genes and temperature [12] and the rates of their biological functions are critically dependent on environmental temperature. The effect of temperature on developmental rate is direct and development is faster at higher temperatures. However, this increase of developmental rate of embryos at higher temperatures occurs only within tolerable thermal limits [13]. In the present study, organogenesis and physiological responses of M. idella idella eggs incubated at different temperatures indicates that higher temperatures increase the rate of embryonic development. Dark brownish structures appearing on the surface of embryos during the development phase of M. idella idella are known as protoplasmic islands. Morphology of protoplasmic islands of embryos was specific at different incubation temperatures. Embryos with round, oval or star shaped protoplasmic islands demonstrated normal development until hatching. But at 36°C the protoplasmic island were not clearly visible. Immediately after appearance of degenerated protoplasmic islands, complete mortality was recorded at 36°C. Further isolation and structural analysis of these protoplasmic islands may indicate a functional significance during early development of M. idella M. idella idella. It is evident that heat shock proteins are induced at early developmental phases in response to various physiological stimuli (growth factors, cell differentiation, and hormonal stimulation) and under the influence of temperature [14]. In M. rosenbergii also at 36°C complete mortality of embryos was reported once the protoplasmic island was degenerated [15].

Morphometric measurements indicated that an early increase in the length of the major axis of embryos at higher temperatures. Finally, length at hatching was increased with increasing temperatures and the larval major axis was highest at 33°C. However, complete mortality was observed at 36°C after 10 days, which indicates that incubation at 36°C is an upper temperature threshold for M. idella idella embryos. Earlier reports on duration of hatching in Macrobrachium spp at different temperatures indicates that eggs hatch out in 25 days at 26°C, 20 days at 28-28.5°C and in 17 days at 32°C [16]. In the present study, in M. idella idella hatching took first at 33°C (241 hrs) than followed by 30°C (292 hrs) and 26°C (340 hrs). Even though hatching took first at 33°C the survival rate was low when compared with 30 and 26°C. In M. rosenbergii the survival was more at 33°C than in 29 and 25°C [15]. In most crustaceans the incubation period is highly dependent on the temperature [17,18]. Incubation periods of Moreton Bay (Australia) population of S. serrata are usually in early spring, at water temperatures of 18-20°C [17]. Water temperatures in estuaries and coastal waters along the north coast of South Africa range between 17-22°C in winter and 23-30°C in summer [19]. Temperature directly influences the developmental rate and development is faster at increasing temperatures. However, this increase of developmental rate of embryos with increase in temperatures occurs only within the acceptable thermal limits [7,13]. In general, thermal limits are narrower for early stages and reduced survival of embryos and juveniles but wider for the adults [7]. Acclimation temperature or thermal history may also affect the temperature tolerance of embryos. In crayfish, time needed for egg development varies with temperature, suggesting the possibility of extending or reducing the incubation period [20]. In crustacean eggs, metabolic rate increases with temperature [21], which affects growth [22], survival [23], and yolk absorption rates [24]. However, high temperature could cause high mortality or serious deformities during egg incubation [25]. Studies on fish [26,27], crustaceans [23,25] and molluscs [28] have shown that there are different approaches to studying the effects of temperature on development. One of the criteria is to measure changes in lipid, protein, and carbohydrate biochemical contents during egg development, which reflect utilization rate [29].

Experiments with Liocarcinus depurator showed that a three-fold decrease in development time could occur in successive batches of eggs incubated during the early spring to mid-summer breeding season in one locality [30]. In L. holsatus and Necora puber hatching success was greater at low temperature-low salinity and high temperature-high salinity combinations. However, L. holsatus was relatively more tolerant to the lower range while N. puber was more tolerant to the higher range [31]. The better survival of L. holsatus eggs and larvae at lower temperatures than that of N. puber may reflect field situation; L. holsatus produces eggs and larvae earlier and at colder temperatures than does N. puber [30,32] noted that within a related group of species there was a direct relationship between egg size and incubation period. In the present study the eggs of the same species (M. idella idella) responded differently when exposed to same environmental conditions [33] studied the effect of temperature on egg extrusion rate in C. quadricarinatus. They found that egg production was more frequent when mature females were maintained in water over 28°C compared to 25°C and lower. In general the survival and incubation time decreased with increased temperature [34]. As expected, higher temperature caused faster development because it had a direct effect on physiological and biochemical processes [35]. This is a reflection of higher metabolic rate [20] and a decrease in the duration of embryonic or larval development, commonly documented for crustaceans [23]. This effect was previously observed in C. quadricarinatus by [34,36] reported approximately 28 days for egg incubation until hatching at 28°C, which is very similar to the present work. The difference in water temperature from 22 to 31°C resulted in 50% shorter development time from egg extrusion to juvenile stage in the work. However, shorter development time in response to higher water temperature is compromised with abnormal development and lower survival [37] found most hatchling crayfish reared at extremely high temperatures had deformed limbs and failed to moult normally. In the present study, deformities and abnormal sizes were not considered, although some deformed hatchlings were observed at 33°C that subsequently died. Crustaceans are highly sensitive to environmental changes during ontogeny, and are therefore at higher risk to reach lethal temperature during this period [24,38] report that the consumption rate of total lipid and protein in Artemia gradually increases as temperature increases. The higher lipid depletion rate at higher temperature occurs because lipids are the main source of energy during ontogeny of aquatic organisms [39,40]. Growth and differentiation processes demand large amounts of energy and all metabolic processes are intensified when the temperature is higher [25]. It was also observed that lipid consumption per day intensified at hatching, which could be related to a higher energy production need during this process [41]. For proteins, the consumption rate during embryogenesis might increase as temperature rises [42]. At high temperatures, tissue synthesis is inefficient due to the high cell proliferation rate, and more protein is used as fuel [39]. In the present study, embryonic development was assessed after acclimating the brooders (carrying newly released eggs in the brood chamber) at the rate of 1°C per day to test the temperatures and maintained until hatching, as described by [43]. A negative slope in the linear regression of development time to hatching, indicates a strong inverse relation with incubation temperature.

Our study indicates a direct linear relationship between development rates of M. idella idella embryos with incubation temperature. Hence, a direct relation between organogenesis and morphological measurements and development was established. A rapid increase in major axis length observed at 30 and 33°C with a concomitant rate of development and earlier hatching. But when compared to 30°C with 33°C the survival rate is very low. Increase in larval length was observed at higher incubation temperatures indicate that such larvae may develop into dominant prawns. Therefore, incubation temperature may prove vital in producing healthy, high quality prawn seeds for successful prawn farming, by taking advantage of "leap frog pattern" of enhanced growth and overall production. However, this hypothesis needs to be investigated by rearing freshwater prawns from embryos until adult stages at higher temperatures.

In the present study the optimal temperature for incubating M. idella idella eggs is recorded at 30°C considering the rate of development and hatching percentage. Total mortality was found at 36°C. High temperatures are related to poor cementing and attachment of eggs [31]. In the case of M. rosenbergii, the development and hatching percentage were maximum at 33°C than in 29°C and 99% mortality was recorded at 36°C [15]. Poor hatching percentage and formation of malformed embryos in later stages at 36°C suggest that this rearing temperature is well above the tolerance limit for development of L. rohita eggs or may be due to the lack of adequate enzymes involved in hatching [44]. This may be due to adaptive response of M. idella idella embryos evolved over the years due to global warming and climatic changes. The complete mortality percentage and gross morphological abnormalities at 36°C suggest that the thermal limit for embryonic development of M. idella idella is below 36°C. However, some of the embryos reared at 33°C reached relatively advanced developmental stages in a short time, inspite of low survival rate and gross abnormalities.

From the point of fertilization until hatching, low temperatures retard and high temperatures accelerate embryonic development [45,46]. According to [26] 29-33°C is acceptable for M. rosenbergii embryonic development, which is higher than earlier reports of 29-31°C [47]. According to [5] the optimal temperature for incubating L. rohita eggs is recorded at 31°C considering the rate of development and hatching percentage, which is higher than earlier reports of L. rohita [48]. But in the present experiment, M. idella idella at 26-30°C were found to be optimum temperature than other experimental temperature. Overall results suggest that 30°C is the ideal temperature for egg incubation of M. idella idella for faster embryonic development, better hatching percentage and least time duration for attaining given ontogenic stages. In L. rohita at 26 and 31°C suggests that these temperature ranges are most suitable for incubation but 31°C is the ideal

temperature for egg incubation. A rise in the optimum temperature for embryonic development over the years may be due to continuous warming in the test region along with a gain of adaptive capability and induced thermal tolerance over the years. This hypothesis needs to be tested at the genetic level. These results may be a prelude to effectively utilize the benefits of temperature on better hatching rate and reduced hatchery man-days and ultimately the cost of production in M. idella idella hatcheries. However, hatchery seed production of M. idella idella is recommended between 26 and 30°C. This study reveals that M. idella idella embryos can accommodate climatic changes due to global warming up to 33°C, without hampering the reproduction and embryonic development.

Acknowledgment

Authors are grateful to the Director of the Centre and the authorities of Annamalai University for providing with facilities. They extend their thanks to Prof. L. Kannan, for critically going through the manuscript and offering comments. The Umino Pet Palace, Chennai is acknowledged for the supply of fishes. We also thankfully acknowledge the valuable comments obtained from three anonymous reviewers, which greatly improved the manuscript.

References

1. Houghton RA, Woodwell GM (1989) Global climatic change. Sci American 260: 36-44.

2. Chatterjee N, Pal AK, Manush SM, Das T, Mukherjee SC (2004) Thermal tolerance and metabolic status of Labeo rohita and Cyprinus carpio early fingerlings acclimated to three different temperatures. J Ther Biol 29: 265-270.

3. Houde ED (1989) Comparative growth, mortality, and energetics of marine fish larvae: temperature and implied latitudinal effects. Fish Bull US 87: 471-495.

4. Brett JR (1969) Temperature and fishes. Chesapeake Science 10: 275-276.

5. Das T, Pal AK, Chakraborty SK, Manush SM, Dalvi RS, et al. (2006) Thermal dependence of embryonic development and hatching ratein Labeo rohita, Hamilton, 1822. Aquacult 255: 536-541.

6. Buddington RK, Hazel JR, Poroshov SI, Vaneehenhaam J (1993) Ontogeny of the capacity for homeoviscous adaptation in white sturgeon (Acipenser transmontanus). J Exp Zool 256: 18-28.

7. Cossins AR, Bowler K (1987) Temperature Biology of Animals. Chapman and Hall, London.

8. Subasinghe RP, Sommerville C (1992) Effects of temperature on hatchability, development and growth of eggs and yolk-sac fry of Orechromis mossambicus (Peters) under artificial incubation. Aqua Fisher Manag 23: 31-90.

9. Allanson BR, Noble RG (1964) The tolerance of Tilapia mossambica (Peters) to high temperature. Trans Amer Fish Soc 93: 323-332.

10. Manush SM, Pal AK, Chatterjee N, Das T, Mukherjee SC (2004) Thermal tolerance and oxygen consumption of Macrobrachium rosenbergii acclimated to three temperatures. J Ther Biol 29: 15-19.

11. Hall BK (1922) Evolutionary Developmental Biology. Chapman and Hall, London.

12. Ojanguren AF, Brana F (2003) Thermal dependence of embryonic growth and development in brown trout. J Fish Biol 62: 580-590.

13. Atkinson D (1996) Ectotherm life-history responses to developmental temperature.

14. Gething M, Sambrook J (1992) Protein folding in the cell. Nat 355: 35-45.

15. Manush SM, Pal AK, Das T, Mukherjee SC (2006) The influence of temperatures ranging from 25 to 36°C on developmental rates, morphometrics and survival of freshwater prawn (Macrobrachium rosenbergii) embryos. Aquacult 256: 529-536.

16. Ogasawara YF (1984) Ecology of prawns and shrimp.

17. Haesman MP, Fielder DE (1983) Laboratory spawning and mass rearing of the mangrove crab Scylla serrata (Forskal), from first zoea to first crab stage. Aquacult 34: 303-316.

18. Hines AH (1986) Larval pattern in the life histories of Brachyuran crabs

(Crustacea, Decapoda, Brachyura). Bull Mar Sci 39: 444-466.

19. Robertson WD, Kruger A (1994) Size at maturity, mating and spawning in the portunid crab, Scylla serrata (Forskal) in Natal, South Africa. Est Coast Shelf Sci 39: 185-200.

20. Reynolds J (2002) Growth and reproduction.

21. Naylor K, Taylor E, Bennett D (1999) Oxygen uptake of developing eggs of Cancer pagurus (Crustacea: Decapoda: Cancridae) and consequent behavior of ovigerous females. J Mar Biol Ass 79: 305-315.

22. Jones DA (1994) Effect of temperature on growth and survival of the tropical freshwater crayfish Cherax quadricarinatus (Von Martens) Decapoda: Parastacidae.

23. Paula J, Mendes R, Paci S, Mc Laughin P, Gheradi F, et al. (2001) Combined effects of temperature and salinity on the larval development of the estuarine mud prawn Upogebia africana (Crustacea, Thalassinidae). Hydrobiol 449: 141-148.

24. Evjemo O, Danielsen T, Olsen Y (2001) Losses of lipid, protein and n-3 fatty acids in enriched Artemia franciscana starved at different temperatures. Aquacult 193: 65-80.

25. Kumlu M, Eroldogan O, Aktas M (2000) Effects of temperature and salinity on larval growth, survival and development of Penaeus semisulcatus. Aquacult 188: 167-173.

26. Ojanguren A, Reyes-Gavilan F, Rodriguez R (1999) Effects of temperature on growth and efficiency of yolk utilization in eggs and pre-feeding larval stages of Atlantic salmon. Aquacult Int 7: 81-87.

27. Keckeis H, Kamler E, Bauer-Nemeschkal E, Schneeweiss K (2001) Survival, development and food energy partitioning of nase larvae and early juveniles at different temperatures. J Fish Biol 59: 45-61.

28. Gilroy A, Edwards S (1998) Optimum temperature for growth of Australian abalone Haliotis rubra (Leach) and greenlip abalone, Haliotis laevigata (Leach). Aquacult Int 29: 481-485.

29. Lemos D, Phan V (2001) Energy partitioning into growth, respiration, excretion and exuvia during larval development of the shrimp Farfantepenaeus paulensis. Aquacult 199: 131-143.

30. Wear RG (1974) Incubation in British decapod crustacea, and the effects of temperature on the rate and success of embryonic development. J Mar Biol Assoc 54: 745-762.

31. Choy SC (1986) Ecological studies on Liocarcinus puber (L.) and L. holsatus (Fabricius) (Crustacea, Brachyura, Portunidae) around the Gower Peninsula, South Wales.

32. Choy SC (1988) Reproductive biology of Liocarcinus puber and L. holsatus (Decapoda, Brachyura, Portunidae) from the Gower Peninsula, South Wales. Mar Ecol 9: 227-241.

33. Yeh H, Rouse D (1995) Effects of water temperature, density, and sex ratio on the spawning rate of red claw crayfish Cherax quadricarinatus (Von Martens). J World Aquacult Soc 26: 160-164.

34. Zhao Y, Meng F, Chen L, GU Z, Xu G, et al. (2000) Effects of different gradient temperatures on embryonic development of the Cherax quadricarinatus (Crustacea, Decapoda). J Lake Sci 12: 59-62.

35. Brown C, Terwilliger N (1999) Developmental changes in oxygen uptake in Cancer magister (Dana) in response to changes in salinity and temperature. J Exp Mar Biol Ecol 241: 179-192.

36. Yeh H, Rouse D (1994) Indoor spawning and egg development of the red claw crayfish Cherax quadricarinatus. J World Aquacult Soc 25: 297-301.

37. Rhodes C (1981) Artificial incubation of the eggs of the crayfish Austropotamobius pallipes. Aquacult 25: 129-150.

38. Agard J (1999) A four dimensional response surface analysis of the ontogeny of physiological adaptation to salinity and temperature in larvae of the palaemonid shrimp Macrobrachium rosenbergii (de Man). J Exp Mar Biol Ecol 236: 209-233.

39. Anderson DT (1982) Embryology. Acadamic Press, New York.

40. Holland DL (1978) Lipid reserves and energy metabolism in the larvae of benthic marine invertebrates.

41. Heras H, Gonzales-Baro M, Pollero R (2000) Lipid and fatty acid composition and energy partitioning during the embryo development in the shrimp Macrobrachium borellii. Lipids 35: 645-651.

42. Conceicao L, Ozorio R, Suurd E, Verreth J (1998) Amino acid profile and amino acid utilization in larval African catfish Clarias gariepinus, effects of ontogeny and temperature. Fish Physiol Biochem 19: 43-57.

43. Beitinger TL, Bennett WA, McCauley RW (2000) Temperature tolerances of North American freshwater fishes exposed to dynamic changes in temperature. Environ Biol Fishes 58: 237-275.

44. Reddy PK, Lam TJ (1991) Effect of thyroid hormones on hatching in the tilapia, Oreochromis mossambicus. Gene Compar Endocri 81: 484-491.

45. Hart PR, Purser GJ (1995) Effects of salinity and temperature on eggs and larvae of the greenback flounder. Aquacult 136: 221-230.

46. Hamel P, Mangan P, East P, Lapointe M, Laurendeau P (1997) Comparison of different models to predict the in situ embryonic development rate of fish with special reference to white sucker (Catostomus commersoni). Canadian J Fish Aquat Sci 54: 190-197.

47. Sebastian CD (1996) A manual on seed production and farming of giant freshwater prawn, Macrobrachium rosenbegii. The Marine products Export Development Authority, (Ministry of Commerce, Govt of India), Pananmpilly Avenue, Cochin.

48. Ponnuraj M, Murugesan AG, Sukumaran N (2002) Effect of temperature on incubation time, fertilization rate and survival of the spawn of Rohu, Labeo rohita.

3

Effect of Decreasing Water Levels on the Spawning Rate and Egg Count of Female Crayfish *Procambarus (Austrocambarus) llamasi* (Villalobos, 1955)

Carmona-Osalde Claudia[1], Puerto-Novelo Enrique[2] and Miguel Rodriguez-Serna[1]*

[1]*National Autonomous University of Mexico (UNAM), Faculty of Science, Multidisciplinary Teaching and Research Unit, Sisal, Aquaculture Biotechnology Area, Mexico*
[2]*Center for Research and Advanced Studies of the IPN (CINVESTAV-IPN) Unit MÉRIDA, Mexico*

Abstract

Three different water levels were assessed to establish their role in *Procambarus llamasi* spawning synchronization. A total of 132 crayfish were used for this assay, 120 were female and 12 were male FI (reproductive male stage), with an average initial size of 45 mm total length and 2.5 g in weight. *P. llamasi* were produced under controlled conditions at CINVESTAV-Merida, Yucatan, Mexico. The experimental system consisted of 12 plastic tanks of 0.60×0.34×0.28 m, equipped with water recirculation, biological filters, individual PVC shelters, a constant water temperature of 26°C and total darkness. Decreasing water levels in the crayfish tanks showed an important negative effect of this factor on most of the biological parameters measured, including the spawning rate and the number of eggs spawned by females. Spawning female sizes in low water levels were smaller and with fewer eggs than in the other levels. These results revealed an important relationship between water level and spawning rate for *P. llamasi*.

Keywords: Water levels; Spawning; Eggs numbers; *Procambarus llamasi*

Introduction

Increasing interest in crayfish culture has lead to develop research on the role that environmental factors play in reproduction. Most attempts in controlling breeding have been made by assessing different environmental factors, mainly, temperature and photoperiod regimens [1-3].

Environmental factors may affect the reproductive performance of crayfish in different ways. Reproduction comprises a series of physiological processes from gonadogenesis to spawning. Synchronization of spawning is critical from a productive point of view, as it offers two main advantages for hatchery operations. First, it reduces the required number of broodstock, as non-productive females would not be maintained in the system. Second, it allows having juvenile crayfish of the same age and size at the same time all year long.

Recent studies have found different spawning results among crayfish species. In some, like *Cherax quadricarinatus* (von Martens), *Procambarus clarkii* (Girard), and *Orconectes virilis* (Hagen), spawning was improved by increasing temperature and photoperiod under laboratory conditions [4-6]. In other species, such as *Procambarus (Autrocambarus)* and *llamasi* (Villalobos), temperature improved only gonad maturation, and there was not noted any positive effect on spawning (Figure 1).

According to [7] and Collins and Mitchell, synchronic spawning cannot be achieved using temperature, photoperiod, or density management. For *P. llamasi* and *C. quadricarinatus*, the best spawning rate reported under control conditions is less than 50%.

Some efforts have been made to improve spawning synchronization, but to date, it is not clear which environmental factor serves as trigger cue in *P. llamasi*. In other aquatic organisms, water level manipulation is used to control spawning by decreasing water depth [8,9]. In crayfish, water level fluctuation has been identified as an important and critical factor in inducing reproductive adaptations [10-13]. An example of this reproductive cycle adaptation in relation to the hydrological regime is found in *P. clarkii*. According to *P. clarkii* is capable of developing different reproductive strategies directly related to the hydroperiod of the region.

In an attempt to better understand the possible role of water level in the spawning response of *P. llamasi*, we performed a study in which three different water depths were assessed on the spawning rate and

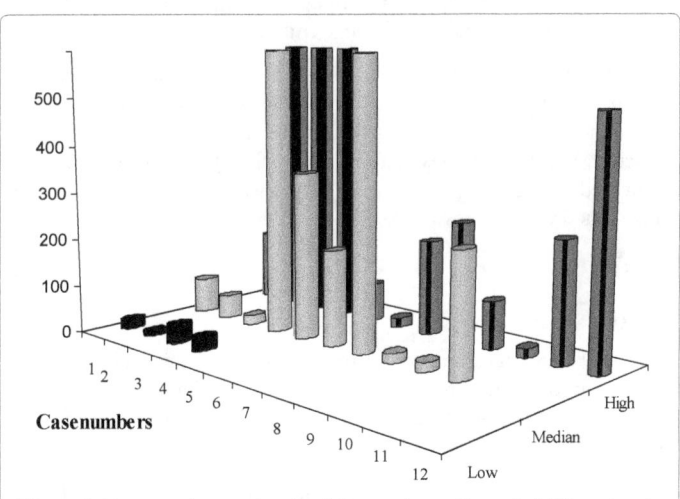

Figure 1: Eggs numberper female of *Procambarus llamasi* at different water levels.

*Corresponding author: Miguel Rodriguez-Serna, National Autonomous University of Mexico (UNAM), Faculty of Science, Multidisciplinary Teaching and Research Unit, Sisal, Aquaculture Biotechnology Area, Mexico
E-mail: mrserna66@gmail.com.

number of eggs per spawned female of *P. llamasi* in order to determine the possible effect of water levels on spawning synchronization.

Materials and Methods

A total of 132 *P. llamasi* were used for this study. 120 were female and 12 were male FI (reproductive male) with an average initial size of 45 mm total length and a weight of 2.5 g. *P. llamasi* were produced under controlled conditions at CINVESTAV- Merida, Yucatan, Mexico. The experimental system consisted of twelve plastic tanks of 0.60×0.34×0.28 m, provided with water recirculation, biological filters, individual PVC shelters, constant water temperature of 26°C and total darkness. Ten females and one reproductive male FI were placed in each plastic tank.

Three different water levels were tested, in quadruplicate, on the spawning rate and the number of eggs per female: a low level of 0.03 m; a medium level of 0.15 m, and a high level of 0.25 m.

During the 120 days of the experiment, animals were fed with a commercial shrimp diet (Purina: 35% protein, 6% lipids) once a day *ad libitum*. All tanks were checked every day for dead animals and spawned females. Spawned females were removed from the system and eggs were counted. To maintain the same density proportion, removed females were replaced with new ones.

Mortality and spawning rate were recorded throughout the experimental period.

Growth was estimated as a function of weight and total length, with measurements taken every 15 days. Specific Growth Rate (SGR) was estimated according to Goolish and Adelman (1984), as well as the following parameters:

Survival [% =100 ((Initial Number–Final Number)/Total Number)] Weigh Gain=[%=100 ((Final Weight–Initial Weight)/Initial Weight)] Length=[%=100 ((Final Length–Initial Weight)/Initial Weight)] SGR (%/day)=100 (loge Final Weight-loge Initial Weight)/Time)

Survival, growth, spawning and egg count/female results were compared using a one-way variance analysis (ANOVA) with a 95% significance level. The difference between means was determined with a Tuckey Multiple Ranges test [14] and all statistical evaluation was performed using Statgraphics ver. 4.1software. Values were arcsine transformed when necessary.

Results

Survival

Survival rate of experimental animals showed statistical differences ($p < 0.05$) among the three water levels assessed (Table 1). Low water levels yielded the lowest survival rate with 77%, while medium water levels yielded the highest with 98%.

According to the present results, decreasing water levels had a negative effect on crayfish survival, with an increase in crayfish mortality as water depth was reduced.

Growth

Table 1 shows the obtained growth parameters. There was a statistical difference in final weight between low and high water levels, with 3.45 g for the former and 5.19 for the latter. Animals in lower water levels experienced a reduction in growth as compared to animals in greater depths. The Specific Growth Rate was statistically similar between low and medium levels, but not between these and the highest

Parameter	Low	Median	High	±SE
Initial number	40	40	40	
Final number Survival (%)	31	39	36	1.5618
	77[a]	98[c]	90[b]	
Initial Length	48.6[a]	46.6[a]	42.25[a]	1.7538
Final Length	61.75[a]	62.55[a]	64.92[a]	1.6494
Initial Weight (g)	2.96[a]	2.61[a]	2.0[a]	0.1572
Final Weight (g)	6.42[a]	6.89[a]	7.11[a]	0.4212
Weight gain (g)	3.45[a]	4.27[ab]	5.19[b]	0.4165
Weight gain (%)	133.74[a]	185.19[a]	343.13[b]	29.233
Specific Growth Rate (%/day)	0.825[a]	1.069[a]	1.527[b]	0.0928
Feed intake (g)	2.90[a]	2.87[a]	2.90[a]	0.1414
Feed Conversion Ratio	1.8[a]	1.11[ab]	1.0[b]	0.3295
Female spawning (%)	13.33[a]	25.6[b]	34.3[c]	1.33
Eggs number per female	26[a]	218.5[b]	280.8[b]	25.9
Spawned female size (mm)	51[a]	57[ab]	62[b]	0.3623

*Numbers with the same superscript are not statistically different ($p > 0.05$).

Table 1: ANOVA results for the effect of water levels on spawning and egg count in *Procambarus llamasi*.

water level. The SGR of crayfish under low water level conditions was almost half of that achieved in high water level conditions.

Food efficiency

Food efficiency parameters are shown in Table 1. Feed intake results were very similar among the three water level treatments, with values reported between 2.87 to 2.90 g.

Feed Conversion Ratio showed efficiency values in all treatments with statistical differences between animals in the low water level treatment and the high water level treatment.

Spawning

Spawning results are shown in Table 1. An important effect of water level was observed on the spawning rate of crayfish among the treatments. Statistical differences were found among low, medium, and high water levels, with the lowest percentage (13.33%) in low water conditions and the highest (34.3%) in high water levels.

Spawning was found to be directly related to water depth. Spawned females in the lowest water level tanks had smaller sizes than those spawned in the other two treatments.

Egg number

The number of eggs spawned per female is presented in Table 1. As with the spawning rate, water level had an important effect on this parameter. Statistical differences were found among water levels with the best results in high water level treatments.

Discussion

Important efforts have been made to manipulate environmental factors in controlling crustacean reproduction. Many studies asses the modulator effect of light and temperature in the reproductive cycle of marine and freshwater species, but the truth is that maturation and, particularly, spawning have not been successful in achieving aquaculture goals with changes in photoperiod and/or temperature regimes.

If we take a look at penaeid shrimp maturation and spawning results, positive results have been reported with long photoperiods and temperatures above 25°C [15-17] but these never promoted faster maturation and better spawning rates than eyestalk ablation [18]. In lobsters, the effects of photoperiod and temperature are so variable that no independent response could be identified. Ovarian development required combined conditions of these factors. Under a long photoperiod (14 L), ovarian development progressed slowly with temperature and the developmental rate increased as temperature increased. On the other hand, ovarian development under a short photoperiod (10 L) depended on temperature. Ovarian development at 13°C progressed similar to that of a long photoperiod, whereas it was prevented or considerably delayed at 19° and 25°C. In *Panulirus japonicus* (von Siebold), both temperature and photoperiod are important factors in controlling ovarian development, but not in spawning [6].

For crayfish, controversial results have been obtained using different photoperiods and temperatures [4-6]. Numerous studies have analyzed the effect of either temperature or photoperiod separately, or in conjunction. When assessed separately, some authors argue that both factors often interact in complex ways (e.g. when both are increased they often affect gonad maturation and spawning synergistically), and may produce different effects depending on the species [19,20].

Several attempts have been made to induce off-season spawning in crayfish from the northern hemisphere and temperate climate by manipulating both temperature and photoperiod with temperature as the predominate cue for egg laying [21] found that increasing temperature by 5°C triggered egg laying in *Astacus astacus* L., whereas photoperiod played only a secondary role throughout the course of two simulated cycles in one full year. According to high temperature following several cold months accelerated egg laying compared to that obtained in natural populations, whereas exposure to a long photoperiod had no effect in *Orconectes limosus* (Rafinesque). In *Orconectes virilis*, raising temperature slightly higher than that in the natural environment following several months of low temperature accelerated egg laying by 5 to 7 weeks earlier than crayfish remaining outdoors. In this case, the photoperiod was not a triggering factor, unlike the increase in temperature.

In *P. llamasi*, previous results reported no effect of light on gonad maturation or spawning. No differences in the reproductive performance were observed between animals maintained under complete darkness and those in natural photoperiod. In contrast, constant temperatures in the same species showed an important effect on gonad maturation, but not on the spawning rate.

P. llamasi gonad maturation is mostly affected by temperature. Unfortunately, the effect of this factor has not been successfully proven on crayfish spawning and to date it is not clear which environmental factor or factors may serve as a key cue for triggering this process [22]. In general terms, temperature alone improves maturation, but not spawning; whereas the effect of light has not been clearly established.

From an aquaculture point of view, synchronization of spawning could allow for some important advantages in broodstock areas and management. Differences in crayfish spawning rates have been reported between natural populations and laboratory conditions, with higher spawning rates and synchronization under natural conditions.

According to [20] the spawning rate of natural populations in Doñona, Spain, of *P. clarkii*, is between 75% and 93%. In contrast, *C. quadricarinatus* and *P. llamasi* maintained under experimental conditions resulted in less than 50% of spawned females with no synchronization [23].

In some aquatic organisms, decreasing the water level has a spawning-promoting effect. Water level is one of the most common factors that successfully induce spawning in the culturing practices of some fish and water snail species. *Pomacea urceus* (Müller) snails mate at the end of the rainy season and burrow into the substratum as the water levels of the rivers and swamps decrease.

Snails, therefore, use a decrease in water level as the cue for spawning. A decrease in water level under controlled conditions triggers mating and spawning. In tanks without water level fluctuations, no mating or spawning took place. This method is inexpensive and reliable, and has important practical applications for snail culture.

In the present study, water level was decreased with the intent to induce spawning in crayfish *P. llamasi*. Water level, however, not only showed a negative influence on the spawning rate of crayfish, but also in the growth and survival of the species. Fecundity decreased drastically in low water treatments as compared to the other two treatments. Mean fecundity reported by [13] for *P. llamasi*, under laboratory conditions, is of 311 eggs/female, with a minimum of 200 and a maximum of 700. In the medium and high level treatments, fecundity and female average size improved. It is possible that animals maintained in low water levels suffer from stress, which is reflected in the limited development of the parameters studied.

Very little is known about the effect of water level and crayfish reproduction. There are no previous results on the effect of water level and crayfish spawning under laboratory conditions, but natural population sites, where crayfish are found to be present, had a water depth of 20 cm and generally no crayfish are found in those with 15 cm or less of water depth [24]. The effect of water level on the reproductive strategies under natural conditions has been reported for *P. clarkii* in Spain by [12]. *P. clarkii*, adjusts its reproductive cycle to the annual and inter-annual hydrologic fluctuations of water levels at the Doñona Park, Spain, with three different strategies. Populations living in freshwater with a temporal seasonal system and a short hydroperiod of 3-4 months reproduce only once per year, when temperature is approximately 20°C and water depth is approximately 30 cm [25].

Populations living in ecosystems with a long hydroperiod (greater than 6 months) have two reproductive periods. Thirdly, crayfish living in permanent ecosystems show a large reproductive period from fall to spring.

In this sense, the Yucatan Peninsula has three main seasons: rainy, dry, and "nortes" (northerlies). During the rainy and north-winds season (May-December), most of the land is flooded. When the dry season arrives, water bodies begin to reduce until only some shallow waters remain and these maintain most of the adult population.

Under this dry condition, in which high individual aggregation and temperatures are combined, females probably mature but do not spawn. Using a different strategy, they wait until the water level rises. This could allow for young crayfish to maximize the utilization of the flooded areas.

The reproductive strategy of this species may be based on the presence of water for the new generations. The dry season lasts from four to five months, a very long time to withstand temperatures above 30°C and high desiccation. Increased water levels may be a more important signal for *P. llamasi* to ensure survival of the offspring in places where water accumulation is very difficult because of the karstic nature of its soil.

In summary, decreasing water levels in crayfish tanks showed an important negative effect on most of the biological parameters measured, including spawning rate and the number of eggs spawned by females. Spawned female sizes in low water levels were smaller and produced fewer eggs than those in the other levels. It is interesting, however, to have found reproduction under this water level condition.

Acknowledgment

To program PAPIIT-UNAM for the financial support by project IN203906-3: Adaptation of the crayfish Procambarus llamasi to contrasting environments in the Yucatan Peninsula

References

1. Yeh HS, Rouse BD (1995) Effects of water temperature, density and sex ratio on the spawning rate of red claw crayfish Cherax quadricarinatus (von Martens). Journal of the World Aquaculture Society 26: 160-164.

2. Carmona-Osalde C, Rodríguez-Serna M, Olvera-Novoa MA (2002) The influence of the absence of light on the onset of first maturity and egg laying of the crayfish Procambarus (Austrocambarus) llamasi (Villalobos, 1955). Aquaculture 212: 289-298.

3. Carmona-Osalde C, Rodríguez-Serna M, Olvera-NovoaMA, Gutierrez-Yurrita PJ (2004) Gondadal development, spawning, growth and survival of the crayfish Procambarus llamasi at three different water temperatures. Aquaculture 232: 305-316.

4. Portelance B, Dubé P (1995) Temperature and photoperiod effects on ovarian maturation, ovarian growth and egg-laying of Orconectes virilis. Freshwater Crayfish 8: 321-330.

5. Provenzano AJ, Handwerker TS (1995) Effects of photoperiod on spawning of red swamp crayfish, Procambarus clarkii, at elevated temperature. Freshwater Crayfish 8: 311-320.

6. Matsuda H, Takenouchi T, Takashi Y (2002) Effects of photoperiod and temperature on ovarian development and spawning of the Japanese spiny lobster Panulirus japonicus. Aquaculture 205: 385-398.

7. Carmona-Osalde C, Rodríguez-Serna M, Olvera-NovoaMA, Gutierrez-Yurrita PJ (2004) Effect of density and sex ratio on gonad development and spawning in the crayfish Procambarus llamasi. Aquaculture 236: 331-339.

8. Kungvankij P (1988) Guide to marine finfish hatchery management. F.A.O., Rome.

9. Ramanrine IW (2003) Induction of spawning and artificial incubation of eggs in the edible snail Pomacea urceus (Muller). Aquaculture 215: 163-166.

10. Huner JV (1988) Procambarus in North America and elsewhere.

11. Momot WT (1993) The role of exploitation in altering the processes regulating crayfish populations. Freshwater Crayfish 9: 101-117.

12. Gutiérrez-Yurrita PJ (1997) he ecological role of crayfish (Procambarus clarkii) in aquatic ecosystems Doñona National Park. An ecophysiological perspective and Bioenergetics.

13. Rodríguez-Serna M (1999) Biology and systematics Cambáridos southeastern Mexico and its potential use in aquaculture.

14. Zar JH (1996) Biostatistical analysis.

15. Crocos PJ, Kerr JD (1986) Factors affecting induction of maturation and spawning of the tiger prawn Penaeus esculentus (Haswell), under laboratory conditions. Aquaculture 58: 203-214

16. Brown A, McVey JP, Middleditch BS, Lawrence AL (1979) The maturation of white shrimp (Penaeus setiferus) in captivity. World Mariculture Society 10: 435-444.

17. Cripe GM (1994) Induction of maturation and spawning of pink shrimp, Penaeus duarorum by changing water temperature, and survival and growth of young. Aquaculture 128: 255-260.

18. Aktas, M., Kumlu M, Eroldogan OT (2003) Off-season maturation and spawning of Penaeus semisulcatus by eyestalk ablation and/or temperature-photoperiod regimes. Aquaculture 228: 361-370.

19. Dube P, Portelance B (1992) Temperature and photoperiod effects of ovarian maturation and egg laying of the crayfish Orconectes limosus. Aquaculture 102: 161-168.

20. Gutiérrez-Yurrita PJ, Montes C (1999) Bioenergetics and phenology of reproduction of the introduced red swamp crayfish, Procambarus clarkii, in DonÄ ana National Park, Spain, and implications for species management. Feshwater Biology 42: 561-574.

21. Westin L, Gydemo RG (1986) Influence of light and temperature on reproduction and moulting frequency of the crayfish, Astacus astacus (L.). Aquaculture 52: 43-50.

22. Karplus I, Gideon H, Barki A (2003) Shifting the natural spring–summer breeding season of the Australian freshwater crayfish Cherax quadricarinatus into the winter by environmental manipulations. Aquaculture 220: 277-286.

23. Barki, A, Levi T, Hulata G, Karplus I (1997) Annual cycle of spawning and molting in the red-claw crayfish, Cherax quadricarinatus, under laboratory conditions. Aquaculture 157: 239-249.

24. Jordan CF (1996) Spatial ecology of decapods and fishes in a northern everglades wetland mosaic.

25. Goolish EM, Adelman IB (1984) Effects of ration size and temperature on the growth of juvenile common carp Cyprinus carpio. Aquaculture 36: 27-35.

4

Effect of Mushroom Beta Glucan (MBG) on Immune and Haemocyte Response in Pacific White Shrimp (*Litopenaeus vannamei*)

Chih-Chiu Yang[1]*, Shiu-Nan Chen[2], Chung-Lun Lu[2], Sherwin Chen[2], Kam-Chiu Lai[1] and Wen-Liang Liao[2]

[1]*College of Life Science, National Taiwan University, Taipei, Taiwan*
[2]*Institute of Fisheries Science, National Taiwan University, Taipei, Taiwan*

Abstract

The total haemocyte count (THC), differential haemocyte count (DHC), respiratory bursts, and phenoloxidase (PO) activity were determined in the Pacific white shrimp *Litopenaeus vannamei* which administrated by feeding with diets containing mushroom beta glucan (MBG) at 0.05% and 0.1%. Results showed that shrimp fed a diet containing 0.05% MGB had significantly increased THC and semi-granular cells ratio at 28 days. Intracellular superoxide anion (O_2^-) production were significantly increased at 14 days which shrimp fed a diet containing 0.05% MBG and PO activity were significantly increased at 14 days which shrimp fed a diet containing 0.1% MGB but equal with control at 28 days. In previously study, semi-granular cells were found as the primary immune activator with their high levels of granules, enzymes and proteins. In this study, the effectively enhance immune responses and compare to high induced semi-granular cells in Pacific white shrimps concluded that MBG is an effective and powerful immunostimulants for shrimp. For application in aquaculture, MBG administration through dietary administration is also a convenient and useful practical technique in the future.

Keywords: *Litopenaeus vannamei*; Mushroom beta glucan; Hyaline cell; Semigranual cell; Granual cell

Introduction

Pacific White shrimp *Litopenaeus vannamei* is the primary penaeid shrimps currently being cultured in Central and South America, as well as Pacific Rim countries. For more than two decades, shrimp farming has suffered viral disease problems like Taura syndrome virus (TSV) and Pacific white spot syndrome virus (WSSV) [1,2], as well as vibriosis due to *Vibrio alginolyticus* and *V. harveyi* [3]. Disease outbreaks are often a result of a deteriorated environment which is associated with intensification of cultivation, and with increases in the proportion of potentially pathogenic species in the Vibrio population of cultured pond waters [4]. Therefore, the immune ability of shrimp and their susceptibility to pathogens are of primary concern when they are subjected to environmental stressors. Penaeid shrimp like other invertebrates rely on an innate immune system for protection against pathogens [5]. Once microorganisms or other foreign particles invade the haemocoel of the host, they encounter a complex system of innate defence mechanisms involving cellular and humoral responses [6]. These include phagocytosis, nodulation, encapsulation, synthesis of antimicrobial peptides, blood coagulation, release of stress-responsive proteins, and the prophenoloxidase (proPO) activating system that leads to melanization [7]. The total haemocyte count (THC), phenol oxidase (PO) activity, respiratory burst (release of superoxide anion), superoxide dismutase (SOD) activity, phagocytic activity and bacterial clearance efficiency are commonly used as indicators for evaluating a host's immunity and health status [8,9] .

The immune stimulatory effects of glucan, chitosan, and other polysaccharides have been widely studied in fish and crustaceans [10]. The shrimp's immune system is activated by pattern recognition proteins including lipopolysaccharide, β-glucan and peptidoglycan binding proteins [11]. Once these proteins are bound to their specific targets, they activate haemocytes to release their contents and trigger different biochemical mechanisms [12]. Immunostimulants have been found to increase the immune responses in several shrimp species by promoting phagocytosis, bactericidal activity, proPO activity, and respiratory bursts, and enhancing resistance against pathogens [13].

β-glucan is polysaccharides of D-glucose monomers linked by β-glycosidic bonds. This structure occur most commonly in algals, the bran of cereal grains, the cell wall of baker's yeast, certain fungi, mushrooms and bacteria. The β-glucan isolated from the mushrooms are poly-glucose of β-1,3 or β-1,3 and β-1,6 cross-linked compounds [14]. Approximately 50% of the fungal species are able to release water-soluble β-glucan. In particular, β-glucans with β-1,3 as the primary chain and β-1,6 as the branched chains demonstrate immune-regulatory effects. Different fungal species contain β-glucans with varying proportions of the main and branched chains. In the recent years, there have been many studies published in the field of aquaculture relating to injection, immersion and supplementation of β-glucans in the feeds to enhance nonspecific immunity of the cultured animals. Chen and Ainsworth noted that injection of β-glucan is effective against *Edwardsiella ictaluri* in catfish *Lctalurus punctatus* [15]. Also, β-glucan enhanced the cytotoxic activities of macrophages, lymphocytes and natural killer cells in fish immune system [16,17] and promoted proPO in the lymphatic fluids of crayfish and horseshoe crabs, leading to a fortified bactericidal effect and phagocytosis, in addition to promotion of haemolymph coagulation [18,19]. Immersion of 0.5 and 1.0 mg/ml β-glucan improved *Penaeus monodon* in enhances the shrimps' immunity against *Vibrio vulnificus* [20]. Furthermore, β-glucan was examined in Pacific white shrimps that can enhance these shrimps' immunity against WSSV [21].

In this paper, diets containing mushroom β-glucan (MBG)

*Corresponding author: Chih-Chiu Yang, College of Life Science, National Taiwan University, Taipei, Taiwan, E-mail: snchen@ntu.edu.tw

(M.W.=559.63 to 6.65 kDa) was administered among *L. vannamei* to study the its impact on THC, differential haemocyte count (DHC), intracellular superoxide anion (O_2^-) production and PO activity.

Materials and Methods

Animal

L. vannamei were obtained from National Taiwan Ocean University (Keelung, Taiwan). Upon arrival, shrimps were acclimated to laboratory conditions for one week in a glass tank (200 L) and fed a basal diet (without MBG). At the beginning of the experiment, nine aquaria (200 L) were each stocked with 50 shrimps with an average initial weight of 13 ± 1 g. Three groups of shrimps were fed with experimental diet. Continuous aeration was maintained for each aquarium and 75% of the water in each aquarium was changed weekly to remove impurities and maintain optimal water quality. Dissolved oxygen concentration was maintained ≧4 mg/L throughout the experimental period. Water temperature ranged from 27-29°C, as pH ranged from 7.3-8.3 and salinity ranged from 30-32‰. A photoperiod of 12 hours light, 12 hours dark (08.00-20.00 h) was implemented. Groups of shrimps were fed their respective diet at a rate of 8% of body weight each day. This daily ration was divided into two equal feedings at 08.30 and 17.30 h. The duration of the study was four weeks.

Preparation of experimental diet

The experimental Mushroom Beta Glucan (MBG) extracted from *Ganoderma lucidum* and *Coriolus versicolor* was supplied by Super Beta Glucan Inc. (Irvine, California, USA) [17]. Based on the recommended nutrient requirements of *L. vannamei* [22], a basal diet was formulated without immunostimulants (Table 1), and 0.05% and 0.1% of MBG was added and mixed to the shrimp feeds as the experimental diet. All dietary ingredients were ground into a fine powder through a 149 µm mesh sieve and mixed thoroughly with fish oil and cold water. The mixture was then passed through an extruder with a 1.5 mm die to produce spaghetti-like strings. After drying, the diets were cut into pellets and stored at −20°C for experiment.

Experimental design

Haemolymph extraction and haemocyte identification: At 0, 14 and 28 days of the feeding trial, five shrimps were randomly selected from each aquarium. Haemolymph was withdrawn from the ventral sinus of each shrimp with a 1-mL sterile syringe (26-gauge) containing 0.9 ml of an anticoagulant solution (0.45 M NaCl, 0.1 M

Ingredients	Percentage
Fish meal	25
Shrimp head meal	5
Peanut meal	14
Squid visceral meal	5
Soybean meal	18
Fish oil	1
Soy lecithin	2
Wheat flour	27.58
Choline chloride (50%)	0.30
Ca(H_2PO_4)_2	0.37
Vitamin permix	0.5
Mineral permix	1
Cellulose	0.25

MBG replaced from cellulose, T1 group contains 0.05% of MBG; T2 group contains 0.1% of MBG

Table 1: Composition of the basal diets (as percentage dry weight).

glucose, 10 mm EDTA, 30 mm Sodium citrate, 26 mm citrate acid, pH 7.3, osmolality 900 mOsm kg^{-1}), a modification from the solution described by Söderhäll and Smith [23]. A drop of the anticoagulante-haemolymph mixture (100 µL) was placed on a haemocytometer to measure hyaline cell, granular cell and semi-granular cell), and the total haemocyte count (THC) and different haemocyte count (DHC) were determined using an inverted phase-contrast microscope (Olympus, IMT-20, Japan). The remainder of the haemolymph mixture was used for subsequent tests.

Intracellular superoxide anion (O_2^-) production assay: Intracellular superoxide anion (O_2^-) production of the haemocytes was quantified using the reduction of nitrobluetetrazolium (NBT) to formazan as a measure of superoxide anion production [20]. The collected haemolymph was diluted with 4 volumes of anticoagulant, and then centrifuged at 700×g for 10 min at 4°C. The resultant haemocyte pellet was then resuspended to 10^6 cells ml^{-1} in a modified complete Hank's balanced salt solution (MCHBSS) containing 10 mM CaCl$_2$, 3 mM MgCl$_2$, 5 mM MgSO$_4$ and 24 mg ml^{-1} HBSS (Sigma). Haemocyte suspension (100 µL) was added to flat-bottomed 96 well microtitre plates (10^5 haemocytes well^{-1}) and cytocentrifuged at 700xg for 20 min at 4°C. After removing the supernatant, O_2^- production of the haemocytes was quantified by a zymosan (2 mg ml^{-1}, Sigma) induced well which compare to non-induced well (MCHBSS added). NBT (100 µL, 0.3% in MCHBSS) was then added to the haemocytes and more incubated for 30 min at 37°C. The staining reaction was terminated by removing the NBT solution and adding absolute methanol. After three washes with 70% methanol, haemocytes were air-dried. 120 µL 2 M KOH and 140 µL dimethyl sulfoxide were added to dissolve the cytoplasmic formazan. Optical densities of the dissolved cytoplasmic formazan were measured at 630 nm with a precision microplate reader. Ratio of OD$_{630}$ of the zymosan stimulated haemocytes to the OD$_{630}$ of the control haemocytes was used as an index for comparing the effect of different level of MBG on O_2^-.

PO activity assay: Phenoloxidase (PO) activity was measured spectrophotometrically by recording the formation of dopachrome produced from L-3,4-dihydroxyphenylalanine (L-DOPA) as previously described by Liu et al. [24,25]. Optical density at 490 nm of the shrimp's PO activity was measured using a spectrophotometer and expressed as dopachrome formation per 50 µL of haemolymph.

Statistical Analysis

Data were analyzed by one-way ANOVA from which multiple comparisons among the different means detected were made with Duncan's new multiple range tests. Statistical significance was determined by setting the aggregate type I error at 1% (P<0.01) for each set of comparison.

Result

Total haemocyte count (THC)

Shrimps fed with diets containing MBG at 0.1% concentrations for 14 day exhibited higher THC than the control group significantly. Statistically significant higher THC was also observed in the shrimps receiving feeds containing 0.05 and 0.1% MBG for 28 days than the control group (P<0.05) (Figure 1).

Differential haemocyte count (DHC)

Shrimps fed with diets containing 0.05% MBG for 14 days exhibited statistically significant lower granular cell counts than the

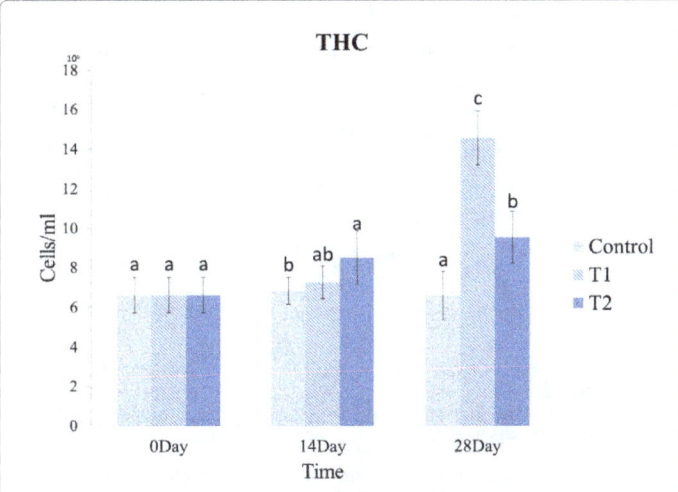

Figure 1: THC of Pacific white shrimp (*L. vannamei*) fed with diets contain.

Days	Population	Control	0.05%	0.1%
0		11.05% ± 2.50%	11.15% ± 2.20%	11.35% ± 2.29%
14	Hyaline cell	13.92% ± 2.85%	13.20% ± 2.94%	9.92% ± 2.28%
28		11.38% ± 2.83%[a]	6.37% ± 2.22%[b]	9.95% ± 1.77%[a]
0		28.05% ± 6.98%	28.05% ± 6.98%	28.05% ± 6.98%
14	Granular cell	31.26% ± 4.33%[a]	26.03% ± 3.95%[b]	30.62% ± 4.18%[a]
28		13.75% ± 4.24%[a]	10.32% ± 3.26%[b]	11.49% ± 2.74%[ab]
0		60.90% ± 7.04%	60.90% ± 7.04%	60.90% ± 7.04%
14	Semi-Granular cell	54.82% ± 4.18%[b]	60.77% ± 2.64%[a]	59.46% ± 4.68%[ab]
28		74.86% ± 6.04%[c]	83.31% ± 3.35%[a]	78.56% ± 2.70%[b]

Values are means_S.D, *n*=5

a,b,c: Values with different superscripts in each column are significantly different (*P*<0.01)

Table 2: Different haemocyte count (DHC) of Pacific white shrimp fed diets with different level of Mushroom beta-glucan (MBG) for 28 days.

groups fed with 0.1% and control diet, respectively (p<0.05), but a statistically significant higher semi-granular cell count than the group fed with control diet (p<0.05). At 28 days, a significant lower hyaline and granular cell count was observed in the group receiving feeds containing 0.05% MBG. In addition, the MBG treated groups for 28 days showed the higher semi-granular cell count than the group fed with control diet (p<0.05) (Table 2).

Intercellular superoxide anion (O_2^-) production

Results obtained from the observation of intercellular superoxide anion production in the shrimp fed with various concentrations of MBG were presented in Figure 2. Both the groups fed with diets containing MBG for 14 days exhibited statistically significant higher activities than the control group (P<0.05), 0.05% MBG showed the highest activity and then was 0.1% MBG. At 28 days, 0.05% MBG and 0.1% MBG treated group showed statistically significant higher activities than the control group, but no statistical difference existed between these two groups(P<0.05).

PO activity

Results obtained from the observation of PO activity in the groups fed with two concentrations of MBG were presented in Figure 3. PO activity of the group fed with MBG for 14 days exhibited statistically significant higher activities than the control group (P<0.05) and showed

a dosage dependent. At 28 days, there was no statistically significant differences in experimental treated groups (P>0.05).

Discussion

Beta glucans derived from microorganisms have been used to enhance innate immunity and improve resistance against pathogens in experimental aquatic animals [26,27]. In modern aquaculture applications, beta glucan derived from the mushrooms has demonstrated its efficacy in enhancing innate immunity and pathogen resistance in cultured shrimps and mollusks successfully [28-31].

Beta glucan not only plays an important role in the nonspecific immune system of the Pacific white shrimps, but it is also associated with shrimp haemocyte's pattern recognition, phagocytosis, reactive oxygen species formation, prophenoloxidase activating system, encapsulation, nodule formation, and release of antimicrobial peptides and lysozymes [6,7,24]. With elevated haemolymph THC, there is an increased number of immune cells to combat infectious sources and subsequently, reduces mortality. Among all haemocytes types in Pacific white shrimps, hyaline cells are the smallest shape with the highest nucleus-to-cytoplasm ratio and minimal granules to execute

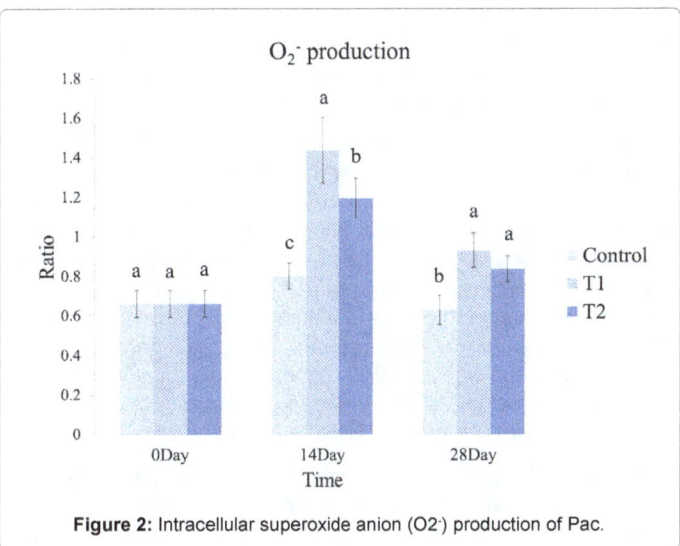

Figure 2: Intracellular superoxide anion (O2-) production of Pac.

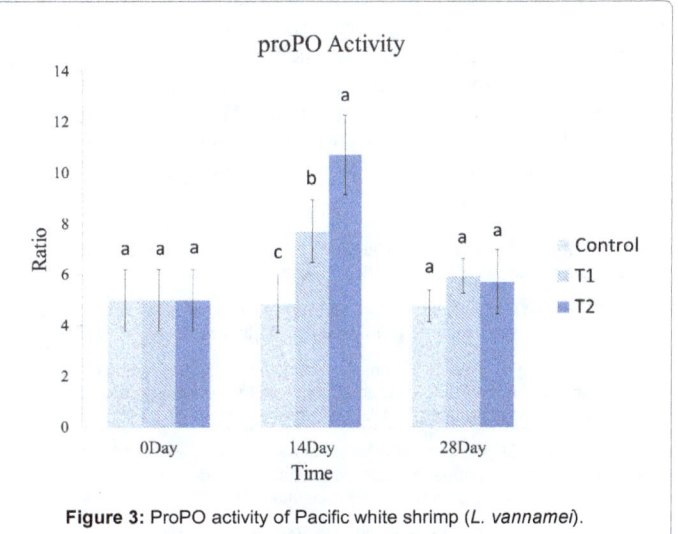

Figure 3: ProPO activity of Pacific white shrimp (*L. vannamei*).

phagocytosis. The size of semi-granular cells and their nucleus-to-cytoplasm ratio is in between that of the granular cells and hyaline cells. Additionally, semi-granular cells contain fewer granules and more cytoplasm and are capable of complement-dependent cytotoxicity, prophenoloxidase (proPO) activation and weak phagocytosis. Granular cells have the largest volume, smallest nucleus and largest granules of the three types of cells [6,7,23,32]. Of the three, semi-granular cells are the most unstable and sensitive haemocytes, which frequently degranulate after being stimulated by foreign particles such as LPS and β-glucan, resulting in cell breakdown if more seriously [7,33,34].

Because each of the three types of haemocytes has its niche of function, the varying proportions of the three cell types at a given time indicate a particular immune function in progress. After administration of MBG for 14 days, the total numbers of all three haemocytes were higher in treated groups than the control group. After 28 days, 0.05% treated group showed the highest total blood cell counts. However, after 14 days, the proportions of haemocytes did not change, indicating an enhancement of immune function from baseline without bias towards a specific function. At 28 days, 0.05% treated group showed an increase in the proportion of semi-granular cells, as these cells are the most sensitive and reactive of the three haemocytes [33]. In another words, semi-granular cells act as the primary immune activator of the Pacific white shrimps with their high levels of granules, enzymes and proteins. Therefore, an increase in the proportions of semi-granular cells can effectively enhance immune response in Pacific white shrimps.

In the present study, both haemocyte intercellular superoxide anion production and PO activity increased after MBG treated at 14 days. After 28 days, these factors were reduced and similar to the control group. On the other hand, THC and DHC levels were significantly different between the treatment and control groups. We speculate that this difference is a result of using immune-naive Pacific white shrimps without previous stimulation by infectious sources as our experimental hosts. Consequently, although immune markers were not elevated, THC and DHC levels showed a significant increase after supplementation of MBG in the feeds, resulting in an increased immune response to infections [13,16,18,21].

Using *G. lucidum* and *C. versicolor* as the sources of origin, mushroom beta glucan was extracted and analyzed, followed by employment of HPLC to determine the specific molecular weight [17]. Supplementation to feeds and determination of immune markers were subsequently performed. This study showed that MBG enhances the innate immune response with surrogate markers of THC, DHC, intercellular superoxide anion production, and PO activity of the Pacific white shrimps. As the use of natural immunomodulators is getting more attention, we suggest that MBG is an effective and powerful Immunostimulants for shrimp aquaculture and can be useful in practical applications during the post larvae stage.

References

1. Lo CF, Chang YS, Peng SE, Kou GH (2003) Major viral diseases of Penaeus monodon in Taiwan. J Fisheries Soc Taiwan 30: 1-13.

2. Yu CI, Song YL (2000) Outbreaks of Taura syndrome in Pacific white shrimp Penaeus vannamei cultured in Taiwan. Fish Pathol 35: 21-4.

3. Liu CH, Cheng W, Hsu JP, Chen JC (2004) Vibrio alginolyticus infection in the white shrimp Litopenaeus vannamei confirmed by polymerase chain reaction and 16S rDNA sequencing. Dis Aquat Organ 61: 169-174.

4. Lavilla-Pitogo CR, Leano EM, Paner MG (1998) Mortalities of pond-cultured juvenile shrimp, Penaeus monodon, associated with dominance of luminescent vibrios in the rearing environment. Aquaculture 164: 337-349.

5. Anderson DP (1992) Immunostimulants, adjuvants, and vaccine carriers in fish: Applications to aquaculture. Ann Rev Fish Diseases 2: 281-307.

6. Takahashi Y, Itami T, Kondo M (1995) Immunodefense system of crustacea. Fish Pathol. 30:141-50.

7. Jiravanichpaisal P, Lee BL, Söderhäll K (2006) Cell-mediated immunity in arthropods: hematopoiesis, coagulation, melanization and opsonization. Immunobiology 211: 213-236.

8. Le Moullac G, Haffner P (2000) Environmental factors affecting immune responses in Crustacea. Aquaculture 191: 121-131.

9. Rodriguez J, Le Moullac G (2000) State of the art of immunological tools and health control of penaeid shrimp. Aquaculture 191: 109-119.

10. Sakai M (1999) Current research status of fish immunostimulants. Aquaculture 172: 63-92.

11. Cheng W, Liu CH, Tsai CH, Chen JC (2005) Molecular cloning and characterisation of a pattern recognition molecule, lipopolysaccharide- and beta-1,3-glucan binding protein (LGBP) from the white shrimp Litopenaeus vannamei. Fish Shellfish Immunol 18: 297-310.

12. Wang XW, Wang JX (2013) Pattern recognition receptors acting in innate immune system of shrimp against pathogen infections. Fish Shellfish Immunol 34: 981-989.

13. Chang CF, Chen HY, Su MS, Liao IC (2000) Immunomodulation by dietary beta-1, 3-glucan in the brooders of the black tiger shrimp Penaeus monodon. Fish Shellfish Immunol 10: 505-514.

14. Yano T, Mangindaan REP, Matsuyama H (1989) Enhancement of the Resistance of Carp Cyprinus-Carpio to Experimental Edwardsiella-Tarda Infection, by Some Beta-1,3-Glucans. Nippon Suisan Gakk 55: 1815-1819.

15. Chen D, Ainsworth AJ (1992) Glucan Administration Potentiates Immune Defense-Mechanisms of Channel Catfish, Ictalurus-Punctatus Rafinesque. J fish diseases 15: 295-304.

16. Engstad RE, Robertsen B, Frivold E (1992) Yeast glucan induces increase in lysozyme and complement-mediated haemolytic activity in Atlantic salmon blood. Fish Shellfish Immunol 2: 287-297.

17. Chang CS, Huang SL, Chen S, Chen SN (2013) Innate immune responses and efficacy of using mushroom beta-glucan mixture (MBG) on orange-spotted grouper, Epinephelus coioides, aquaculture. Fish Shellfish Immunol 35: 115-125.

18. Söderhäll K (1981) Fungal cell wall beta-1,3-glucans induce clotting and phenoloxidase attachment to foreign surfaces of crayfish hemocyte lysate. Dev Comp Immunol 5: 565-573.

19. Cerenius L, Söderhäll K (2011) Coagulation in invertebrates. J Innate Immun 3: 3-8.

20. Song YL, Hsieh YT (1994) Immunostimulation of tiger shrimp (Penaeus monodon) hemocytes for generation of microbicidal substances: analysis of reactive oxygen species. Dev Comp Immunol 18: 201-209.

21. Bai N, Gu M, Zhang W, Xu W, Mai K (2014) Effects of ß-glucan derivatives on the immunity of white shrimp Litopenaeus vannamei and its resistance against white spot syndrome virus infection. Aquaculture 426-427: 66-83.

22. Shiau SY (1998) Nutrient requirements of penaeid shrimps. Aquaculture 164: 77-93.

23. Söderhäll K, Smith VJ (1983) Separation of the haemocyte populations of Carcinus maenas and other marine decapods, and prophenoloxidase distribution. Dev Comp Immunol 7: 229-239.

24. Hemández-López J, Gollas-Galván T, Vargas-Albores F (1996) Activation of the Prophenoloxidase System of the Brown Shrimp (Penaeus californiensis Holmes). Comp Biochem Physiol 113: 61-66.

25. Liu CH, Chen JC (2004) Effect of ammonia on the immune response of white shrimp Litopenaeus vannamei and its susceptibility to Vibrio alginolyticus. Fish Shellfish Immunol 16: 321-334.

26. El-Boshy ME, El-Ashram AM, AbdelHamid FM, Gadalla HA (2010) Immunomodulatory effect of dietary Saccharomyces cerevisiae, beta-glucan and laminaran in mercuric chloride treated Nile tilapia (Oreochromis niloticus) and experimentally infected with Aeromonas hydrophila. Fish Shellfish Immunol 28: 802-808.

27. Harikrishnan R, Balasundaram C, Heo MS (2011) Diet enriched with mushroom

Phellinus linteus extract enhances the growth, innate immune response, and disease resistance of kelp grouper, Epinephelus bruneus against vibriosis. Fish Shellfish Immunol 30: 128-134.

28. Encarnacion AB, Fagutao F, Hirayama J, Terayama M, Hirono I, et al. (2011) Edible mushroom (Flammulina velutipes) extract inhibits melanosis in Kuruma shrimp (Marsupenaeus japonicus). J Food Sci 76: C52-58.

29. Gu M, Ma HM, Mai KS, Zhang WB, Bai N, et al. (2011) Effects of dietary beta-glucan, mannan oligosaccharide and their combinations on growth performance, immunity and resistance against Vibrio splendidus of sea cucumber, Apostichopus japonicus. Fish Shellfish Immunol 31: 303-309.

30. Yeh SP, Hsia LF, Chiu CS, Chiu ST, Liu CH (2011) A smaller particle size improved the oral bioavailability of monkey head mushroom, Hericium erinaceum, powder resulting in enhancement of the immune response and disease resistance of white shrimp, Litopenaeus vannamei. Fish Shellfish Immunol 30: 1323-1330.

31. Zhu F, Zhang X (2012) Protection of shrimp against white spot syndrome virus (WSSV) with Î²-1,3-D-glucan-encapsulated vp28-siRNA particles. Mar Biotechnol (NY) 14: 63-68.

32. Guzman MA, Ochoa JL, Vargas-Albores F (1993) Haemolytic activity in the brown shrimp (Penaeus californiensis Holmes) haemolymph. Comp Biochem Physiol A Comp Physiol 106: 271-275.

33. Cerenius L, Jiravanichpaisal P, Liu HP, Söderhill I (2010) Crustacean immunity. Adv Exp Med Biol 708: 239-259.

34. Sritunyalucksana K, Söderhäll K (2000) The proPO and clotting system in crustaceans. Aquaculture 191: 53-69.

5

Effect of Stocking Large Channel Catfish in a Biofloc Technology Production System on Production and Incidence of Common Microbial Off-Flavor Compounds

Bartholomew W Green[1]* and Kevin K Schrader[2]

[1]US Department of Agriculture, Agriculture Research Service, Harry K. Dupree Stuttgart National Aquaculture Research Center, Stuttgart, Arkansas USA
[2]US Department of Agriculture, Agricultural Research Service, Natural Products Utilization Research Unit, Thad Cochran National Center for Natural Products Research, University, Mississippi, USA

Abstract

Density-dependent production and incidence of common microbial off-flavors caused by geosmin and 2-methylisoborneol were investigated in an outdoor biofloc technology production system stocked with stocker-size (217 g/fish) channel catfish at 1.4, 2.1, or 2.8 kg/m³. Individual weight at harvest ranged from 658-829 g/fish and was inversely related to stocking density. Net fish yield ranged from 3.8-5.4 kg/m³, and increased linearly as stocking density increased. The percentage of sub-marketable fish (<0.57 kg/fish) increased linearly with increasing stocking rate. Mean total feed consumption increased linearly with stocking density, but feed consumed per fish was inversely related to stocking density. Feed conversion ratio did not differ significantly among treatments. Concentrations of geosmin and 2-methylisoborneol in biofloc water were low throughout the study. All sampled fillets contained low concentrations of geosmin and 2-methylisobornel, but these fillets likely would not be deemed as having objectionable "earthy" or "musty" off-flavors when evaluated by trained processing plant flavor testers because of the low concentrations present. Data from this study combined with data from our two previous studies provide strong evidence that the incidence of geosmin- and 2-methylisoborneol-related off-flavor episodes is low in the BFT production system.

Keywords: Biofloc technology; Channel catfish; Ictalurus Punctatus; Stockers; Density; Geosmin; 2-Methylisoborneol; MIB; Market-Size fish

Introduction

High yields are obtained from the biofloc technology (BFT) production system in response to high stocking and feeding rates because the biofloc, which is maintained in suspension by continuous aeration, metabolizes excreted feed nitrogen [1,2]. Net yield of market-size channel catfish (*Ictalurus punctatus*) as high as 9.3 kg/m³ has been reported for BFT production [3]. In addition to channel catfish, the BFT production system is used to grow the Pacific white shrimp (*Litopenaeus vannamei*) [4] and [5] and Nile tilapia (*Oreochromis niloticus*) [2] and [6].

Stocking rate is known to affect channel catfish production at different life stages and in a variety of production environments, including the BFT production system [7-9]. Although, individual fish growth and final fish size are inversely related to stocking rate, yield can increase with stocking rate because of the greater number of fish. In investigating density effects on channel catfish production in BFT production, stocking rate was inversely related to individual weight at harvest, but positively related to net fish yield [3,9,10]. Stocker-size catfish (115-150 g/fish) are being stocked increasingly by farmers in food-fish ponds so that harvested fish are within the 0.57-2.04 kg/fish size range preferred by processing plants. In 2009, 56.6% of fish stocked into production ponds by farmers were stocker catfish [11].

The effect of stocking rate on rearing stocker-size catfish to market size in BFT production has not been researched. Only one study addresses production of market size catfish in ponds stocked with stocker catfish. In that study, up to 98.5% of the channel catfish population was within the preferred size range when ponds were stocked with 0.26 kg/fish average-size stocker catfish [12]. Thus, it is important to determine how stocking rate of stocker-size catfish affects production of market size fish in BFT production.

Geosmin and 2-methylisoborneol (MIB) are the compounds responsible for the "earthy" and "musty" off-flavors, respectively, and these compounds can accumulate in fish flesh and temporarily render them unmarketable [13] and [14]. Harvest delays caused by off-flavor episodes are a persistent problem for catfish farmers: 69.6% of operations and 53.3% of food-fish ponds experienced delayed harvest in 2002, and 80.7% of operations and 48.1% of food-fish ponds experienced delayed harvest in 2009 [11,15]. Geosmin and MIB have been detected in channel catfish BFT culture units, but aqueous concentrations generally are low and in preceding studies only 11% of culture units contained fish that would be judged as having "earthy" or "musty" off-flavors when evaluated by trained processing plant flavor testers [3,10]. In contrast, concentrations of geosmin and MIB in food-fish pond waters can exceed 2,000 ng/L and 700 ng/L, respectively [16-18], and as many as 76% of ponds may contain off-flavored fish from July-September [19]. Thus, reduced incidence of episodes of "earthy" or "musty" off-flavors is a potential advantage of the BFT production system compared to static-water pond systems.

In this study we sought to determine the effect of initial biomass of stocker-size channel catfish on production characteristics, water

---

*Corresponding author: Green BW, US Department of Agriculture, Agricultural Research Service, Harry K. Dupree Stuttgart National Aquaculture Research Center, USA, E-mail: bart.green@ars.usda.gov

quality, and microbial off-flavor compounds in an outdoor BFT production system.

Materials and Methods

Biofloc technology production system

This study was carried out in nine 15.6-m³ tanks located outdoors at the USDA Agricultural Research Service, Harry K. Dupree Stuttgart National Aquaculture Research Center (HKDSNARC), Stuttgart, AR, USA. Triplicate tanks (described in detail by Green et al. [10]) were assigned using a completely randomized design to initial fish biomass treatments of 1.4, 2.1, or 2.8 kg/m³ (5.4, 8.1, or 10.8 fish/m²; designated LO, MED, HI, respectively). Animal care and experimental protocols were approved by the HKDSNARC Institutional Animal Care and Use Committee and conformed to ARS Policies and Procedures 130.4 and 635.1.

Between 23 April-9 May 2012, tanks were filled with well water (total alkalinity=228.4 mg/L as CaCO₃), and each was seeded with 2.3 m³ of water from a HKDSNARC pond containing a phytoplankton bloom, fertilized with 1.5 kg 11-37-0 (N-P-K) and 2.0 kg dried molasses (Sweet45, Westway Feed Products, New Orleans, LA, USA), and treated with 4.5 kg stock salt to ensure that chloride concentration exceeded 100 mg/L. Pond water was added to the tanks to expedite development of a phytoplankton bloom to aid in the removal of total ammonia-N (TAN). An additional mean of 2.4, 2.0, and 4.7 kg dried molasses was added to LO, MED, and HI treatment tanks, respectively, from 15 May-1 June. Dried molasses was added to tanks as a carbon source to stimulate bacterial transformation of TAN [20,21]. No water was exchanged, but well water was added as needed to replace evaporative loss and losses to draining settling chambers. Sodium bicarbonate (1.13 kg/tank) was added as needed to maintain pH above pH 7.0; mean total sodium bicarbonate added was 5.7, 7.6, and 8.8 kg/tank for the LO, MED, and HI treatments, respectively.

Each tank was equipped with a 130-L (117-L working volume) conical-bottom settling chamber; a 2.5-cm diameter air lift moved water to the settling chamber at 5.6 L/min. Settling chambers were operated, on average, 5.9 h/d on 11 d during the period 26 July-14 August in order to reduce TSS concentration to approximately 300-400 mg/L as recommended by Green et al. [10].

Catfish stocking and feeding rates

Stocker channel catfish were harvested from a holding pond, and fish from the population retained by a No. 70 bar grader (217 ± 74 g/fish; mean ± SD; coefficient of variation, CV,=34.1%) were stocked randomly into tanks on 10 May 2012. A random sample of 310 fish from the initial population was weighed individually (Figure 1). Fish were fed once daily with a commercially produced 32% protein floating

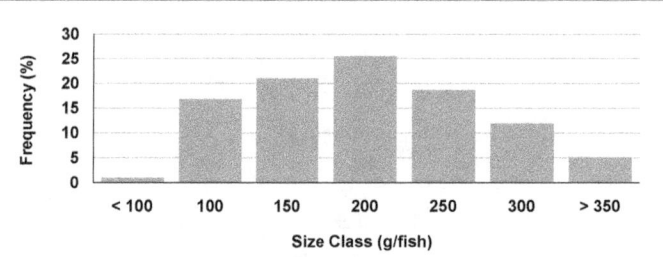

Figure 1: Mean size class distribution of the population of stocker channel catfish stocked into an outdoor biofloc technology production system. Mean (± SD) initial weight was 217 ± 74 g/fish.

extruded feed (premium formulation, Delta Western Feed Mill, Indianola, Mississippi) to apparent satiation (10 min) and the amount was recorded. Feed conversion ratio (FCR) was calculated for each tank as the total quantity of administered feed (wet weight) divided by the net total yield. The weight of dead fish recovered was recorded, but not all mortalities were recovered; weight of dead fish was not included in FCR calculation. All tanks were harvested by draining; two replicate tanks per treatment were harvested on 11 October and the remaining replicate per treatment was harvested on 12 October. At harvest, all fish per tank were weighed individually.

Water quality analyses

Water samples were collected weekly from each tank. Sample pH was measured electrometrically. Total alkalinity (titration to pH 4.5), settleable solids (SS), total suspended solids (TSS), and total volatile solids (TVS) in raw samples were measured using methods given by Eaton et al. [22]. Water was filtered through 0.2-um pore size membrane filter and analyzed for nitrite-nitrogen (NO₂-N, diazotization), nitrate-nitrogen (NO₃-N, cadmium reduction), and soluble reactive phosphorus (PO₄-P, ascorbic acid method) using flow injection analysis according to manufacturer instructions (FIAlab 2500; FIAlab Instruments, Bellevue, WA, USA). Flow injection analysis also was used to quantify total ammonia-nitrogen (TAN) fluorometrically in filtered samples using the o-phthaldialdehyde method of Genfa and Dasgupta [23]. Water samples were filtered through a 0.45-um pore size glass fiber filter for chlorophyll a analysis. Chlorophyll a was extracted in 2:1 chloroform:methanol from the phytoplankton (planktonic algae and cyanobacteria as well as those associated with the biofloc) retained on the filter, and the chlorophyll a concentration in the extract was determined by spectroscopy [24].

Dissolved oxygen (DO) and temperature in each tank were monitored continuously (10-sec scan rate) by a galvanic oxygen sensor (Type III, Oxyguard, Birkerød, Denmark) and a thermister (Model 109, Campbell Scientific, Logan, UT, USA) connected to a datalogger (Model CR206 or CR1000, Campbell Scientific, Logan, UT, USA).

Determination of microbial off-flavor compounds

Water samples were collected from each tank on 20 June and at approximately 4-wk intervals thereafter through 10 October for analysis of geosmin and 2-methylisoborneol (MIB). A sample of the biofilm that accumulated on the tank liner at the water surface also was collected from each tank on 20 June. Sample handling and shipment to the USDA-ARS Natural Products Utilization Research Unit (NPURU), Oxford, MS, USA, for analysis followed Schrader et al. [3]. Four samples from 20 June were lost accidently during shipment due to vial breakage.

Five catfish were selected at random from each tank at harvest, euthanized by cranial percussion, and filleted. Catfish fillets (one fillet/fish) were placed in individual plastic bags, vacuum sealed, and immediately frozen until overnight shipment to the USDA-ARS-NPURU for analysis. Fish fillets were stored frozen until further processing to obtain microwave distillates. For analysis of each fillet, a single 20-g sample was resected from the anterior end of the fillet by cutting 1-cm wide portions (2-3 portions per fillet) vertically from the dorsal to ventral side of the fillet and then each 1-cm wide sample was cut into approximately 1-cm cube-like pieces to undergo microwave distillation according to the method of Lloyd and Grimm [25].

Prior to analysis, water samples and microwave distillates of catfish fillet samples were processed by placing 0.6-mL aliquots into separate

2-mL glass crimp-top vials each containing 0.3 g sodium chloride. The methodology of Lloyd et al. [26] as modified by Schrader et al. [27] was used to quantify geosmin and MIB using solid phase microextraction and gas chromatography-mass spectrometry (SPME-GC-MS). Samples were analyzed using an Agilent 6890 gas chromatograph (Agilent, Palo Alto, CA, USA) and Agilent 5973 mass selective detector with attached CombiPal autosampler and solid phase microextraction assembly (LEAP Technologies, Inc., Carrboro, NC, USA). The GC-MS conditions were the same as those outlined by Schrader et al. [28] and each sample was run in triplicate. The instrumental detection limit for each compound was 1 part per trillion

Data analysis

Datasets were analyzed using the mixed models analysis of variance (MIXED), frequency (FREQ), and regression (REG) procedures of SAS version 9.4 (SAS Institute, Cary, NC, USA). Mean geosmin and MIB concentrations in water and geosmin concentration in fillets were not normally distributed and an appropriate data transformation was not found; therefore, data were analyzed by nonparametric one-way analysis of variance (NPAR1WAY).

Results

Fish production and feed consumption

Stocker-size channel catfish attained mean final weights that ranged from 658-829 g/fish (Table 1) and decreased linearly with increased stocking density (R^2=0.617, P=0.012). Final weight CV averaged 31.7,

32.6, and 33.4% for the LO, MED, and HI treatments, respectively, and did not differ significantly among treatment (P=0.515). Additionally, no significant difference (P>0.05) was detected between initial and final weight CV within each treatment. Chi-square analysis indicated a significant (P<0.001) association between stocking rate and fish size classes (Figure 2). Specifically, there were fewer fish in the <0.57-0.57 kg/fish size classes and more fish in the 0.79 kg/fish and larger size classes than expected in the LO treatment. In the MED treatment, there were fewer fish in the 0.68 kg/fish and smaller size classes, and more fish in the 0.79-1.02 kg/fish size classes than expected. There were more fish in the <0.57-0.68 kg/fish size classes and fewer fish in the 0.79 kg/fish and larger size classes than expected in the HI treatment. The percentage of sub-marketable fish (<0.57 kg/fish) increased linearly with increasing stocking rate (R^2=0.699, P=0.005). The percentage of fish larger than 0.57 (R^2=0.699, P=0.005) decreased linearly with increased stocking rate. Fish growth differed significantly among treatments and was linearly related to stocking density (R^2=0.698, P=0.005).

Gross and net yields ranged from 5.2-8.2 and 3.8-5.4 kg/m³, respectively, and increased linearly as initial biomass increased (R^2=0.900, P<0.001, and R^2=0.715, P=0.004, respectively). Catfish survival was not affected by stocking rate and averaged 97.2% across treatments.

Feed consumption was affected significantly by treatment (Table 1). Following a period of rapid increase that occurred during the first month following stocking, consistently high daily feed consumption was sustained from 10 June-15 September. Mean daily feed consumption during this peak feeding period increased linearly with increased

Initial	Individual			Fish		Feed			
Biomass	weight	Yield (kg/m³)		growth	Survival	Daily	Peak[†]	Total	
(kg/m³)*	(g/fish)	Gross	Net	(g/d)	(%)	(g/m³/d)		(kg/m³)	FCR[††]
1.4	828.9[a]	5.2[b]	3.8[b]	4.0[a]	97.3	50[b]	60[b]	7.7[b]	1.5
2.1	771.0[ab]	7.1[a]	5.0[a]	3.6[ab]	96.7	69[a]	82[a]	10.6[a]	1.5
2.8	658.4[b]	8.2[a]	5.4[a]	2.9[b]	97.7	76[a]	92[a]	11.7[a]	1.4
Pooled SE	34.1	0.3	0.2	0.2	0.01	3	3	0.4	0.1
ANOVA, P>F	0.032	0.001	0.001	0.022	0.658	0.001	0.001	0.001	0.615

*n=3 replicates per treatment.
[†]Period of peak feed consumption, weeks 24-37 (10 June-15 September).
[††]FCR=wet weight of feed/net fish yield.

Table 1: Least squares means (± SE) for individual weight at harvest, gross and net yields, net daily yield, survival, daily and peak feed consumption, total feed, and feed conversion ratio (FCR) for stocker-size (217 g/fish) channel catfish stocked at 1.4-2.3 kg/m³ in an outdoor biofloc technology production system and grown for 154 d.

Treatment	NH₄-N[a]	NO₂-N[a]	NO₃-N[a]	PO₄-P[a]	pH	T Alk[a]	Chl a[a]	SS[a]	TSS[a]
1.4 kg/m³									
Initial	1.01	0.02	0.00	7.85	8.7	217.1	661.8	3	105.4
Final	0.02	0.03	94.91	23.58	7.5	84.4	1,694.0	57	667.8
Pooled SE	0.38	0.01	5.54	0.63	0.0	4.5	197.9	1	16.5
Pr > F	0.139	0.671	0.007	0.002	0.002	<0.001	0.026	<0.001	<0.001
2.1 kg/m³									
Initial	0.84	0.03	0.00	8.32	8.7	214.1	675.9	4	114.8
Final	0.02	0.02	112.60	28.35	7.3	76.8	1,471.5	75	760.0
Pooled SE	0.37	0.01	11.51	2.01	0.1	4.4	231.0	7	41.4
Pr > F	0.260	0.895	0.020	0.002	<0.001	<0.001	0.072	0.002	0.005
2.8 kg/m³									
Initial	0.41	0.03	0.00	7.48	8.7	218.4	602.8	4	102.2
Final	0.02	0.03	130.88	32.67	7.3	81.6	1,048.2	68	753.3
Pooled SE	0.14	0.01	6.45	0.70	0.1	6.8	234.9	1	25.0
Pr > F	0.128	0.982	0.005	<0.001	<0.001	0.002	0.251	<0.001	<0.001

[a] Total ammonia nitrogen (mg/L NH₄-N), nitrite-nitrogen (mg/L NO₂-N), nitrate-nitrogen (mg/L NO₃-N), soluble reactive phosphorus (mg/L PO₄-P), total alkalinity (mg/L as CaCO₃ T Alk), chlorophyll a (mg/m³ Chl a), settleable solids (mL/L SS), and total suspended solids (mg/L TSS).

Table 2: Within treatment comparison of least squares mean (± SE) initial and final water quality variable concentrations for outdoor biofloc technology tanks stocked with large (217 g/fish) channel catfish at 1.4–2.8 kg/m³.

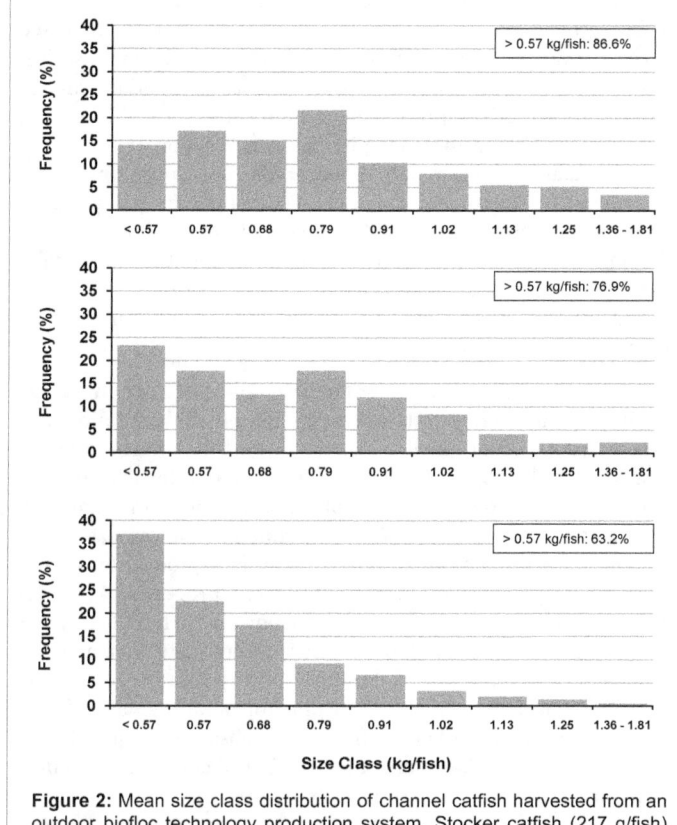

Figure 2: Mean size class distribution of channel catfish harvested from an outdoor biofloc technology production system. Stocker catfish (217 g/fish) were stocked at 1.4 kg/m³ (top), 2.1 kg/m³ (middle), or 2.8 kg/m³ (bottom). Fish < 0.57 kg/fish are sub-marketable.

stocking biomass (R^2=0.850, P<0.001). Daily feed consumption during the peak feeding period was 18% higher, on average, than mean daily feed consumption for the entire experiment. Mean total feed consumption was linearly related to initial biomass (R^2=0.822, P<0.001). However, feed consumed per fish decreased linearly (R^2=0.659, P=0.008) with increased stocking density and averaged 1.23, 1.15, and 0.94 kg/fish for the LO, MED, and HI treatments, respectively. Feed conversion ratio did not differ significantly among treatments.

Water quality

Mean daily DO concentrations did not differ significantly among treatments, averaging (± SE) 6.4 ± 0.2 mg/L (P=0.697) (79.5 ± 2.6% saturation, P=0.758). Mean daily water temperature ranged from 13.9 –32.0°C over the course of the experiment, did not differ significantly among treatments (P=0.364), and averaged 26.7 ± 0.2°C.

No significant differences (P>0.05) were detected among treatments in initial concentration of any water quality variable. Mean final PO4-P concentration differed significantly (P=0.020) among treatments, and increased linearly (R^2=0.728, P=0.003) as stocking rate increased. No other significant differences (P>0.05) among treatments were detected for final water quality variable concentration. Significant differences were detected between mean initial and final concentrations of all water quality variables within treatment except for TAN and NO_2-N (Table 2).

Microbial off-flavor compounds

Geosmin and MIB concentrations in water were low in all tanks, and ranged from 0-11 and 0-31 ng/L, respectively (Table 3). Geosmin concentration did spike in one tank on one sample date, but decreased to near zero by the next sample date. Concentrations of geosmin and MIB were at or below the instrument detection threshold of 1 ng/L in 59% and 61%, respectively, of the water samples analyzed. No significant treatment differences were detected for mean geosmin (P=0.415) or mean MIB (P=0.125) concentration in water. Geosmin (P=0.169) and MIB (P=0.726) concentrations in the biofilm did not differ significantly among treatments. No significant treatment differences were detected for geosmin (P=0.866) or MIB (P=0.283) concentrations in fillets. Mean aqueous and fillet concentrations of geosmin and MIB consistently have been low during three consecutive years of research on catfish production using the BFT production system (Table 4).

	Sampling Date					
	6/20		7/11	8/15	9/12	10/10	10/11-12
Tank†	Water	Biofilm	Water	Water	Water	Water	Fillet
Geosmin							
R2	0 (0)	2 (0)	0 (0)	0 (0)	10 (0)	1 (0)	5.7 (1.6)
R3	0 (0)	467 (150)	0 (0)	1 (0)	2 (0)	2 (1)	66.6 (13.7)
R8	·	542 (28)	0 (0)	0 (0)	7 (0)	1 (0)	23.9 (5.2)
R4	0 (0)	97 (3)	0 (0)	7 (1)	1 (0)	2 (0)	52.9 (20.2)
R6	·	1 (0)	0 (0)	0 (0)	8 (1)	2 (2)	13.5 (2.4)
R7	·	78 (4)	0 (0)	0 (0)	11 (0)	1 (1)	22.1 (4.6)
R1	0 (0)	34 (2)	0 (0)	1 (0)	10 (1)	4 (2)	25.6 (9.3)
R5	4	4 (1)	0 (0)	482 (26)	4 (0)	1 (0)	19.4 (2.1)
R9	·	152 (8)	3	0 (0)	5 (0)	1 (0)	23.1 (6.2)
MIB							
R2	0 (0)	0 (0)	0 (0)	0 (0)	6 (1)	1 (0)	10.4 (3.5)
R3	0 (0)	78 (35)	0 (0)	4(1)	0 (0)	1 (0)	25.8 (6.2)
R8	·	0 (0)	0 (0)	0 (0)	0 (0)	2 (1)	18.0 (2.2)
R4	7 (1)	13 (2)	0 (0)	31 (5)	0 (0)	1 (0)	19.8 (4.7)
R6	·	0 (0)	0 (0)	15 (3)	0 (0)	5 (1)	16.3 (5.8)
R7	·	0 (0)	5 (1)	0 (0)	0 (0)	2 (1)	17.1 (4.0)
R1	0 (0)	0 (0)	6 (1)	0 (0)	0 (0)	2 (0)	12.6 (3.7)
R5	6 (1)	0 (0)	0 (0)	19 (1)	0 (0)	6 (1)	9.4 (3.0)
R9	·	69 (8)	3 (3)	10 (3)	0 (0)	2 (0)	13.9 (6.3)

†Initial biomass: 1.4 kg/m³ (R2, R3, R8); 2.1 kg/m³ (R4, R6, R7); 2.8 kg/m³ (R1, R5, R9).

*Sample lost during shipment.

Table 3: Mean (±SD) geosmin and 2-methylisoborneol (MIB) concentrations during 2012 in water (ng/L), biofilm (ng/L), and fillet (ng/kg) in freshwater biofloc technology production system tanks stocked with channel catfish at three densities.

	Initial	Aqueous		Fillet		
	Biomass	Geosmin	MIB	Geosmin	MIB	
Year	(kg/m³)	(ng/L)		(ng/kg)		Source
2012	1.4	1.7 (0.3)	1.0 (0.0)	32.1 (18.1)	18.1 (4.4)	Present experiment
	2.1	2.7 (0.3)	5.0 (1.7)	29.5 (12.0)	17.7 (1.0)	
	2.8	34.3 (31.8)	4.0 (1.2)	22.7 (1.8)	12.0 (1.3)	
2011	1.4	64.5 (33.7)	53.0 (25.0)	123.3 (35.3)	18.5 (2.3)	Green et al. [10]
	1.8	5.4 (41.3)	12.7 (30.7)	29.3 (2.6)	55.6 (10.1)	
	2.3	34.1 (33.7)	27.1 25.0)	85.9 (10.3)	56.6 (14.5)	
2010	0.4	6.5 (1.6)	21.2 (15.9)	216.1 (137.1)	243.4 (200.6)	Schrader et al. [3]
	0.9	9.1 (1.2)	18.3 (10.0)	17.7 (13.3)	30.9 (5.3)	
	1.4	6.2 (2.2)	5.9 (1.5)	35.8 (4.6)	25.0 (1.4)	
	2.5	58.8 (36.9)	15.1 (4.7)	60.8 (11.4)	31.6 (4.6)	

Table 4: Comparison of mean (± SE) treatment concentrations of geosmin and 2-methyisoborneol (MIB) in water and fillet meat of channel catfish reared in an outdoor biofloc technology production system in the current and previous studies.

Discussion

Density-dependent effects were observed for channel catfish growth and yield in the BFT production system. Although mean individual weight at harvest was inversely related to stocking rate, fish yield increased linearly because of the increasing number of fish. However, the proportion of market-size fish decreased linearly as fish yield increased, which will impact production economics. The common size range accepted by processing plants without penalty is 0.57-2.04 kg/fish. Despite a 10-15% decrease in numbers of market-size fish and a 7% decrease in mean final individual weight compared to the LO treatment, the MED treatment yield of market-size fish was 21% higher. Fish in the HI treatment were 14.6% smaller, the proportion of harvested fish within the two processing plant size ranges decreased 17-32%, and yield of market-size fish decreased as much as 14% in comparison to the MED treatment. Thus, it appears that MED treatment performed the best. However, an economic analysis, which was beyond the scope of this study, would be required to identify the best stocking density.

Stocker catfish (217 g/fish) were grown in the BFT production system for the first time in the current study. No data exist for performance of stocker catfish raised to market size in the BFT production system, and only one study reports on earthen pond production. Stocker catfish (0.26 kg/fish) stocked into earthen ponds at approximately 0.22 kg/m³ (11,115/ha) grew at 4.0 g/d, achieved a mean final weight of 0.91 kg/fish in 164 d, and consumed 1.17 kg feed/fish [12]. This performance is similar to that of fish in the LO treatment of the current study. However, because of the increased stocking density, the 3.8 kg/m³ NFY in the LO treatment was substantially higher that the estimated 0.43 kg/m³ NFY in the pond study.

Channel catfish exhibit density-dependent growth during different phases of production and in multiple production environments [3,7] and [29-33]. Social interactions can affect growth of individual fish within the population [34] and [35]. Competition for food among individuals of a fish population is the most-common form of social interaction and results in growth dispensation [36]. If food is assumed to be distributed according to a size hierarchy, then size variation (as measured by the CV) would be exacerbated as competition intensifies. Increased stocking rate could be one factor that increases competition and variation in final fish size. Although stocking rate in the present study affected fish growth, variability in final fish size did not differ significantly among treatments and did not differ from variability in initial fish size. Few studies have examined the impact of stocking rate on variation in channel catfish final individual weight. Kilambi et al. [37] report that growth of channel catfish stocked in cages at 2.5-6.5 kg/m³ is unaffected by stocking rate as is variability in final individual weight, with CVs ranging from 26.5-41.1%. In a large study on growth variation of channel catfish reared in cages, Konikoff and Lewis [38] report that final individual weight CVs converge towards 30-40%. In another BFT study [3], channel catfish individual weight CV decreased from 43.2% at stocking to 29.0-36.9% at harvest, whereas in a flow-through system, individual weight CV decreased from 45.5% at stocking to 40.7% at harvest [39]. Results of the present experiment were consistent with these reports in that final weight CVs decreased only slightly from the initial weight CV and were within the 30-40% range.

Competition for food in the current experiment likely was negligible because fish were fed daily to apparent satiation. Consequently, final weight CVs at the different stocking rates would not be expected to differ. However, despite the linear increase in feed consumption in response to increased stocking density, feed consumption per fish and fish growth decreased linearly as stocking density increased, but feed conversion was unaffected. This inverse individual fish feed consumption-stocking rate relationship also was observed for channel catfish in an earlier BFT production system study [3] and in earthen pond studies [40] and [41]. Despite being fed to apparent satiation a diet formulated to meet nutritional needs, feed consumption was restricted by increased stocking rate. Differential mortality cannot explain these differences because fish survival in the current study was high and did not differ significantly among treatments. Water quality variable concentrations all were within ranges considered acceptable for rapid fish growth and would not be expected to affect individual fish feed consumption or growth.

Water quality in the static water BFT tanks was driven by feed input in response to high fish biomass. High phytoplankton biomass (as indicated by chlorophyll a concentration) and nitrification (as indicated by high NO_3-N concentrations) converted the excreted feed nitrogen. As the amount of feed application increased in response to increasing fish biomass, NO_3-N, PO_4-P, and TSS increased, which is consistent with results from other studies on BFT production systems dominated by photo- and chemo-autotrophic processes [3,5,10,42]. Nitrification caused a significant reduction in pH, but this was moderated by periodic addition of sodium bicarbonate. Increased TSS limited light penetration, which inhibited phytoplankton growth more in the MED and HI treatments. Thus, phytoplankton uptake of TAN likely was greater in the LO treatment. The absence of soil, which is a major sink for P [43], explains the high PO_4-P concentrations in all treatments.

Aqueous concentrations of geosmin and MIB consistently were low throughout the present study, and these results were consistent with those for other BFT culture of channel catfish [3,10]. Based on results of previous BFT studies [3,10], the presence of off-flavor-producing microorganisms likely was transitory because the turbulent mixing of the water in BFT tanks favors faster-growing diatoms and chlorophytes over cyanobacterial bloom-forming genera, which lose the cell buoyancy regulation competitive advantage they enjoy in quiescent waters [13,44].

Catfish fillets sampled in the current study all had analytical instrument detectable concentrations (≥ 1 ng/kg) of geosmin and MIB, but no sampled fillet exceeded the previously reported sensory threshold detection levels for geosmin (250-500 ng/kg) and MIB (100-200 ng/kg) of trained catfish processing plant flavor testers [45]. It is unlikely that fish in the present study would be classified as having objectionable "earthy" or "musty" off-flavors when evaluated by trained processing plant flavor testers because aqueous geosmin and MIB concentrations to which they had been exposed were low.

Three consecutive years (including the present study) of aqueous and fillet geosmin and MIB data [3,10] provides strong evidence of a reduced incidence of common microbial off-flavor episodes (i.e., "earthy," "musty") for channel catfish grown in the BFT production system compared to earthen ponds. Additionally, other types of off-flavors that can occur in catfish related to their foraging for food (e.g., "grassy," "vegetable") [46] are unlikely to occur in the BFT production units lined with high-density polyethylene. In the two earlier studies [3,10], 11.1% of tanks each year contained off-flavored fish, whereas in the current study no tank contained off-flavored fish. In catfish pond waters in the southern U.S., in contrast, geosmin and MIB concentrations that exceed 2,000 ng/L and 700 ng/L, respectively, are observed [16-18] and episodes of off-flavored (gesomin- or MIB-tainted) fish can be correlated with the presence of an off-flavor-producing cyanobacterium [19]. Geosmin and MIB off-flavors are prevalent in commercial catfish ponds from July-September, during which time up to 76% of ponds can contain off-flavored fish [19]. Only one report was found in which the incidence of off-flavor episodes for pond-raised channel catfish approached levels we observed in the BFT production system. Torrans and Lowell [47] report that 0-20% of channel catfish ponds co-stocked with blue tilapia (*Oreochromis aureus*) contained off-flavored catfish whereas 58-67% of catfish ponds in monoculture contained off-flavored fish.

In summary, density-dependent growth of stocker catfish was observed, but did not lead to increased variability in fish final weights at the different stocking rates. The highest yield of market-size fish was obtained by stocking fish at 2.1 kg/m³. However, an economic analysis would be required to identify the best stocking density. Although competition for food likely was negligible because fish were fed to apparent satiation, feed consumption per fish and fish growth were inversely related to stocking rate. The reason as to why individual fish feed intake decreased as stocking rate increased remains unknown, but likely is related to some aspect of social interaction. Concentrations of geosmin and MIB in tank waters were low throughout the study and while fillets from sampled fish contained analytically detectable geosmin and MIB concentrations, no fish would be classified as having "earthy" or "musty" off-flavors when evaluated by trained processing plant flavor testers. Data from this study combined with data from our two previous studies provide strong evidence that the incidence of common microbial off-flavor episodes is low in the BFT production system. However, there remains a continued need to identify the microbial sources of and elucidate the dynamics of geosmin and MIB production in the BFT production system.

Acknowledgment

The technical assistance of Greg O'Neal, Matt McEntire, Dewayne Harries, and Phaedra Page is greatly appreciated. This study was funded by the USDA/ARS under project number 6225-31630-006-00D. No sources of funding external to any author institution were used for this study. None of the authors has any conflict of interest. Mention of trade names or commercial products in this article is solely for the purpose of providing specific information and does not imply recommendation or endorsement by the U.S. Department of Agriculture. USDA is an equal opportunity provider and employer.

References

1. Hargreaves JA (2006) Photosynthetic suspended-growth systems in aquaculture. Aquacultural Engineering 34: 344-363.

2. Avnimelech Y (2007) Feeding with microbial flocs by tilapia in minimal discharge bio-flocs technology ponds. Aquaculture 264: 140-147.

3. Schrader KK, Green BW, Perschbacher PW (2011) Development of phytoplankton communities and common off-flavors in a biofloc technology system used for the culture of channel catfish (Ictalurus punctatus). Aquacultural Engineering 45: 118-126.

4. Browdy CL, Venero JA, Stokes AD, Leffler J (2009) Superintensive production technologies for marine shrimp Litopenaeus vannamei: technical challenges and opportunities. In: Burnell G, Allan GI, editors. New Technologies in Aquaculture: Improving Production Efficiency, Quality and Environmental Management. Woodhead Publishing Ltd., Cambridge. p 1010-1028.

5. Ray AJ, Lewis BL, Browdy CL, Leffler JW (2010) Suspended solids removal to improve shrimp (Litopenaeus vannamei) production and an evaluation of a plant-based feed in minimal-exchange, superintensive culture systems. Aquaculture 299: 89-98.

6. Rakocy JE, Bailey DS, Thoman ES, Shultz RC (2004) Intensive tank culture of tilapia with a suspended, bacterial-based treatment process: new dimensions in farmed tilapia. In: Bolivar R, Mair G, Fitzsimmons K, editors. Proceedings of the Sixth International Symposium on Tilapia in Aquaculture. Philippines Bureau of Fisheries and Aquatic Resources, Manila. p. 584-596.

7. Dunham RA, Brummett RE, Ella MO, Smitherman RO (1990) Genotype-environment interactions for growth of blue, channel and hybrid catfish in ponds and cages at varying densities. Aquaculture 85: 143-151.

8. Baumgarner BL, Schwedler TE, Eversole AG, Brune DE, Collier JA (2005) Production characteristics of channel catfish, Ictalurus punctatus, stocked at two densities in the Partitioned Aquaculture System. Journal of Applied Aquaculture 17: 75-83.

9. Green BW (2010) Effect of channel catfish stocking rate on yield and water quality in an intensive, mixed suspended-growth production system. North American Journal of Aquaculture 72: 97-106.

10. Green BW, Schrader KK, Perschbacher P (2014) Effect of stocking biomass on solids, phytoplankton communities, common off-flavors, and production parameters in channel catfish biofloc technology production system. Aquaculture Research 45: 1442-1458.

11. United States Department of Agriculture (USDA) (2010) Catfish 2010 Part II: Health and Production Practices for Foodsize Catfish in the United States, 2009. National Animal Health Monitoring System, Fort Collins.

12. Green BW, Engle CR (2004) Growth of stocker channel catfish to large market size in single-batch culture. Journal of the World Aquaculture Society 35: 25-32.

13. Paerl HW, Tucker CS (1995) Ecology of blue-green algae in aquaculture ponds. Journal of the World Aquaculture Society 26: 109-131.

14. Tucker CS (2000) Off-flavor in aquaculture. Reviews in Fisheries Science 8: 45-88.

15. United States Department of Agriculture (USDA) (2003) Catfish 2003 Part II: Reference of Health and Production Practices for Foodsize Catfish in the United States, 2003. National Animal Health Monitoring System, Fort Collins.

16. Schrader KK, Blevins WT (1993) Geosmin-producing species of Streptomyces and Lyngbya from aquaculture ponds. Canadian Journal of Microbiology 39: 834–840.

17. Zimba PV, Grimm CC (2003) A synoptic survey of musty/muddy odor metabolites and microcystin toxin occurrence and concentration in southeastern USA channel catfish (Ictalurus punctatus Ralfinesque) production ponds. Aquaculture 218: 81-87.

18. Schrader KK, Dennis ME (2005) Cyanobacteria and earthy/musty compounds found in commercial catfish (Ictalurus punctatus) ponds in the Mississippi Delta and Mississippi-Alabama Blackland Prairie. Water Research 39: 2807-2814.

19. Van der Ploeg M, Tucker CS (1994) Seasonal trends of flavor of channel catfish, Ictalurus punctatus, from commercial ponds in Mississippi. Journal of Applied Aquaculture 3: 121-140.

20. Avnimelech Y, Diab S, Kochba M, Mokady S (1992) Control and utilization of inorganic nitrogen in fish culture ponds. Aquaculture and Fisheries Management 23: 421-430.

21. Avnimelech Y (1999) Carbon/nitrogen ratios as a control element in aquaculture systems. Aquaculture 176: 227-235.

22. Eaton AD, Clesceri LS, Rice EW, Greenberg AE (2005) Standard Methods for the Examination of Water and Wastewater, 21st Edition. American Public Health Association, Washington, DC.

23. Genfa Z, Dasgupta PK (1989) Fluorometric measurement of aqueous ammonia ion in a flow injection system. Analytical Chemistry 61: 408-412.

24. Lloyd SW, Tucker CS (1988) Comparison of three solvent systems for extraction of chlorophyll a from fish pond phytoplankton communities. Journal of the World Aquaculture Society 19: 36-40.

25. Lloyd SW, Grimm CC (1999) Analysis of 2-methylisoborneol and geosmin in catfish by microwave distillation-solid-phase microextraction. Journal Agricultural and Food Chemistry 47: 164-169.

26. Lloyd SW, Lea JM, Zimba PV, Grimm CC (1998) Rapid analysis of geosmin and 2-methylisoborneol in water using solid phase micro extraction procedures. Water Research 32: 2140-2146.

27. Schrader KK, Nanayakkara NPD, Tucker CS, Rimando AM, Ganzera M, et al. (2003.) Novel derivatives of 9, 10-anthraquinone are selective algicides against the musty-odor cyanobacterium Oscillatoria perornata. Applied and Environmental Microbiology 69: 5319-5327.

28. Schrader KK, Davidson JW, Rimando AM, Summerfelt ST (2010) Evaluation of ozonation on levels of the off-flavor compounds geosmin and 2-methylisoborneol

in water and rainbow trout *Oncorhynchus mykiss* from recirculating aquaculture systems. Aquacultural Engineering 43: 46-50.

29. Engle CR, Valderrama D (2001) Effect of stocking density on production characteristics, costs, and risk of producing fingerling channel catfish. North American Journal of Aquaculture 63: 201-207.

30. Pomerleau S, Engle CR (2003) Production of stocker-size channel catfish: Effect of stocking density on production characteristics, costs, and economic risk. North American Journal of Aquaculture 65: 112-119.

31. Li MH, Manning BB, Robinson EH, Bosworth BG (2003) Effect of dietary protein concentration and stocking density on production characteristics of pond-raised channel catfish *Ictalurus punctatus*. Journal of the World Aquaculture Society 34: 147-155.

32. Allen KO (1974) Effects of stocking density and water exchange rate on growth and survival of channel catfish *Ictalurus punctatus* (Rafinesque) in circular tanks. Aquaculture 4: 29-39.

33. Baumgarner BL, Schwedler TE, Eversole AG, Brune DE, Collier JA (2005) Production characteristics of channel catfish, *Ictalurus punctatus*, stocked at two densities in the Partitioned Aquaculture System. Journal of Applied Aquaculture 17: 75-83.

34. Jobling M, Wandsvik A (1983) Effect of social interactions on growth rates and conversion efficiency of Arctic charr, *Salvelinus alpinus* L. Journal of Fish Biology 22: 577-584.

35. Huss M, Byström P, Persson L (2008) Resource heterogeneity, diet shifts and intra-cohort competition: effects on size divergence in YOY fish. Oecologia 158: 249-257.

36. Brett JR (1979) Environmental factors and growth. In: Hoar WS, Randall DJ, Brett JR, editors. Fish Physiology, Volume VIII. Academic Press, New York, London. p. 599-675.

37. Kilambi RV, Adams JC, Brown AV, Wickizer WA (1977) Effects of stocking

38. Konikoff M, Lewis WM (1974) Variation in weight of cage-reared channel catfish. Progressive Fish-Culturist 36: 138-144.

39. Carmichael GJ (1994) Effects of size-grading on variation and growth in channel catfish reared at similar densities. Journal of the World Aquaculture Society 25: 101-108.

40. Southworth BE, Engle CR, Stone N (2006) Effect of multiple-batch channel catfish, *Ictalurus punctatus*, stocking density and feeding rate on water quality, production characteristics and costs. Journal of the World Aquaculture Society 37: 452-463.

41. Robinson EH, Li MH (2008) Effect of feeding diets with and without fish meal on production of channel catfish, *Ictalurus punctatus*, stocked at varying densities. Journal of Applied Aquaculture 20: 233-242.

42. Ray AJ, Dillon KS, Lotz JM (2011) Water quality dynamics and shrimp (*Litopenaeus vannamei*) production in intensive, mesohaline culture systems with two levels of biofloc management. Aquacultural Engineering 45: 127-136.

43. Masuda K, Boyd CE (1994) Phosphorus fractions in soil and water of aquaculture ponds built on clayey utisols at Auburn, Alabama. Journal of the World Aquaculture Society 25: 379-395.

44. Reynolds CS (1984) The Ecology of Freshwater Phytoplankton. Cambridge University Press, Cambridge.

45. Grimm CC, Lloyd S, Zimba PV (2004) Instrumental versus sensory detection of off-flavors in farm-raised channel catfish. Aquaculture 236: 309–319.

46. Schrader KK, Tucker CS (2012) Evaluation of off-flavor in pond-raised channel catfish following partial crop harvest. North American Journal of Aquaculture 74: 385-394.

47. Torrans L, Lowell F (1987) Effects of blue tilapia/channel catfish polyculture on production, food conversion, water quality and channel catfish off-flavor. Proceedings Arkansas Academy of Science 41: 82-86.

Development and Performance Evaluation of an Automatic Fish Feeder

Ogunlela AO* and Adebayo AA

Department of Agricultural and Biosystems Engineering, University of Ilorin, Ilorin Nigeria

Abstract

Aquaculture, the process of raising aquatic animals in ponds, is gaining more attention in recent times. The feeding system is an important aspect of aquacultural practice. A simple, relatively inexpensive automatic fish feeder was designed, constructed and evaluated. The operation of the feeder does not require highly technical expertise. This paper reports the design considerations, materials used and the effectiveness of the device, based on analysis of manual feeding and automatic feeding. The main features of the device are: hopper (stainless steel), bi-directional motor, feed platform and electrical control box. The design was based on specific parameters which included capacity of culture tank, stocking density, fish biomass, diameter of the feed, angle of repose and bulk density (of the feed). The total cost of the device was 17,000 naira (approx. 106 U.S. dollars). The device was tested under two culture tanks (0.75 m³ each) with 10 kg-33 juvenile cat fish (*Clarias gariepinus*) placed in each tank with one feeding automatically and the other, manually. The feeder evaluation was based on feed conversion ratio (FCR) and feeding efficiency (FE).

The total average gain in weight per fish was higher in the automatic feeding (89.50 g) than in manual (78.50 g). An FE of 20.9% was obtained in the automatic feeding and 18.6% in manual, in relation to their FCRs. A t-test, conducted at 5% significance level, indicated a significant difference in the two feeding methods.

Keywords: Automatic feed dispensers; Aquaculture; Fish ponds; Feeding; Catfishg

Introduction

Aquaculture (fish cultivation), a rapidly- growing entrepreneurial activity, contributes to food security and poverty alleviation in many developing nations. Feeding is one of the most important aspects of fish growth and production. A major challenge facing aquaculture development is the management of feeding systems. Feed adjustment to meet fish requirement is very important for income/benefit maximization. Feeding frequency is thus an essential consideration. Aderolu et al., [1] studied the effect of feeding frequency on growth performance, feed utilization and economic viability of African catfish (*Clarias gariepinus*).

The efficiency and profitability of aquacultural practice could be enhanced with improved technology. This has necessitated the design, development and construction of automatic feeding devices to meet feeding needs and to reduce labor requirements, thereby reducing the cost of fish production.

Mohapatra et al., [2] developed and tested a demand fish feeder, fabricated with Fibre Reinforced Plastic (FRP) material. The feeder was specifically for carp, and was tested in outdoor culture systems. Demand feeders, controlled by the fish needs, could be bait- rod (pendulum)-type or submerged plate-type [3]. Tadayoshi [4] developed an automatic fish feeder which had the capability of sensing uneaten feed. Noor et al., [5] designed an automatic fish feeder using PIC microcontroller. The basic components of the feeder are pellet storage, former, stand, DC motor and microcontroller.

While several automatic fish feeders are available in developed nations, they are scarce in Nigeria and other developing countries (e.g. Anyadike et al., [6]), mainly attributable to the cost of importation. In their design, Anyadike et al., [6] utilized a plastic hopper, with a galvanized-metal discharge chute and a valve attached. The device is capable of discharging 240 g of pelleted feed in 120 seconds. The objective of this work was to develop and evaluate the performance of an automatic fish feeder- to enhance aquacultural practice.

Materials and Methods

Design considerations

Some properties of the feed pellet considered were: angle of repose, specific gravity and bulk density. Also, parameters considered were:

1. Culture system
2. Capacity of the pond (culture tank)
3. Stocking density
4. Average feed requirement
5. Capacity and shape of the hopper
6. Discharge rate through the outlet of the hopper
7. Power requirement by the motor
8. Operation time and operation interval.

Fish biomass=capacity of the tank × stocking density $\quad$ (1)

Daily Feed Need=fish biomass x % of the body weight feeding $\quad$ (2)

$$\textit{Amount of feed needed per operation} = \frac{\textit{Daily feed need}}{\textit{number of operation per day}} \quad (3)$$

Discharge rate through the outlet of the hopper [7], $Q = \frac{\pi g}{16k}\rho D^3 \quad$ (4)

*Corresponding author:Ogunlela AO, Department of Agricultural and Biosystems Engineering, University of Ilorin, Ilorin Nigeria
E-mail: aogunlela@yahoo.com

Where:

Q=volumetric flow rate, m³/s

D=orifice diameter, m

g=acceleration due to gravity, m/s²

k=coefficient of drag

ρ=bulk density, kg/m³

Mass flow rate=volumetric flow rate x average density of the pellet (5)

Time of operation=amount of feed needed per operation/mass flow rate   (6)

$$\text{Operation interval(OP)} = \frac{number\ of\ hours\ per\ day}{number\ of\ feedings\ per\ day} \qquad (7)$$

Design of control box (Timer)

The 555 timer IC can be configured in three different modes: astable, monostable and bistable. The astable and monostable were adopted for this project. These devices are precision timing circuits capable of producing accurate time delays or oscillation. In the time-delay or monostable mode of operation, the time interval is controlled by a single external resistor and capacitor network. In the astable mode, the 555 timer acts as a "one-shot" pulse generator. The time of operation was calculated to be 3 sec, and the range of the operation was assumed to be 1 to 10 sec. The variable resistor that can delay for this period was calculated from the equation.

Monostable (Timer)=1.1RC   (8).

Description of the device

Figures 1 and 2 shows the general features of the automatic fish feeder. The component parts of the machine (device) include: the hopper, top cover (LID), the base (comprising the motor and feed platform) and the electrical control box. The hopper is made of stainless steel (1mm thickness) and it is of composite shape (cylindrical and fulcrum). The top cover is made of the same material as the hopper and it protects the feed from rain and contaminants. The base consists of 6V, 3W bi-directional motor and feed platform attached to it. The feed platform opens and closes the discharge outlet as the motor rotates. The electrical control box controls and regulates the feeding operation and the frequency.

Operation of the machine

The hopper contains the feed which comes out through the discharge outlet. When the machine is switched on and reset, the feed platform moves in bi-directional (to and fro) motion, during which there is opening and closing of the discharge outlet for pre-determined period. The desired amount of the feed would be dispensed into the pond and this completes an operation. After the operation is completed, the machine will automatically stop for preset hours (1, 2, 3............. hrs) based on the number of operations needed per day. When the hours are completed, the machine will start again and dispense the same amount of feed as in the previous operation, and the operation continues.

The machine is powered by electricity and it has a back-up (6V battery) which can last for at least 3 days (72 hrs) when fully charged. The device can be used for both local and imported dry pellet of size 0.5 mm-9 mm. The cost estimate for the production of the machine was ₦17,000 (approx. US $106). The construction materials and the Bill of

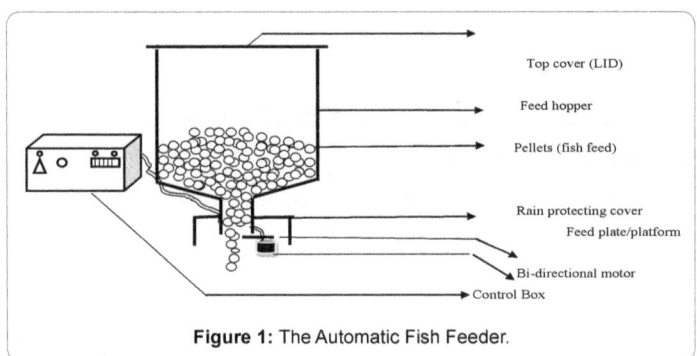

Figure 1: The Automatic Fish Feeder.

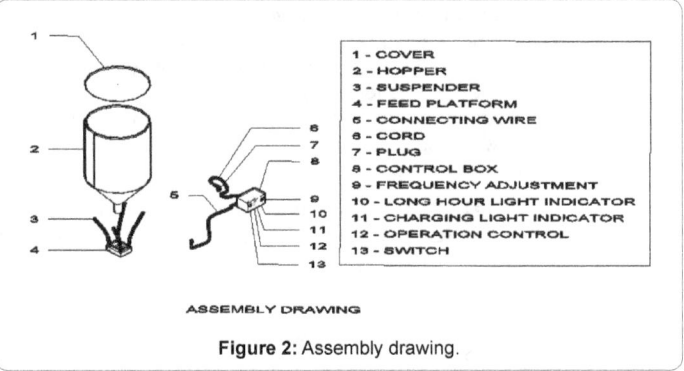

1 - COVER
2 - HOPPER
3 - SUSPENDER
4 - FEED PLATFORM
5 - CONNECTING WIRE
6 - CORD
7 - PLUG
8 - CONTROL BOX
9 - FREQUENCY ADJUSTMENT
10 - LONG HOUR LIGHT INDICATOR
11 - CHARGING LIGHT INDICATOR
12 - OPERATION CONTROL
13 - SWITCH

ASSEMBLY DRAWING

Figure 2: Assembly drawing.

Engineering Measurement and Evaluation (BEME) are shown in Tables 1 and 2, respectively.

Performance evaluation

The performance evaluation of the device was conducted using a recirculatory aquaculture system (RAS) located behind the Department of Agricultural and Biosystems Engineering, University of Ilorin; Ilorin, Nigeria.

Ilorin (longitude 4°35'E, latitude 8° 30'N), the capital of Kwara State of Nigeria, has two main seasons: wet (March-October) and dry (November-February). The experiments were conducted from April to June, 2013, involving two culture tanks, each of 0.75 m³ volume, with 10 kg-33 juvenile catfish (*Clarias gariepinus*) placed in each tank with one feeding automatically and the other, manually.

"Durante" fish feed, weighing balance and meter rule were also used in the investigation. The feeder was placed over a stand fixed at the corner of the culture tank. Growth rate of fish was estimated by sampling 10 fishes from the rearing tank every week (7 days interval). The feed conversion ratio (FCR) and the feeding efficiency (FE) were used for the performance evaluation:

$$FCR = \frac{total\ amount\ of\ the\ feed\ given\ (\text{gram})}{total\ gain\ in\ weight\ by\ the\ fish\ (\text{gram})} \qquad (9)$$

$$FE = \frac{1}{FCR} \qquad (10)$$

Results and Discussion

Tables 3 and 4 show the catfish growth rate for automatic and manual feeding, respectively.

For the automatic feeding (Table 3);

Components	Materials Used	Dimensions	Remarks
1. Lid	Stainless steel (1 mm thickness)	260 mm (diameter)	Cut out and folded to make a cover
2. Upper Cylinder	Stainless steel (1 mm thickness)	300 mm x 790 mm	Cut out and folded to form a cylinder of 250 mm diameter
3. Frustum	Stainless steel (1 mm thickness)	195 mm x 790 mm	Cut and folded to form a frustum of upper diameter of 250 mm and base diameter of 30 mm, height of 150 mm. The frustum joined to the upper cylinder at an angle of 50°.
4. Base Cylinder (outlet)	Stainless steel	50 mm x 95 mm	Cut and folded to form a cylinder of 30 mm diameter. Then welded to the base of the frustum.
5. Base	PVC, motor and wire		PVC used to cover the motor, also as feed platform. The wire connects the base to the control box. The base was suspended to the feed hopper using copper wire.
6. Control Box	PVC, veroboard, resistors, capacitors, transistors, relays, 555 timers, transformer, diodes and regulator.		PVC used for casing. The components were laid on the veroboard.

Table 1: Construction materials.

Materials	Quantity	Unit cost (₦)	Total cost (₦)
Stainless steel	¼ sheet	4,000 (1/4 sheet)	4,000
Acrylonitrile plastic steel	½ sheet	1,000 (1/2 sheet)	1,000
Transformer (220-9V)	1 piece	500	500
Resistors (R1,R2,...R9)	....	500	500
Capacitor (C1, C2, ...C9)	....	500	500
Bi-directional motor (6V-3W)	1 piece	2000	2000
LED (D1, D2, D3)	....	200	200
Integrated Circuit (IC1, IC2 and IC3)	....	1500	1500
Regulator (RG1,RG2)	....	500	500
Relay (RL1, RL2 and RL3)		1000	1000
Variable Resistor	....	300	300
Hopper construction workmanship and other costs			5,000
Total			17,000

Table 2: Bill of Engineering Measurement and Evaluation (BEME).

S/N	Date	Number of Fish	Average weight of fish (g)	Total feed consumed per fish (g)	Average gain in weight per fish (g)
1	16/04/2013	10	300.00	0.00	0.00
2	23/04/2013	10	305.00	42.00	5.00
3	30/04/2013	10	313.50	43.00	8.50
4	7/05/2013	10	323.00	44.50	9.50
5	14/05/2013	10	335.50	46.00	10.5
6	21/05/2013	10	345.50	47.00	10.30
7	28/05/2013	10	355.53	49.00	9.80
8	04/06/2013	10	365.80	51.10	10.5
9	11/06/2013	10	377.20	52.00	11.40
10	18/06/2013	10	389.20	53.40	12.00

The total average of feed consumed per fish during the period of the experiment = 428.00 g
The total average gain in weight per fish during the period of the experiment = 89.50 g

Table 3: Growth rate of catfish for the automatic feeder.

$$FCR = \frac{428}{89.5} = 4.78$$

$$FE = \frac{1}{4.78} = 20.9\%$$

For manual feeding (Table 4);

$$FCR = \frac{421.50}{78.50} = 5.37$$

$$FE = \frac{1}{5.37} = 18.6\%$$

A t-test was also used in analyzing the data. Table 5 shows the result of the statistical analysis, at 5% significance level.

Conclusion

An automatic fish feeder was designed, constructed and evaluated. Its main components are: hopper, bi-directional motor, feed platform and electrical control box. The device was incorporated into a recirculatory aquaculture system (RAS) and the evaluation was based on feed conversion ratio (FCR) and feeding efficiency (FE) using juvenile catfish (*Clarias gariepinus*). The feeding efficiency was higher in the automatic feeding (20.9%) than in manual (18.6%). Use of the automatic feeder will improve aquacultural practice.

S/N	Date	Number of Fish	Average weight of fish (g)	Total feed consumed per fish (g)	Average gain in weight per fish (g)
1	16/04/2013	10	300.00	0.00	0.00
2	23/04/2013	10	304.00	42.00	4.00
3	30/04/2013	10	312.50	44.00	8.50
4	7/05/2013	10	320.20	45.00	7.70
5	14/05/2013	10	329.00	46.00	8.80
6	21/05/2013	10	335.00	46.50	8.00
7	28/05/2013	10	344.50	47.00	9.50
8	04/06/2013	10	354.70	49.00	10.20
9	11/06/2013	10	365.50	50.00	10.80
10	18/06/2013	10	376.50	52.00	11.00

The total average of feed consumed per fish during the period of the experiment=421.50 g

The total average gain in weight per fish during the period of the experiment=78.50 g

Table 4: Growth rate of catfish for manual feeding.

Feeding Method	Mean gain in weight per fish(g)	Standard deviation (g)	t-test value
Automatic feeding	8.95	3.57	1.077
Manual feeding	7.85	3.23	0.973

The mean gain in weight per fish in automatic feeding was higher than in manual.

Table 5: Result of statistical analysis.

References

1. Aderolu AZ, Seriki BM, Apatira AL, Ajaegbo CU (2010) Effects of feeding frequency on growth, feed efficiency and economic viability of rearing African catfish (Clarias gariepinus, Burchell 1822) fingerlings and juveniles. African Journal of Food Science 4: 286-290.

2. Mohapatra BC, Sarkar B, Sharma KK, Majhi D (2009) Development and Testing of Demand Feeder for Carp Feeding in Outdoor Culture System. Agricultural Engineering International, the CIGR EJournal.

3. Varadi L (1984) Mechanized feeding in Aquaculture. Inland Aquaculture Engineering. Food and Agriculture Organization (FAO). Fisheries and Aquaculture Department.

4. Tadayoshi, Nagatomi (2003) Automatic Fish Feeder with Uneaten Feed Sensor for Environmental Preservation. Yamaha Mot Tech Rev.

5. Noor MZH, Hussain AK, Saad MF, Ali MSAM, Zolkap M (2012) The design and development of automatic fish feeder system using PIC microcontroller. Control and System Graduate Research Colloquium (ICSGRC), 2012 IEEE.

6. Anyadike CC, Eze M, Ajah GN (2010) Development of an automatic fish feeder. Journal of Agricultural Engineering and Technology 18: 29-36.

7. Gregory JM, Feldler CB (1987) Equation describing granular flow through circular orifices. Transactions of the ASAE 30: 529-532.

7

Effects of Acute and Chronic Nitrite Exposure on Rabbitfish *Siganus rivulatus* Growth, Hematological Parameters, and Gill Histology

Patrick Saoud I[1]*, Naamani S[2], Ghanawi J[1] and Nasser N[1]

[1]*Department of Biology, American University of Beirut, Lebanon*
[2]*Department of Biology, Beirut Arab University, Beirut, Lebanon*

Abstract

Nitrite is toxic to fishes and is often encountered in recirculation aquaculture systems. Accordingly, the nitrite tolerance of potential aquaculture candidates needs to be assessed before the fish can be farmed in land-based recirculation systems. In the present work, we studied the susceptibility of the marbled rabbitfish *Siganus rivulatus* to nitrite. In the first experiment, we placed fish at 0, 40, 50, 60, 70, 80, 90, 100, 110, 120 and 130 mg l^{-1} NO_2-N and evaluated 96 h LC_{50}. In the second experiment we measured survival and growth of fish reared at 0, 10, 20, 30, 40, and 50 mg l^{-1} NO_2-N for eight weeks. Blood parameters of fish in the various treatments were also measured and gill histology studied. Finally, methemoglobinemia in fish reared at various nitrite conditions was assessed. The NO_2-N 96 h LC_{50} of *S. rivulatus* juveniles was 105 mg l^{-1}. In the growth experiment, fish mortality was greater than in the control at NO_2-N concentrations 30 mg l^{-1} and greater. Growth in all treatments was less than in the control but there were no significant differences among treatments. Aqueous nitrite affected various hematological parameters such as hematocrit and total hemoglobin. Compared to other aquacultured marine fishes, the marbled rabbitfish is considered tolerant to environmental nitrite.

Keywords: Rabbitfish; *Siganus rivulatus*; Nitrite; Methemoglobin

Introduction

As locations suitable for aquaculture become scarcer and more costly and diseases more prevalent, modern aquaculture facilities are opting for intensive and biosecure rearing systems, mostly Recirculation Aquaculture Systems (RAS). Intensive aquaculture could provide greater economic profits but intensification of fish culture frequently leads to increased loads of nitrogenous and other toxic metabolites, resulting in water quality deterioration. One of the toxic nitrogenous wastes often encountered in aquaculture systems is nitrite (NO_2^-) which often reaches toxic levels in intensive RAS [1,2]. In most intensive marine fish tank culture, some nitrite is tolerated as it is difficult to maintain a nitrite-free RAS. Consequently, aquaculturists need to understand the tolerance of the organism they are farming to chronic exposure to low concentrations of nitrite.

Nitrite is an intermediate product formed by bacterial nitrification of ammonia. Nitrite in the water competes with chloride on the chloride-bicarbonate exchanger present in the apical membranes of chloride cells of fish gills. Nitrite also competes with chloride for transfer across erythrocyte membranes leading to the oxidation of hemoglobin to met-hemoglobin [3]. Consequently, excessive nitrite levels in culture systems can cause depressed growth [4], increased susceptibility to disease, and eventual mortality [5]. However, this competition with chloride decreases the detrimental effects of nitrite in marine waters and makes nitrite more dangerous in freshwater aquaculture.

Acute and chronic toxicities of nitrite have been extensively studied in freshwater species such as rainbow trout *Oncorhynchus mykiss* [6], Siberian sturgeon *Acipenser baerii*, Brandt [7], *matrinxã Brycon cephalus* [8], tambaqui *Colossoma macropomum* [9], Walleye *Sander vitreus* [10], mrigal carp *Cirrhinus mrigala* [11], European eel *Anguilla anguilla* [12], common carp *Cyprinus carpio* L. [13], and silver perch *Bidyanus bidyanus* [14]. Additionally, a few studies have evaluated responses of marine fish such as pompano *Trachinotus marginatus* [15], silver sea bream *Sparus sarba* [16,17], dark-banded rockfish, *Sebastes inermis* [18], Atlantic cod *Gadus morhua* [19], flounder *Platichthys flesus* [2], and cobia *Rachycentron canadum* [20]

to ambient nitrite, but results vary greatly among species. Thus, we cannot use conclusions from previous work to estimate rabbitfish *S. rivulatus* tolerance to environmental nitrite.

Fish gills are multifunctional organs needed for respiration, osmo-regulation, acid-base balance and nitrogenous excretion [21]. The large surface area and direct continuous contact with the surrounding water make the gills the first target to waterborne chemicals [22]. Pollutants enter the organism through the gills and exert their primary toxic effects on the bronchial epithelium [23]. Thus, morphological changes in fish gills are among the most commonly recognized responses to environmental stressors and are indicative of physical and chemical stress in marine as well as freshwater habitats [21,24]. Histopathological changes in gills such as epithelial lifting, hypertrophy, hyperplasia, epithelial necrosis, edema, and fusion of secondary lamellae are the major effects reported in fish exposed to various types of pollutants and toxic substances [24]. There are presently no published reports on the effects of ambient nitrite on rabbitfish gills.

Marbled rabbitfish *Siganus rivulatus* is a euryhaline, herbivorous marine fish widely distributed along the Eastern Mediterranean and East Indian Ocean [25]. This teleost is a valuable fishery species that is relatively easy to farm and thus considered of great potential for warm-water aquaculture diversification [26-28]. However, for the rabbitfish aquaculture industry to succeed, tolerance to metabolites that the fish will be exposed to has to be understood. The present

*Corresponding author: Patrick Saoud I, Department of Biology, American University of Beirut, Beirut, Lebanon, E-mail: is08@aub.edu.lb

work was performed to evaluate the tolerance of marbled rabbitfish to nitrite exposure. We determined acute nitrite tolerance of juvenile rabbitfish by establishing 96-h LC_{50} of NO_2-N. We then evaluated the effects of chronic exposure of *S. rivulatus* to nitrite on survival, growth performance, hematological and biochemical parameters. Finally, we determined levels of total hemoglobin and methemoglobin in the fish upon chronic nitrite exposure.

Materials and Methods

Fish acquisition and experimental conditions

Marbled rabbitfish *S. rivulatus* were caught in traps off the beach south of Beirut and immediately transported to the aquaculture laboratory at the American University of Beirut (AUB). The system was housed in an environmentally controlled facility. Photoperiod was maintained by a timer at 14:10 (Light: Dark) throughout the experiment. Dissolved oxygen concentrations were maintained above 5 mg l^{-1}. Salinity was maintained at 35‰ and water temperature at 27°C. Dissolved oxygen, salinity and temperature were measured daily using an YSI-85 salinometer. pH was maintained between 8.0 and 8.2 using sodium bicarbonate and measured using a handheld pH meter. Nitrite-N levels in the tanks were adjusted via additions of $NaNO_2$ of a known volume from a 20 g l^{-1} NO_2-N stock solution which was prepared before the experiment by dissolving 98.57 g of 99.9% $NaNO_2$ in 1 liter of de-ionized water. Nitrite concentrations were measured daily as described by Parsons et al. [29].

Fish were acclimated in 1 m^3 circular quarantine tanks for two weeks and offered 35% protein, 8% lipid commercial diet (Rangen EXTR 350, Rangen Inc., Buhl, Idaho, USA) twice daily to apparent satiation. After two weeks of acclimatization, fish were size sorted and transferred to an indoor tank system consisting of 33 insulated 55-L (30×60×30 cm; W×L×H) glass aquaria. All water used was stored in large closed tanks and treated prior to use. Seawater treatment included chlorination to remove any possible bacterial contamination, de-chlorination using sodium thiosulfate and addition of EDTA to remove possible heavy metal contamination.

Experimental design

Three experiments were performed based on modifications of methods described by Clesceri et al. where applicable [30]. The protocols are USEPA approved and are standard methods for toxicity testing in aquatic environments. LC_{50} was considered to be the concentration at which at least half the fish in a treatment died within a set time frame.

Experiment 1: Acute Nitrite Exposure (96 h LC_{50})

In the first experiment juvenile rabbitfish (n=528, average weight =8.1 g ± 0.5; mean ± SD) were stocked at densities of 16 fish per aquarium in 33 aquaria for the 96 h toxicity test. Each tank was randomly assigned one of 11 treatments with three replicate tanks per treatment. Treatments were 0, 40, 50, 60, 70, 80, 90, 100, 110, 120 and 130 mg l^{-1} NO_2-N. The nitrite-N tested was chosen after a preliminary trial. Nitrite challenge was started by replacing ambient water with clean, nitrite free seawater and adding to each tank a calculated amount of stock solution (20 g l^{-1} NO_2-N) to obtain the desired NO_2-N concentrations. Salinity, temperature and pH were controlled throughout the experiment.

Prior to exposure, all fish were fasted for 24 h. During the test, fish were not fed and were counted twice daily at 08:00 and 19:00. The fish's behavior was observed and loss of equilibrium recorded when observed. Dead fish were removed and recorded at each counting event.

Nitrite-N concentrations in the tanks were measured daily. After 96 h, fish were counted, behavior recorded and the experiment terminated.

Experiment 2: Chronic nitrite exposure

Aqueous nitrite concentrations that did not cause any fish mortality for 96 h were used in the second part of the project. Fish were size sorted (weight=8.6 ± 0.6 g; length=9.2 ± 0.3 cm) and stocked 16 per tank with three replicates per treatment. The initial Fulton's condition index K=100×(weight in g)/(total length in cm)³ of the fish was 1.08. Each tank was aerated using submersible diffusers connected to a regenerative air blower. Six plastic barrels (200-liter) were filled with seawater and $NaNO_2$ added to five of them to raise NO_2-N concentrations to 10, 20, 30, 40, and 50 mg l^{-1}. Water from each barrel was pumped into three tanks using a submersible power-head pump connected to flow restrictors so that the volume of each tank was completely replaced each 12 h. Fish were offered the Rangen commercial diet at 3% body weight divided into two daily feedings, mornings and evenings. All leftover feed and feces were siphoned out of each tank daily. All fish were group-weighed weekly and feed ration adjusted accordingly. Ammonia concentration in the tanks was measured twice a week and remained less than 0.01 mg l^{-1}. The experiment was terminated after eight weeks, and all surviving fish were harvested, counted, group weighed and individually weighed. Blood was collected for haematological tests as described below.

Experiment 3: methemoglobin determination

The fish in experiment 2 were too small to allow extraction of sufficient blood for total haemoglobin and methemoglobin tests. Accordingly, larger fish were challenged with nitrite for 19 days in order to determine blood proportions of methemoglobin and total haemoglobin. Rabbitfish (weight=33.7 ± 0.5 g; length=14.1 ± 0.5 cm) were acclimated for 1 week to laboratory conditions in 1 m^3 circular tanks. Nine fish were then transferred to each of 15 glass tanks used in previous experiment and maintained under experimental conditions for 19 days. Treatments were 0, 10, 20, 30 and 40 mg l^{-1} NO_2-N. Water quality parameters were measured daily. Temperature was maintained at 27°C, salinity at 35‰, oxygen remained above 5 mg l^{-1}, pH was 8.0 ± 0.2, and NH_3-N was below 0.02 mg l^{-1}. Fish were offered the Rangen commercial feed at 3% body weight divided into two daily feedings, mornings and evenings. On the morning of the twentieth day after induction, fish were caught and placed in 100 mg l^{-1} MS-222 treated water to anaesthetize them and blood was extracted for haematology.

Blood collection

Fish were fasted for 24 h prior to blood collection. Blood was collected by cardiac puncture using heparin coated needles and 1 ml syringes. Blood was collected from two to four fish per tank and pooled in order to secure a minimum volume of blood needed for analyses and transferred to heparin coated micro centrifuge tubes. Collected blood was held on ice until all samples were obtained. Hematological analyses were performed within 2 h of blood collection.

Hematological and biochemical tests

Blood samples with anticoagulant (sodium heparin) were used for hematological examination. Blood was diluted 1:200 using modified Natt Herrick's solution [31] and total blood cell numbers, Total Erythrocyte Counts (TECs) and White Blood Cells (WBCs) counted using a modified Neubauer hemocytometer. Differential blood cell counts were performed on blood films fixed with absolute methanol, and stained with modified Wright's- Giemsa stain. A total of 800 White

Blood Cells (WBCs) per slide were identified and counted as described by Ellis [32] and Ainsworth [33]. Each type of blood cell was expressed as a percentage of the total number of blood cells examined.

Hematocrit measurements were made in duplicate by drawing well-mixed blood into heparin-coated microhematocrit tubes (75 mm length, inside diameter 1 mm, ABCO, Dealers, Inc., IL, USA) and centrifuged at 10,000g for 5 min in a microhematocrit centrifuge. Total hemoglobin was determined by the cyanmethemoglobin technique using Drabkin's reagent (Sigma) and light absorption at 540 nm [34]. Total Plasma Protein (TPP) was determined by placing a drop of blood plasma onto a plasma protein veterinary refractometer (RHC-200ATC, Westover Scientific, Inc., WA, USA). Lactate was measured using a hand-held device (Accutrend' Plus System, Roche Diagnostics GmbH). Plasma nitrite levels were determined according to Shechter et al. [35]. Nitrite is not stable in an acidic environment, so an alkaline extraction in six parts zinc sulphate (4.31%) and five parts of NaOH (0.8%) was performed and the resultant solution was maintained at 0°C for 60 min then centrifuged for 15 min at 1000 rpm. An aliquot of the resulting supernatant was used for nitrite determination by adding sulphanilic acid and Cleve's acid. This produced a final red-violet product, and nitrite concentration was calculated from absorbance at 520 nm.

Percentage Methemoglobin (MetHb) in the blood of fish exposed to nitrite was measured using a modification of the method described by Horecker and Brackett [36]. Briefly, 300 µl of whole blood was added to 3 ml of hemolyzing solution and vortexed to promote lysis of erythrocytes. The resulting solution was divided into two equal portions. To one portion, 20 µl of 10% potassium ferricyanide $K_3Fe(CN)_6$ solution was added. All samples were then centrifuged for 3 min at 10,000 g and 4°C. One milliliter of each supernatant was placed into cuvettes. Absorbance of the untreated [without $K_3Fe(CN)_6$] and treated [with $K_3Fe(CN)_6$] samples at 820 nm were recorded as A_1 and B_1, respectively. 10 µl of non-neutralized cyanide solution was added to each cuvette, and the new absorbance recorded as A_2 and B_2. The percentage MetHb was calculated as follows: % MetHb = 100 $(A_1-A_2)/(B_1-B_2)$.

Histology of gills

In Experiment 2, three fish from each tank were taken for histological examination. Gills were excised and samples were immediately fixed in Bouin's fixative, dehydrated using gradients of alcohol concentrations and embedded in paraffin. Sections (5 µm) were stained with Ehrlich's haematoxylin and counter stained with eosin. Then the tissue specimens were dehydrated, cleared in xylene and embedded in paraffin. Slides were examined using a light microscope.

Statistical analysis

All statistical analyses were performed using SAS for Windows (V8e, SAS Institute Inc., Cary, North Carolina USA), with a significance level α=0.05. LC_{50} values with 95% confidence limits were determined using probit analysis of Log_{10} transformed concentration values. Growth and hematology variables were analyzed using one way analysis of variance (ANOVA) to test for significance assuming a completely randomized experimental design. Student-Newman-Keuls mean separation tests were used to determine differences among treatment means in all experiments.

Results

Acute nitrite exposure (96h LC_{50})

The NO_2-N 96 h LC_{50} of *S. rivulatus* juveniles was 105 mg l^{-1} with

95% confidence limits between 97.8 and 114.3 mg l^{-1}. All fish survived for 96 h in the treatments containing nitrite concentrations from 0 to 60 mg l^{-1}. Some mortality was observed in treatments containing nitrite concentrations from 70 to 100 mg l^{-1} (Figure 1). Mortalities increased significantly at 110 mg l^{-1} where half of the fish were dead after a 96 h exposure. Almost all fish in treatments 120 and 130 mg l^{-1} NO_2-N were dead at 96 h.

Mortality of rabbitfish juveniles after a 24 hour exposure was 0% for nitrite concentrations ranging from 0 to 120 mg l^{-1}, and was only 2% at 130 mg l^{-1}. After 48 h of nitrite exposure, mortality of juveniles was 0% for nitrite-N concentrations of 0 to 80 mg l^{-1}, and increased with nitrite-N increase between 90 and 130 mg l^{-1}. Significant differences among treatments started appearing after 48 h of exposure and became more apparent with time. After 96 h of exposure, most of the fish had died in 120 and 130 mg l^{-1} NO_2-N (Figure 1). Fish exposed to nitrite concentrations ranging from 0 to 70 mg l^{-1}, exhibited no abnormal behavior. Fish swam normally with no loss of balance or orientation. However, at greater nitrite concentrations, fish exhibited erratic swimming, and were occasionally observed lying still at the bottom of the containers. At NO_2-N concentrations of 110, 120 and 130 mg l^{-1}, all fish were lethargic and dark in color, thus exhibiting usual signs of stress in marbled rabbitfish.

Chronic nitrite exposure

Growth and survival: All fish in treatment 50 mg l^{-1} NO_2-N died by the end of the fourth week (Table 1). All fish at 0 mg l^{-1} NO_2-N survived. Mortality was observed in treatments 10 and 20 mg l^{-1} NO_2-N but results were not significantly different from the control. In treatments

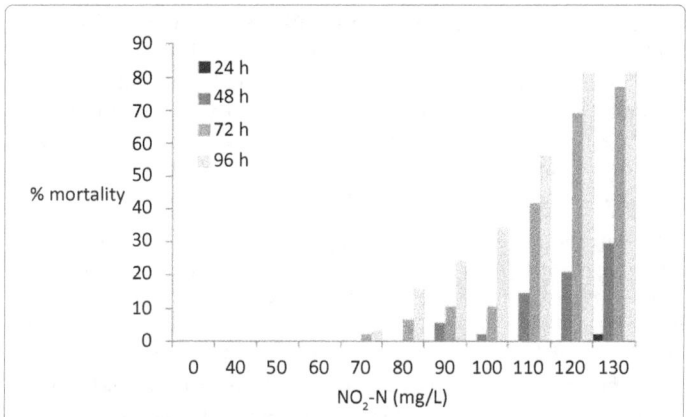

Figure 1: Rabbitfish (*Siganus rivulatus*) mortality during acute exposure to various NO_2-N concentrations (0-130 mg l^{-1}).

NO_2-N (mg l^{-1})	S (%)	Wt_f (g)	TL_f (cm)	K_f	HSI_i	HSI_f
0	100[a]	14.8[a]	11.15[a]	1.06	1.67	1.66
10	84.7[a]	9.6[b]	9.65[b]	1.02	1.67	1.90
20	76.5[a]	7.6[b]	9.09[b]	1.00	1.67	1.91
30	47.0[b]	7.2[b]	8.99[b]	0.98	1.67	1.51
40	40.0[b]	7.1[b]	8.97[b]	0.96	1.67	1.22
PSE*	6.42	0.72	0.22	0.03	-	0.19

Table 1: Survival (S), final weight (Wt_f), final total length (TL_f), final Fulton-type condition index (K_f), initial hepatosomatic (HSI_i) and final hepatosomatic index (HSI_f) of rabbitfish (*Siganus rivulatus*) exposed to various nitrite-N concentrations (0-50 mg l^{-1}) for 56 days.

Values in the same column sharing the same letter are not significantly different from each other.

*PSE=Pooled standard error

30 and 40 mg l^{-1} NO$_2$-N survival were 47.0% and 40.0% respectively, significantly less than in lower concentration treatments (Table 1). The final weight and final length of fish in treatment 0 mg l^{-1} were greater than final weight and final length of fish in all other treatments. There were no significant differences in final weight and final length among treatments 10, 20, 30 and 40 mg l^{-1} although a decreasing trend in growth was observed as nitrite concentration increased (Table 1).

Although final condition index (Kf) decreased as nitrite concentration increased, results were not significantly different from each other among treatments (Table 1). There were also no significant differences in initial Hepato-Somatic Index (HSI) or in final HSI among treatments (Table 1).

Hematological and biochemical parameters: Total Erythrocyte Count (TEC) decreased progressively with increasing nitrite concentrations (Table 2). Compared to the initial value (3.86×10^6 μl^{-1}), a significant reduction (P<0.05) in TECs was observed at all nitrite concentrations from 10 mg l^{-1} (2.86×10^6 μl^{-1}) to 40 mg l^{-1} (1.67×10^6 μl^{-1}) (Table 2). Hematocrit (Ht) values in treatments 20, 30, and 40 mg l^{-1} NO$_2$-N were significantly less (P<0.05) than Ht values at 0 and 10 mg l^{-1} NO$_2$-N (Table 2). Significantly lower total hemoglobin (Hb) levels were found in fish exposed to 10, 20, 30, and 40 mg l^{-1} NO$_2$-N compared to the control (Table 2).

Total Leukocyte Count (TLC) of fish in treatments 10, 20 and 30 mg l^{-1} NO$_2$-N were significantly greater than TLC in the control group (P<0.05) (Table 2). However, TLC in fish at 40 mg l^{-1} NO$_2$-N was significantly less than TLC in other treatments (P<0.05). TLC in fish at 40 mg l^{-1} NO$_2$-N was even less than TLC in control fish although the difference was not statistically significant (Table 2).

Average Total Plasma Protein (TPP) content in control fish was 3.92 g dl^{-1}. A reduction in TPP was observed with increasing nitrite concentrations (Table 2). At 40 mg l^{-1} NO$_2$-N, TPP was significantly less than in fish in all other treatments. There was no significant difference in TPP among all other treatments (Table 2). Lactate levels increased as nitrite concentrations increased, but there were no significant differences among treatments (Table 2).

Neutrophil numbers increased progressively with increasing nitrite concentrations. The proportion of neutrophils in the blood of fish at 20, 30, and 40 mg l^{-1} NO$_2$-N were significantly greater than in the control (Table 3). The proportion at 30 mg l^{-1} NO$_2$-N was greater than at 10 mg l^{-1} and the proportion at 40 mg l^{-1} was greater than the neutrophil proportion of the fish blood at all other treatments (Table 3). Eosinophil proportion in blood also increased with increasing ambient nitrite concentrations and in fish maintained at 40 mg l^{-1} NO$_2$-N it was

NO$_2$-N (mg l^{-1})	Ht (%)	Total Hb (mg dl^{-1})	TEC (x10^6 μl^{-1})	TLC (x10^6 mm^{-3})	TPP (g dl^{-1})	Lactate (mmol l^{-1})
0	36.65^a	11.67^a	3.86^a	77.81^b	3.92^a	2.33
10	32.38^a	8.68^b	2.86^b	108.19^a	3.86^a	2.72
20	22.33^b	6.06^b	2.33^{bc}	121.67^a	3.17^a	2.03
30	22.63^b	5.77^b	2.31^{b,c}	108.88^a	3.18^a	2.23
40	20.50^b	5.32^b	1.67^c	65.13^b	2.10^b	3.30
PSE*	2.31	0.47	0.28	8.21	0.28	0.59

Table 2: Hematological and Biochemical parameters of rabbitfish *Siganus rivulatus* exposed to various nitrite-N concentrations (0-50 mg l^{-1}) for 56 days.

Values in the same column sharing the same letter are not significantly different from each other.

Ht=Hematicrit; Hb=hemoglobin; TEC=Total erythrocyte count; TLC=Total leukocyte count; TPP=total plasma protein.

*PSE=Pooled standard error

NO$_2$-N (mg l^{-1})	Thrombocytes (%)	Neutrophils (%)	Lymphocytes (%)	Monocytes (%)	Eosinophils (%)
0	80.00^a	8.78^d	10.89^b	0.56	0.11^b
10	68.11^{a,b}	17.22^{c,d}	13.89^b	0.56	0.22^b
20	53.14^b	28.43^{b,c}	16.43^b	1.14	0.71^b
30	35.25^c	34.00^b	29.25^a	1.25	1.25^b
40	11.00^d	54.00^a	30.50^a	2.00	2.75^a
PSE*	5.25	4.20	2.40	0.42	0.31

Table 3: Average percentages of differential blood counts in *Siganus rivulatus* rabbitfish after exposure to 56 days chronic nitrite experiment.

Values in the same column sharing the same letter are not significantly different from each other.

*PSE=Pooled standard error

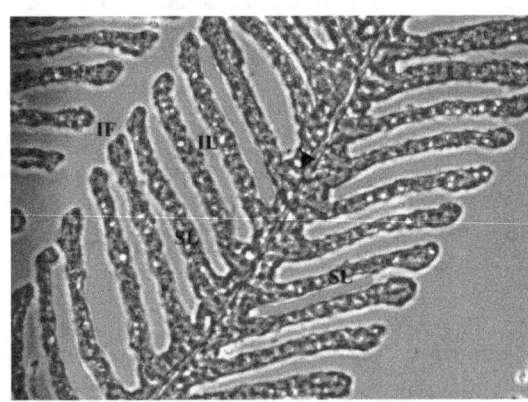

Figure 2: Light micrographs of gills of fish reared in nitrite free water. Gills have normal architecture, good integrity of the central axis (▶), parallel and well organized secondary lamellae (SL) and evident interfilament (IF) space and interlamellar (IL) space. [Formalin fixed-H&E stained preparation, X 400].

significantly different from all other treatments (Table 3). Lymphocytes proportion in blood of control fish was 10.89%, increasing to 13.89% and 16.43% in blood of fish at 10 and 20 mg l^{-1} NO$_2$-N respectively (Table 3). However, the differences were not significantly different from each other. In blood of fish maintained at 30 and 40 mg l^{-1} NO$_2$-N, lymphocyte proportions increased significantly to reach 29.25 and 30.50% of WBCs, respectively (Table 3). Thrombocytes proportion in blood of control fish was 80.0% and decreased significantly as ambient nitrite increased. In blood of fish maintained at 40 mg l^{-1} NO$_2$-N, thrombocytes comprised only 11% of total WBCs (Table 3). Monocytes proportion in blood increased with an increase in ambient nitrite concentrations but values were not significantly different among treatments (Table 3). No basophils were found in any of the hematological slides that were examined.

Histology: Gill preparations of control fish had normal gill architecture and typical structural organization of gill filaments (Figure 2). The structure of the lamellae observed was comparable to gills described by Gisbert et al. [7] and Kroupova et al. [6] for healthy fish. The gill arch was composed of healthy looking long gill filaments (primary lamellae) bearing distinct secondary lamellae on both sides and well separated with evident interlamellar space. Secondary lamellae were evenly arranged, running parallel to each other, and also showed clear interlamellar spaces.

Gill preparations of fish in 10 mg l^{-1} NO$_2$-N revealed distinguishable deformations and apparent gill damage such as twisted uneven secondary lamellae, fusion and hyperplasia of secondary lamellae. Fish exposed to 20 mg l⁻¹ nitrite nitrogen had dilations of central axes and infiltration of red blood cells with hyperplasia of spiked and shortened secondary lamellae. Fish exposed to 30 mg l⁻¹ nitrite nitrogen had blunt ended secondary lamellae with aneurysms, and hypertrophy and hyperplasia of epithelial cells (Figure 3). Gills of fish exposed to 40 mg l⁻¹ NO_2-N revealed significant disorganization and twisting of secondary lamellae with severe hyperplasia leading to lamellar fusion. Additionally, aneurysms and vascular congestions of lamellae were observed. Fish exposed to 50 mg l⁻¹ NO_2-N exhibited complete loss of gill structure resulting from lysis at several points (Figure 4a and 4b). Gill alteration included irregular, twisted and curled secondary lamellae, and severe gill hypertrophy, hyperplasia, lamellar telangiectasia and ruptures of lamellar epithelia, and lamellar fusion.

Plasma nitrite and methemoglobin determination: Blood hemoglobin (total Hb) of the rabbitfish decreased as ambient nitrite increased. Significantly lesser total hemoglobin levels were found in fish exposed to 10, 20, 30, and 40 mg l⁻¹ NO_2-N compared to the control (Figure 5a). Moreover, the percentage of MetHb increased significantly in fish exposed to 30 and 40 mg l⁻¹ NO_2-N concentrations compared with controls, and reached proportions between 28.55 and 71.40% of the total Hb respectively (Figure 5b). Plasma nitrite concentration increased significantly with increasing aqueous nitrite concentrations (Figure 5c). At 0 mg l⁻¹ ambient NO_2-N, plasma nitrite was 0.05 mg l⁻¹ and increased nearly ten-fold to 0.47 mg l⁻¹ when fish were maintained at 40 mg l⁻¹ NO_2-N for 19 days.

Discussion

Acute nitrite exposure (96 h LC₅₀)

Results of the acute nitrite toxicity study indicated that *S. rivulatus* juveniles have a 96 h LC_{50} of 10⁵ mg l⁻¹ NO_2-N. Fish challenged with excessive nitrite exhibited erratic swimming, disorientation and lack of balance within 48 h of exposure. Similar behavior was reported in cobia exposed to nitrite by Rodrigues et al. [20]. The impaired swimming performance is possibly because of increased MetHb in the blood that causes a decrease in O_2 available to tissues thus limiting activity [37]. Margiocco et al. found that nitrite concentrations in liver, brain, gills and muscles do not reach the same levels as in the blood, which suggests that blood is the primary target of nitrite toxicity and that in turn affects all the other organs by reducing oxygen delivery [38].

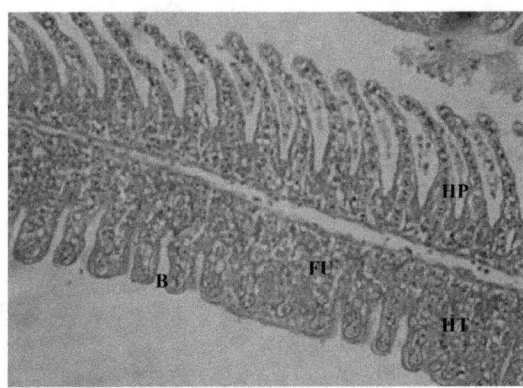

Figure 3: Light micrographs of gills of fish reared in 30 ppm nitrite-N. Gill damage is apparent. Notice lamellar fusion (FU) and hypertrophy (HT), and hyperplasia (HP) of the twisted irregular secondary lamellae with blunt ends (B). [Formalin fixed-H&E stained preparation, X 200].

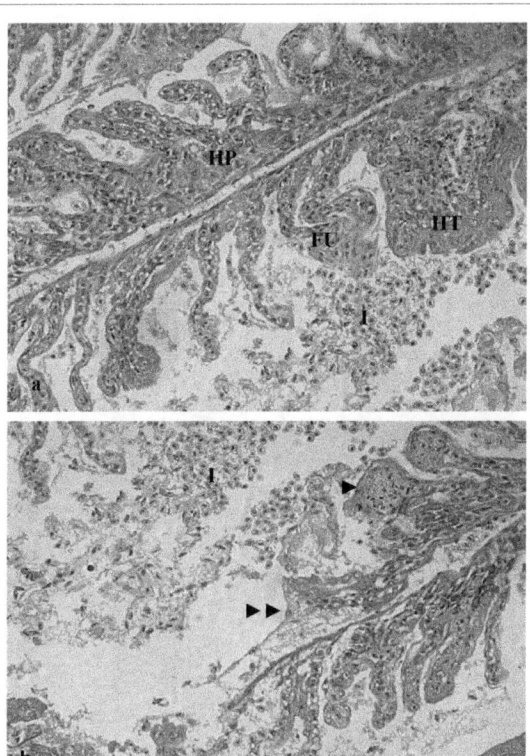

Figure 4: a and b: Light micrographs. Gills of group 6 (50 ppm nitrite). a) Showing marked histological alterations and severe deformations with severe irregular and curled secondary lamellae, extensive hypertrophy (HT) leading to severe lamellar fusion (FU) of secondary lamellae with severe blood infiltration (I). b) Revealing remarkable and widespread tissue damage and lysis at several points (►►) with severe telangiectasia (►) and infiltration of blood cells (I). [Formalin fixed- H&E stained preparation, X 400].

No data about the tolerance of marbled rabbitfish to nitrite is reported in the literature. However, when compared to various other marine species, *S. rivulatus* shows middle-of-range tolerance to nitrite-nitrogen. The flounder *Paralichthys orbignyanus* is the most sensitive marine fish on record with a 96 h LC_{50} of 30 mg l⁻¹ NO_2-N [39] and the dark-banded rockfish *Sebastes inermis* is the least sensitive fish reported, with a 96 h LC_{50} of 700 mg l⁻¹ NO_2-N [18]. Other species show intermediary toxicity: 85 mg l⁻¹ NO_2-N for red drum *Sciaenops ocellatus* [40] and 199 mg l⁻¹ NO_2-N for the marine pejerrey *Odontesthes argentinensis* [41]. Tolerance of *S. rivulatus* to short-term nitrite exposure is closest to that of Siberian sturgeon *Acipenser baeri* having LC_{50} of 130 mg l⁻¹ NO_2-N [16]. Although nitrite tolerance changes with age, our interest was in tolerance of juveniles because these are the age groups important to aquaculturists.

Chronic nitrite exposure

Survival and growth of juvenile marbled rabbitfish exposed to NO_2-N concentrations of 10 to 40 mg l⁻¹ over 56 days decreased as the nitrite concentration increased. However, nitrite uptake, toxicity, and effects vary among fish species and life stages [3,42,43], as well as with environmental conditions such as temperature [44], thus making comparison of toxicity values among studies difficult.

Although studies on long-term effects of aqueous nitrite on growth and survival of marine fish are relatively scarce, there are reports that increasing nitrite levels lead to net reductions in growth and/or survival of various fish species [4,6,14]. Our results indicate growth

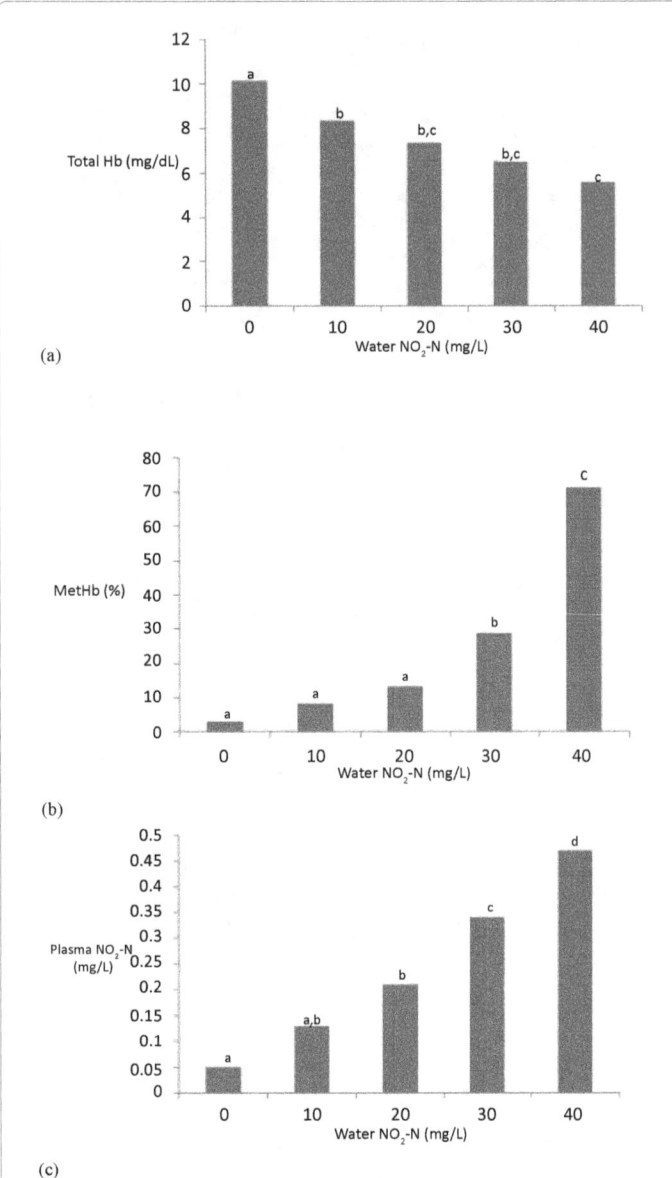

Figure 5: (a,b,c): Plasma nitrite (NO$_2$-N), total hemoglobin (total Hb) and methemoglobin values (MetHb) of rabbitfish *Siganus rivulatus* exposed to various nitrogen nitrite concentrations (0-40 mg l^{-1}) for 19 days.

suppression and mortality but these results were only significant at the greater nitrite concentrations (30 and 40 mg l^{-1} NO$_2$-N) used in the present experiment. A similar trend of reduction in growth of marine fish concomitant with increasing nitrite concentrations was reported by Siikavuopio and Saether working with juvenile cod *Gadus morhua* [19]. However, growth suppression was not always correlated with increasing nitrite concentrations. Wedemeyer and Yasutake found no growth suppression of steelhead trout *Salmo gairdneri* exposed to low nitrite concentrations for six months [45]. Similarly, Kamstra et al., observed no growth reduction in European eel *Anguilla anguilla* exposed to sub-lethal nitrite concentrations [46].

Finally, no significant differences in HSI among fish exposed to the various nitrite concentrations were observed in the present work. Similar results were observed by Deane and Woo working with the bream *Sparus sarba* [17]. Apparently, although nitrite is partly

detoxified by the liver of some fishes such as rainbow trout [47], most of the mitigation is performed by methemoglobin reductase in oxygenated RBCs [3,48], thus possibly sparing the liver. However, nitrite has been reported to cause some liver damage in various fishes [49-51]. Huang and Chen suspected liver necrosis in eel *Anguilla anguilla* following exposure to very low levels of nitrite [12]. Nitrite does not seem to have similar deleterious effects on rabbitfish livers as it has on livers of some other fishes.

A significant reduction in total erythrocyte count as observed in the present work was reported in various freshwater fishes [9,11] and Park et al. [18] reported a decrease in hemoglobin content with increasing nitrite concentrations in the marine dark-banded rockfish. Jensen suggested that significant changes in the blood parameters of fish caused by nitrite exposure, such as reduction in total erythrocyte count, total Hb concentration, and Ht levels, may be a consequence of passage of nitrite into the blood stream causing RBC hemolysis [52]. The problems are further compounded because hemolysis results in efflux of RBC potassium, leading to RBC shrinkage and creating a situation of oxygen shortage in the fish [52-54].

Leukocytes are involved in the regulation of immunological functions and their numbers increase as fish respond to stressful conditions [55-57]. Although some authors consider fish thrombocytes to be equivalent to platelets in mammals, we consider thrombocytes as complete white blood cells as suggested by Tavares-Dias and Oliveira [58]. In the present study, TLC increased with increase in nitrite concentration up to 20 mg l^{-1} NO$_2$-N then decreased at greater nitrite-N concentrations (30 and 40 mg l^{-1}). Similar results were found in *Cirrhinus mrigala* (Ham) where TLC increased with increasing nitrite concentrations and then decreased with further exposure [11]. Possibly, the excessive stress caused failure/exhaustion of leukopoiesis, resulting in reduction of TLC at higher nitrite concentrations [11]. It is also possible that decreased oxygen content of the blood affected hemopoietic tissues such as the kidney.

Differential leukocyte counts of juvenile *Siganus rivulatus* in the present study showed an increase in the percentages of neutrophils, lymphocytes, monocytes, and eosinophils concomitant with an increase in nitrite concentrations. Such results are not surprising considering stress generally causes neutrophilia in fish [59]. Moreover, Das et al. reported that lymphocyte and monocyte prevalence is affected by hypoxic stress [11]. No basophils were observed in blood of fish in the present experiment. Apparently, such results are common in Mediterranean and Red Sea fishes. No basophils were found in blood of six Mediterranean fish species examined by Pavlidis et al. [60] or in 50 species of fishes from the Red Sea [61].

Total plasma protein values of *Siganus rivulatus* showed a decreasing trend with increasing nitrite concentrations, with a significant reduction of TPP in fish maintained at 40 mg l^{-1} NO$_2$-N. Such results corroborate results observed in dark-banded rockfish *Sebastes inermis*, which when subjected to nitrite concentrations between 200-700 mg l^{-1} NO$_2$-N for 96 h portrayed a significant reduction in serum protein [18]. Similarly, Das et al. reported significant decrease of serum protein level in C. mrigala following exposure to 4-10.4 mg l^{-1} nitrite [11]. Possibly, fish require extra energy to maintain metabolism when subjected to toxicant induced stress and requirement is met by increasing protein catabolism, thereby reducing plasma protein [11]. In addition, hemolysis and removal of RBCs could cause a dilution of plasma which is also observed as a small reduction in plasma protein levels [11]. The reduction in plasma protein in the present study, concomitant with increasing nitrite concentrations is probably a result of both protein catabolism and plasma dilution.

One effect of hypoxia on fish is an increase in lactate production [62,63]. Woo and Chiu reported a significant increase in blood lactate of *L. calcarifer* exposed to various aqueous nitrite levels [64]. Although in the present work no significant differences were observed among blood lactate levels of fish maintained at various nitrite levels, results do suggest an increase in blood lactate of fish reared in the 40 mg l^{-1} NO$_2$-N waters compared to fish in other treatments. The fact that lactate increase was only observed at high nitrite levels suggests that there might be a threshold for oxygen content in the blood, below which lactate starts increasing. Similar observations of plasma lactate being only discernible in fish exposed to extremely high nitrite concentration were reported by Jensen et al. working with carp [62].

Histological changes in fish gills are among the most commonly recognized responses to environmental pollutants [21,24,65]. In the present study, light micrographs of gills revealed that chronic exposure to nitrite resulted in distinguishable tissue deformations and apparent gill damage as compared to gills of fish not exposed to nitrite. These observations were similar to those of Bullock [66], Larmoyeaux and Piper [67], Takashima and Hibiya [68], and Fernandes et al. [23] to name but a few.

In several slides, epithelial lifting and lysis of some blood sinuses were observed. Epithelial lifting possibly served as a defense mechanism, because separation of lamellar epithelia increase the distance across which waterborne pollutants must diffuse to reach the bloodstream [69,70]. However, as Peters et al. suggested, epithelial lifting causes an increased diffusion distance across the membrane reducing the gill's functional surface area and capacity for efficient gas exchange [71].

The most characteristic features observed in nitrite exposed fish were lamellar aneurysms. The lamellar aneurysms are believed to result from the collapse of pillar cell systems and the breakdown of vascular integrity with a release of large quantities of blood that push the lamellar epithelium outward [72]. In the present study, excessive nitrite caused severe histopathological alterations in the gills of *Siganus rivulatus* and may be the reason for the mortality observed. Fish exposed to lesser nitrite concentrations exhibited histological signs of stress but were able to survive.

Plasma nitrite and methemoglobin determination

Methemoglobin levels in siganid blood increased as environmental nitrite concentration increased. At an aqueous concentration of 40 mg l^{-1} NO$_2$-N, blood methemoglobin reached an average of 71.4% of total hemoglobin. Similar observations are reported for sea bass *Dicentrarchus labrax* by Scarano et al. [50]. As the proportion of methemoglobin in the blood increased, the proportion of functional hemoglobin decreased but also the total amount of hemoglobin in the blood decreased. Such results are corroborated by Brown and Mcleay [73] and Margiocco et al. [38] suggesting that although anoxia resulting from methemoglobinemia is the primary mechanism of nitrite toxicity [74], nitrite also causes a decrease in total hemoglobin in the blood which compounds the problem. However, this observation has not been reported in all fish species exposed to nitrite. Hemoglobin decrease was not observed in sea bass exposed 24 h to nitrite, and was not reported in channel catfish exposed 24 h to 5 mg l^{-1} NO$_2$-N [75]. Moreover, methemoglobinemia has been questioned as the sole mechanism of nitrite toxicity in fish [76,77]. Regardless of the mode of action of environmental nitrite, the marbled rabbitfish is quite tolerant

of nitrite if reared at a salinity of 35‰ making the species quite suitable for intensive aquaculture in recirculation systems.

Acknowledgment

The present work was funded by a grant from the Lebanese National Council for Scientific Research.

References

1. Tomasso JR (1994) Toxicity of nitrogenous wastes to aquaculture animals. Reviews in Fisheries Science 2: 291-314.

2. Grosell M, Jensen FB (2000) Uptake and effects of nitrite in the marine teleost fish Platichthys flesus. Aquat Toxicol 50: 97-107.

3. Jensen FB (2003) Nitrite disrupts multiple physiological functions in aquatic animals. Comp Biochem Physiol 135: 9-24.

4. Colt J, Ludwig R, Tchobanoglous G, Cesh Jr JJ (1981) The effects of nitrite on the short-term growth and survival of channel catfish, Ictalurus punctatus. Aquaculture 24: 111-122.

5. Svobodova Z, Machova1 J, Poleszczuk G, Huda J, Hamaakova J, et al. (2005) Nitrite poisoning of fish in aquaculture facilities with water-recirculating systems. Acta Vet Brno 74: 129-137.

6. Kroupova H, Machova J, Piackova V, Blahova J, Dobsikova R, et al. (2008) Effects of subchronic nitrite exposure on rainbow trout (Oncorhynchus mykiss). Ecotoxicol Environl Saf 71: 813-820.

7. Gisbert E, Rodriguez A, Cardona L, Huertas M, Gallardo MA, et al. (2004) Recovery of Siberian sturgeon yearlings after an acute exposure to environmental nitrite: changes in the plasmatic ionic balance, Na+ - K+ ATPase activity, and gill histology. Aquaculture 239: 141-154.

8. Avilez IM, Altran AE, Aguiar LH, Moraes G (2004) Hematological responses of the neotropical teleost matrinxa (Brycon cephalus) to environmental nitrite. Comp Biochem Physiol 139: 135-139.

9. Da Costa OTF, Ferreira DJDS, Mendonca FLP, Fernandes MN (2004) Susceptibility of the Amazonian fish, Colossoma macropomum (Serrasalminae), to short-term exposure to nitrite. Aquaculture 232: 627-636.

10. Madison BN, Wang YS (2006) Haematological responses of acute nitrite exposure in walleye (Sander vitreus). Aquat Toxicol 79: 16-23.

11. Das PC, Ayyappan S, Jena JK, Das BK (2004) Nitrite toxicity in Cirrhinus mrigala (Ham): acute toxicity and sub-lethal effect on selected haematological parameters. Aquaculture 235: 633-644.

12. Huang CY, Chen JC (2002) Effects on acid-base balance, methaemoglobinemia and nitrogen excretion of European eel after exposure to elevated ambient nitrite. Journal of Fish Biology 61: 712-725.

13. Svobodova Z, Machova J, Drastichova J, Groch L, Luskova V, et al. (2005) Haematological and biochemical profile of carp blood following nitrite exposure at different concentration of chloride. Aquaculture Research 36: 1177-1184.

14. Frances J, Allan GL, Nowak BF (1998) The effects of nitrite on the short-term growth of silver perch (Bidyanus bidyanus). Aquaculture 163: 63-72.

15. Costa LDF, Filho KCM, Severo MP, Sampaio LA (2008) Tolerance of juvenile pompano Trachinotus marginatus to acute ammonia and nitrite exposure at different salinity levels. Aquaculture 285: 270-272.

16. Huertas M, Gisbert E, Rodriguez A, Cardona L, Williot P, et al. (2002) Acute exposure of Siberian sturgeon (Acipenser baeri, Brandt) yearlings to nitrite: median-lethal concentration (LC$_{50}$) determination, haematological changes and nitrite accumulation in selected tissues. Aquatic Toxicology 57: 257-266.

17. Deane EE, Woo NYS (2007) Impact of nitrite exposure on endocrine, osmoregulatory and cytoprotective functions in the marine teleost Sparus sarba. Aquat Toxicol 82: 85-93.

18. Park I-S, Lee J, Hur J-W, Song Y-C, Na HC, et al. (2007) Acute toxicity and sublethal effects of nitrite on selected hematological parameters and tissues in dark-banded rockfish, Sebastes inermis. Journal of the World Aquaculture Society 38: 188-199.

19. Siikavuopio SI, Saether BS (2006) Effects of chronic nitrite exposure on growth in juvenile Atlantic cod, Gadus morhua. Aquaculture 255: 351-356.

20. Rodrigues RV, Schwarz MH, Delbos BC, Sampaio LA (2007) Acute toxicity

and sublethal effects of ammonia and nitrite for juvenile cobia Rachycentron canadum. Aquaculture 271: 553-557.

21. Au DWT (2004) The application of histo-cytopathological biomarkers in marine pollution monitoring: a review. Mar Pollut Bulletin 48: 817-834.

22. Perry SF, Laurent P (1993) Environmental effects on fish gill structure and function: recent advances and future directions. In: Jensen F, Rankin C edn. Fish Ecophysiology. Chapman and Hall, London 231-264.

23. Fernandes C, Fontainhas-Fernandes A, Monteiro SM, Salgado MA (2007) Histopathological gill changes in wild leaping grey mullet (Liza saliens) from the Esmoriz-Paramos coastal lagoon, Portugal. Environ Toxicol 22: 443-448.

24. Mallat J (1985) Fish gill structural changes induced by toxicants and other irritants: a statistical review. Canadian Journal of Fisheries and Aquatic Sciences 42: 630-648.

25. Woodland DJ (1983) Zoogeography of the Siganidae (Pisces): an interpretation of distribution and richness patterns. Bulletin of Marine Science 33: 713-717.

26. Saoud IP, Kreydiyyeh S, Chalfoun A, Fakih M (2007) Influence of salinity on survival, growth, plasma osmolality and gill Na+-K+-ATPase activity in the rabbitfish Siganus rivulatus. Journal of Experimental Marine Biology and Ecology 384: 183-190.

27. Saoud IP, Ghanawi J, Lebbos N (2008) Effects of stocking density on survival, growth, size variation and condition index of the rabbitfish Siganus rivulatus. Aquaculture International 16: 109-116.

28. Saoud IP, Mohanna C, Ghanawi J (2008) Effects of temperature on survival and growth of juvenile rabbitfish (Siganus rivulatus). Aquaculture Research 39: 491-497.

29. Parsons TR, Maita Y, Lalli, CM (1985) A Manual of Chemical and Biological Methods for Seawater Analysis. Pergamon Press, Elmsford, New York, USA.

30. Clesceri LS, Greenberg AE, Trussell RR (1989) Standard methods for the examination of water and wastewater, 7th edn. American Public Health Association, Washington, USA.

31. Natt MP, Herrick CA (1952) A new blood diluent for counting erythrocytes and leukocytes of the chicken. Poultry Science 31: 735-738.

32. Ellis AE (1977) The leucocytes of fish: A review. Journal of Fish Biology 11: 453-491.

33. Ainsworth AJ (1992) Fish granulocytes: morphology, distribution and function. Annual Review Fish Disease 2: 123-148.

34. Drabkin DL, Austin JH (1932) Spectrophotometric constants for common hemoglobin derivatives in human, dog, and rabbit blood. Journal of Biological Chemistry 98: 719-733.

35. Shechter H, Gruener N, Shubal HI (1972) Micromethod for the determination of nitrite in blood. Analytica Chimica Acta 60: 93-99.

36. Horecker BL, Brackett FS (1944) A rapid spectrophotometric method for the determination of methemoglobin and carbonylhemoglobin in blood. Journal of Biological Chemistry 152: 669-677.

37. Brauner CJ, Val AL, Randall DJ (1993) The effect of graded methaemoglobin levels on the swimming performance of Chinook salmon (Oncorhynchus tshawytscha). J Exp Biol 185: 121-135.

38. Margiocco C, Arill A, Mensi P, Schenone G (1983) Nitrite bioaccumulation in Salmo gairdneri Rich and hematological consequences. Aquatic Toxicology 3: 261-270.

39. Bianchini A, Wasielesky Jr W, Miranda KC (1996) Toxicity of nitrogenous compounds to juveniles of flatfish Paralichthys orbignyanus. Bulletin of Environmental Contamination and Toxicology 56: 453-459.

40. Wise DJ, Tomasso JR (1989) Acute toxicity of nitrite to red drum Sciaenops ocelattus: effect of salinity. Journal of World Mariculture Society 20: 193-198.

41. Sampaio LA, Pisseti TL, Morena M (2006) Acute toxicity of nitrite on larvae of the marine pejerrey Odontesthes argentinensis (Teleostei, Atherinopsidae). Ciência Rural 36: 1008-1010.

42. Lewis Jr WM, Morris DP (1986) Toxicity of nitrite to fish: A review. Transactions of the American Fisheries Society 115: 183-195.

43. Martinez CBR, Souza MM (2002) Acute effects of nitrite on ion regulation in two neotropical fish species. Comp Biochem Physiol 113: 151-160.

44. Saroglia MG, Scarano G, Tibaldi E (1981) Acute toxicity of nitrite to sea bass (Dicentrarchus labrax) and European eel (Anguilla anguilla). Journal of the World Mariculture Society 12: 121-126.

45. Wedemeyer GA, Yasutake WT (1978) Prevention and treatment of nitrite toxicity in juvenile steelhead trout (Salmo gairdneri). Journal of the Fisheries Research Board of Canada 35: 822-827.

46. Kamstra A, Span JA, Van Weerd JH (1996) The acute toxicity and sublethal effects of nitrite on growth and feed utilization of European eel, Anguilla anguilla (L). Aquaculture Research 27: 903-911.

47. Doblander C, Lackner R (1996) Metabolism and detoxification of nitrite by hepatocytes. Biochimica Biophys Acta 1298: 270-274.

48. Freeman L, Beitinger TL, Huey DW (1983) Methemoglobin reductase activity in phylogenetically diverse piscine species. Comparative Biochemistry and Physiology 75: 27-30.

49. Arillo A, Gaino E, Margiocco C, Mensi P, Schenone G (1984) Biochemical and ultrastructural effects of nitrite on rainbow trout: liver hypoxia at the root of the acute toxicity mechanism. Environ Res 34: 135-154.

50. Scarano G, Saroglia MG, Gray RH, Tibaldi E (1984) Hematological responses of sea bass Dicentrarchus labrax to sublethal nitrite exposure. Transactions of the American Fisheries Society 113: 360-364.

51. Michael MI, Hilmy AM, El-Domiaty NA, Wershana K (1987) Serum transaminases activity and histopathological changes in Clarias lazera chronically exposed to nitrite. Comp Biochem Physiol 86: 255-262.

52. Jensen FB (1990) Nitrite and red cell function in carp: control factors for nitrite entry, membrane potassium ion permeation, oxygen affinity and methaemoglobin formation. J Exper Biol 152: 149-166.

53. Knudsen PK, Jensen FB (1997) Recovery from nitrite-induced methaemoglobinemia and potassium balance disturbance in carp. Fish Physiology and Biochemistry 16: 1-10.

54. Vedel NE, Korsgaard B, Jensen FB (1998) Isolated and combined exposure to ammonia and nitrite in rainbow trout (Oncorhynchus mykiss): effects on electrolyte status, blood respiratory properties and brain glutamine/glutamate concentrations. Aquatic Toxicology 41: 325-342.

55. Wlasow T, Dabrowska H (1990) Haematology of carp in acute intoxication with ammonia. Polskie Archiwum Hydrobiologii 37: 419-428.

56. Svobodova Z, Vykusova B, Machova J (1994) The effects of pollutants on selected haematological and biochemical parameters in fish. In: Muller R, Lloyd R edn. Sublethal and Chronic Effects of Pollutants on Freshwater Fish. FAO Fishing news books, Oxford 39-52.

57. Nussey G, van Vuren JHJ, du Preez HH (2002) The effect of copper and zinc at neutral and acidic pH on the general haematology and osmoregulation of Oreochromis mossambicus. African Journal of Aquatic Sciences 27: 61-84.

58. Tavares-Dias M, Oliveira SR (2009) A review of the blood coagulation system of fish. Brazilian Journal of Biosciences 7: 205-224.

59. Blaxhall PC (1972) The haematological assessment of the health of freshwater fish: A review of selected literature. Journal of Fish Biology 4: 593-604.

60. Pavlidis M, Futter WC, Katharios P, Dianach P (2007) Blood cell profile of six Mediterranean fish species. Journal of Applied Ichthyology 23: 70-73.

61. Saunders DC (1968) Variations in thrombocytes and small lymphocytes found in circulating blood of marine fishes. Transactions of the American Microscopical Society 87: 39-43.

62. Jensen FB, Andersen NA, Heisler N (1987) Effect of nitrite exposure on blood respiratory properties, acid-base and electrolyte regulation in the carp (Cyprinus carpio). Journal of Comp Physiol 157: 533-541.

63. Stormer J, Jensen FB, Rankin JC (1996) Uptake of nitrite, nitrate and bromide in rainbow trout, Onchorhynchus mykiss: effects on ionic balance. Canadian Journal of Fisheries and Aquatic Sciences 53: 1943-1950.

64. Woo NYS, Chiu SF (1996) Metabolic and osmoregulatory responses of the Sea Bass Lates calcarifer to nitrite exposure. Environmental Toxicology and Water Quality 12: 257-264.

65. Laurent P, Perry SF (1991) Environmental effects on fish gill morphology. Physiology and Zoology 64: 4-25.

66. Bullock GL (1972) Studies on selected myxobacteria pathogenic for fishes and

on bacterial gill disease in hatchery-reared salmonids. US Bureau of Sport Fisheries and Wildlife, Technical Paper 60.

67. Larmoyeaux JD, Piper RG (1973) Effects of water reuse on rainbow trout in hatcheries. Progressive Fish-Culturist 35: 2-8.

68. Takashima F, Hibiya T (1995) An atlas of fish histology. Normal and pathological features, 2nd edn., Kodansha Ltd., Tokyo.

69. Karan V, Vitorovic S, Tutundzic V, Poleksic V (1998) Functional enzymes activity and gill histology of carp after copper sulfate exposure and recovery. Ecotoxicol Environ Saf 40: 49-55.

70. De Boeck G, Grosell M, Wood C (2001) Sensitivity of the spiny dogfish *Squalus acanthias* to waterborne silver exposure. Aquat Toxicol 54: 261-275.

71. Peters G, Hoffmann R, Klinger H (1984) Environmental induced gill disease of cultured rainbow trout (*Salmo gairdner*). Aquaculture 38: 105-126.

72. Alazemi BM, Lewis JW, Andrews EB (1996) Gill damage in the freshwater fish *Gnathonemus petersii* (family: Mormyridae) exposed to selected pollutants: an ultrastructural study. Environmental Technology 17: 225-238.

73. Brown DA, McLeay DJ (1975) Effect of nitrite on methemoglobin and total hemoglobin of juvenile rainbow trout. The Progressive Fish-Culturist 37: 36-38.

74. Cameron JN (1971) Methemoglobin in erythrocytes of rainbow trout. Comparative Biochemistry and Physiology 40: 743-749.

75. Huey DW, Beitinger TL (1980) Hematological responses of larval *Rana catesbeiana* to sublethal nitrite exposures. Bulletin of Environmental Contamination and Toxicology 25: 574-577.

76. Smith CE, Williams WG (1974) Experimental nitrite toxicity in rainbow trout and Chinook salmon. Transactions of the American Fisheries Society 103: 389-390.

77. Crawford RE, Allen GH (1977) Seawater inhibition of nitrite toxicity to chinook salmon. Transactions of the American Fisheries Society 106: 105-109.

8

Effect of Probiotic on Microbiological and Haematological Responsiveness of Cat fish (*Heteropnuestes fossilis*) Challenged with Bacteria *Aeromonas hydrophila* and Fungi *Aphanomyces invadans*

Meeran Mohideen[1, 2]*, and Haniffa MA[1]

[1]*Centre for Aquaculture Research and Extension, St. Xavier's (Autonomous) College, Palayamkottai, 627002, Tamil Nadu, India*
[2]*Institute for Research in Molecular Medicine, University Sains Malaysia, Pulau Penang, 11800, Malaysia*

Abstract

The use of probiotic for disease prevention and improved nutrition in aquaculture is becoming popular due to an increasing demand for environment friendly aquaculture. Here we used *Bacillus subtilis* as a probiotic to fish to evaluate the effect of probiotic on microbiological and haematological responsiveness of cat fish (*Heteropnuestes fossilis*) challenged with bacteria *Aeromonas hydrophila* and fungi *Aphanomyces invadans*. *Heteropnuestes fossilis* were collected from local market at Tirunelveli, Tamil Nadu, India. Fish were subjected into microbiological, haematological, physiological observation. In *H. fosilis*, probiotic accepted fishes gained more weight than that of the control fishes fed with control diet. The gut micro flora of *H. fossilis* was found to be 6.3×10^6, 5.7×10^7, 5.4×10^5 and 5.1×10^5 cells in D1 treatment fishes on $10^4, 10^5, 10^6$ and 10^7 dilutions respectively. The microbiological estimation also showed a dual increase in trial count in T2 injected fish than that of the T3 injected fish. Many factors can influence the immune response of fish. Among them are stressors and environmental factors are natural. In the present investigation behavioural symptoms to pathogenicity such as imbalance, restlessness and avoidance of food were observed. Pathological symptoms include fin necrosis and tail rot which were also observed. In some cases septicaemic ulceration was noticed. Haematological parameters elicited changes which are able to reveal some clues for diagnosis and prognosis of the disease state. T2 fishes were inflicted alterations in TEC, TLC, DLC, and Hb content which indicated decrease state of immunity, when compared with T3 fishes. Bacteria injected fishes showed good healthy status whereas fungal injected fishes showed non healthy status of fish.

Keywords: Probiotic; Microbiological; Haematological; Cat fish; Bacteria; Fungi

Introduction

Only in the last few years, aquaculture has undergone rapid advancement. The principle reasons for the increased interest and development of fish farming are due to the recent advances in the development of fish culture techniques in the world particularly in the field of husbandry and management of culture system and the development of standardized artificial breeding technologies such that supply of seeds is guaranteed and controlled by the fish farmer. A large number of reports are available on general biology in relation to food, feeding habits and breeding [1]. However, information in relation to nutrient requirements of the fish protein is limited. Development of economical feed mixture is an important factor in fish culture in which the growth of fish is influenced by the quality and quantity of the diet. Reports are available on the growth of cultivable fishes using animal and plant sources of protein diets [2]. For optimum fish growth the use of fish meal (25-65%) as higher dietary protein in fish feeds causes more expensive. Hence, the development of low cost and nutritionally balanced diet is in urgent need. The alternative sources of protein either by partial or complete replacement of the fish meal have been studied using various ingredients [3]. This study focussed on investigation attempts have been made to produce the fish feed using fish meal like anchovy, jawala and flour like soy flour, tapioca flour and wheat flour and sunflower oil, aquasavour, vitamin C.

Infectious diseases are considered as one of the main barriers to the successful development and continuation of molluscan and shrimp aquaculture as they limit production in terms of quality, quantity and regularity [4]. Although disease control is an inherent component of any intensive animal production system, controlling disease in the aquatic environment is further complicated by the intimate relationship that exists between pathogens and their host and the frequent use of open production system [5]. However, excessive antimicrobial use can lead to the emergence of bacterial resistance [6,7]. Hence the use of probiotics for disease prevention and improved nutrition in aquaculture is becoming popular due to an increasing demand for environment friendly aquaculture [6]. Probiotics act as growth promoter and reduce the substrate of pathogenic microbes. Commonly available probiotics are *Lactobacillus acidophilus*, *Bacillus subtilis* and EfinolG (mixture of microbes). Several studies have shown that probiotics improves the growth rate of fishes by improving the immune status of fishes [1,4,8-10]. The use of Probiotics to displace pathogenic bacteria by competitive process is a better remedy than administering antibiotics. *Pseudomonas fluorescens* (AH2) was shown to be strongly inhibitory against *Vibrio anguillarum* and it reduced the mortality rate of rainbow trout injected by *Vibrio anguillarum* [5]. Improved disease resistance has also been observed in cod fry fed with dry feed containing *Carnobacterium divergens* [11,12] showed that the survival and growth of the black tiger shrimp (*Penaeus monodom*) [12]. Thus, probiotics have been shown to be effective in a wide range of

---

*Corresponding author: Meeran Mohideen, Institute for Research in Molecular Medicine, University Sains Malaysia, Pulau Penang, Malaysia
E-mail: meeran_micro@yahoo.co.in

species for the promotion of growth, enhanced nutrition, immunity and survival.

Materials and Methods

Heteropneustes fossilis were collected from local market in Tirunelveli, TamilNadu, India and transported to CARE aqua farm. They were acclimatized to laboratory condition for a week. Ambient temperature 29±1°C and pH 7, 1±0.5mg/lr, was maintained, throughout the experiment and 1/3rd of water was renewed daily.

Feed formulation

Fishes were fed regularly with an artificial balanced diet made up of wheat flour, tapioca flour, soya flour, vegetable oil, anchovy, jawala to control fishes and *Bacillus subtilis* was added in the diet to experimental fishes (*B. Subtilis* was purchased from Xi'an Lyphar Biotech co., Ltd, China). Fishes were reared in plastic troughs (capacity 40 litres). The following ingredients were used to prepare feed pellet for *Heteropneustes fossilis* for 100 mg fish feed.

Anchovy	-	26.9 gm
Soy bean	-	25 gm
Jawala Acetes	-	20 gm
Tapioca meal	-	10.9 gm
Wheat flour	-	10 gm
Sunflower oil	-	5.8 ml
Mono sodium phosphate	-	0.5 gm
Aquasavour	-	0.3 gm
Ascorbic acid	-	0.02 gm
Probiotic concentration	-	2 mg

Fish meal

Anchovy and Jawala prawn were purchased from local fish market. They were powdered and sieved to required size. The protein content of the fish meal was 60%.

Flour: Wheat, soybean and tapioca flour were used. Tapioca flour acts as a binder and source of carbohydrate. Gelatinization of tapioca flour improves the stability of the feed.

Oil: 5.8 ml of sunflower oil was used in the food formulation.

Constituents

Energy	-	884 cal
Total fatty acids	-	100 gm
i. SFA	-	12 gm
ii. MUFA	-	28 gm
iii. PUFA	-	60 gm
iv. Trans fatty acids	-	BDL
i. Saturated Fatty Acid		
ii. Mono Saturated Fatty Acid		
iii. Poly Saturated Fatty Acid		
iv. Below Detectable limit (0.05%)		

Vitamin: 0.02 gm Vitamin C was used to formulate fish feed.

Feed preparation

All powdered ingredients were weighed separately and mixed with required amount of hot water to make semi moist dough and sterilized .Then 5.8ml sunflower oil was added under appropriate temperature

and the probiotic *B. subtilis was* also added under aseptic condition. This mixture was pelleted and dried.

Experimental design

The experimental fishes *H. fossilis* were reared in plastic trough of 40lr capacity. Twelve troughs containing six fishes in each were taken and filled with water. This was divided into two groups as Diet 1 (D_1) treatment (4 troughs) and Diet 2 (D_2) probiotic treatment (8 troughs). The control fishes were fed with control basal diet while the probiotic treatment fishes were fed with *B. subtilis* added diet for 21 days feeding trial. Before the feeding trial the gut micro flora and the haematology of the experimental fishes were analysed.One third of water was changed daily. After completion of feeding trial the fishes were divided into three treatments (D_1) group fishes was kept as treatment 1 (T1) The D_2 group was further divided into two treatments with four troughs of fish for each. The treatment II fishes were injected *A. hydrophila* in treatment III were *A. invadans* injected fishes. This disease challenge was done for 15 days and the microbiological and haematological changes were observed and recorded.

LD$_{50}$

The LD$_{50}$ *Aeromonas hydrophila* (10^6 dilution) by one fish/dose, six fishes each replicate, were injected intramuscularly, in 0.1 ml concentration. *Aphanomyces invadans* (10^5 dilution) by one fish/dose, six fishes each replicate, were injected intramuscularly, in 0.1 ml concentration.

Disease challenge

Before the treatment the control fishes were fed with control feed and the experimental fishes were fed with probiotic containing feed. After the treatment all the experimental fishes were fed with control feed including control fishes. The treatment (T1) fishes were injected with physiological saline. The treatment II (T2) and treatment III (T3) fishes were injected with 0.1 ml of *A. hydrophila* and *A. invadans* respectively.

Disease challenge was done on 31st day by intramuscular injection of *A. hydrophila* and *A. Invadans* to (T2) and (T3) group fishes respectively. At the start of experiment (T=0) fishes in (T2) and (T3) of the fishes were fed with feed containing probiotic, while the remaining 1st group was fed with non probiotic feed. The immunological parameters (RBC count, WBC count, DLC, PCV, MCV, MCHC, Haemoglobin content of blood) of the fish were recorded on (31, 30, 35 and 40 days.) in all the three treatments.

Microbiological estimation

The samplings were made on 0, 30, 35, 40and45 days for microbial investigation of gut of experimental fish *H. fossilis* in order to check the changes in gut micro flora after probiotic feeding.

Enumeration of total heterotrophic bacterial count

The experimental fish *H. fossilis* were taken and wiped with alcohol to remove the surface bacteria. The fish were dissected and the intestine was removed and homogenized with sterile distilled water and grinded with morter and pestle. 1ml of homogenized solution was taken and added into sterile blank water representing as 10^{-1} dilution. Then it was serially diluted upto 10^{-10} dilution. For Heterotrophic bacterial count 0.1 ml was taken from 10^{-4} to 10^{-7} dilution and spread plated into sterile nutrient agar plates (10^{-4} to 10^{-7}). Duplicates were maintained for each dilution. The plates were incubated at 37°C for 24 hrs. After incubation colonies were observed and recorded.

Collection of blood sample

Blood was obtained by puncturing the heart by using 1ml insulin syringe. Before that the syringe and blood collecting vials are coated with anti-coagulant heparin. Anticoagulated blood was used for the analysis of the blood variables except differential leucocyte count and serological parameter.

Total erythrocyte count

Enumeration of total erythrocytes was done with Haemocytometer. Blood was diluted 200 times in the standard RBC pipette with RBC diluting fluid.

Total leukocyte count

Enumeration of total leucocytes was done using Haemocytometer.

Differential leucocyte count

For differential leucocyte count minimum of 100 leucocytes were classified and counted from each smear. They were indentified under 40X magnification.

Haemoglobin content

Sahlis's method had been employed for estimating haemoglobin content of blood.

Mean Corpuscular Haemoglobin (MCH)

It's used to determine the average haemoglobin content in single red cell in micro gram.

Haematocrit Value (Packed Cell Volume-PCV)

The haematocrit value (Hk) determines the ratio of the volume of the blood cells to that of blood plasma. The haematocrit value (Hk) is expressed as the percentage fraction of blood cells, in the total volume.

Mean Corpuscular Volume

The average volume of a single red cell in cubic microns.

Mean Corpuscular Haemoglobin Concentration (MCHC)

To determine the haemoglobin content of 20 μm of the packed cells as percentage as opposed to the percentage of haemoglobin of whole blood.

Results

The experimental fish *H. fossilis* readily accepted the probiotic diet (*B. subtilis*) and basal diet (control). No mortality occurred during the feeding trial. Length and weight gain of *H. fossilis* fed with probiotic diet was higher than the fishes fed with control diet. The gut micro flora of *H. fossilis* was found to be $6.3 \times 10^6, 5.7 \times 10^7, 5.4 \times 10^5$ and 5.1×10^5 cells in D1 treatment fishes on $10^{4}, 10^5, 10^6$ and 10^7 dilutions respectively (Table 1) on initial day (day 0). On 30th day the gut micro flora of *H. fossilis* fed with control diet and probiotic diet was found to be 5.7×10^7, 4.8×10^7, 5.1×10^6 and 4.3×10^5 cells on $10^4, 10^5, 10^6$ and 10^7 dilutions respectively (Table 2). On day 35, 40 and 45 the microbial analysis was done in three treatments such as T1, T2 and T3. The gut micro flora of T2 fishes was found to be $7.4 \times 10^6, 8.1 \times 10^5, 7.9 \times 10^5$ and 4.7×10^5 on 35th day in $10^4, 10^5, 10^6$ and 10^7 dilutions respectively. This showed a gradual decrease in T3 fishes (Table 3). On 40th day its showed gradual decrease of $7.2 \times 10^6, 7.3 \times 10^5, 6.9 \times 10^5$ and 4.3×10^5 in $10^4, 10^5, 10^6$ and 10^7 dilutions respectively. But in T3 fishes showed fluctuation (Table 4) when compared to T3 fishes on day 35.

The fungal and bacterial load in T2 and T3 fishes was decreased on day 45 (Table 5) when compared to days 40. The fungal colonies showed only a slight limitation in the growth recording 4.1×10^6, $5.7 \times 10^5, 6.2 \times 10^5$ and 4.1×10^5 on day 45 in $10^4, 10^5, 10^6$ and 10^7 dilutions respectively. The bacterial load and fungal load were found to be least in T2 and T3 on day 45. The results of gut microflora were showed in Tables 1-5.

Intramuscular injection of *Aeromonas hydrophila* and *Aphanomyces invadans* showed slight to severe dermomuscular lesions, in the experimental fish *H. fossilis*. Among the three treatments external lesion initially developed as a blanched area with superficially abrading lesion and excavated lesion with swelling in both probiotic treated (T2 and T3) fishes. But in the case of probiotic fed *A. hydrophila* injected (T2) group the infections gradually decreased during the experimental period from day 9(+++) until day 15(+). Whereas in *A. invadans* injected group (T3) lesion slightly decreased on day 9(+++) and further decreased on day 15(++). In the *A. hydrophila* injected group external lesion was reduced in the probiotic feeding than *A. invadans* injected group. The control fish did not develop any lesion.

The values of various Haematological indices of T1,T2,T3 are shown in Tables 6-13 i.e, TLC, TEC, DLC, PCV, Hb content, MCH, MCV, MCHC respectively.

TLC of T1 did not show much variation and found to be 3.3 (Cells/mm³) on 30th day and 3.4 (Cells/mm³), 3.1 (Cells/mm³), 3.0 (Cells/mm³) on day 35,40,45 respectively .The TLC of *A. hydrophila* (T2) injected fishes showed a decrease on the day 35 and gradually increased on day 45. In *A. invadans* (T3) injected fishes too the leucocytes showed a decrease on 35th day 2.6 (Cells/mm³) and a slight increase on day 45th day 4.0 (Cells/mm³) were observed (Table 6).

The TEC in T1 fishes were found to be normal from the 30th day 3215 (Cells/mm³) to 45th day 3100(Cells/mm³). In T2 fishes the red blood cells showed increased level on day 35 and it gradually decreased on day 45. In case of T3 also red blood cells increased on day 35 and slightly decreased on day 45 (Table 7).

DLC in T1 fishes were found to be from the 30thday to 45th day showed fluctuation in T2 fishes when compared with T1 fishes (Table

Dilution Factor	D1	D2
10^4	6.3×10^6	5.4×10^6
10^5	5.7×10^7	5.2×10^6
10^6	5.4×10^5	5.0×10^5
10^7	5.1×10^5	5.1×10^4

Table 1: Heterotrophic Bacterial Count (Day 0).

Dilution Factor	Colonies in number
10^4	5.7×10^7
10^5	4.8×10^7
10^6	5.1×10^6
10^7	4.3×10^5

Table 2: Heterotrophic Bacterial Count (Day 30).

Dilution Factor	T1	T2	
104	6.7×106	7.4×106	4.1×105
105	6.1×106	8.1×105	4.0×105
106	8.3×105	7.9×105	5.1×104
107	8.0×105	4.7×105	3.2×104

Table 3: Heterotrophic Bacterial Count (Day 35).

Dilution Factor	T1	T2	T3
104	6.4 × 106	7.2×106	4.3 ×105
105	6.2 × 106	7.3 × 105	4.0 ×105
106	5.8 × 106	6.9 × 105	7.9 ×104
107	7.3 × 105	4.3 ×105	5.7 ×104

Table 4: Heterotrophic Bacterial Count (Day 40).

Dilution Factor	T1	T2	T3
104	6.7 × 106	4.1×106	4.0 ×105
105	6.3 × 106	5.7 × 105	7.9 ×104
106	7.1× 105	6.2 × 105	4.1 ×104
107	5.8 × 105	4.1 ×105	3.2 ×105

Table 5: Heterotrophic Bacterial Count (Day 45).

Day	Sample	Mean	SD
30th Day	T1	3.3	1.2
	T2	4.1	1.7
	T3	3.9	1.5
35th Day	T1	3.4	1.3
	T2	3.9	0.9
	T3	2.6	0.7
40th Day	T1	3.1	0.8
	T2	3.8	1.3
	T3	3.1	0.8
45th Day	T1	3.0	1.2
	T2	4.3	1.4
	T3	4.0	0.7

Table 6: Comparison of TLC (Cells/mm3) in *H. f ossilis* administered w ith *A. hydrophila* and *A. invadans*.

Day	Sample	Mean	SD
30thDay	T1	3.2	0.7
	T2	3.2	1.5
	T3	2.8	1.0
35th Day	T1	3.1	0.8
	T2	3.5	1.2
	T3	3.6	1.0
40th Day	T1	3.4	0.5
	T2	3.7	1.7
	T3	3.7	1.3
45th Day	T1	3.1	0.2
	T2	3.7	1.4
	T3	3.9	0.9

Table 7: Comparison of TEC(Cells/mm3) in *H. f ossilis* administered w ith *A. hydrophila* and *A. invadans*.

Cells	Duration							
	Bacteria Injected							
	Control (T1)				Test (T2)			
	30th day	35th day	40th day	45th day	30th day	35th day	40th day	45th day
Neutrophils %	60	40	40	39	60	50	56	42
Eosinophils %	2	8	10	10	5	5	5	9
Basophils %	6	12	10	9	3	2	1	1
Lymphocytes %	30	40	39	41	32	39	38	48
Monocytes %	2	0	1	1	0	4	0	0

Table 8: DLC(%) in *H. f ossilis* administered w ith *A. hydrophilia* (106 CFU/0.1ml)

Cells	Duration							
	Fungus Injected							
	Control (T1)				Test(T3)			
	30th day	35th day	40th day	45th day	30th day	35th day	40th day	45th day
Neutrophiles %	52	40	45	48	43	62	51	50
Eosinophiles %	3	8	10	9	5	3	4	5
Basophiles %	5	10	5	6	10	2	1	1
Lymphocytes %	33	40	40	37	42	30	41	44
Monocytes %	7	2	0	0	0	3	3	0

Table 8.1 DLC(%) in *H. f ossilis* administered w ith *A. invadans* (105 CFU/0.1 ml)

Duration	Bacteria Administered				Fungus Administered	
	Control (T1)		Test (T2)		Test (T3)	
	PCV%	Plasma	PCV%	Plasma	PCV%	Plasma
30th day	25	75	25	75	26	74
35th day	25	75	22	78	14.7	85.3
40th day	24	76	20	80	13.2	86.8
45th day	25	75	18.2	81.8	11.1	88.1

Table 9: Haemetocrit value of *H. f ossilis* administered w ith *A. hydrophilia* (106 CFU/0.1ml) and *A. invadans* (105 CFU/0.1 ml).

Duration	Bacteria Injected		Fungus Injected
	Control (T1)	Test (T2)	Test (T3)
30th day	19.8	19.7	20
35th day	19.7	14.5	15
40th day	19.8	10	13
45th day	19	11.1	10

Table 10: Comparison of Haemoglobin content (in gram) of blood *in H. f ossilis* administered w ith *A. hydrophilia* (106 CFU/0.1ml) and *A. invadans* (105 CFU/0.1ml)

Duration	Bacteria Injected		Fungus Injected
	Control (T1)	Test (T2)	Test (T3)
30th day	0.0049	0.0047	0.0050
35th day	0.0051	0.0037	0.0054
40th day	0.0050	0.0034	0.0050
45th day	0.0064	0.0044	0.0045

Table 11: Comparision of MCH (in microgram) *in H. f ossilis* administered w ith *A. hydrophilia* (106 CFU/0.1ml) and *A. invadans* (105 CFU/0.1ml).

Duration	Bacteria Injected		Fungus Injected
	Control (T1)	Test (T2)	Test (T3)
30th day	0.00622	0.00622	0.00652
35th day	0.00638	0.00556	0.00525
40th day	0.00600	0.00671	0.00510
45th day	0.00801	0.00723	0.00501

Table 12: Comparision of MCV (in cubicmicrons) *in Heteropnuestes f ossilis* administered w ith *Aeromonas hydrophilia* (106 CFU/0.1ml) and *Aphanomyces invadans* (106 CFU/0.1ml).

8). In T3 fishes also showed a decrease level of DLC from 30th day to 45th day (Table 8.1).

The PCV and total plasma level was found to be higher in T2 fishes were higher than compared to T1 and T3 fishes. Total plasma increased from 30th day to 45th day in T3 (Table 10) when compared to T1 and T2 (Table 9).

Haemoglobin content of T3 fishes were higher than T1 and T2 (Table 10). The MCH level of T3 was higher than that of the T1 and T2

Duration	Bacteria Injected		Fungus Injected
	Control (T1)	Test (T2)	Test (T3)
30th day	79.2	78.8	76.92
35th day	78.80	65.90	102.04
40th day	82.50	50.00	98.48
45th day	76.00	60.98	90.09

Table 13: Comparision of MCHC (%) in Heteropnuestes f ossilis administered w ith Aeromonas hydrophilia (106 CFU/0.1ml) and Aphanomyces invadans (105 CFU/0.1 ml).

(Table 11). The MCV level showed fluctuations (Table 12); MCHC level of T3 showed higher than that of T1 and T2 (Table 13). No mortality was observed in T1, whereas in T2 2% and T3 6% were noticed. Lesions induced by A. invadans were higher than A. hydrophila on H. fossilis (Table 14).Because of probiotic feeding H. fossilis gained weight in D1 was about 10.25 gm, and in D2 about 8.0 gm (Table 15).

Discussion

The recent intensification of aquaculture has lead to a number of fish diseases due to environmental and physiological stress in fish population. Overcrowding, handlings of fish and water quality have resulted in disease outbreaks. The present investigation reports a study of microbiological and haematological responsiveness of H. fossils fed with probiotic diet during experimental infection by A. hydrophila and A. invadans. This study reports A. hydrophila and A. invadans caused infection and severe lesions were noticed in T2 and T3.

A. hydrophila is disseminated as a cosmopolitan especially in aquatic environments which provides ample opportunity to fishes and leads to infection and then causing mortality [13]. So, the pathological symptoms made by A. hydrophila in this study, were similar to the studies made earlier [14].

TEC in A. hydrophila infected H. fossilis and A. invadans infected H. fossilis exhibited a significant decrease on prolonged exposure. Previous studies reported a hike in TEC during the unhealthy state of fish [13,15]. Accordingly, high counts are associated with the abnormal conditions of fish. Hence, a sudden increase in the TEC is indication of a severe infection by the opportunistic bacteria and fungi. This might have been accomplished by a rapid mobilization of RBC from the haemopoietic tissue, which may transport higher amounts of Oxygen particularly to withstand stress factor caused by A. hydrophila[1,16].

TLC also showed an increasing trend in the infected the H. fossilis with A. hydrophila and A. invadans. Increase in the number of WBC's has been reported by a several investigators. The results of the present study agrees with the works of Innocent research group [17]. DLC studies showed that lymphocytes constituted maximum percentage followed by neutrophils, Ainsworth has suggested that neutrophils and macrophages are responsible for bacterial uptake [18]. Similar observation of neutrophilia an inflammatory response was record by Innocent et al., [19] in M. montanus infected by A. Hydrophila [19].

Varieties of leukocyte types are involved in innate cellular defense of fish including macrophages, granulocytes and non-specific cytotoxic cells [20]. The above findings support the present study. Monocytes and macrophages are probably the single most important cell mediated immune reponse of fish [21]. They are the primary cells involved in phagocytosis and the killing of pathogens upon first recognition and subsequent infection [22]. Secombes has reported that in fish granulocytes especially neutrophils are the primary cells involved in the initial stages by inflammation. Granulocytes are highly motile, phagocytic and produce reactive oxygen species. The results of the present studies also show significant increase in neutrophils and lymphocytes. Lymphocyte numbers decreased significantly in the post challenge sample while neutrophils and monocyte remain unchanged. Hb content showed a decrease in H. fossilis experimentally infected by A. Hydrophila leading to anaemia.

The administration of probitics feed in experimentally infected H. fossilis elicited alteration in haematological parameters such as TEC, TLC, DLC, MCV, MCH, MCHC, PCV. An analysis of results of present study reveals that B. sublilis plays a vital role in increasing length and weight of fish. And also, it did not cause any changes in gut micro flora.

A. hydrophila and A. Invadans are potentially effective to the H. fossilis in 10^{-4} CFU/ml concentration. The challenging of A. hydrophila and A. invadans in H. fossilis, in terms of immunological responsiveness shows good healthy status. Post challenging of these pathogen causes disease in fishes. But B. subtilis is not having any vital role in immune mechanism of fish after post challenging.

Lesions developed by A. invadans were higher than that of the A. hydrophila induced lesions. So, the probiotic was not efficient in fungal infection. During probiotic feeding weight was also gained when compared to control. So probiotic acts as a one of the growth promoter in H. fossilis.

Conclusion

In H. fosilis, probiotic accepted fishes gained more weight than that of the control fishes. The microbiological estimation also showed a dual increase in trial count in T2 injected fish than that of the T3 injected fish. Many factors can influence the immune response of fish. Among them are stressors and environmental factors are natural. In the present investigation behavioural symptoms to pathogenicity such as imbalance, restlessness and avoidance of food were observed. Pathological symptoms include fin necrosis and tail rot which were also observed. In some cases septicaemic ulceration was noticed. Haematological parameters elicited changes which are able to reveal some clues for diagnosis and prognosis of the disease state. T2 fishes were inflicted alterations in TEC, TLC, DLC, and Hb content which indicated decrease state of immunity, when compared with T3 fishes. Bacteria injected fishes showed good healthy status whereas fungal injected fishes showed non healthy status of fish.

	30th Day	33rd Day	39th Day	42th Day	45th Day
T1	-	-	-	-	-
T2	-	+	++	++++	+++
T3	-	++	+++	+++	++++

- Normal Intact skin
-/+ Intact but melanized skin at injection site
+ Blanching Slight sw elling of injection site
++ Blanching furuncle – like lesion w ith dermal erosion w ith or w ithout hemorrhagic periphery
+++ Extensive blanching lesion w ith furuncle like ulcerated core
++++ Ulcerated lesion w ith underlying necrotic musculature

Table 14: Lesions induced by A. hydrophila and A. invadans on H.Fossilis.

Diet	Initial weight	Final weight	WG
D1	27.25	37.50	10.25
D2	25.08	33.08	8.0

Table 15: During probiotic feeding Weight Gained (WG).

References

1. Manju RA, Haniffa M, Singh SA, Ramakrishnan CM, Dhanaraj M, et al. (2011) Effect of dietary administration of Efinol® FG on growth and enzymatic activities of *Channa striatus*. Journal of Animal and Veterinary Advances 10: 796-801.

2. Khan MA, Jafri AK, Chadha NK, Usmani N (2003) Growth and body composition of rohu (*Labeo rohita*) fed diets containing oilseed meals: partial or total replacement of fish meal with soybean meal. Aquaculture Nutrition 9: 391-396

3. Gomes EF, Rema P, Kaushik SJ (1995) Replacement of fish meal by plant proteins in the diet of rainbow trout (*Oncorhynchus mykiss*): digestibility and growth performance. Aquaculture 130: 177-186.

4. Nakai T, Park SC (2002) Bacteriophage therapy of infectious diseases in aquaculture. Research in microbiology 153: 13-18.

5. Gram L, Løvold T, Nielsen J, Melchiorsen J, Spanggaard B (2001) In vitro antagonism of the probiont *Pseudomonas fluorescens* strain AH2 against *Aeromonas salmonicida* does not confer protection of salmon against furunculosis. Aquaculture 199: 1-11.

6. Balcázar JL, De Blas I, Ruiz-Zarzuela I, Cunningham D, Vendrell D, et al, (2006) The role of probiotics in aquaculture. Veterinary microbiology 114: 173-186.

7. Verschuere L, Rombaut G, Sorgeloos P, Verstraete W (2000) Probiotic bacteria as biological control agents in aquaculture. Microbiology and Molecular Biology Reviews 64: 655-671.

8. Gildberg A, Johansen A, Bøgwald J (1995) Growth and survival of Atlantic salmon (*Salmo salar*) fry given diets supplemented with fish protein hydrolysate and lactic acid bacteria during a challenge trial with *Aeromonas salmonicida*. Aquaculture 138: 23-34.

9. Sundary (2007) Role of Probiotics on Water Quality Management Sangram Keshari Rout Asm Nandi. Environmental Biotechnology.

10. Watanabe T, Kiron V (1994) Prospects in larval fish dietetics Aquaculture 124: 223-251.

11. Olafsen JA (2001) Interactions between fish larvae and bacteria in marine aquaculture. Aquaculture 200: 223-247.

12. Rengpipat S, Rukpratanporn S, Piyatiratitivorakul S, Menasaveta P (2000) Immunity enhancement in black tiger shrimp (*Penaeus monodon*) by a probiont bacterium (*Bacillus S11*). Aquaculture 191: 271-288.

13. Tendencia EA, Dela Peña MR, Fermin AC, Lio-Po G, Choresca CH, Inui Y (2004) Antibacterial activity of tilapia *Tilapia hornorum* against *Vibrio harveyi*. Aquaculture 232: 145-152.

14. Cahill MM (1990) A review virulence factors in motile *Aeromonas* species. Journal of Applied Bacteriology 69: 1-16.

15. Fathima KMSA, Annalakshmi T, Innocent BX (2012) Immunostimulant Effect of Vitamin-A in *Channa Punctatus* Challenged with *Aeromonas Hydrophila*: Haematological Evaluation. Journal of Applied Pharmaceutical Science 2: 123-126.

16. Nielsen ME, Hoi L, Schmidt AS, Qian D, Shimada T, et al. (2001) Is *Aeromonas hydrophila* the dominant motile Aeromonas species that causes disease outbreaks in aquaculture production in the Zhejiang Province of China? Diseases of aquatic organisms 46: 23-29.

17. Innocent BX, Fathima MSA, Dhanalakshmi (2011) Studies on the immouostimulant activity of *Coriandrum sativum* and resistance to *Aeromonas hydrophila* in *Catla catla*. Journal of Applied Pharmaceutical Science 1: 132-135.

18. Ainsworth AJ (1992) Fish granulocytes: morphology distribution and function. Annual Review of Fish Diseases 2: 123-148.

19. Innocent BX, Martin P (2004) Haematological studies in Mystus montanus exposed to gram negative bacteria Aeromonas hydrophila. Indian Journal of environmental protection.

20. Galindo-Villegas J, Hosokawa H (2004) Immunostimulants: towards temporary prevention of diseases in marine fish Advances.

21. Clement S, Lovell R (1994) Comparison of processing yield and nutrient composition of cultured Nile tilapia (*Oreochromis niloticus*) and channel catfish (*Ictalurus punctatus*) Aquaculture 119: 299-310.

22. Shoemaker CA, Evans JJ, Klesius PH (2000) Density and dose: factors affecting mortality of Streptococcusiniae infected tilapia (Oreochromisniloticus). Aquaculture 188: 229-235.

9

Effects of Bamboo Charcoal Added Feed on Reduction of Ammonia and Growth of *Pangasius hypophthalmus*

Quaiyum MA, Jahan R, Jahan N, Akhter T and Islam M Sadiqul*

Department of Fisheries Biology & Genetics, Bangladesh Agricultural University, Mymensingh-2202, Bangladesh

Abstract

A 50-day feeding trial was conducted to determine the effects of dietary bamboo charcoal (BC) on ammonia (NH_3-N) excretion and growth performances of *Pangasius hypophthalmus*. Four levels of BC (0%, 0.5%, 1% and 2%) were supplemented to the diet composition and fed to fish (initial body weight 1.18 ± 0.04 g) twice a day. At the end of the trial, mean of final weight (g), final length (cm), weight gain (g), length gain (cm), percent weight gain, percent length gain, specific growth rate (% per day), feed conversion ratio, survival rates and water quality parameters i.e, ammonia (NH_3-N), pH, and dissolved oxygen were measured and found that fish fed 2% BC diet showed significantly ($P<0.05$) higher growth enhancement than those of fish fed the control diet (0% BC). Ammonia concentration over the experimental period decreased with increasing dietary BC. Moreover, in histological observation it was found that the villus height and villus area in all intestinal segments tended to increase with increasing dietary BC supplementation. The present results indicate stimulating effects of dietary BC on intestinal villi and the diet supplemented with 2% BC was found to have a suitable level to fulfill the maximum growth performances of *P. hypophthalmus* and to decrease the ammonia concentration.

Keywords: Bamboo charcoal; Ammonia; Growth; *Pangasius hypophthalmus*

Introduction

There is a great potential for *Pangasius hypophthalmus* in Bangladesh. It is very much demandable in local markets because of its lower market price. The vast majority of poor people consume *P. hypophthalmus* as this fish is delicious and tasty due to its high fat content. Moreover, the climate, water and soil conditions of Bangladesh have proved totally suitable for *P. hypophthalmus* production and it is one of the most suitable catfishes for rearing in ponds [1]. *Pangasius* culture has proved itself as a profitable enterprise due to year round production, quick growth and high productivity. In addition, it can be stocked at a much higher density in ponds compared to other cultivable species [2].

However, costly feed and low market price has slowed progress in farming of this fish. In addition, due to high accumulation of nitrogenous waste products that is toxic to fish considered as a limiting factor for growth and survival of fish are affecting the culture of this species [3]. One of the sources of this nitrogenous waste (e.g. ammonia) is from the supplemented feed that fed to fish. An effective way to reduce the waste load is to modify aqua feeds with the aim of reducing excretion of nitrogen, phosphorus and total solids relative to fish growth [4,5]. Therefore, several efforts have been made to produce high quality animal products without using medicines and to reduce environmental contamination by efficient utilization of natural substances. Some of these natural substances (e.g. wood charcoal, bamboo charcoal, coconut shell charcoal etc.) are not cited in the scientific literature, but are used locally. Like wood charcoal bamboo charcoal (BC) is also an activated charcoal made by dry distillation of a thick-stemmed bamboo and powder of which is known as a universal adsorbent, because it can bind with variety molecules since it contains a complex network of pores of various shapes and sizes [6]. Now-a-days, BC has been used in animal feed formulation as an additive because they absorb ammonia and nitrogen, and activates the intestinal function through eliminating the poisons and impurities from the gastrointestinal tract of land animals [7,8]. Utilization of charcoal from wood or bamboo may provide an economical way to eliminate noxious substances because of their cheaper cost [9]. Moreover, BC is considered to have a higher adsorption capacity than wood charcoal because it has about 4 times more cavities, 3 times more mineral content and 4 times better absorption rate [6]. Reports have clarified the ammonia adsorption effect of BC in aqueous solution [10], and dietary addition of BC effects on digestion, nitrogen retention and excretion of growing goats [8]. They also found that goats fed a diet containing 0.5 g of BC per kg of body weight grew faster than the controls. Recently it was found that bamboo charcoal boosts tilapia growth [11]. However, very limited studies about BC in aquatic animal nutrition as a feed ingredient have been conducted.

Therefore, the major purpose in the present study was to assess the effectiveness of dietary BC supplementation on growth performance of *P. hypophthalmus* and elimination of ammonia nitrogen excretion from the water during study period.

Materials and Methods

Experimental site

The experiment was carried out in the Backyard Laboratory of the Faculty of Fisheries, Bangladesh Agricultural University, Mymensingh, Bangladesh. The experiment was conducted for a period of 50 days. The research work was undertaken in 13 glass aquaria (average capacity 50L). An adequate level of dissolved oxygen in each aquarium was maintained through artificial aeration during the experimental period.

*Corresponding author: M Sadiqul Islam, Department of Fisheries Biology and Genetics, Faculty of Fisheries, Bangladesh Agricultural University, Mymensingh-2202, Bangladesh, E-mail: sadiqul1973@yahoo.com

Preparation of BC

Bamboo was cut into small pieces and put into a tightly sealed container made of iron. This is then placed on a hot fire for at least an hour. Once the fire was out, the container left to cool down completely before it opened. The BC then pounded into a fine powder and the composition of BC is as follows in Table 1.

Feed formulation

Standard fish feed (Table 2) obtained from commercial feed company (Mega Feed Co. Ltd., Bangladesh) was used as a basal feed supplemented with BC powder at 0, 0.5, 1.0 and 2% in control, T_1, T_2 and T_3, respectively. Diets were prepared by mixing the dry ingredients and water (35% of the dry weight of ingredients) and then pellet-type diets were produced through a meat grinder with a diameter disc (size, 1.9-2.2 mm). The diets were later oven dried (60°C for 2 h) to approximately 11% moisture. After preparation, the diets were stored at refrigerator until used.

Fingerling collection and stocking of fish: Fingerling of *P. hypophthalmus* was collected from the local fish seed retailer and carefully transported to avoid physical injury. One fish per liter stocking density and three aquaria were used for each treatment.

Feeding rates: Fingerlings were fed with experimental diets twice a day in the morning at 9.00 am and in the afternoon at 3.00 pm throughout the study period. Fingerling in each aquarium were fed daily at the rate of 7% of their body weight; the amount was fixed after observing that it was not interested to take more than that amount of feed.

Sampling procedure: Every 10 d, the body weight of fish were measured. For weighing 5 fingerlings were collected from each aquarium. All animals were placed on paper towels to remove excess water and then weighed using an electric balance (Mettler PJ3000). After recording the length and weight of fingerlings were released in the aquarium.

Measurement of water quality parameters

All the following water quality parameters were recorded at 10 days interval.

Item	%
Ash	6.35
Nitrogen	0.57
Phosphate	1.06
Potassium	2.10
Silicon dioxide	1.20
pH	8.50

Table 1: Composition of BC powder.

Item	%
Moisture	10
Protein	28
Crude fat	6
Ash	18
Crude fiber	7
Calcium	1.8
Phosphorus	0.7
Gross energy (Kcal/Kg)	3400

Table 2: Ingredient and chemical composition of the basal commercial diet.

Ammonia: Ammonia (NH_3-N) of the aquarium water was recorded in mg/l with the help of ammonia test kit (Hanna Instrument Ammonia Test Kit for Fresh Water).

Dissolved oxygen: Dissolved oxygen of the aquarium water was recorded in mg/l with the help of a dissolved oxygen meter (Model Oxi 3150i, Germany).

pH: pH of the water was recorded with the help of a pH meter (pH meter L20 ME 1-1 LER TOLEDO, Switzerland).

Histological observation of intestine

For the histological observation intestines were quickly removed from fish and place in 10% formaldehyde fixative solution. Then the anterior and middle central portions of the intestine of 0.5 cm thickness were put into the cassettes separately for histological examination. They were dehydrated in graded alcohol series, embedded in paraffin, sectioned for 5-7μm in thickness using a microtome (MICROM HM355S, Germany) and stained with Haematoxylin and Eosin, then mounted in DPX mountant and photographed with an OLYMPUS-CX41 microscope which was equipped with the SONY DSC-W220 camera. At least two glass slides were prepared from each portion of the intestine.

Data analysis

The collected data were statistically analyzed by one way ANOVA (analysis of variance) with the help of SPSS (Statistical Package for Social Science) to see whether the influence of different treatments on these parameters were significant or not.

Results

Water quality parameters

Ammonia: The ammonia (NH_3-N) values of water ranged from 0.02 ± 0.01 to 2.06 ± 0.12 mg/l during the study period. Significantly (P < 0.05) higher value of ammonia which is not good for fish was found in the control while lowest value which is suitable for fish was found in T_3 (2% BC) (Figure 1).

pH

The pH values of water ranged from 6.89 ± 0.11 to 7.81 ± 0.09 during the study period. The highest level of pH was found in the control while the lowest level of pH was found in T_3 (2% BC). There were significant differences (P<0.05) between the control and T_3. A proportional relationship between pH and ammonia (NH_3-N) concentration were also observed during the study period (Figure 2).

Dissolved oxygen

During the study period dissolved oxygen content of water ranged from 4.58 ± 0.32 to 5.97 ± 0.09 mg/l. The highest value 5.97 ± 0.09 mg/l was found in T_3 which was significantly (P<0.05) higher than the control. An inverse relationship between dissolved oxygen and ammonia (NH_3-N) concentration were observed during the study period (Figure 3).

Growth parameters

As shown in Table 3 the maximum growth enhancement was noticed at the 2% BC supplementation level. However, fish groups that received dietary BC from 0.5 to 2% level showed higher values of weight gain, SGR and FCR than the control. No significant differences in final length and length gain were observed between fish fed BC containing

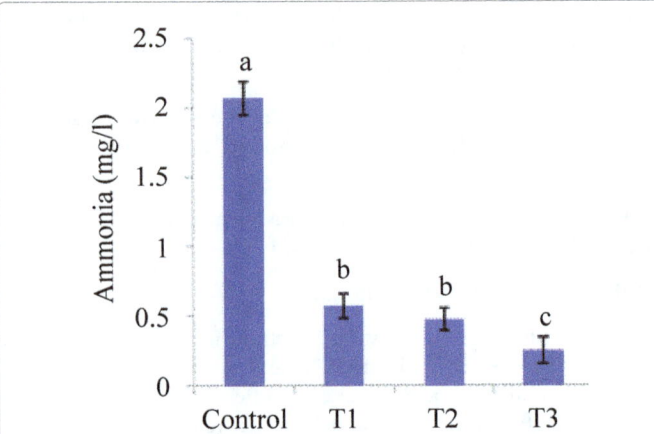

Figure 1: Ammonia (NH$_3$-N) concentration (mg/l) in different treatments during the study period. a, b, c means with different superscripts are significantly different from each other (P<0.05).

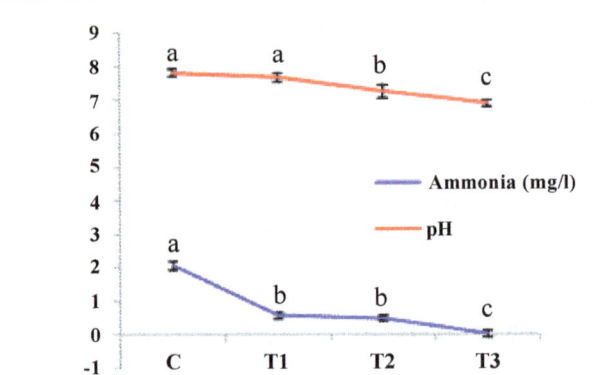

Figure 2: pHin different treatments during the study period. a, b, c means with different superscripts are significantly different from each other (P<0.05).

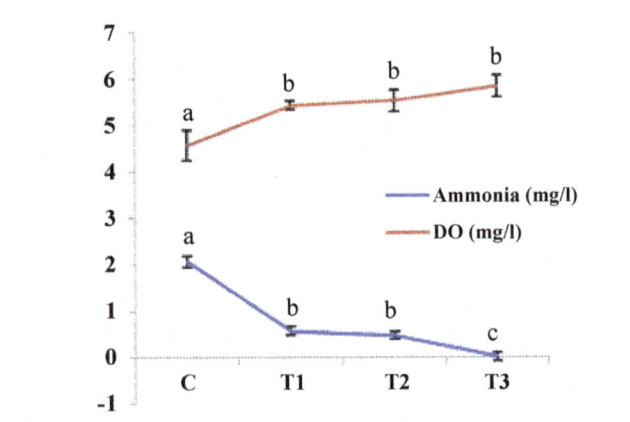

Figure 3: Dissolved oxygen (mg/l)in differenttreatments. Different superscript alphabets in each treatment group are significantly different at P<0.05.

diet. It was also seen that the survival rates of *P. hypophthalmus* under the trail treatments were significantly (P<0.05) higher than the control. In an experiment in the earthen pond our study showed that 1.5% BC supplementation level was suitable for the maximum growth enhancement (data not shown) (Table 3).

Values are presented as mean ± SE. Values in the same row having different superscript letters are significantly different (P<0.05). The lack of superscript letter indicates no significant differences among treatments.

Mean weight gain=mean final weight (g)-mean initial weight (g);

Mean length gain=mean final length (cm)-mean initial length (cm);

% Weight gain=(final body weight–initial body weight)×100/initial body weight;

% Length gain=(final body length–initial body length)×100/initial body length;

Specific growth rate (SGR)=(Log final body weight–Log initial body weight)×100/feeding period;

Feed conversion ratio (FCR) = feed fed/live weight gain;

Survival rate (%)=No. of fish harvested ×100/No. of fish stocked.

Histological observation

Intestine of *P. hypophthalmus* of all treatments had almost normal structure but there was slight change in the villus height and lumen area of those intestines. It was observed that villus height was increased and consequently lumen areas of those intestines were decreased with increasing BC supplementation level (Figure 4).

Parameters	Control	T$_1$	T$_2$	T$_3$
Initial weight (g)	1.17 ± 0.40	1.17 ± 0.40	1.17 ± 0.40	1.17 ± 0.40
Initial length (cm)	3.63 ± 0.15	3.63 ± 0.15	3.63 ± 0.15	3.63 ± 0.15
Final weight (g)	2.88 ± 0.03^c	3.02 ± 0.06^c	3.21 ± 0.03^b	3.67 ± 0.05^a
Final length (cm)	6.36 ± 0.29	6.79 ± 0.43	6.82 ± 0.16	7.10 ± 0.34
Weight gain (g)	1.71 ± 0.03^c	1.85 ± 0.06^c	2.03 ± 0.03^b	2.50 ± 0.05^a
Length Gain (cm)	2.73 ± 0.29	3.16 ± 0.43	3.19 ± 0.16	3.46 ± 0.34
% Weight gain (%)	145.32 ± 2.70^c	157.22 ± 5.58^c	172.80 ± 2.73^b	212.46 ± 4.71^a
% Length gain (%)	75.22 ± 7.99^c	87.06 ± 12.07^b	87.79 ± 4.63^b	95.41 ± 9.53^a
SGR (% per day)	0.73 ± 0.01^c	0.78 ± 0.02^c	0.84 ± 0.01^b	0.97 ± 0.01^a
FCR	3.44 ± 0.06^d	3.18 ± 0.11^c	2.89 ± 0.04^b	2.35 ± 0.05^a
Survival rate (%)	77.33 ± 0.88^b	84 ± 1.15^a	82 ± 0.57^a	84 ± 1.15^a

Table 3: Growth parameters of *P. hypophthalmus*.

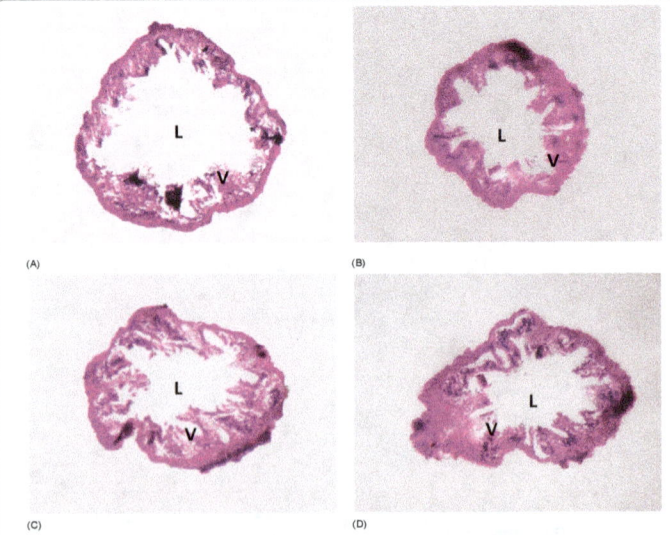

Figure 4: Histological section of the intestine of *P. hypophthalmus*collected from the (A) control; (B)T$_1$; (C)T$_2$; and (D) T$_3$. (V=villus, L=lumen)

Discussion

Water quality parameters

One important objective of this study is to determine whether the ammonia nitrogen could be reduced by dietary BC. Some studies have reported on the ammonia nitrogen excretion of Japanese flounder (*Paralichthys olivaceus*) [12-15]. In the present experiment, the maximum ammonia (NH_3-N) concentration was found in the control while minimum concentration was recorded at T_3 and concentration increased with decreasing BC supplementation level. This result showed the dose-related effect of dietary BC on ammonia concentration.

During the study period a relationship among ammonia, pH, and dissolved oxygen were observed. In the control, ammonia and pH values were high and DO was low. It was found that ammonia and pH were decreased but DO was increased with increasing BC supplementation level and the lowest value of ammonia was found in T_3 where the higher level of BC (2%) was supplied. These results indicated that the dietary BC supplementation could be a potential feed additive to eliminate the ammonia from the culture system.

Growth performances

Based on the growth data, the optimum dietary BC supplementation level for the *P. hypophthalmus* was found to be 2% of the diet. The maximum mean final weight was 3.68g in T_3 where higher level of BC was supplied. The minimum mean final weight was 2.89 g in the control where 0% BC was supplied. Again the maximum mean final length was 7.1 cm observed in T_3 and the minimum mean final length was 6.37 cm observed in the control. However, our experiment in the earthen pond indicated a little bit lower concentration of BC (about 1.5%) than the aquarium was the optimum for the growth of *P. hypophthalmus* and elimination of ammonia (data not shown). The present finding was less than that found in study where the highest weight gain was obtained at 4% BC supplementation in tiger puffer fish [15]. Although it was found 0.5% BC was suitable level for optimum growth of juvenile flounder [16]. These results indicated the species-related effect of dietary BC on growth, and it might be because of the differences in digestion and feeding behaviors of these species. In the Aigamo ducks [17] it was found that the highest mean final weight gain was in 1% SB (mixture of bamboo charcoal powder and bamboo vinegar solutions) group.

In the present experiment the mean weight gain of *P. hypophthalmus* in different treatments varied from 1.71 to 2.50 g. The highest mean weight gain was found in T_3 while the lowest mean weight gain was observed in the control. Again the highest mean length gain (3.46 cm) was found in T_3 and the lowest mean weight gain (2.73 cm) was found in the control which is almost similar to the findings of [16]. In white leghorn hens (*Gallus domesticus*) the highest weight gain was observed in case of 0.5% of BC supplementation [17] whereas in Aigamo ducks the mean weight gain of 1% SB group was 8% heavier than the control group [18].

The higher percent weight gain was found in T_3 and it was 212.46% while the lower percent weight gain was found in the control which was 145.32%. Again the higher percent length gain was 95.41% which was found in T_3 and the lower percent length gain was 75.23% which was found in the control. More or less similar type of percent weight gain was observed in Japanese flounder by [16].

The specific growth rates of our studied fish under different treatments were varied from 0.73 to 0.97%. SGR value was higher in T_3 where BC level was higher (2%) and SGR value was lower in control

where fish reared without BC. SGR were observed 0.02 to 0.68% in *P. hypophthalmus* using 35-40% protein containing feed [19]. The result obtained in our study was much lower than that (3.09 to 3.51%) was found by [20] in Thai pangas. The FCR under different treatments ranged from 2.36 to 3.44. The improved performance of FCR was observed at 2% BC containing diet that is similar to the findings of [16,18].

Our findings indicated that the dietary BC supplementation could be a potential feed additive to enhance the growth of *P. hypophthalmus* and supports research in tiger puffer [15] and other studies that reported growth in goats [8], in broiler chicks [21], and in Japanese flounder [22].

Histological studies

During the experiment it was observed that, intestine of *P. hypophthalmus* of all treatments had almost normal structure. The villus height and villus area of intestines of same treatment were almost in similar structure but there was a slight change in the internal configurations of intestines of different treatments. It was also observed that villus height was increased and lumen area was decreased with increasing BC supplementation level. Increased height of intestinal villi means a greater surface area for nutrient absorption [23]. Greater villus height and increased cell mitosis numbers in the intestine are indicators of activation of the function of the intestinal villi [24,25]. Furthermore, increased villus size was also associated with activated cell proliferation in the crypt [26] and provided more surface area for nutrient absorption and thus improved nutrient digestibility [23]. These reports suggest our findings that the increased villus height and decreased lumen area in BC supplemented fish would be multiplicatively stimulated by the influence of BC.

Acknowledgment

The authors wish to thank Bangladesh Agricultural University Research System (BAURES) for financial support to MS Islam for making it possible to carry out this research. The assistance of Professor Dr. Harunur Rashid of the Department of Fisheries Management, Bangladesh Agricultural University for the histology work is gratefully acknowledged.

References

1. Nesar A, Hasan MR (2007) Sustainable livelihoods of pangus farming in rural Bangladesh. Sustainable Aquacult 2: 5-11.

2. Ali MZ, Hossain MA, Mazid MA (2005) Effect of mixed feeding schedules with varying dietary protein levels on the growth of sutchi catfish, Pangasius hypophthalmus (Sauvage) with silver carp, Hypophthalmichthys molitrix (Valenciennes) in ponds. Aquacult Res 36: 627-634.

3. Person-Le Ruyet J, Chartois H, Quemener L (1995) Comparative acute ammonia toxicity in marine fish and plasma ammonia response. Aquacult 136: 181-194.

4. Cowey CB, Cho CY (1991) Nutritional Strategies and Aquaculture Waste, Proceedings of the first international symposium on nutritional strategies in management of aquaculture waste, University of Guelph, Guelph, Ontario, Canada.

5. Talbot C, Hole R (1994) Fish diets and the control of eutrophication resulting from aquaculture. J Applied Ichth 10: 258-270.

6. Zhao R, Yuan J, Jiang T, Shi J, Cheng C (2008) Application of bamboo charcoal as solid-phase extraction adsorbent for the determination of atrazine and simazine in environmental water samples by high-performance liquid chromatography-ultraviolet detector. Talanta 76: 956-959.

7. Mekbungwan A, Yamakuchi K, Sakaida T (2004) Intestinal villus histological alterations in piglets fed dietary charcoal powder including wood vinegar compound liquid. Anat Histol Embryol 33: 11-16.

8. Van DTT, Mui NT, Ledin I (2006) Effect of method of processing foliage of

Acacia mangium and inclusion of bamboo charcoal in the diet on performance of growing goats. Animal Feed Sci Tech 130: 242-256.

9. Prasad KSN, Aruna C, Chhabra A (2000) Effect of addition of charcoal in the concentrate mixture on rumen fermentation, nutrient utilization and blood profile in cattle. Ind J Dairy Biosci 11: 116-119.

10. Asada T, Ohkubo T, Kawata K, Oikawa K (2006) Ammonia adsorption on bamboo charcoal with acid treatment. J Health Sci 52: 585-589.

11. Yang S, Wen Y, Liu F (2009) Bamboo charcoal boosts tilapia growth. Aquafeed, 29.

12. Alam MS, Teshima S, Ishikawa M, Koshio S (2002) Arginine requirement of juvenile Japanese flounder Paralichthys olivaceus. Estimated by growth and biochemical parameters. Aquacult 205: 127-140.

13. Kikuchi K, Takeda S, Honda H, Kiyono M (1991) Effect of feeding on nitrogen excretion of Japanese flounder, Paralichthys olivaceus. Nippon Suisan Gakkaishi 57: 2059-2064.

14. Kikuchi K, Takeda S, Honda H, Kiyono M (1992) Nitrogen excretion of juvenile and young Japanese flounder, Paralichthys olivaceus. Nippon Suisan Gakkaishi 58: 2329-2333.

15. Moe T, Ishikawa M, Koshio S, Yokoyama S (2009) Effects of dietary bamboo charcoal on growth parameters and nutrient utilization of tiger puffer fish, Takifugu rubripes. Aquacult Sci 57: 53-60.

16. Moe T, Koshio S, Ishikawa M, Yokoyama S (2010) Effects of supplementation of dietary bamboo charcoal on growth performance and body composition of juvenile Japanese flounder, paralichthys olivaceus. J World Aquacult Soc 41: 255-262.

17. Ruttanavut J, Yamauchi K, Goto H, Erikawa T (2009) Effects of dietary bamboo charcoal powder including vinegar liquid on growth performance and histological intestinal change in aigamo ducks. Int J Poultry Sci 8: 229-236.

18. Yamauchi K, Ruttanavut I, Takenoyama S (2010) Effects of dietary bamboo charcoal powder including vinegar liquid on chicken performance and histological alterations of intestine. J Animal Feed Sci 19: 257-268.

19. Alam MS (2004) Growth performances and morphological variations among Thai pangas, P. hypophthalmus collected from four different hatcheries in Mymensingh. Bang J Prog Agr 15: 141-149.

20. Azimuddin KM (1998) Effect on Stocking Density on the Growth of Thai pangus, Pangasius sutchi (Folwer) in Net Cage by Using Formulated Diet, MS Thesis, Department of Aquaculture, Bangladesh Agricultural University, Mymensingh, Bangladesh.

21. Kutlu HR, Unsal L, Gorgulu M (2001) Effects of providing dietary wood (oak) charcoal to broiler chicks and laying hens. Animal Feed Sci Tech 90: 213-226.

22. Yoo JH, Ji SC, Jeong GS (2005) Effect of dietary charcoal and wood vinegar mixture (CV82) on body composition of olive flounder, Paralichthys olivaceus. J World Aquacult Soc 36: 203-208.

23. Onderci M, Sahin N, Sahin K, Cikim G, Aydin A, et al. (2006) Efficacy of supplementation of alpha-amylase-producing bacterial culture on the performance, nutrient use and gut morphology of broiler chickens fed a corn-based diet. Poultry Sci 85: 505-510.

24. Langhout DJ, Schutte JB, Van leeuwen P, Wiebenga J, Tamminga S (1999) Effect of dietary high and low methyllated citrus pectin on the activity of the ileal microflora and morphology of the small intestinal wall of broiler chickens. Brit Poultry Sci 40: 340-347.

25. Yasar S, Forbes JM (1999) Performance and gastrointestinal response of broiler chicks fed on cereal grain-based foods soaked in water. Brit Poultry Sci 40: 65-67.

26. Lauronen J, Pakarinen MP, Kuusanmaki P, Savilahti E, Vento P, et al. (2000) Intestinal adaptation after massive proximal small bowel resection in the pig. Brit Poultry Sci 41: 416-423.

10

Effect of Dietary Carbohydrate Levels and Feeding Frequencies on Growth and Carbohydrate Digestibility by White Shrimp *Litopenaeus vannamei* Under Laboratory Conditions

Zainuddin*, Haryati and Siti Aslamyah

Department of Fisheries, Faculty of Marine Science and Fisheries, Hasanuddin University, Sulawesi Selatan 90245, Indonesia

Abstract

This study aims to determine the level of carbohydrate and feeding frequency on the growth and the carbohydrate digestibility of juvenile shrimp vanamei. This study used a factorial design patterns completely randomized design with two factors and three replications of each factor given time. The treatments tested were factor A (carbohydrate feeding different levels, namely 26, 32, 38, and 44%) and factor B (feeding frequency 2 times, 4 times, and 6 times per day). Shrimp juvenile were used have an average individual weight of 0.3 g. Feeding dose is 10% of the body weight and the feeding frequency adapted to the treatment. The results showed that combination treatment with carbohydrate levels by 38% and the feeding frequency 4 times a day are the best combination of treatments to the specific growth rate and digestibility of carbohydrates juvenile white shrimp.

Keywords: Level; Carbohydrates; Frequency; Feeding; Growth, Digestibility

Introduction

Shrimp is the one commodity that contributes significantly to the increase in local revenues in South Sulawesi. In 2006 South Sulawesi shrimp production reached 19414 tonnes and a decline into 16361.4 tonnes in 2007 [1]. Decreased production of tiger shrimp in recent time due to virus attacks WSSV causes the need for diversification of species that are more resistant to disease. Shrimp Litopenaeus vannamei vanamei is one type of penaeid shrimp that have endurance higher than the species of tiger shrimp against viral attack.

In a system of intensive shrimp farming in ponds vanamei, feed is one of the strategic components that determine the success of the business. Feed is a very large part of the operational costs in the cultivation of crustaceans [2]. At the event, almost 60-70% of total production costs for the purchase of feed [3,4]. But the last few years of cultivation of these commodities often fail. One of the factors that lead to failure in shrimp farming in Indonesia vanamei cultivation technology is the application that does not comply with the carrying capacity of the waters, the farming technologies among others including feeding technology [5]. The high organic matter derived from the feed that is not consumed and is derived from the metabolism, is one of the triggers declining water quality.

The success of shrimp farming vanamei among others determined by the quality of feed used. To produce optimal growth, feed with shrimp require a fairly high protein content . Optimal growth is achieved when vanamei shrimp shrimp feed with protein content of 40-50% [6]. However, the protein content is too high can lead to decreased water quality cultivation media, which comes from the feed that can not be consumed, faeces and feed protein metabolism. [7] suggested that the protein needs of the shrimp can be reduced if the energy needs can be met from other sources of non-protein , such as carbohydrates

Carbohydrates and lipids are important nutritional components in food shrimp [8]. Increasing the proportion of carbohydrates and not protein feed to meet energy needs can ultimately feed costs. The addition of vegetable protein to reduce the cost of feed has been widely studied [9-12]. In addition to required as an energy source, shrimp also need carbohydrates for synthesis of chitin. Used by shrimp chitin in the growth process to form and replace eksoskleton during the molting process.

Carbohydrates are a source of cheap energy, but the ability of aquatic organisms, including shrimp to utilize limited. This is due to lack of ability to digest and regulate plasma glucose concentrations. The use of carbohydrate by fish and shrimp are less efficient than land animals [13]. The ability of shrimp in utilizing the limited carbohydrate digestibility due to low [14] and low concentrations of plasma glucose regulation [15]. The low digestibility of carbohydrates associated with the availability of the enzyme α-amylase, whereas low concentrations of plasma glucose regulation allegedly caused by a deficiency of the hormone insulin. Based on the recommendation of the humans who suffer from diabetes, suggested that the feeding frequencyis much more then the ability to utilize carbohydrates can be improved. This is in line with research that the continuous feeding may increase the use of carbohydrates and increase fat reserves through increased lipogenesis process. In addition to the feeding frequencymore frequently, the possibility of higher feed can be consumed, so the rest of which will feed into the cultivation medium, which in turn will affect the water quality can be eliminated. Under these conditions, the purpose of this study was to determine the level of carbohydrate feed and feeding frequency on the growth rate and the best carbohydrate digestibility of juvenile shrimp vanamei.

---

*Corresponding author: Zainuddin, Department of Fisheries, Faculty of Marine Science and Fisheries, Hasanuddin University, Sulawesi Selatan 90245, Indonesia
E-mail: zainuddin_latif2013@yahoo.co.id

Materials and Methods

Experimental condition and animal

The container used in this study is 60 cm×50 cm×50 cm sized glass aquarium with a total of 36 pieces each capacity 20 liters. The water used is sea water which has been diluted upto 20 ppt salinity.

Test animals used in this study were white shrimp juvenile (*Litopenaeus vannamei*) postlarva 25 local stadia. The prawns were taken from the nursery people in Maros. Shrimp is adapted to the type of artificial feed is tested. The stocking density of test animals used were 20 individual/container. Percentage of prescribed daily feeding as much as 10%.

Feed and husbandry

Feed used in this study is the artificial feed composition has been determined and is shown in Tables 1 and 2. Especially for observation digestibility of dry matter in the feed chromium oxide added as much as 0.5%.

In order to achieve the desired objectives of the research, white shrimp juvenile maintained for about two months At the time of maintenance, it is worth noting that the feeding frequency. The daily feeding frequency is performed twice a day, four times a day and six times per day according to the tested treatment. As for sampling weights and measurements performed once every week.

The design of experiments

The experimental design used in this study was a factorial design with completely randomized design basis. The first factor is the level of carbohydrates in the diet are: (A1) Carbohydrate content of feed by 44%, (A2) Carbohydrate content of feed by 38%, (A3) Carbohydrate content of feed by 32%, (A4) Carbohydrate content of feed by 26%.

Each level karbrohidrat given repeat 3 times. The second factor is the feeding frequency, respectively: (B1) feeding frequency of twice per day, (B2) feeding frequency of four times per day, (B3) feeding frequency of six times per day. Each treatment was given repetition

Ingredient	Diet (g/100 g as-fed basis)			
	A1	A2	A3	A4
Local fish meal	5	16	27	35
Head shrimp meal	10	10	10	10
Soybean meal	30	27	25	30
Corn meal	10	10	10	10
Bran meal	24	24	19	11
Wheat meal	17	9	5	-
Fish oil	2	2	2	2
Vitamin mix	1	1	1	1
Mineral mix	1	1	1	1
Cromium oxide	0.5	0.5	0.5	0.5

Table 1: Ingredient composition of experimental diet.

	% Dry basis			
	A1	A2	A3	A4
Crude protein	30.09	35.88	41.56	49.71
Crude lipid	6.42	6.73	6.61	8.37
Nitrogen-free extract	49.65	40.41	32.68	18.44
Crude fiber	5.14	4.93	4.48	6.68
Ash	8.7	12.05	14.67	16.8
Gross energy (KKal/kg)	4068.3	4016.1	3980.1	3976.4

Table 2: Proximate analysis (g/100 dry wt) of experimental diets used.

Feed	Feeding frequency		
	B1	B2	B3
A1	6.288 ± 0.443[b]	6.796 ± 0.938[b]	6.756 ± 0.443[b]
A2	7.198 ± 1.294[b]	9.888 ± 0.083[a]	6.494 ± 0.615[b]
A3	9.244 ± 0.877[a]	8.453 ± 0.946[a]	6.279 ± 1.172[b]
A4	6.363 ± 0.721[b]	6.082 ± 0.498[b]	6.594 ± 0.705[b]

Note: different letters in the same row and column indicates treatment significantly different (P<0.05)

Table 3: Daily specific growth rate (%) of shrimp juvenile the combined treatment of feed carbohydrate levels and feeding frequency.

feeding frequency 3 times. Thus obtained 12 combined treatment replicates were each given 3 times to obtain 36 experimental units.

Variables

Specific growth rate (SGR): SGR=(ln Wt-ln Wo)/t×100

Where: Wt=average individual weight at the final experiment (g), Wo=average individual weight at the initial, experiment (g) t=length of culture (days)

Carbohydrate digestibility: Carbohydrate digestibility will be calculated using the formula in Strickland and Parsons et al. [7] as follows:

Digestibility of dry matter (%)=(1-% Cr_2O_3 in feed/% Cr_2O_3 in feces)×100

Data analysis: Data were analyzed using analysis of variance. If the results of the analysis proved that significant treatment followed by Tukey's W-test to determine the treatment that produced the best response.

Results and Discussion

Specific Growth Rate (SGR)

Specific growth rate of juvenile white shrimp A2B2 highest in the combination treatment, followed by A3B1 and A3B2 with values respectively 9.888 ± 0.083%, 9.244 ± 0.877 and 8.453 ± 0.946%. All three treatment combinations were not significantly different (P>0.05) but significantly different from other treatment combinations (P<0.05) (Table 3).

The results showed that the specific growth rate of white shrimp juvenile daily highest achieved in the combined treatment 38% carbohydrate content of feed and feeding frequency of 4 times per day. This suggests that the juveniles were able to take advantage of white shrimp feed with feed carbohydrate content to 38% by increasing the feeding frequency of upto 4 times per day. Based on the recommendation of the humans who suffer from diabet, suggested that the feeding frequency is much more then the ability to utilize carbohydrates can be improved. This is in line with research that the continuous feeding may increase the use of carbohydrates and increase fat reserves through increased lipogenesis process. In addition to the feeding frequency more frequently, the possibility of higher feed can be consumed, so the rest of which will feed into the cultivation medium, which in turn will affect the water quality can be eliminated. Increased use of carbohydrates by shrimp is expected to increase levels of carbohydrate and reducing protein content in the composition of artificial feed. Growth performance and feed utilization efficiency is significantly influenced by the level of carbohydrate feed.

However, an increased feeding frequency up to 6 times per day instead give effect to the reduction in the rate of growth (Table 3).

Although shrimp fed with high frequency by consuming large amounts of feed but narrow intervals of frequencies, then feed more quickly through the digestive tract and cause digestive ineffective [16] reported the feeding frequency 2 times per day with a span of 12 hours resulted in the best growth in Lates calcarifer juvenile feeding frequency compared to 3 and 4 times per day with a shorter time span. Based on this it can be explained that the maintenance of white shrimp juvenile feeding frequency 4 times a day is enough to improve growth performance.

Carbohydrate digestibility

The results showed that the combined treatment of carbohydrate level of feed and feeding frequency significantly (<0.05) the digestibility of carbohydrates . Single factor of feed carbohydrate levels had no significant effect (P>0.05), but the frequency factor of feeding a significant effect (P<0.05) the digestibility of carbohydrates. Feeding frequency four and six times per day was not significantly different (P>0.05), but both were significantly different (P<0.05) with the feeding frequency two times a day (Table 4 and Figure 1).

Carbohydrates are a source of cheap energy, but the ability of aquatic organisms, including shrimp to utilize limited. This is due to lack of ability to digest and regulate plasma glucose concentrations. The low digestibility of carbohydrates associated with the availability of the enzyme α-amylase, whereas low concentrations of plasma glucose regulation allegedly caused by a deficiency of the hormone insulin. The results showed that the digestibility of carbohydrates by white shrimp juvenile increases with increasing feeding frequency. Despite this feeding frequency four and six times a day gives the same response. This is in line with research that the continuous feeding may increase the use of carbohydrates and increase fat reserves through increased lipogenesis process. In addition to the feeding frequency more frequently, the possibility of higher feed can be consumed, so the rest of which will feed into the cultivation medium, which in turn will affect the water quality can be elimination. It is seen from ammonia –N excretion were lower in the feeding frequency four times in this study.

The ability to use carbohydrates as an energy source varies between fish, shrimp, and terrestial animals. The use of carbohydrate by fish and shrimp are less efficient than terrestial animals [13]. The use of hot water in the manufacture of pellets was also influential, as it is known that carbohydrates are gelatinized, which increases the digestibility of shrimp [14-16]. The results showed that an increase in starch gelatinization for commercial feed manufacturing is expected to have a positive effect on digestibility [17]. Carbohydrates and lipids are important nutritional components in food shrimp [8]. From a practical standpoint, it is worth understanding how carbohydrates are used to provide information in design a better feed for the different growth phases of the shrimp.

The results of the study [17] showed that salinity variations do not affect the digestibility of carbohydrates and lipids by white shrimp juvenile on the container control. The quantity and quality of waste excreted by shrimp depends on consumption, digestion, and metabolism of the compound feed [18,19] reported that the highest energy of feed consumed by white shrimp juvenile Litopenaeus vannamei was obtained when the salinity of the water is maintained at 26 ppt. White shrimp reared in low salinity shown to promote the growth and survival by adjusting the level of nutrients in the feed [20,21]. When the shrimp are exposed to low salinity, shrimp should resist the loss of Na + and Cl- by the active uptake of Na + from the water in exchange for H+, which occurs in the apical membrane of the cell osmoregulatory to increase capacity osmoregulatory [22-24].

During the ongoing research of water quality parameter values remain at the limit eligibility for the growth and survival of white shrimp juvenile especially salinity was maintained in the range of 20 ppt. The values of physical parameters of the water quality is within acceptable limits for maintenance indoor shrimp production [10].

Conclusion

Based on the research that has been done can be concluded that the best level of carbohydrate feed was 38% with a feeding frequency four times a day. Level of 38% carbohydrate feed produces the highest specific growth rate and high carbohydrate digestibility.

References

1. Department of Fisheries and Marine Resources of South Sulawesi (2008) Annual Report of Fisheries Development Goals Realization and South Sulawesi. Department of Fisheries and Marine Resources of South Sulawesi province.

2. Cortés-Jacinto, Villarreal-Colmenares EH, Civera-Cerecedo R, Martínez-Cordova L (2003) Effect of dietary protein level on growth and survival of juvenil freshwater crayfish Cherax quadricarinatus (Decapoda: Parastacidae). Aquacult Nutr 9: 207-213.

3. Haryati, Saade E, Zainuddin (2009) Formulation and feed aplication for brustock and cultivation: Feed suplement aplication for the increase quality of tiger shrimp local brustock. University of Hasanuddin, Makassar.

4. Haliman RW, Dian AS (2005) White shrimp Litopenaeus vannamei: cultivation and market prospect of white shrimp disease resistence. Swadaya Press, Jakarta.

5. Zainuddin, Abustang, Siti Aslamyah (2009) Probiotic utilization in commercial feed for cultivation of tiger shrimp. University of Hasanuddin, Makassar.

6. FAO (1987) Feed and feeding of fish and shrimp. A manual on the preparation and presentation of compound feeds for shrimp and fish aquaculture.

7. Koshio S, Teshima TS, Kanazawa A, Watase T (1993) The effect of dietary protein content on growth, digestion efficiency and nitrogen excretion of juvenil kuruma prawns, Penaeus japonicus Aquaculture 113: 101-114.

8. Gaxiola G, Cuzon G, Garcia T, Tabeada G, Brito R, et al. (2005) A factorial effects

Feed	Feeding frequency		
	B1	B2	B3
A1	86.26 ± 0.728ᵃ	87.95 ± 0.560ᵇ	88.89 ± 0.475ᵇ
A2	84.64 ± 1.206ᵃ	89.27 ± 0.340ᵇ	90.43 ± 0.273ᵇ
A3	84.79 ± 0.371ᵃ	88.99 ± 0.696ᵇ	90.77 ± 0.431ᵇ
A4	84.61 ± 0.517ᵃ	89.16 ± 0. 423ᵇ	89.33 ± 0.325ᵇ

Note: different letters in the same row indicate treatments significantly different (P<0.05)

Table 4: Carbohydrate digestibility (%) of white shrimp juvenilein the combined treatment of feed carbohydrate levels and feeding frequency.

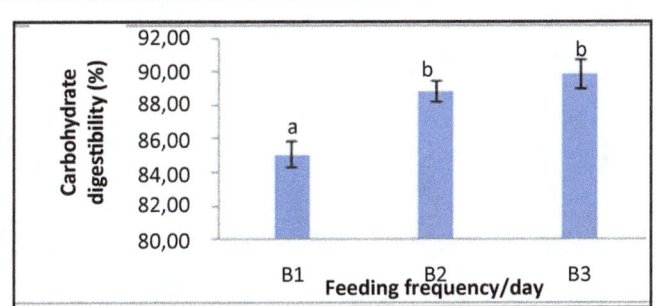

Figure 1: Graph digestibility of carbohydrates in the treatment of shrimp juvenile by feeding frequency per day.

of salinity, dietary carbohydrate and molt cycle on digestive carbohydrases and hexokinases in Litopenaeus vannamei (Boone, 1931). Comp Biochem Physiol 140: 29-39.

9. Cuzon G, Rosas C, Gaxiola G, Taboada G, Van Wormhoudt A (2000) Utilization of carbohydrates by shrimp

10. Campaña Torres A, Martinez Cordova LR, Villarreal Colmenares, Cortés-Jacinto E (2010) Evaluation of different concentrations of adult live Artemia (Artemia franciscana, Kellogs 1906) as natural exogenous feed on the water quality and production parameters of Litopenaeus vannamei (Boone 1931) pre-grown intensively. Aquacult Res 42: 40-46.

11. Radford CA, Marsden ID, Jeffs AG (2008) Specific dynamic action as an indicator of carbohydrate digestion in juvenil spiny lobsters, Jasus edwardsii. Mar Freshwat Res 59: 841-848.

12. Olmos J, Ochoa L, Paniagua-Michel J, Contreras R (2011) Functional feed assessment on Litopenaeus vannamei using 100% fish meal replacement by soybean meal, high levels of complex carbohydrates and Bacillus probiotic strains. Mar Drugs 9: 1119-1132.

13. Guo R, Liu YJ, Tian LX, Huang JW (2006) Effect of dietary cornstarch levels on growth performance, digestibility and microscopic structure in the white shrimp, Litopenaeus vannamei reared in brackish water. Aquacult Nutr 12: 83-88.

14. Mohapatra M, Sahu NP, Chaudhari A (2003) Utilization of gelatinized carbohydrates in diets of Labeoro hita fry. Aquacult Nutr 9: 189-196.

15. Shiau SY (1997) Utilization of carbohydrates in warmwater fish - with reference to tilapia, Oreochromis niloticus X O. aureus. Aquaculture 151: 79-96

16. Salama AJ (2008) Effects of different feeding frequency on the growth, survial and feed conversion ratio of the Asian sea bass Lates calcarifer juvenils reared under hypersaline seawater of the Red Sea. Aquacul Res 39: 561-567.

17. Thomas M, Van Vliet T, Van Der Poel AFB (1998) Physical quality of pelleted animal feed. Contribution of feed stuff components. Anim Feed Sci Technol 70: 59-78

18. Amirkolaie AK (2011) Reduction in the environmental impact of waste discharged by fish farms through feed and feeding. Rev. Aquacul 3: 19-26.

19. Valdez GF, Diaz F, Re AD, Sierra E (2008) Effect of salinity on physiological energetic of white shrimp Litopenaeus vannamei (Boone). Hidrobiologica 18: 105-115.

20. Gong H, Jiang DH, Lightner D, Collins C (2003) Improving osmoregulatory capacity of Litopenaeus vannamei cultured in brackish water through dietary modifications. Aquaculture America, Louisville, Kentucky, USA.

21. Pérez-Velázquez M, González-Félix ML, Jaimes Bustamente F, Martínez-Córdova LR, Trujillo-Villalba DA (2007) Investigation of the effects of salinity and dietary protein level on growth and survival of Pacific white shrimp, Litopenaeus vannamei. J World Aquac Soc 38: 475-485.

22. Gucic M, Jacinto EC, Cerecedo RC, Marie DR, Martínez-Córdova LR (2013) Apparent carbohydrate and lipid digestibility of feeds for whiteleg shrimp, Litopenaeus vannamei (Decapoda: Penaeidae), cultivated at different salinities. Rev Biol Trop 3: 1201-1213.

23. Bückle LF (2006) Osmoregulatory capacity of the shrimp Litopenaeus vannamei at different temperatures and salinities, and optimal culture environment. Rev Biol Trop 54: 745-753.

24. Hurtado MA, Racotta IS, Arjona O, Hernández-Rodríguez M, Goytortúa E, et al. (2006) Effect of hypo-and hyper-saline conditions on osmolarity and fatty acid composition of juvenil shrimp Litopenaeus vannamei (Boone, 1931) fed low-and high-HUFA diets. Aquacult Res 37: 1316-1326.

11

Determination of Protein, Lipid and Carbohydrate Contents of Conventional and Non-Conventional Feed Items used in Carp Polyculture Pond

Islam MA[1]*, Asadujjaman M[2], Biswas S[1], Manirujjaman M[3], Rahman M[1], Hossain MA[2], Uddin AMM[1], Asaduzzaman M[1], Rahman MS[1] and Munira S[1]

[1]*Department of Biochemistry and Molecular Biology, University of Rajshahi, Rajshahi-6205, Bangladesh*
[2]*Department of Fisheries, University of Rajshahi, Rajshahi-6205, Bangladesh*
[3]*Department of Biochemistry, Gonoshasthaya Samaj Vittik Medical College and Hospital, Gono University, Savar, Dhaka-1344, Bangladesh*

Abstract

This study was conducted with a view to comparing the protein, lipid and carbohydrate contents in conventional and non-conventional feed items and to recommend suitable strategy in selecting feed item for the development of weed based fish farming in carp polyculture pond. Six different conventional and non-conventional fish feed items like rice bran, wheat bran, mustard oilcake, Azolla, grass and banana leaves were tested to determine the nutrient contents under 6 treatments as T_1, T_2, T_3, T_4, T_5 and T_6, respectively. In this study, nutrient contents (protein, lipid and carbohydrate) were monitored monthly. Significant variations ($P<0.05$) were found in the mean values of nutrient contents with different treatments of feed items but in case of same feed item no significant difference was found in the nutrient content at different months. Among the non-conventional feed items treatment T_4 (Azolla) varied more significantly ($P<0.05$) for the mean values of protein content. Findings indicated that Azolla was more nutritive and low cost effective diets for fish farming in Bangladesh.

Keywords: *Azolla;* Conventional and non-conventional feed; Carp polyculture; Bangladesh

Introduction

The technique of polyculture of fish is based on the concept of utilization of different trophic and spatial niches of a pond in order to obtain maximum fish production per unit area. Different compatible species of fish of different trophic and spatial niches are raised together in the same pond to utilize all sorts of natural food available in the pond [1]. Supplementary feed plays an important role in achieving higher fish production. Unfortunately lack of low cost supplementary feed is found as one of the major problems in aquaculture in Bangladesh [2]. It was thus considered necessary to look for cheaper and locally available materials as substitutes.

The optimal protein requirements of carp are affected by the nutritional value of the dietary protein and level of non-protein energy in the carp diet. When sufficient energy sources such as lipids and carbohydrates are available in the diet, most of the ingested protein goes to protein synthesis. Adult Indian major carps require 30% dietary protein for proper growth and survival. Lipids or fats are required as sources of energy and essential fatty acids, and serve as carriers for fat-soluble vitamins. The gross lipid requirement of Indian major carp is 7-8% of the diet, and young fish require relatively more fat and protein than adults. Carbohydrate is the least-expensive nutrient and also a less expensive energy source for carp. Indian major carp, being herbivorous/ omnivorous feeders, easily digest appreciable quantities of carbohydrates in their diets. A dietary level up to 30% carbohydrate does not affect the growth of carp and growth retardation and reduced feed efficiency are observed, however, when carbohydrate levels exceeded 35% of diet. Fish culture is induced primarily by the need for increased protein supply. One of the most essential prerequisites for the successful management of fish culture programme is a comprehensive understanding of feeding [3]. The increase in cost and demand of feed protein from conventional sources necessitates fish culturists of the developing countries to incorporate cheap and locally available ingredients in fish feeds. The utilization of aquatic plants having high food value are used to supplement fish food has taken a new dimension for producing the much required animal protein at low cost [4].

Aquatic macrophytes have been known to have potential food value [5]. A perusal of the available literature shows that some of the aquatic weeds are highly nutritive and, therefore, one alternative solution to check the massive population of these weeds might be their utilization through incorporation as components of feedstuff for fish. In fact, significant effort has been directed towards evaluating the nutritive value of different non-conventional feed resources, including terrestrial and aquatic macrophytes, to formulate nutritionally balanced and cost-effective diets for fish and poultry [6-9]. Most of these nutritional studies were carried out abroad and no comprehensive studies are found in comparing the nutritional quality of both conventional and non-conventional feeds for fish farming in Bangladesh. However, before advocating the utilization of these aquatic weeds for supplementation of fish feeds, there is an urgent need to explore their nutritional quality, throughout the major culture season in ponds under carp polyculture system. Therefore, the present study aimed at evaluating the protein, lipid and carbohydrate content in conventional and non-conventional feed items used for carp polyculture system in Bangladesh.

Materials and Methods

Duration and location of the study

The study was conducted for a period of six months from April 2010 to September 2010. Feed items were collected from the fish farming study site located at Alampur village under Kushtia district

*Corresponding author: Islam MA, Department of Biochemistry and Molecular Biology, University of Rajshahi, Rajshahi-6205, Bangladesh
E-mail: maislam06@gmail.com

of Bangladesh. Whereas nutrient analysis was done at the Protein and Enzyme Research Laboratory under the department of Biochemistry and Molecular Biology, Rajshahi University, Rajshahi, Bangladesh.

Experiment design

The current experiment was carried out under six treatments of feed items each with three replications. The treatment assignments were designated as T_1, T_2, T_3, T_4, T_5 and T_6 for rice bran, wheat bran, mustard oilcake, Azolla, grass and banana leaves, respectively. Conventional feed items (rice bran, wheat bran, mustard oilcake) were collected from local market during the experimental period. Non-conventional feed item like Azolla was collected from Azolla ponds adjacent to the research area whereas grass and banana leaf were collected from adjacent grass field and banana garden. Both conventional and non-conventional feed items were collected once a month for nutritional analysis throughout the experimental period.

Nutrient analysis of the collected samples

Total protein, total lipid and total carbohydrate of the collected samples were determined by the micro-kjeldahl method [10,11] methods and Anthrone method [12] respectively.

Statistical analysis

All the data were subjected to ANOVA (analysis of Variance) using computer software SPSS (Statistical Package of Social Science). The mean values were also compared to see the significant difference from the DMRT (Duncan Multiple range Test) [13].

Results

Monthly variations

Protein content significantly varied from 6.05 ± 0.45% with T_6 (banana leaf) at 6th month (September, 2010) to 31.20 ± 0.32% with treatment T_3 (mustard oilcake) at 2nd month (May, 2010). Lipid content significantly varied from 2.95 ± 0.21% with treatment T_6 (banana leaf) at 5th month (August, 2010) to 13.72 ± 0.36% with treatment T_3 (mustard oilcake) at 4th month (July, 2010). Carbohydrate significantly varied from 32.85 ± 0.14% with treatment T_3 (mustard oilcake) at 4th month (July, 2010) to 66.35 ± 0.32% with T_2 (wheat bran) at 3rd month (June, 2010). In the same feed item no significant difference in the nutrient content was found during the study period (Tables 1-6).

Mean variations

The variations in the mean values of nutrient contents (protein, lipid and carbohydrate) with different treatments of feed items are presented in Table 7 and Figure 1. Protein content significantly varied from 6.18 ± 0.13% with treatment T_6 (banana leaf) to 30.53 ± 0.40% with treatment T_3 (mustard oilcake). Lipid content significantly varied from 3.06 ± 0.09% with treatment T_6 (banana leaf) to 13.33 ± 0.10% with treatment T_3 (mustard oilcake). Carbohydrate significantly varied from 32.95 ± 0.29% with treatment T_3 (mustard oilcake) to 66.12 ± 0.47% with treatment T_2 (wheat bran).

Discussion

Monthly variations of the nutrient contents

Protein content varied from 6.05 ± 0.45% with (T_6 at 6th month) to 31.20 ± 0.32% (T_3 at 2nd month). Lipid content ranged from 2.95 ± 0.21% (T_6 at 5th month) to 13.72 ± 0.36% (T_3 at 4th month). Carbohydrate content ranged from 32.85 ± 0.14% (T_3 at 4th month) to 66.35 ± 0.32%

(T_2 at 3rd month). Suresh and Mandal worked on the determination of nutritive value of rice bran, mustard oil cake and Azolla for a period of 4 months from July to October. In rice bran they found crude protein and crude fibre as 12.6% and 21.9%, respectively. In mustard oilcake, crude protein and crude fibre was 38.6% and 6.8%, respectively and in Azolla, crude protein and crude fibred was 26.5% and 20.4%, respectively. Sithara and Kamalaveni [14] worked on the formulation of low cost fish feed using Azolla as a protein supplement during September to March and reported 20-25.5% protein in Azolla. Ebrahim [15] used Azolla as tilapia diet for a period of 90 days in summer season and reported 20% protein in Azolla. Fasakin and Balogan [16] worked on the nutritional aspects of Azolla in August, 1997 and reported 20.9% protein in Azolla.

Present findings also indicated that in case of same feed item, no significant difference was found in the nutrient content at different months (Tables 1-6). This might be due to no major change in the temperature was found to affect the growth and composition of Azolla during the study period. This statement was almost agreed with Lumpkin and Plucknett [17] who reported that change in Azolla composition was subjected to change in environment. Statement also agreed with Van-Hove [18] and Ebrahim [15] who reported that change in Azolla composition was subjected to change in species.

Mean variation of the nutrient contents

In the present study the protein content varied from 6.18 ± 0.13% (T_6, banana leaf) to 30.53 ± 0.40% (T_3, mustard oilcake), lipid content varied from 3.06 ± 0.09% (T_6, banana leaf) to 13.33 ± 0.10% (T_3, mustard oilcake) and carbohydrate content varied from 32.95 ± 0.29% (T_3, mustard oilcake) to 66.12 ± 0.47% (T_2, wheat bran). The highest protein and lipid content was found in treatment T_3 (mustard oilcake) whereas the highest carbohydrate content was found in treatment T_2, wheat bran (66.12 ± 0.47%) followed by T_4, Azolla (50.21 ± 0.54%), T_6, banana leaf (48.50 ± 0.51%), T_5, grass (46.36 ± 0.16%), T_1, rice bran (44.09 ± 0.67%), T_3, mustard oilcake (32.95 ± 0.29%). Hepher [19] reported the protein content of ricebran, wheat bran, oil cake and Azolla as 11.88%, 14.57%, 30-33% and 19.27%, respectively. Banerjee and Matai [20] determined the nutritive status of *Azolla pinnata* and reported protein as 21.9% and Lipid as 3.8%. Gavina [21] reported crude protein of 20.98%, crude fat of 5.17% and crude fiber of 19.30% in Azolla. Tavares [22] observed 38.8% crude protein, 3.8% crude fat and 13.2% crude fiber in dried duck weed. They also reported that the protein content of duckweeds growing on nutrient poor and nutrient rich water varied between 15-25% and 35-45% (Dry matter basis), respectively. In case of conventional feed items the major nutrient like protein varied from 14.40 ± 0.32% (rice bran) to 30.53 ± 0.40% (mustard oilcake). Whereas in case of non-conventional feed items the protein varied from 6.18 ± 0.13% (banana leaf) to 18.58 ± 0.09% (Azolla). Being an omnivore, the fish can also feed on vegetation [23] and may be able to assimilate Azolla in the diets.

The chemical composition of Azolla species varies with ecotypes and with the ecological conditions and the phase of growth. The crude protein content is about 19-30 percent dry matter basis during the optimum conditions for growth [24,25]. The protein contents of Azolla species are comparable to or higher than that of most other aquatic macrophytes. Aquatic weeds' are highly nutritious with protein content of 20-30%, when cultivated in nutrient rich waters [26]. Importantly, they are preferred food of a wide range of herbivorous fish such as grass carp (*Ctenopharyngodon idella*), silver barb (*Barbonymus gonionotus, Puntius jerdoni*), tilapias (*Oreochromis niloticus, Tilapia rendalli, Tilapia zillii*) and rohu (*Labeo rohita*) [27,28].

Nutrients	Months					
	April	May	June	July	August	September
Protein (%)	14.60 ± 0.22[a]	13.92 ± 0.19[a]	14.65 ± 0.19[a]	14.50 ± 0.36[a]	14.22 ± 0.28[a]	14.50 ± 0.24[a]
Lipid (%)	10.42 ± 0.31[a]	10.50 ± 0.25[a]	10.64 ± 0.25[a]	10.20 ± 0.21[a]	10.24 ± 0.15[a]	10.45 ± 0.26[a]
Carbohydrate (%)	44.25 ± 0.41[a]	43.72 ± 0.19[a]	43.85 ± 0.19[a]	44.20 ± 0.24[a]	44.32 ± 0.20[a]	44.20 ± 0.16[a]

Figures bearing common letter(s) in a row as superscript do not differ significantly ($P < 0.05$)

Table 1: Monthly variations in nutrient (protein, lipid and carbohydrate) contents with treatment T_1 (Rice, *Oryza sativa* bran).

Nutrients	Months					
	April	May	June	July	August	September
Protein (%)	17.20 ± 0.05[a]	17.05 ± 0.12[a]	17.25 ± 0.12[a]	16.95 ± 0.24[a]	17.10 ± 0.34[a]	17.22 ± 0.18[a]
Lipid (%)	6.75 ± 0.41[a]	6.66 ± 0.69[a]	6.80 ± 0.69[a]	7.12 ± 0.46[a]	6.47 ± 0.32[a]	6.32 ± 0.38[a]
Carbohydrate (%)	66.20 ± 0.36[a]	65.75 ± 0.32[a]	66.35 ± 0.32[a]	66.32 ± 0.26[a]	66.12 ± 0.15[a]	65.99 ± 0.23[a]

Figures bearing common letter(s) in a row as superscript do not differ significantly ($P < 0.05$)

Table 2: Monthly variations in nutrient (protein, lipid and carbohydrate) contents with treatment T_2 (Wheat, *Trticum aestivum* bran).

Nutrients	Months					
	April	May	June	July	August	September
Protein (%)	30.65 ± 0.18[a]	31.20 ± 0.32[a]	30.50 ± 0.32[a]	30.25 ± 0.15[a]	30.15 ± 0.11[a]	30.45 ± 0.17[a]
Lipid (%)	13.34 ± 0.31[a]	13.24 ± 0.47[a]	13.25 ± 0.47[a]	13.72 ± 0.36[a]	13.22 ± 0.18[a]	13.20 ± 0.19[a]
Carbohydrate (%)	32.86 ± 0.18[a]	32.90 ± 0.25[a]	33.10 ± 0.25[a]	32.85 ± 0.14[a]	32.98 ± 0.31[a]	33.02 ± 0.46[a]

Figures bearing common letter(s) in a row as superscript do not differ significantly ($P < 0.05$)

Table 3: Monthly variations in nutrient (protein, lipid and carbohydrate) contents with treatment T_3 (Mustard, *Brassica napus* Oilcake).

Nutrients	Months					
	April	May	June	July	August	September
Protein (%)	18.65 ± 0.08[a]	18.45 ± 0.41[a]	18.35 ± 0.41[a]	18.45 ± 0.32[a]	18.75 ± 0.24[a]	18.80 ± 0.26[a]
Lipid (%)	3.25 ± 0.09[a]	3.15 ± 0.12[a]	3.12 ± 0.12[a]	3.35 ± 0.18[a]	3.14 ± 0.34[a]	3.10 ± 0.41[a]
Carbohydrate (%)	50.36 ± 0.75[a]	50.45 ± 0.61[a]	50.20 ± 0.61[a]	50.15 ± 0.54[a]	50.20 ± 0.17[a]	49.88 ± 0.27[a]

Figures bearing common letter(s) in a row as superscript do not differ significantly ($P < 0.05$)

Table 4: Monthly variations in nutrient (protein, lipid and carbohydrate) contents with treatment T_4 (*Azolla pinnata*).

Nutrients	Months					
	April	May	June	July	August	September
Protein (%)	7.28 ± 0.35[a]	7.32 ± 0.25[a]	7.45 ± 0.25[a]	7.15 ± 0.14[a]	7.25 ± 0.19[a]	7.12 ± 0.23[a]
Lipid (%)	6.35 ± 0.05[a]	6.28 ± 0.06[a]	6.45 ± 0.06[a]	6.23 ± 0.12[a]	6.21 ± 0.18[a]	6.32 ± 0.28[a]
Carbohydrate (%)	46.58 ± 0.12[a]	46.30 ± 0.41[a]	45.95 ± 0.41[a]	46.85 ± 0.38[a]	46.70 ± 0.19[a]	45.76 ± 0.14[a]

Figures bearing common letter(s) in a row as superscript do not differ significantly ($P < 0.05$)

Table 5: Monthly variations in nutrient (protein, lipid and carbohydrate) contents with treatment T_5 (Grass, *Cynodon dactylon*).

Nutrients	Months					
	April	May	June	July	August	September
Protein (%)	6.25 ± 0.11[a]	6.20 ± 0.21[a]	6.32 ± 0.21[a]	6.12 ± 0.31[a]	6.14 ± 0.36[a]	6.05 ± 0.45[a]
Lipid (%)	3.05 ± 0.04[a]	3.12 ± 0.11[a]	3.10 ± 0.11[a]	3.20 ± 0.17[a]	2.95 ± 0.21[a]	2.96 ± 0.41[a]
Carbohydrate (%)	48.85 ± 0.36[a]	47.98 ± 0.26[a]	48.10 ± 0.26[a]	48.30 ± 0.31[a]	48.90 ± 0.35[a]	48.85 ± 0.24[a]

Figures bearing common letter(s) in a row as superscript do not differ significantly ($P < 0.05$)

Table 6: Monthly variations in nutrient (protein, lipid and carbohydrate) contents with treatment T_6 (Leaf of banana, *Musa acuminata*).

Overall findings indicated that inspite of having variations in nutrient contents, monthly supply of nutrients was almost same respective feed item under non-conventional feeds as with conventional feeds. Mean values of the nutrient contents under non-conventional feed items are found potentials for the development of low cost aquaculture.

Fish feed generally constitutes 60-70% of the operational cost in intensive and semi- intensive aquaculture system [29]. The fish feed used in aquaculture is quite expensive, irregular and short in supply in many third world countries. These feeds are sometimes adulterated, contaminated with pathogen as well as containing harmful chemicals for human health. Naturally there is a need for the development of healthy, hygienic fish feed which influences the production as well as determines the quality of cultured fish. Considering the importance of nutritionally balanced and cost-effective alternative diets for fish, almost similar expression to evaluate the nutritive value of different non-conventional feed resources, including terrestrial and aquatic macrophytes was found with Wee and Wang [9] and Mondal and Ray [30]. However potentials roles of aquatic and terrestrial macrophytes as supplementary feeds in fish farming were also found to be expressed with Bardach [31] and Edwards [32].

Treatments	Nutrient content		
	Protein (%)	Lipid (%)	Carbohydrate (%)
T₁ (Rice bran)	14.40 ± 0.32^d	10.41 ± 0.31^b	44.09 ± 0.67^e
T₂ (Wheat bran)	17.13 ± 0.07^c	6.69 ± 0.30^c	66.12 ± 0.47^a
T₃ (Oilcake)	30.53 ± 0.40^a	13.33 ± 0.10^a	32.95 ± 0.29^f
T₄ (*Azolla pinnata*)	18.58 ± 0.09^b	3.19 ± 0.10^d	50.21 ± 0.54^b
T₅ (Grass- *Cynodon dactylon*)	7.26 ± 0.18^e	6.31 ± 0.13^c	46.36 ± 0.16^d
T₆ (Leaf of *Musa acuminata*- Banana leaf)	6.18 ± 0.13^f	3.06 ± 0.09^d	48.50 ± 0.51^c
F value	16.42	13.88	114.85
P value	0.002	0.004	0.0000008

Figures bearing common letter(s) in a column as superscript do not differ significantly (P<0.05)

Table 7: Variations in the mean values of protein, lipid and carbohydrate contents in different fish feed items.

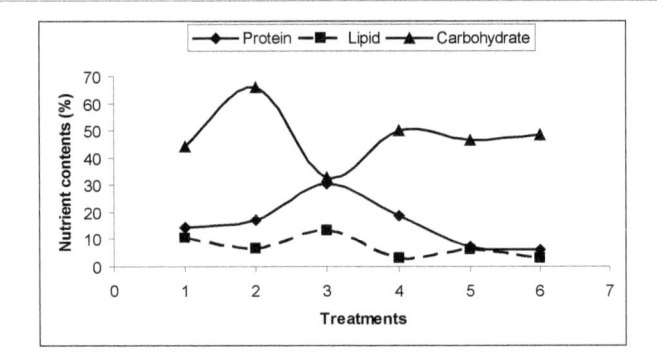

Figure 1: Variations in the mean values of nutrient contents under different fish feed items

Conclusion

In case of conventional feed items, protein, lipid and carbohydrate varied from 14.40 ± 0.32% to 30.53 ± 0.40%, 6.69 ± 0.30% to 13.33 ± 0.10% and 32.95 ± 0.29% to 66.12 ± 0.47%. In case of non-conventional feed items, protein, lipid and carbohydrate varied from 6.18 ± 0.13% to 18.58 ± 0.09%, 3.06 ± 0.09% to 6.31 ± 0.13% and 46.36 ± 0.16% to 50.21 ± 0.54%. Inspite of variations weeds are moderately nutritive and low cost effective diets for fish. However, the present study did not evaluate the fish production and economy of feed and weed based systems.

Recommendation

Present findings explored the nutritive aspects of both conventional and non-conventional feed items and question raised about the response of utilizing the feed especially of aquatic weeds to fish growth and economy. Therefore, it is recommended to conduct further study on the evaluation of fish production and economy under different feed and weed based systems in polyculture ponds.

Acknowledgement

The research work was conducted under a financial support by the Ph. D. Fellowship Programme of Ministry of Science and Technology, Govt. of the People's Republic of Bangladesh which is gratefully acknowledged.

References

1. Rahman MM, Varga I, Chowdhury SN (1992) Manual on polyculture and integrated fish farming in Bangladesh. Food and Agriculture Organization (FAO), Rome, Italy.

2. DoF (2011) National Fish Week 2011 Compendium (in Bengali).

3. Halver JE (1972) Fish nutrition. Academic Press, New York.

4. Lakshmanan MAV, Murthy DS, Pillai KK, Banerjee (1967) On a new artificial feed for carp. fry. FAO Fisheries Report 44: 373-387.

5. Edwards P (1980) Food potential of aquatic macrophytes. ICLARM Studies and Reviews 5: 51

6. Edwards P, Kamal M, Wee KL (1985) Incorporation of composted and dried water hyacinth in pelleted feed for the tilapia Oreochromis niloticus (Peters). Aquaculture and Fisheries Management 16: 233-248.

7. Patra BC, Ray AK (1988) A preliminary study on the utilization of the aquatic weed Hydrilla verticillata Rayle as feed by the carp, Labeo rohita (Hamilton): growth and certain biochemical composition of flesh. Indian Biology XX: 44-50.

8. Ray AK, Das I (1995) Evaluation of dried aquatic weed, Pistia stratiotes meal as a feedstuff in pelleted feed for rohu, Labeo rohita fingerlings. J Applied Aquaculture 5: 35-44.

9. Wee KL, Wang SS (1987) Nutritive value of Leucaena leaf meal in pelleted feed for Nile Tilapia. Aquaculture 62: 97-108.

10. Rangama S (1977) Manual of analysis of Fruits and vegetable products, Tata Mc Graw- Hill Publishing Company Ltd, New Delhi.

11. Bligh EG, Dyer W (1989) Total Lipid Extraction and Purification. Can J Biochem Physiol 37: 911-917.

12. Boel E, Huge-Jensen B, Christensen M, Thim L, Fiil NP (1988) Rhizomucor miehei triglyceride lipase is synthesized as a precursor. Lipids 23: 701-706.

13. Gomez KA, Gomez AA (1984) Statistical Procedure for Agricultural Research.

14. Sithara K, Kamalaveni K (2008) Formulation of low-cost feed using Azolla as a protein supplement and its influence on feed utilization in fishes. Current Biotica 2: 212-219.

15. Ebrahim MSM, Zeinhom MM, Abou-Seif RA (2007) Response of Nile tilapia (Oreochromis niloticus) fingerlings to diets containing Azolla meal as a source of protein. J Ara Aquaculture Society 2: 54-68.

16. Fasakin EA, Balogun AM (2001) Nutritional and anti-nutritional analyses of Azolla africana Desv. and Spirodela polyrrhiza L. Schleiden as feedstuffs for fish production.

17. Lumpkin TA, Plucknett L (1982) Azolla as a green manure: use and management in crop production.

18. Van-Hove C, de Waha Baillonville T, Diara HF, Godard P, Mai Kodomi Y, et al. (1987) Azolla collection and selection. Azolla Utilization.

19. Hepher B (1988) Nutrition of Pond Fishes. Cambridge University Press, London, UK.

20. Banerjee A, Matai S (1990) Composition of Indian aquatic plants in relation to utilization as animal forage. J. Aquat. Plant Manage 28: 69-73.

21. Gavina LD (1994) Pig-Duck-Fish-Azolla integration LA Union Philippines.

22. Tavares FA, Roudrigues JSR, Fracalossi DM, Esquivel J, Roubach R (2008) Dried duckweed and commercial feed promote adequate growth performance of tilapia fingerlings. Biotemas 21: 91-97.

23. Santhanam R, Sukumaran N, Natarajan P (1990) A Manual of Fresh Water Aquaculture. Oxford and IBH Publishing Go. Pvt. Ltd, New Delhi.

24. Peters GA, Mayne BC, Ray TB, Toia RE (1979) Physiology and biochemistry of the Azolla-Anabaena symbiosis.

25. Becking JH (1979) Environmental requirements of Azolla for use in tropical rice production. In Nitrogen and Rice, Los Banos, Laguna, International Rice Research Institute.

26. Culley DD, Rejmankova E, Koet J, Prye JB (1981) Production, chemical quality and use of duckweeds (Lemnaceae) in aquaculture, waste management and animal feeds. J World Maricult Soc 12: 27-49.

27. Singh SB, Pillai KK, Chakraborty PC (1967) Observation on the efficacy of grass carp in controlling and utilizing aquatic weeds in ponds in India. Proc. Indo-Pacific Fish Counc 12: 220-235.

28. Gaiger IG, Porath D, Granoth G (1984) Evaluation of duckweed (Lemna gibba) as feed for tilapia (Oreochromis nilotieus cross Oreochromis aureus) in a recirculating unit. Aquacultre 41: 235-244.

29. Singh PK, Gaur SR, Chari MS (2006) Growth Performance of Labeo rohita (Ham.) Fed on Diet Containing Different Levels of Slaughter House Waste. J Fish Aquat Sci 1: 10-16.

30. Mondal TK, Ray AK (1999) The nutritive value of Acacia auriculiformis leaf meal in compounded diets for Labeo rohita fingerlings.

31. Bardach JE, Ryther JH, MeLarney WO (1974) Aquaculture: The Farming and Husbandry of Freshwater and Marine Organisms, Wiley-Interscience, New York.

32. Edwards P (1990) Use of terrestrial vegetation and aquatic macrophytes in aquaculture.

12

Effects of Different Levels of Copper Sulfate on Growth and Reproductive Performances in Guppy (P. *reticulate*)

Mahsa Javadi Moosavi[1]* and Vali-Allah Jafari Shamushaki[2]

[1]*M.Sc Graduated of Aquaculture, Gorgan University of Agricultural Science and Natural Resources, Faculty of Fishery and Environmental Science, Golestan, I.R. Iran*
[2]*Assistance Professor, Department of Fisheries, Gorgan University of Agricultural Sciences and Natural Resources, Gorgan, Iran*

Abstract

Adult Guppies (Poecilia reticulate) were exposed to copper sulfate ($CuSO_4$ $5H_2O$) to evaluate the effects on growth, survival and reproduction performances. Total 480 individuals (mean age of 2.5-3 months) were employed in 5 experimental groups containing 16 fish per group and exposed to 4 sub-lethal levels of copper (0 as control, 0. 004, 0. 013, 0. 019 and 0. 026 mg $CuSO_4.l^{-1}$) for a period of 56 days. Control group had relative advantage than experimental ones at both growth and reproduction performances. As the copper concentration increased, Relative fecundity, gonadosomatic index, surviving rate, offspring production and feed conversion ratio decreased but specific growth rate increased significantly ($P<0.05$). Focused energy consumption in liver for Cu detoxification process and lack of energy for other physiological demands confirm low SGR and high FCR values in this study. It's obvious that copper has its toxic effects for guppy, even at lower concentrations than LC50 value (0. 46 mg $Cu.l^{-1}$)

Keywords: Guppy; Copper sulfate; Growth performance; Reproductive performance

Introduction

Copper (Cu) like other trace elements [zinc (Zn), iron (Fe), manganese (Mn), etc. serves important functions in living cells and is essential for fish [1]. Copper sulfate ($CuSO_4$) is the sulfated form of copper which routinely used as an algicide in commercial and recreational fish ponds. It is generally recognized that copper can be highly toxic to teleosts [2,3]. In aquaculture, copper is being used as eternal parasites, bacterial and fungal disease preventer and also weeds cleaning in sulfated form [4]. The ideal concentration of copper sulfate for weed termination has been suggested to be as much as 1mg.l^{-1} in which has lower poisonous effect on fish but highly affects the invertebrates [5]. According to the records of United States Environmental Protection Agency (EPA), copper sulfate is an ordinary compound which is being used broadly in aquaculture. Moreover, the Food and Drug Administration (FDA) has barred the medicinal use of copper in aquaculture.

It should be considered that copper could be accumulated in water body and gradually increases the concentration will raise to lethal concentration for fishes [4]. Toxicity of copper to aquatic species depends on factors such as organism sensitivity, concentration of copper and its bioavailability [6], total hardness, pH [7], organic particles or various other inorganic cations and anions [8]. So it is notable to avoid using copper in waters with lower alkalinity than 50 mg $CaCo_3.l^{-1}$. The range of copper sulfate is in aquaculture as much as 0.025-2 mg.l^{-1} according to alkalinity and total hardness [9] but the usual effective use in aquaculture is reported as much as 0.01 of total alkalinity [10].

Guppy (P. *reticulata*) (or rainbow fish) is one of the most widely distributed tropical fish in the world. It is a member of the Poeciliidae family and, like all other members of the family, is live-bearing. Northeast South America is the native habitat for guppies but now, they could be found all around the world. High adaption ability makes them live in many different environmental and ecological conditions. Guppies exhibit sexual dimorphism and omnivorous feed habit (algal remains, diatoms, invertebrates, plant fragments, mineral particles, aquatic insect larva, etc.). Females produce offspring from first 10 weeks to 34 months of age, but first reproduction occur in 10-20 weeks of age. They are used as a model organism in the field of ecology, evolution, and behavioral studies [11]. Due to copper sulfate therapeutic trait which made it an ordinary drug for use in ornamental fish's hatcheries and personal aquariums despite its toxic effects and shortage of focused study on reproductive performance in guppy (P. *reticulate*), this study carried out.

Methods and Materials

This study performed at the aquatic laboratory of Shahid Fazli Bar Abadi located in Gorgan university of Agricultural Science and natural Resources, Golestan, Iran. Experimental fishes were bought from a local hatchery (Shast Kolah road, Gorgan province, Iran). Upon arrival, 480 individuals of 2.5-3 months aged guppies were acclimated to laboratory conditions for 2 weeks in a 1000 L round fiberglass tank measuring 1 m in diameter and fed a commercial diet (0.5-0.8 mm in size, Pars Kilka Corp, Babolsar, Iran) as much as 3% of body weight twice a day with two equal meals at 0800 and 1700 h. The experiment was conducted in a completely randomized design with six replications per treatment for 56 days. At the beginning of the experiment, 16 fish (N: 12 female and 4 male; mean length: 3.59 ± 0.11 cm; mean weight: 0.36 ± 0.01 g) were stocked in each aquarium (50×35×30 cm) designed to contain 35 L tap water (pH: 7.4 ± 0.12; salinity; 0.35 ± 0.12 ppt; DO: 7.36 ± 0.98 mg.l^{-1}; total hardness: 270-300 mg.l^{-1}CaCO$_3$) and equipped with airstones to maintain dissolved oxygen levels as much as possible. The water temperature was kept 28 ± 1°C by electrical heaters. Water quality variables were checked daily. Handmade plastic Happas were

*Corresponding author: Mahsa Javadi Moosavi, M. Sc Graduated of Aquaculture, Gorgan University of Agricultural Science and Natural Resources, Faculty of Fishery and Environmental Science, Golestan, I.R. Iran
E-mail: javadimoosavi@gmail.com

put in aquariums to act as shelters for new born offsprings from predation by parents.

In this study, five levels - one as blank and four as experimental groups - of copper sulfate penta-hydrate (CuSO$_4$ 5H$_2$O, Merck, Germany) concentrations including 0 (as control), 0.004, 0.013, 0.019 and 0.026 mg CuSO$_4$.l^{-1}. Prior to start the trial, the mean 96 h LC50 in guppies determined as much as 0.046 mg Cu.l^{-1} then, four concentrations were selected randomly. According to each level of copper sulfate, selected concentration prepared [12] into four reserved tanks to renew the replaced water volume. All experimental groups have been monitored daily by atomic absorption [13] to ensure the determined concentrations achieved [14].

Acute median lethal concentration (LC50) and their 95% confidence limits for all tests obtained by Finney's method were calculated with the formula of Mohapatra and Rengarajan. LC50 values of 24, 48, 72 and 96h were determined using Finney's method of probit analysis" and with SPSS computer statistical software. The mean lethal concentration LC50 for an exposure period of 24, 48, 72 and 96 hr was designated by trial and error. Amounts of LC1, LC10, LC30, LC50, LC70, LC90, LC99 were calculated by probit tables, mortality and probit regression.

At the end of the trial, percentage of body Specific Growth Rate (SGR) [ln(final weight)-ln(initial weight)/experimental period)]×100, Feed Conversion Ratio (FCR) [dry feed consumed (g)/gain in wet weight (g) [15], relative fecundity (RF) [absolute fecundity/body weight] [16], Gonadosomatic Index (GSI) [ovary weight/body weight]×100 [17], Surviving Rate (SR) [final fish number/initial fish number]×100 [18] and Offspring Production (OP) [new born offspring/number of adult females] calculated.

All data were reported as mean ± standard deviation. Statistics were performed by using one-way analysis of variance (ANOVA) followed by Duncan multiple comparisons test if significant differences were found. A Kolmogrov–Smirnov test was used to assess normality of distribution and abnormal data were log transformed. Significance was set at P<0.05 level.

Results and Discussion

Liver, brain, heart, kidney, and muscle are the main storing places for copper; in tissues and blood cells, it bounds to proteins, including many enzymes. It exerts a wide range of physiological effects on vital), and other hematopoietic tissues in which threats the present fishes in area [19]. Copper sulfate toxicity differs among fish species [20]; For instance, the amounts of 8, 8.97 and 16 mg Cu.l^{-1} reported to have negative effects on tilapia (Oreochromis niloticus × O-aureus), grupper (Epinephelus malabaricus) and channel catfish (Ictalurus punctatus) respectively [21,22].

In 8 weeks, toxic effects of copper sulfate exposure were recorded (Table 1). As seen, the best growth performance resulted in control group (0 mg CuSO$_4$.l^{-1}) and growth diminished as copper sulfate increased.

Despite the same initial weight, age and feed consumption, rising amounts of copper had negative effects on weight gain, specific growth rate and feed conversion ratio (Figure 1). There are hypothesis for reduction in growth performance including:

1) Intestine tissue interruption [23]: Heavy metals have their special recipients in which act as a stimulator to them. Intestine can be considered as a target tissue for metals. Copper which enter into

the intestine can damage the tissue and by means of that, reduces the nutrition intake from intestine. These researchers found out that, the existence of copper sulfate around catfishes reduced their growth performance due to interruption in intestine tissue. Furthermore, copper sulfate decreases the Zn [24] and Se [25] intake from intestine.

2) High energy consumption for copper detoxification in liver [26] reported that, in freshwater prawn (Macrobrachium rosenbergii) copper usually transported by the haemolymph to other organs primarily hepatopancreas for storage and detoxification. Ignoring the physiological differences, at the same polluted condition (Here by copper sulfate) guppies transfer copper to liver for detoxification and for fulfilling the detoxification process, body needs extra energy which is provided by more food consumption. Lack of energy for other metabolic processes result in low growth performance.

3) Chronic stress [27-29]: Chronic stress reduced the growth too [30]. The increase in blood glucose and cortisol concentrations is known as a general secondary response to stress of fish to toxic effects [31]. Griffin et al. [32] reported that, copper sulfate raised blood cortisol level (as stressor factor) in channel catfish (I. punctatus). These researches showed that the initial signs of stress (cortisol level) decreased consequently as copper sulfate diminished in water. At present study, a combination of three mentioned hypothesis is reliable reason for low growth performance. These results were in accordance with Berntssen et al. [33] on Atlantic salmon (Salmo salar), [34] on channel catfish (I. Punctatus). These researchers noted that increase in copper sulfate lead to low growth in fish.

Copper sulfate also had negative effects on reproduction performance (Table 1). In sensitive species of teleosts, copper adversely affects reproduction and survival from 0.01-0.02 mg Cu.l^{-1}. At present study, surviving rate, gonadosomatic index, relative fecundity and finally offspring production decreased by increase in copper sulfate levels (P<0.05) and best performance of mentioned factors resulted in control group (with no copper sulfate). Although relative fecundity between first two treatments (0 and 0.004 CuSO$_4$. l^{-1}) were not analytically significant; but it would be due to rare concentration of experimental treatment (0.004 CuSO$_4$. l^{-1}). Copper may affect reproductive success of fish through disruption of hatch coordination with food availability or through adverse effects on larval fishes. Chronic exposure of representative species of teleosts to low concentrations (0.005 to 0.04 mg.l^{-1}) of copper in water containing low concentrations of organic material adversely affects survival and spawning. Dethloff et al. reported that cortisol is released to the blood via stimulation of the Hypathalamo-Pituitary-Adrenal (HPI) axis by heavy metal exposure. Cortisol has depressive effects on a number of immune responses in fish, including phagocytosis and lymphocyte mitogenesis [35]. Despite the valuable role of HPI hormones in reproductive processes of fishes, cortisol hormone which made in response to stress resistance in fish would interrupt HPI hormones. It will be probably the reason for lowering slope gonadosomatic index in response of rise in copper sulfate concentrations.

Pulsford et al. [36] showed that, cortisol restricts the metabolic activities of macrophage and lymphocyte cells which especially spread in kidney and spleen; on one side, negative effects of copper on intestine inhibited absorption of nutrients like electrolytes and fatty acid intake [37]; on the other side, inducing hypertrophy in gill cells [38], blocking calcium transport in gills through interference with chloride cells and ionic and gas exchanges [39] make fishes weak to stand with situation. That's why low survival rate appears in high doses of copper sulfate.

Parameters [*]	Copper sulfate concentrations (mg CuSO$_4$. l^{-1})				
	0 (Control)	0.004	0.013	0.019	0.026
Initial length (cm)	3.61 ± 0.034[a]	3.58 ± 0.026[a]	3.57 ± 0.014[a]	3.6 ± 0.031[a]	3.59 ± 0.04[a]
Final length (cm)	3.76 ± 0.02[a]	3.71 ± 0.037[ab]	3.65 ± 0.017[ab]	3.7 ± 0.026[ab]	3.6 ± 0.01[b]
Initial weight (gr)	0.36 ± 0.005[a]	0.36 ± 0.017[a]	0.37 ± 0.011[a]	0.37 ± 0.023[a]	0.36 ± 0.011[a]
Final weight (gr)	0.82 ± 0.023[a]	0.78 ± 0.023[a]	0.72 ± 0.011[b]	0.63 ± 0.005[c]	0.65 ± 0.003[c]
Specific growth rate	0.91 ± 0.013[a]	0.86 ± 0.02[a]	0.71 ± 0.015[b]	0.69 ± 0.041[b]	0.59 ± 0.018[b]
Feed conversion ratio	4.36 ± 0.16[c]	4.76 ± 0.065[bc]	5.88 ± 0.00[b]	6.62 ± 0.31[ab]	7.88 ± 0.89[a]
Relative fecundity	56.82 ± 1.82[a]	51.42 ± 3.52[a]	32.45 ± 1.63[b]	34.89 ± 1.51[b]	25.57 ± 1.33[b]
Gonadosomatic index	6.1 ± 0.115[a]	5.61 ± 0.034[b]	5.32 ± 0.005[c]	5.29 ± 0.023[c]	5.31 ± 0.011[c]
Surviving rate	88.88 ± 2.77[a]	55.55 ± 7.34[b]	52.77 ± 7.34[b]	30.55 ± 10.01[c]	27.77 ± 2.77[c]
Offspring production	3.88 ± 0.22[a]	3.33 ± 0.17[b]	1.94 ± 0.07[c]	1.83 ± 0.096[c]	1.55 ± 0.07[c]

exposure to several concentrations of Cooper

[*]Mean ± SD values with different superscript letters within a raw for a parameter are significantly different (P<0.05).

Table 1: The mean results of growth and reproduction performances of guppy (*P. reticulate*) in response of

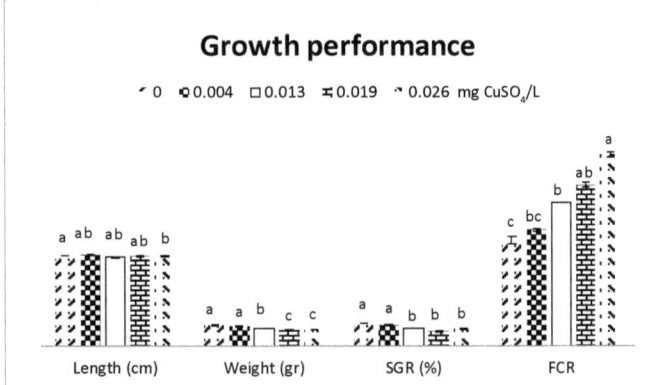

*Different superscript letters within a raw for a parameter are significantly different (P<0.05).

Figure 1: The mean results of growth performance of guppy (*P. reticulate*) in response of exposure to several concentrations of Cooper during 56 days of trial

Acknowledgments

Authors thank Gorgan University of Agricultural Science and Natural Resources for laboratory facilities and financial support to carry out this experiment.

References

1. Lorentzen M, Maage A (1995) Trace element status of juvenile Atlantic salmon Salmo salar L. fed a fish-meal based diet with or without supplementation of zinc, iron, manganese and copper from first feeding Aquaculture Nutrition 5: 163-171.

2. Perschbacher PW, Wurts WA (1999) Effects of calcium and magnesium hardness on acute copper toxicity to juvenile channel catfish. Ictalurus punctatus Aquaculture. 172: 275-280.

3. Straus DL (2003) The acute toxicity of copper to blue tilapia in dilutions of settled pond water. Aquaculture 219: 233-240.

4. Reddy R, Pillai BR, Adhikari S (2006) Bioaccumulation of copper in post-larvae and juveniles of freshwater prawn Macrobrachium rosenbergii (de Man) exposed to sub-lethal levels of copper sulfate. Aquaculture 252: 356-360.

5. Boyd CE (1990) Water quality in ponds for aquaculture. Alabama Agricultural Experiment Station, Auburn university, Birmingham Publishing.

6. De Oliveira-Filho EC, Lopes RM, Paumgartten FJ (2004) Comparative study on the susceptibility of freshwater species to copper-based pesticides. Chemosphere 56: 369-374.

7. Rábago-Castro J, Sanchez J, Pérez-Castañeda R, González-González A (2006) Effects of the prophylactic use of Romet®-30 and copper sulfate on growth, condition and feeding indices in Channel catfish (Ictalurus punctatus). Aquaculture 253: 343-349.

8. Di Giulio RT, Hinton DE (2008) The toxicology of fishes. Taylor and Francis Group, United States of America.

9. Tucker CC, Robinson EH (1990) Channel catfish farming handbook. Springer.

10. Boyd CE (2005) Copper treatments control phytoplankton Global Aquaculture Advocate.

11. Magurran AE (2005) Evolutionary ecology: the Trinidadian guppy. Oxford University Press.

12. AOAC (2002) Official methods of analysis. (16thedn). Association of Official Analytical Chemists., Arlington, VA, USA.

13. Perkin-Elmer (1964) Analytical methods for atomic absorption spectrophotometry. Corporation, USA.

14. Bruland KW, Franks RP (1979) Sampling and analytical methods for the determination of copper, cadmium, zinc, and nickel at the Nanogram per liter level in sea water. Analytica Chimica Acta 105: 233-245.

15. Yengkokpam S, Sahu N, Pal A, Mukherjee S, Debnath D (2007) Gelatinized carbohydrates in the diet of Catla catla Fingerlings: effect of levels and sources on nutrient utilization, body composition and tissue enzyme activities. Australasian Journal of Animal Sciences 20: 89.

16. Khodadoust A, Rasta M, Khara H, Rahbar M (2013) Determination of Some Biometry and Fecundity Indicators in Female Khramulia (Capoeta capoeta gracilis, Keyserling 1861) in the Sefidroud River. World Journal of Fish and Marine Sciences 5: 392-397.

17. Ling S, Hashim R, Kolkovski S, Chong Shu-Chien A (2006) Effect of varying dietary lipid and protein levels on growth and reproductive performance of female swordtails Xiphophorus helleri (Poeciliidae). Aquaculture Research 37: 1267-1275.

18. Lin YH, Shie YY, Kent M, Shiau SY (2010) Dietary copper requirements of juvenile grouper, Epinephelus malabaricus, with an organic copper source. Aquaculture 310: 173-177.

19. Eisler R (1998) Copper hazards to fish, wildlife, and invertebrates: A synoptic review. DTIC Document.

20. Sorensen EM (1991) Metal poisoning in fish. CRC press.

21. Gatlin DM, Wilson RP (1986) Dietary copper requirement of fingerling channel catfish Aquaculture 54: 277-285.

22. Shiau S, Ning Y (2003) Estimation of dietary copper requirements of juvenile tilapia, Oreochromis niloticus x O-aureus. Animal Science 77: 287-292.

23. Handy RD (1992) The assessment of episodic metal pollution. II. The effects of cadmium and copper enriched diets on tissue contaminant analysis in rainbow trout (Oncorhynchus mykiss). Arch Environ Contam Toxicol 22: 82-87.

24. Cousins RJ (1985) Absorption, transport, and hepatic metabolism of copper and zinc: special reference to metallothionein and ceruloplasmin. Physiol Rev 65: 238-309.

25. Lorentzen M, Maage A, Julshamn K (1998) Supplementing copper to a fish meal based diet fed to Atlantic salmon parr affects liver copper and selenium concentrations. Aquaculture Nutrition 4: 67-72.

26. Slatinská I, Smutná M, Havelková M, Svobodová Z (2008) Review article:

biochemical markers of aquatic pollution in fish - glutathione s-transferase. Folia veterinaria 52: 129-134.

27. Dethloff GM, Schlenk D, Khan S, Bailey HC (1999) The effects of copper on blood and biochemical parameters of rainbow trout (Oncorhynchus mykiss). Arch Environ Contam Toxicol 36: 415-423.

28. Rawles S, Kocabas A, Gatlin DM, Du W, Wei C (1998) Dietary Supplementation of Terramycin and Romet-30 Does Not Enhance Growth of Channel Catfish But Does Influence Tissue Residues. Journal of the World Aquaculture Society 28: 392-401.

29. Truchot J, Rtal A (1998) Effects of long-term sublethal exposure to copper on subsequent uptake and distribution of metal in the shore crab Carcinus maenas. Journal of Crustacean Biology.

30. Kjartansson H, Fivelstad S, Thomassen JM, Smith MJ (1988) Effects of different stocking densities on physiological parameters and growth of adult Atlantic salmon (Salmo salar L.) reared in circular tanks. Aquaculture 73: 261-274.

31. Farat O, Cogun HY, Yuzereroglu TA, Gok G, Firat O, et al. (2011) A comparative study on the effects of a pesticide (cypermethrin) and two metals (copper, lead) to serum biochemistry of Nile tilapia, Oreochromis niloticus. Fish Physiol Biochem 37: 657-666.

32. Griffin BR, Davis KB, Schlenk D (1999) Effect of simulated copper sulfate

therapy on stress indicators in channel catfish. Journal of Aquatic Animal Health 11: 231-236.

33. Berntssen MH, Lundebye AK, Maage A (1999) Effects of elevated dietary copper concentrations on growth, feed utilisation and nutritional status of Atlantic salmon (Salmo salar L.) fry. Aquaculture 174: 167-181.

34. Murai T, Andrews JW, Smith Jr RG (1981) Effects of dietary copper on channel catfish. Aquaculture 22: 353-357.

35. Harris J, Bird DJ (2000) Modulation of the fish immune system by hormones. Vet Immunol Immunopathol 77: 163-176.

36. Pulsford A, Crampe M, Langston A, Glynn P (1995) Modulatory effects of disease, stress, copper, TBT and vitamin E on the immune system of flatfish. Fish and Shellfish Immunology 5: 631-643.

37. Irianto A, Austin B (2002) Probiotics in aquaculture. Journal of Fish Diseases 25: 633-642.

38. Van Heerden D, Vosloo A, Nikinmaa M (2004) Effects of short-term copper exposure on gill structure, metallothionein and hypoxia-inducible factor-1alpha (HIF-1alpha) levels in rainbow trout (Oncorhynchus mykiss). Aquat Toxicol 69: 271-280.

39. Evans DH (1987) The fish gill: site of action and model for toxic effects of environmental pollutants. Environ Health Perspect 71: 47-58.

13

Effect of Flow Rate and Length of Gully on Lettuce Plants in Aquaponic and Hydroponic Systems

El-Sayed G Khater* and Samir A Ali

Agricultural Engineering Department–Faculty of Agriculture–Benha University 13736, Egypt

Abstract

The main objective of this research is to study the effect of source of nutrients, water flow rate and length of gully to know the possibility of producing lettuce plants depending on the nutrients existing in effluent fish farm as compared with the lettuce production using standard nutrient solutions. To achieve that was studied the effect of source of nutrients (effluent fish water and nutrient solution), flow rate (1.0, 1.5 and 2.0 L min^{-1}) and length of gully (2, 3 and 4 m) on the following parameters: nutrient uptake, dry weight and NO_3-N content in plant. The obtained results indicated that the fresh and dry weight of shoots increased in nutrient solution over those of effluent fish farm. The fresh and dry weight of shoots decreased with increasing the flow rate and the length of gully. The dry weight of roots increased in nutrient solution over those of effluent fish farm. The dry weight of roots decreased with increasing the flow rate and the length of gully. The NO_3-N content significantly increased in nutrient solution over those of effluent fish farm. The NO_3-N content decreased with increasing the flow rate and length of gully. The NO_3/protein ratio increased in nutrient solution over those of effluent fish farm.

Keywords: Aquaponics; Hydroponics; Aquaculture; Fish farm

Introduction

Aquaponics is the integration of aquaculture (fish farming) and hydroponics (growing plants without soil). In aquaponic system the fish consume food and excrete waste primarily in the form of ammonia. Bacteria convert the ammonia to nitrite and then to nitrate [1-14].

Aquaponics has several advantages over other recirculating aquaculture systems and hydroponic systems that use inorganic nutrient solutions. The hydroponic component serves as a biofilter, and therefore a separate biofilter is not needed as in other recirculating systems. Aquaponic systems have the only biofilter that generates income, which is obtained from the sale of hydroponic produce such as vegetables, herbs and flowers [15].

Small proportion of ammonia is toxic to fish, when as nitrate is not toxic to fish. If nitrate increased over a specific limit it will be toxic to fish eaters (human being) and cause nitrate pollution and the eaters will suffer from methamoglobnia disease. The blood of the affected people became brown and will not be able to carry oxygen to the rest of human organs [16]. To avoid this problem in aquaculture, part of water should be discharged daily and add fresh water instead. Another solution to this problem is establishing hydroponic system attached to the aquaculture and cultivates plants in the hydroponics in order to save discharged-water and gets use of existing nitrate.

Benefits of aquaponics are conservation of water resources and plant nutrients, intensive production of fish protein and reduced operating costs relative to either system in isolation. Water consumption in integrated systems including tilapia production is less than 1% of the required in pond culture to produce equivalent yields [4].

Lettuce is one of the best crops for aquaponic systems because it can be produced in a short period and, as a consequence, pest pressure is relatively low. Unlike tomato and cucumber, a high proportion of the harvested biomass is edible. With lettuce, income per unit area per unit time is very high. Other fast growing and high income generating crops are herbs such as basil and chive, which are being grown commercially in aquaponic systems [15].

The objective of the current investigation was to study the effect

of source of nutrients, water flow rate and length of gully to know the possibility of producing lettuce plants depending on the nutrients existed in effluent fish farm as compared with the lettuce production using standard nutrient solutions.

Materials and Methods

The experiment was carried out at El-Nenaiea farm, Ashmon, El-Minufiya Governorate, during the period of February to April, 2013.

System description

Figure 1 illustrates the experimental setup. It shows the recirculating aquaculture system which consists of fish tanks, screen filter, biological filter, oxygen generator, oxygen maxing and hydroponic units.

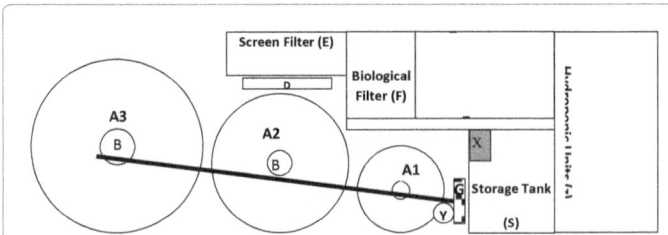

Figure 1: Fish tank, A; particle trap, B; channel collector, D; screen filter, E; biological filter, F; storage tank, S; pumps, G; heat exchanger, X; Downflow oxygen contactor, Y; Hydroponic units, Z.

*Corresponding author: El-Sayed G. Khater, Agricultural Engineering Department – Faculty of Agriculture – Benha University 13736, Egypt
E-mail: alsayed.khater@fagr.bu.edu.eg

The system consists of three circular concrete tanks were used for fish culture. Dimensions of tanks are (5 m diameter x 1.25 m height), (8 m diameter x 1.25 m height), and (10 m diameter x 1.25 m height). The water volumes used in tanks were 25, 50, and 100 m³ respectively. Each tank was provide to a particle trap in the center for water drain waste solids, settleable solids flow under a plate, in a flow of water that amounts to only 5 percent of the total flow leaving the center of the tank. The larger flow (95 percent of the total) exits the tank through a larger discharge strainer mounted at the top of the particle trap.

The drum screen filter used in this system which has dimensions was 1.7 m in diameter and 2.0 m long. The filter was made from stainless steel at private company for steel industry. The fine mesh silk 60 micron was used a media of screening.

Rotating Biological Contactor (RBC) used in this system, has 1.5 m in diameter and 2.0 m long. The filter was made from stainless steel. Used polyethylene tubes were used as a media. The filter was driven by one motor of 1.5 kW power and 1500 rpm and a gearbox to reduce the speed 500 times to give the recommended rotating speed (3 rpm).

Pure oxygen used in this system source of oxygen gas was oxygen generator. Adding pure oxygen gas to water by oxygen mixer. The water and oxygen enter the top of the oxygen mixer, as the water and oxygen move downward.

The hydroponic units (NFT) in this study consisted of two sources of nutrient solution were used (1) Stock nutrient solution and (2) Effluent fish farm, three lengths of gully (2, 3 and 4 m) and three water flow rates 1, 1.5 and 2 L min⁻¹. Intermitted flow (1 minute 'on' and 4 minute 'off') as described by [17].

Figure 2 shows the design of hydroponic units. The gullies were 50 cm wide, slope 2% and stand 1 m high above the ground with row spacing of 20 cm. The gullies frames were made from iron, lined by plastic sheet and covered with foam boards to support the plants.

The solution was pumped from the tank to the upper ends of the gullies. Small tubes were used to supply each gully by nutrient solution or effluent fish farm. Nutrient solution is circulated in closed system. The tank of the nutrient solution system with a capacity of 200 liter capacity was used for collecting the drained solution by gravity from the ends of the gullies. The nutrient solutions were prepared manually once per ten days dissolving appropriate amounts of $Ca(NO_3)_2$, KNO_3, K_2SO_4, KH_2PO_4, $MgSO_4$ and chelates for trace elements into preacidified groundwater, pH was further adjusted to 6.0-7.0 after salt addition.

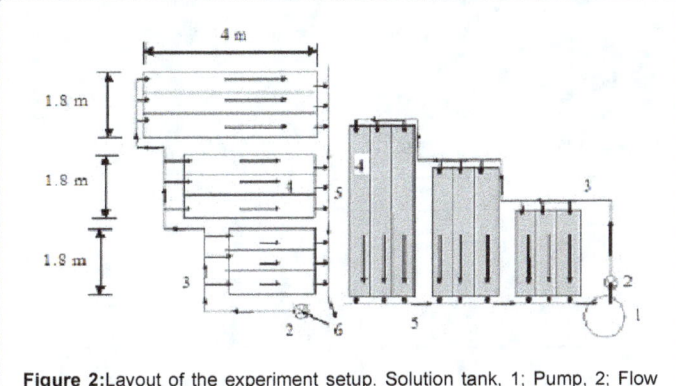

Figure 2: Layout of the experiment setup. Solution tank, 1; Pump, 2; Flow pipe, 3; Gullies, 4; Catchment pipe, 5; Water farm, 6; ▬ Nutrient solution ▭ Effluent fish water.

Lettuce germination

Lettuce seeds (*Romein type*) were sown in the plastic cups (5 cm diameter and 5 cm height) filled with peatmoss. The cups were watered daily using water with nutrient solution. Three weeks old lettuce seedlings were planted in the experimental trays [18].

Measurements

Water samples were taken, at inlet and outlet of the hydroponic units for measured Ammonia (NH_3), Nitrite (NO_2), Nitrate (NO_3), Phosphorus (P), Potassium (K), Calcium (Ca) and Magnesium (Mg). Ammonia (NH_3), Nitrite (NO_2), Nitrate (NO_3) and Phosphorus (P) measured by a Spekol 11 (Model SPEKOL 11–Range 0.1–1000 concentration ± 1 nm λ, UK). Potassium (K) measured by flame photometer (Model Jenway PFP7–Range 0.1–999.9 ppm ± 0.2 ppm, USA). Calcium (Ca) and Magnesium (Mg) measured by using disodium versenate method as described by [19]. The dry weight was measured at the end of the experiment. After measured fresh weight the plants were oven dried at 70°C until constant weight was reached. The NO_3-N content was evaluated after being digested. Nitrate (NO_3-N) content was measured by using salsalic acid as described by [20].

Statistical analysis

The statistical analysis for the data obtained was done according to [21] and the treatments were compared using Least Significant Differences (LSD) test at 99% confidence level [22].

Results and Discussion

Nutrients uptake

Any removal of nutrients from the solution can be equated with uptake by plants, provided that the system is free from leaks, algae and regardless of precipitation. Figures 3a-3e show N, P, K, Ca and Mg uptake by lettuce plants at the end of the growing period. The nutrients uptakes were significantly increased in nutrient solution over those of effluent fish water. The N, P, K, Ca and Mg uptakes were 300.46, 69.01, 434.86, 153.66 and 254.78 mg plant⁻¹, respectively in nutrient solution

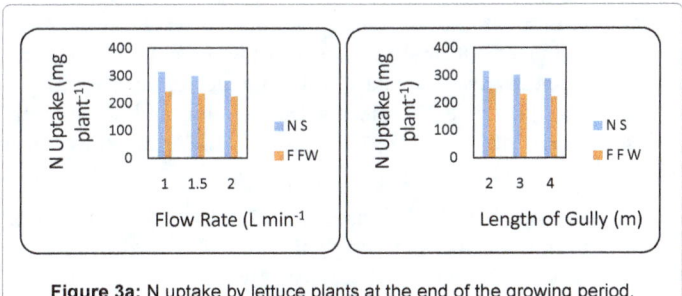

Figure 3a: N uptake by lettuce plants at the end of the growing period.

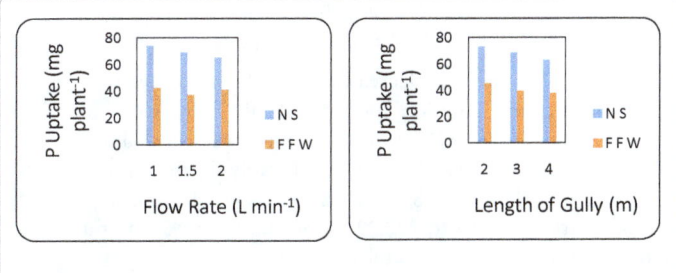

Figure 3b: P uptake by lettuce plants at the end of the growing period.

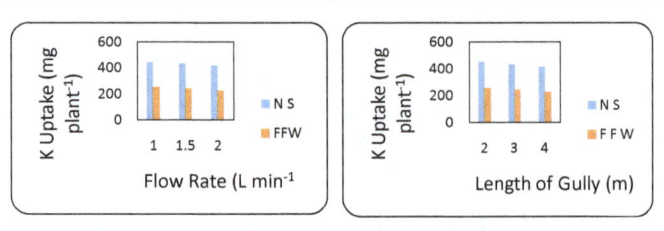

Figure 3c: K uptake by lettuce plants at the end of the growing period.

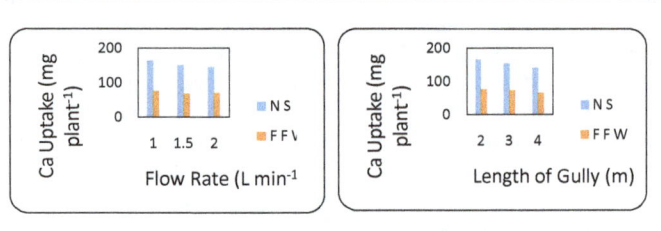

Figure 3d: Ca uptake by lettuce plants at the end of the growing period.

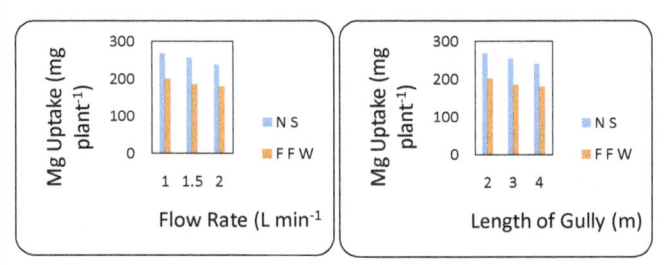

Figure 3e: Mg uptake by lettuce plants at the end of the growing period.

and the N, P, K, Ca and Mg uptakes were 233.73, 40.70, 241.34, 71.08 and 188.78 mg plant^{-1}, respectively in effluent fish water. The nutrients uptakes were decreased with increasing the flow rate and length of gully. N uptake significantly decreased from 316.22 to 282.62 mg plant^{-1} (10.63%) in nutrient solution and significantly decreased from 242.78 to 222.72 mg plant^{-1} (8.26%) in effluent fish water at 1 and 2 L min^{-1} flow rate, respectively. N uptake significantly decreased from 316.22 to 288.48 mg plant^{-1} (8.77%) in nutrient solution and significantly decreased from 250.18 to 221.95 mg plant^{-1} (11.28%) in effluent fish water at 2 and 4 m length of gully, respectively.

P uptake significantly decreased from 74.21 to 65.47 mg plant^{-1} (11.78%) in nutrient solution and significantly decreased from 42.62 to 41.09 mg plant^{-1} (3.59%) in effluent fish water at 1 and 2 L min^{-1} flow rate, respectively. P uptake significantly decreased from 73.34 to 63.17 mg plant^{-1} (13.87%) in nutrient solution and significantly decreased from 45.31 to 38.21 mg plant^{-1} (15.70%) in effluent fish water at 2 and 4 m length of gully, respectively.

K uptake significantly decreased from 444.58 to 416.16 mg plant^{-1} (6.39%) in nutrient solution and significantly decreased from 250.46 to 225.98 mg plant^{-1} (9.77%) in effluent fish water at 1 and 2 L min^{-1} flow rate, respectively. K uptake significantly decreased from 456.00 to 418.37 mg plant^{-1} (8.25%) in nutrient solution and significantly decreased from 258.14 to 228.29 mg plant^{-1} (11.56%) in effluent fish water at 2 and 4 m length of gully, respectively.

Ca uptake significantly decreased from 164.06 to 145.80 mg plant^{-1}

(11.13%) in nutrient solution and significantly decreased from 76.51 to 70.04 mg plant^{-1} (8.46%) in effluent fish water at 1 and 2 L min^{-1} flow rate, respectively. Ca uptake significantly decreased from 165.70 to 141.51 mg plant^{-1} (19.59%) in nutrient solution and significantly decreased from 75.84 to 63.71 mg plant^{-1} (14.04%) in effluent fish water at 2 and 4 m length of gully, respectively.

Mg uptake significantly decreased from 268.72 to 238.91 mg plant^{-1} (11.09%) in nutrient solution and significantly decreased from 200.22 to 179.97 mg plant^{-1} (10.11%) in effluent fish water at 1 and 2 L min^{-1} flow rate, respectively. Mg uptake significantly decreased from 268.99 to 240.68 mg plant^{-1} (10.52%) in nutrient solution and significantly decreased from 201.70 to 179.68 mg plant^{-1} (10.92%) in effluent fish farm at 2 and 4 m length of gully, respectively.

The lowest values of plant consumption were found in treatment of effluent fish water at a flow rate of 2 L min^{-1} with length of gully 4 m and the highest values were recorded at a flow rate of 1 L min^{-1} with length of gully 2 m. Increasing the velocity of water in gullies with increasing the flow rate decreased the rate of nutrient consumption. These results agreed with those obtained by [23-25].

The lowest values of plant consumption were found at 4 m length of gully and the highest values were found at 2 m length of gully. This may be due to the number of plants in case of 4 m length as compared with those of 2 m length. Worthy to note that pumping either nutrient solution or effluent fish water to the growing gullies was adjusted at 1 min pumping and 4 m rest. This was performed with 1.0, 1.5 and 2 L min^{-1} discharge in 2, 3 and 4 m of the gullies. Thus, the nutrient stayed longer under the 4 m length and the total intake periods of nutrients were longer than those achieved with the shorter gullies (2 m). The refreshment of nutrients under the longer gullies were restricted as compared with the shorter ones. These results agreed with those obtained by [26].

Fresh and dry weight

Fresh and dry weight of shoots: Figures 4a and 4b show the effect of source of nutrients, flow rates and lengths of gully on the fresh and dry weights production of lettuce plants at the end of growing period. The fresh and dry weights of shoots w significantly increased in nutrient solution over those of effluent fish water. The fresh weights of shoots were 207.12 and 190.32 g plant^{-1} for nutrient solution and effluent fish water, respectively. The dry weights of shoots were 21.34 and 19.74 g

Figure 4a: Fresh weight of shoots at the end of the growing period.

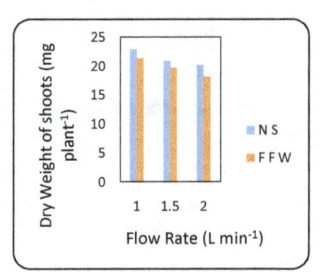
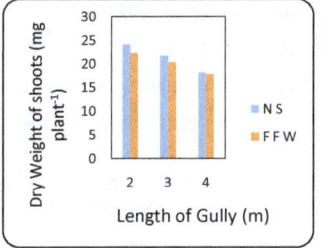

Figure 4b: Dry weight of shoots at the end of the growing period.

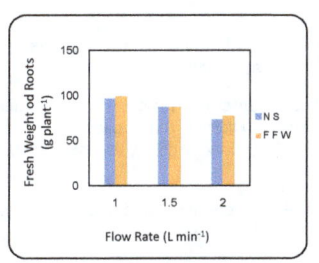
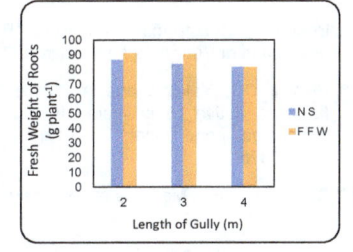

Figure 5a: Fresh weight of roots at the end of the growing period.

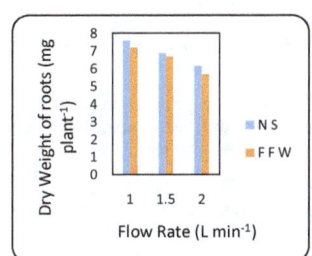
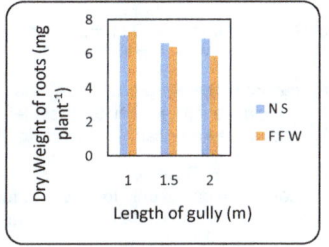

Figure 5b: Dry weight of roots at the end of the growing period.

plant^{-1} for nutrient solution and effluent fish water, respectively. The highest value of fresh and dry weights of 215.88 and 26.36 g plant^{-1} was obtained with a flow rate of 1.0 L min^{-1} and 2 m length of gully, while, the lowest value of fresh and dry weights of 170.68 and 19.05 g plant^{-1} occurred at a flow rate of 1.5 L min^{-1} and 4 m length of gully in nutrient solution. On the other hand, the highest value of fresh and dry weights of 217.18 and 23.72 g plant^{-1} was obtained at a flow rate of 1.0 L min^{-1} with 2 m length of gully, while, the lowest value of fresh and dry weights 163.46 and 15.99 g plant^{-1} was obtained at a flow rate of 2.0 L min^{-1} with 4 m length of gully in effluent fish water.

Decreasing the fresh and dry weights of shoots with increasing the flow rate from 1 to 2 L min^{-1} and increased the length of gully from 2 to 4 m may be attributed to decrease in nutrient consumption rate. These results agreed with those obtained by [27] who found that the dry weight decreased with increasing the flow rate from 0.5 to 1.5 L hour^{-1}.

Fresh and dry weight of roots

Figures 5a and 5b show the fresh and dry weights of roots production of lettuce plants at the end of growing period (50 days). The fresh and dry weights of roots were significantly increased in nutrient solution over those of effluent fish water. The fresh and weight of roots was 83.70 and

87.29 g plant^{-1} for nutrient solution and effluent fish water, respectively. The fresh and dry weights of roots were 6.89 and 6.52 g plant^{-1} for nutrient solution and effluent fish water, respectively. The fresh and dry weights were decreased with increased the flow rate and length of gully. The fresh and dry weights significantly decreased from 96.13 to 72.51 g plant^{-1} (24.57%) and 7.59 to 6.17 g plant^{-1} (18.71%), respectively, in nutrient solution and significantly decreased from 98.46 to 76.68 g plant^{-1} (22.12%) and 7.20 to 5.69 g plant^{-1} (20.20%), respectively, in effluent fish water at 1 and 2 L min^{-1} flow rate, respectively. The fresh and dry weights significantly decreased from 86.13 to 81.55 g plant^{-1} (5.32%) and 7.10 to 6.92 g plant^{-1} (25.35%), respectively, in nutrient solution and significantly decreased from 90.57 to 81.32 g plant^{-1} (10.21%) and 7.29 to 5.87 g plant^{-1} (19.48%), respectively, in effluent fish water at 2 and 4 m length of gullies, respectively.

Decreasing the fresh and dry weights of roots with increasing the flow rate from 1 to 2 L min^{-1} and increasing the length of gully from 2 to 4 m may be attributed to decrease in nutrient consumption rate.

Furthermore, the fresh and dry weights of roots were higher in nutrient solution than in effluent fish water. This helps to explain differences yield and growth of root in between various solutions. Generally, Nutrient solution has provided optimum conditions to the root system of the plant with regard to the amount of nutrients available to the roots and their balance in addition to sufficient, oxygen supply, the appropriate osmotic pressure of solution and its temperature. These results agreed with those obtained by [26].

No$_3$-N content in plant

Figure 6 shows the No$_3$-N content by lettuce plants at the end of the growing period as estimated from the dry weight of the entire plant and nutrient concentration. The No$_3$-N content was significantly increased in nutrient solution over those of effluent fish water. The No$_3$-N content was 227.23 and 107.81 g plant^{-1} for nutrient solution and effluent fish water, respectively. The No$_3$-N content was decreased with increasing the flow rate and length of gully. The No$_3$–N content significantly decreased from 241.07 to 208.62 mg plant^{-1} (13.46%) in nutrient solution and significantly decreased from 113.55 to 96.74 mg plant^{-1} (14.80%) in effluent fish farm at 1 and 2 L min^{-1} flow rate, respectively. The No$_3$-N content significantly decreased from 245.82 to 212.05 mg plant^{-1} (13.74%) in nutrient solution and significantly decreased from 115.66 to 103.17 mg plant^{-1} (10.80%) in effluent fish farm at 2 and 4 m length of gully, respectively.

That is to say, No$_3$–N content in plants grown in nutrient solution was almost 2 times of that in those grown in effluent fish water. This may be due to high concentrations of nutrient solution as compared with those of effluent fish farm.

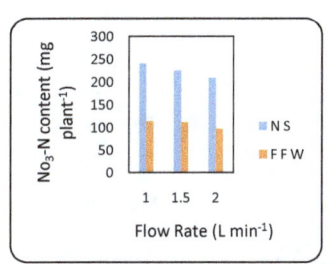
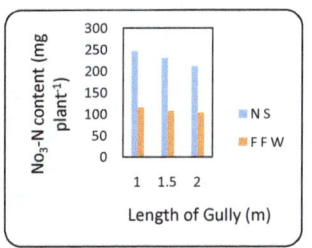

Figure 6: No$_3$-Ncontent in plant at the end of the growing period.

Conclusions

The experiment was carried out to study the effect of source of nutrient, flow rate and length of gully on the following parameters: nutrients uptake, dry weight of shoots and roots and NO_3-N content in plant. The treatments under study are: source of nutrient (waste fish farm and stock nutrient solution), flow rate (1.0, 1.5 and 2.0 L min^{-1}) and length of gully (2, 3 and 4 m).The obtained results can be summarized as follows:

- The nutrients uptakes were decreased with increasing the flow rate and the length of gully. The total nutrients uptake values were higher in nutrient solution than those in effluent fish water.

- The fresh weight of shoots and roots were decreased with increasing the flow rate and the length of gully. The fresh weight of shoots was higher in nutrient solution than in effluent fish water by 8.09%.

- The dry weight of shoots and roots were decreased with increasing the flow rate and the length of gully. The dry weight of shoots and roots were higher in nutrient solution than in effluent fish water by 11.64 and 6.68%, respectively.

- The NO_3-N content was decreased with increasing the flow rate and the length of gully. The NO_3-N content was higher in plants grown in nutrient solution than those of plants grown in effluent fish water by 110.79%.

The best flow rate for 2 m length of gully was 1.5 L min^{-1}, the best flow rate for 3 m length of gully was 1.5 L min^{-1} and the best flow rate for 4 m length of gully was 2 L min^{-1}.

It could be concluded that aquaponic system is one of the economical solutions for getting benefits from the water-waste from the fish farms as it save nutrients and produce fresh vegetables, i.e., lettuce. With using this system successively its cost will be decreased and became more economic. The produced plants via this system considered as an organic product which is more safe for human consumption.

References

1. Johnson DM, Wardlow GW (1997) A prototype recirculating aquaculture–hydroponic system.

2. Diver S (2000) Aquaponics–integration of hydroponics with aquaculture.

3. Bromes B (2002) Aquaponics.

4. Rakocy JE (2002) Aquaponics: vegetable hydroponics in recirculating systems.

5. Selock D (2003) An introduction to aquaponics: the symbiotic culture of fish and plants.

6. Lee CS (2004) Aquaponics, An integrated fish culture and vegetable hydroponics production system.

7. Okimoto DK (2004) Aquaponics export conducts workshops in American Samoa

8. Karen I (2005) Aquaponics from global aquatics turning waste into profits.

9. Nelson RL (2006a) Aquaponics–Hydroponics-Aquaculture.

10. Nelson RL (2006b) Information on aquaponics and aquaculture.

11. Nelson RL (2006c) The source for information on aquaponics and aquaculture.

12. Nelson RL (2008) Aquaponic Food Production. Nelson and Pade Inc. Press, USA.

13. Graber A, Junge R (2009) Aquaponic systems: nutrient recycling from fish wastewater by vegetable production. Desalination 246: 147-156.

14. Suits B (2010) Access to personal agriculture.

15. Rackocy JE, Hargreaves JA (1993) Integration of vegetable hydroponics with fish culture: a review.

16. Tucker CS, Boyd CE (1985) Water quality, channel catfish culture. Elsevier Sci. Publ. Co., Amesterdam, Netherlands.

17. Benoit F, Ceustermans N (1989) Recommendations for the commercial production of Butterhead lettuce in NFT. Soilless Culture 5: 1-12.

18. Sikawa DC, Yakupitiyage A (2010) The hydroponic production of lettuce (Lactuca sativa L) by using hybrid catfish (Clarias macrocephalus × C. gariepinus) pond water: Potentials and constraints. Agri Water Manag 97: 1317-1325.

19. Black CA (1965) Methods of soil analysis. Part 2, chemical and microbiological properties.

20. Chapman HD, Partt FP (1961) Methods of analysis of soils, plant and water.

21. Snedecor GW, Cochran WG (1980) Statistical methods. Iowa State Univ. Press, Ames, Iowa, USA.

22. Gomez KA (1984) Statistical procedures for agricultural research. John Wiely and sons, New York, USA.

23. Graves CJ, Hurd RG (1983) Intermittent solution circulation in the nutrient film technique. Acta Hort 133: 47-52.

24. Rackocy JE, Hargreaves JA, Bailey DS (1993) Nutrient accumulation in a recirculating aquaculture system integrated with vegetable hydroponic production.American Society of Agricultural Engineering,USA.

25. Rackocy JE, Bailey DS, Shultz KA, Cole WM (1997) Evalution of a commercial-scale aquaponic unit for the production of tilapia and lettuce. International Symposium on Tilapia in Aquaculture, USA.

26. Khater EG (2006) Aquaponics: the integration of fish and vegetable culture in recirculating systems.

27. Fahim MM (1989) Design of nutrient film system for agriculture under green-house.

14

Economic Analyze of Costs and Return of Fish Farming in Saki-East Local Government Area of Oyo State, Nigeria

Adeniyi Bashir Tunde[1]*, Kuton MP[2], Ayegbokiki Adedayo Oladipo[3] and Lawal Hakeem Olasunkanmi[4]

[1]*Aquaculture and Fisheries Research Programme, Institute of Food Security, Environmental Resources and Agricultural Research (IFSERAR), Federal University of Agriculture, Abeokuta, Nigeria*
[2]*Department of Marine Sciences, University of Lagos, Akoka, Lagos, Nigeria*
[3]*Food Security and Socio-Economic Research Programme, Institute of Food Security, Environmental Resources and Agricultural Research (IFSERAR), Federal University of Agriculture, Abeokuta, Nigeria*
[4]*Department of Agricultural Economics and Farm Management, Federal University of Agriculture, Abeokuta, Nigeria*

Abstract

Aquaculture is regarded as a lucrative and important endeavour in terms of income generation and supply of animal protein to the majority of population in the country. The study examined economic analysis of fish farming in Saki-East Local Government Area (LGA) of Oyo State, Nigeria. Structured questionnaire was administered to randomly selected respondents to represent the fish farming community in the study area. Data collected were analyzed using descriptive statistics, costs and budgetary analysis and multiple regression analysis.

The results of a Cost and Return Analysis of the fish farming in the study area showed that the total revenues was N244364.30 k per cycle, whereas total cost was N129379.52 k per cycle. This implies that fish farming was profitable and is expected to continue to operate. In addition, Benefit Cost Ratio (BCR) was 1.9, the fish farming is therefore considered to be profitable. The rate of Return on Investment was 0.8887, meaning, for every N1 invested; there will be a return of 88.8 k.

Keywords: Economic analysis; Fish production; Revenue; Total cost; Profitability

Introduction

Aquaculture is the practice of rearing, growing or producing products in water or in managed water systems. Products of aquaculture include plants, insects, crustaceans, bivalves and pearls, fish, and anything else grown in water. Mires, also defined aquaculture as the commercial rearing of fish in conditions where all the basic means of production can be controlled within their respective limitations and from which producers aim to obtain optimal economical results.

Fish farming, an aquaculture practice, which involves the raising of fish, has a long history. Fish were raised in ponds in the ancient civilizations of Egypt and Mesopotamia around 3000BC and carp have been reared in china since about 2000BC. During the middle Ages in Europe, fish ponds were of great important for the supply of fresh during lent and of fast days, and fish farming was generally associated with the monasteries. Fish farming almost disappeared in England after the reformation and in Germany during the thirty years war. However, there was a considerable revival in Germany and central and Eastern Europe during the last century.

Today China is by far the most important producer of farmed fish, accounting for over half the total world output. Aquaculture was introduced to Nigeria in the late 1930's and early 1940s, with the first fish pond built around Onikan, Lagos in 1941. The trend of development followed the same pattern as in other African countries, which have not been very encouraging. According to Fagade [1] in spite of the fact that over 1.5 million hectares of surface water area is available for fish culture, no appreciable result has been recorded in the culture sub-sector.

Fishing like other hunting activities has been a major source of food for the human race and has put an end to unsavoury outbreak of anaemia, Kwashiorkor and so on. Fish accounts for about one fifth of world total supply of animal protein and this has risen five folds over the last forty years from 20million metric tones to 98 million metric tones in 1993. According to Tobor stated that protein of animal origin is in short supply in Nigeria as increase in livestock population is being limited by several causes including virus diseases, drought, scarcity, and high cost of feeds and low genetic potential of indigenous livestock breeds. Short supply and increase in human population have the combined effect of raising the cost of animal protein to a level beyond the reach of low-income groups. The situation has given rise to considerable increase in the demand for fish to supplement animal protein. Also, [2] revealed that fish allows for protein improved nutrition in that it has a high biological value in terms of high protein retention in the body. Furthermore, Slang stated that fish has higher protein assimilation as compared to other animal protein sources, low cholesterol content and one of the safest sources of animal protein. Also, the FAO [3] revealed that West African Countries, of which Nigeria is one, obtain at least 50 percent of their animal protein needs from fish and fish products.

Moreover, Adekoya [4] stated that marine fish products account for at least 60 per cent of the non-plant protein consumed in Nigeria. According to Aromolaran and Akintunde, the residents of Warri in

*Corresponding author: Adeniyi Bashir Tunde, Aquaculture and Fisheries Research Programme, Institute of Food Security, Environmental Resources and Agricultural Research (IFSERAR), Federal University of Agriculture, Abeokuta, Nigeria, E-mail: adeniyibt@funaab.edu.ng

Nigeria preferred fish and actually consumed it more frequently than any other animal protein source. The author found that the average household in the area consumed fish at least once a day for 16 days in one month.

In Nigeria, the population of fish is about 0.7 million metric tons annually which results in a shortfall of about 1.0 million metric tons annually. Only 5% of this 0.7 million metric tone produced locally is from aquaculture. The remaining 95% is from the capture fisheries, which are dominated by the artisanal fishermen. Out of 35 grams of animal protein per day per person recommended by food and agricultural organization, less than 7 grams is consumed on the average [5]. As a result of this, many Nigerians suffer from protein deficiency due to low protein intake.

Economic importance of aquaculture and fishery management in Nigeria

1. It increases food production, especially of animal proteins, and achieving self-sufficiency in aquatic products supplies.

2. It contributes to improvement of human nutrition.

3. It generates new source of employment in rural areas, including part-time employment of farmers and small-scale fishermen, and arresting the migration of people from rural to urban areas.

4. Earning foreign exchange through export or saving foreign exchange through import substitution.

5. It promotes agro-industrial development, which could include processing and marketing of fishery products, feeds and equipment for aquaculture, and seaweed culture for the production of marine colloids, pearl oyster culture etc.

6. It creates and maintains leisure-time activities, including sport fishing and home and public aquaria.

7. Overall development of rural areas through integrated projects, including aquaculture.

Aquaculture management system

Fish farming may range from large scale industrial enterprises to 'backyard' subsistence ponds. Farming systems can be distinguished in terms of input levels.

1. Extensive system: In extensive fish farming, (economic) inputs are usually low. Natural food production plays a very important role, and ponds productivity is relatively low. Fertilizer may be used to increase pond fertility and thus fish production.

2. Semi-intensive system: a moderate level of inputs is used and fish production is increase by the use of fertilizer and/or supplementary feeding. This means higher labour and food costs but higher fish yields more than compensate for this usually.

3. Intensive system: a high level of inputs is used and the ponds are stocked with as many fish as possible. The fish are fed supplementary food, and natural food production play a minor role [6]. In this system, the high feeding costs and risks, due to high fish stocking densities and thus increased susceptibility to diseases and dissolved oxygen shortage, can become difficult management problems. Because of the high production costs you are forced to fetch a high market price in order to make the fish farming economically feasible.

Problem statement

Nigeria has become one of the largest importers of fish in the developing world, importing about 600,000 metric tons annually. To solve the country's problem of high net supply deficit for fish, Nigeria must turn to their under-utilized inland water for improved fish production and aquaculture. Aquaculture expansion, moreover, as been a slow process, as private sector fish farmers have faced major constraints, including lack of seed and quality feed, limited access to institutional credit, lack of skilled personnel as well as inadequate database on biological and ecological requirements of indigenous species with aquaculture potential [7].

Since fish is the most perishable among the good quality food protein sources and usually the cheapest source of animal protein. It is therefore necessary to utilize our limited scarce resources judiciously. In spite of the ever-increasing growth being witnessed by other major sources of animal protein such as livestock and poultry industries, this problem of protein deficiency has continued unabated. The need therefore arise, to explore aquaculture as a means of addressing the problem of protein inadequate protein intake in Nigeria.

The study attempted to achieve the following objectives

a.   To describe the socio-demographic characteristics of fish farmers in the study area;

b.   To determine the cost, profitability and viability of fish farming;

c.   To determine the factors influencing fish production on the study areas;

d.   To make relevant recommendations based on the findings of the study.

Methodology

Saki-East local government was created alongside 182 others created by the federal government during General Sanni Abacha administration in December, 1996. The council was carved out of the defunct Ifedapo local government area. Saki-East comprises of five major towns namely: Agbonle, Ago-Amodu, Ogbooro, Oje-Owode, and Sepeteri. The zoological Old Oyo National Park is located in Sepeteri and the landmass is about 200 km² while the terrain is undulating and paling Savannah grassland with dotted rocks and hills. The common trees are sheabutter and locust bean trees [8]. The local government council is bounded in the north by Oorelope local government, to the south by Atisbo and Olorunsogo to the East and Saki-West to the west. Saki-East local government covers a land area of 1,569 km² and a population of 110,223 based on 2006 census [9].

Method of data collection

Saki-East local government was purposely selected as the study area based on the convenience and familiarity of the researchers to the environment; this was also done to reduce research cost [10]. Primary data were obtained through a random sampling of 50 active fish farmers across the five major towns of Saki-East local government namely: Agbonle, Ago-Amodu, Ogbooro, Oje-Owode, and Sepeteri using structured questionnaires on various aspects of fish farming such as socio-demographic characteristics of the farmer, production income and financial investment. Ten fish farmers were randomly selected in each of the five major towns in the study area.

Analytical techniques

To satisfy the objectives of the study, the following analytical tools were used:

a.   Descriptive analysis such as frequency and percentage were used to describe socio-demographic characteristics of fish farmers.

b.   Cost and returns analysis were used to examine the cost and returns of fish farming.

c.   Profitability rations such as Benefit-Cost Ratio, Return on investment were used to analyze profitability and viability of fish farming.

BCR = TR/TC

Where;

BCR = Benefit Cost Ratio

TR = Total Revenue

TC = Total Cost

Return on Investment (ROI) = Profit/Total Cost

d.   Regression Techniques was employed to identify factors that influence fish production in the study area.

$Y = f (X1, X2, X3,...X8, U)$

where:

Y = total value of production

X1 = age of farmers in years

X2 = size of pond (m²)

X3 = total fish seed (naira)

X4 = cost of feed (naira)

X5 = value of total fertilizer and lime input (naira)

X6 = value of total labour input (naira)

X7 = educational status of the farmers

X8 = sex of the farmers (Dummy: 1 = male, 0 = female)

U = error term assume to have a zero mean and constant variance.

The following functional forms were analyzed in order to choose the lead equation:

$Y = b0 + b1X1+b2X2+b3X3+b4X4+b5X5+b6X6+b7X7+b8X8$ (Linear form)

$Y = b0 + lnb0 + b1lnX1 + b2lnX2 + b3lnX3 + b4lnX4 + b5lnX5 + b6lnX6 + b7lnX7 + b8lnX8$

$LnY = b0 + b1lnX1+b2lnX2+b3lnX3+b4lnX4+b5lnX5+b6lnX6+b7lnX7+b8lnX8$ (Double Log)

Results and Discussion

Socio-economic characteristics of the respondents

The Table 1 below shows the socio-economic characteristics of respondents in the study area. The result showed that about 94 percent of the respondents were male while the female were just 3, representing 6 percent of the respondents [11]. This is an indication that the production of fish is widely popular among the males in the study area while the women might likely be more involved in the marketing aspect of the enterprise.

The age distribution of the respondents showed that most of the fish farmers are still in their active youth age; hence, high productivity should be expected in the study area as about 36 percent and 30 percent of the respondents fall between the age range of 20-30 years and 31-40 years respectively.

The marital status of reflects that 38 respondents, representing about 76 percent were married while about 22 percent were still single. A negligible amount no of the respondents, representing about 2 percent was divorced.

As observed in the table below, the minimum household size of the respondent is 1 and the maximum been 15 people [12]. The observation shows that majority of the respondents (about 56 percent) have an average household size of between 1-5 while about 26 percent and 18 percent of the respondents have household size of between 6-10 and 11-15 respectively.

The population of fish farmers with tertiary education is very high in Saki-East Local Government. 52 percent of the farmers have tertiary qualification while 20 percent have secondary education. The high level of education status recorded among the farmers might not be unconnected with the high percentage of civil servants among the respondents as about 32 percent are majorly civil servants who do fish farming as secondary occupation. Result showed that 34 percent of the respondents have fish farming as their major occupation [13]. Trading was also noticed to be the main occupation of some respondents (12 percent) in the study area while 22 percents of the respondents have various other jobs such tailoring, barbing, auto mechanic, etc., as their major occupation.

Cost and return analysis

The cost of various input used in fish production is described in Table 2 below. Gross return is the monetary value that fish farmers derive from fish catch (output), the gross return or total revenues from fish production in the study area was N244,363.30k per cycle, whereas total fixed costs were N39,773.79k, total variable cost was N89605.73k per cycle. The result showed a profit margin of N114984.78K per cycle. This implies that fish farming in Saki-East local government area is a profitable enterprise [14].

Benefit Cost Ratio (BCR) explains how much the owner gets from every N1 expended on the project. Fish farming in Saki-East LGA seems highly feasible since BCR recorded a value of 1.9 [15]. This result is similar to the work carried out by Ngazy titled 'Appraising Aquaculture of ZALA Park fish cultivation and Makoba Integrated Mariculture Pond System' in which the BCR was 1.5. The Return on Investment (ROI) showed a value of 0.888 which implies that for every N1 invested; there will be a return of 88.8 kobo.

Factors influencing fish production in the study area

Table 3 below summarized the regression results showing the parameters that are significance to fish production in the study area for the purpose of capturing objective number four. Estimated double log was chosen as the lead equation from the three functional forms based on the economic criteria (signs of the regression coefficient), statistical criteria (significance of t-value) and the value of R^2.

The results of double log production function models estimate showed six parameters to be positive compared to linear and semi log with five parameters each [16]. The result reveals that only quantity of feed, quantity of lime and respondents sex are significant

Variable	Frequency	Percentage
Sex		
Male	47	94
Female	3	6.0
Total	50	100
Age (Years)		
20-30	18	36.0
31-40	15	30.0
41-50	12	24.0
51-60	4	8.0
60 and above	1	2.0
Total	50	100.0
Marital Status		
Married	38	76.0
Single	11	22.0
Divorced	1	2.0
Total	50	100.0
Household size		
1-5	28	56.0
6-10	13	26.0
11-15	9	18.0
Total	50	100.0
Education status		
No formal education	3	6.0
Primary education	11	22.0
Secondary education	10	20.0
Tertiary education	26	52.0
Total	50	100
Primary occupation		
Farming	17	34.0
Trading	6	12.0
Civil servant	16	32.0
Artisan & others	11	22.0
Total	50	100

Source: Field Survey, 2013.

Table 1: Socio-economic characteristics of respondents in the study area

factors affecting fish farming output in Saki-East LGA at 5 percent level of significance. This implies that a percentage increase in these variables will lead to certain percentage increases in the dependent variable (total value of production) assuming all other factors are held constant. Parameters such as fish seeds, age and level of education

of the respondents were found to be positive, indicating a positive increase with a unit increase in the factors, though not significant at any probability level. These factors also support the findings of Ahmed et al. [5] which concluded that stocking density and pond size are the factors that significantly influenced tilapia output in small water bodies in Bangladesh.

The F-value of 2.09 being significant is also an indicator that the model has a good fit to justify the factors influencing the fish farming operations in the study area. The values of parameter estimates of the double log equation showed an elasticity of production. The summation of the parameter estimates (b1 + b2 +...b8) gives a value of 3.961 indicating that farmers are operating in stage 1 of production function. This suggests that fish production in the study area had an increasing positive return to scale.

The explicit form of the lead equation is;

$$Y = 0.18 + 0.082X1 - 0.374X2 + 0.318X3 + 2.832X4^* - 2.009X5^* - 0.302X6 + 0.737X7 + 2.677X8^* + Ui$$

(0.858) (0.935) (0.711) (0.752) (0.04) (0.041) (0.764) (0.466) (0.05)

$R^2 = 17.5\%$ F-values=2.09   *coefficient significant at 5% level.

Out of all the eight explanatory variable captured in this study, only three (quantity of feed, X4; quantity of lime, X5; and sex, X8) were found significant at 5% level. This means that only these three variables exerted significant influence on the total revenue of fish produced in the study area. Quantity of feed (X4) showed a positive relationship with the total revenue, this implies that for every percentage increase in quantity of feed at a given level of other variable inputs will increase total value of production by 2.83%. Quantity of lime (X5) showed a negative relationship with the level of income, which implies that

ITEM	Total Values (₦ : K)
Total Revenue	244364.30
Fixed Cost	
Land	26656.25
Drag Net	1740.08
Water Pump	2147.86
Weighing Scale	1576.42
Knives/Cutlasses	638.89
Other Fixed Input	7014.29
Total Fixed Costs (TFC)	**39773.79**
Variable Costs	
Labour	19335.02
Fingerlings (Catfish)	32503.40
Fingerlings (Tilapia)	1144.34
Poultry Dung (kg)	9905.71
Maggot (kg)	1050.11
Concentration Diet (kg)	10445.95
Other Feeds (kg)	2801.11
Limes	12420.09
Total Variable Costs (TVC)	89605.73
Total Cost (TFC+TVC)	129379.52
Net profit (1-2)	**114984.78**
Benefit-Cost Ratio (BCR)	1.9
Return on Investment (ROI)	0.8887

Source: Field survey cost analysis.

Table 2: Cost and return analysis of fish production in saki-east local government

	Linear											
Constant	b_1	b_2	b_3	b_4	b_5	b_6	b_7	b_8	R^2	$\bar{R}^2$	F	Sig.
-0.044	0.420	-0.908	0.653	1.018	-0.499	-0.245	0.869	0.702	0.121	-0.05	0.708	.683
(0.965)	(0.676)	(0.369)	(0.518)	(0.314)	(0.621)	(0.807)	(0.39)	(0.486)				
	Semi log											
2.568	0.499	-0.189	0.533	-0.053	-1.055	-0.061	0.378	0.155	0.114	-0.059	0.661	.722
(0.014)	(0.62)	(0.851)	(0.597)	(0.958)	(0.298)	(0.952)	(0.707)	(0.155)				
	Double log											
0.18	0.082	-0.374	0.318	2.832*	-2.009*	-0.302	0.737	2.677*	0.175	0.014	2.09	.039
(0.858)	(0.935)	(0.711)	(0.752)	(0.04)	(0.041)	(0.764)	(0.466)	(0.05)				

Source: Analysis from field data.

Table 3: Regression result on the total value of production

for every percentage increase in quantity of lime at a given level of other variable inputs, will reduce total value of production by 2.01%. This could be as a result of over liming that can lead to water acidity [17]. Sex (X8) showed a positive relationship with the total value of production, this could be said that male farmers will have a higher labour productivity than female farmers, if all other variables are held constant.

Limitation of the study

Due to the problem of poor farm record keeping, the farmers could not give accurate record of their production and sales as most of the information obtained were recalled from memories [18]. Thus, the production and income data obtained might have been under-reported in this study.

Conclusion and Recommendation

The result of this study showed that fish farming in Saki-East Local Government Area is highly feasible and profitable. The Benefit-Cost Ratio recorded a good value of 1.9 and the Return on Investment also recorded a good value of 0.888 indicating a return of about 88 kobo should be expected on every N1 spent in the enterprise. The result also reflected that age and level of education of the fish farmers have positive relationship with the farmers output, indicating a positive increase with a unit increase in the factors. It can therefore be recommended from the study that government and farm friendly institutions should support the fish farmers in Saki-East LGA in term of more inputs supply and also educating the farmers on the need to form cooperative societies as to attract any government intervention opportunities, this is believed will help the farmers increase their profit margin in the study area and also would help the teaming unemployed youths (including graduates) to develop more interests in agriculture, especially aquaculture.

References

1. Fagade SO (1992) Production, Utilization and Marketing in fisheries; Status and opportunities. 20th Annual Conference of FISON, Abeokuta, Nigeria.

2. Anthonio OR, Akinwumi JA (1991) Supply and Distribution of Fish in Ibadan Nigeria. Geog J 14: 16

3. FAO (1996) Fisheries Technical Paper, Rome

4. Adekoya BB (1993) Fueling in Crisis. Fish Network: a quarterly Publication of the Fisheries Society of Nigeria.

5. Ahmed M, Bimpao MP, Gupta MV (1996) Economics of Tilapia Aquaculture in small Water bodies in Bangladesh.

6. Adekoya BB (2001) Enhancing the income generating capacity of cooperative societies and individuals in Ogun and Lagos States.

7. Borgese EM (1977) Sea Farm: The Story of Aquaculture. Henry N. Abrams Inc, New York.

8. Delgado LD, Wanda M, Rosegrant MW, Meier S (2003) Fish Outlook to 2020. International Food policy Research Institute, Washington.

9. FAO (1991) Fish for food and Employment, Food and Agricultural Organization, Rome.

10. FAO (2007) Aquaculture only way to meet global demand for fish-UN agency, FAO Bulletin.

11. Gertjan DE G, Johannes J (1996) Handbook on the artificial reproduction and Pond rearing of the African Catfish Clarias gariepinus in Sub-Saharan African Godwin Etuk (2005).

12. Igene (1991) Food Science and Technology in Nigeria Agriculture and Industrial Development. Agricultural Science Tech.

13. Igene JO (1997) Food Production and Nutrition in Nigeria. Proceedings of the National Workshop of Nigeria's Position at the World Food Summit, Abuja.

14. Olagunju FI, Adesiyan IO, Ezekiel AA (2004) Economc viability of Fish farming in Oyo State, Nigeria. J. Hum Ecol 21: 121-124.

15. Othman MF, Sadek SS (2004) Industrial Fish Feed Production Development in Egypt. National Consultation on Fish Nutrition Research and Feed Technology, Egypt.

16. Sadek S, Osman MF, Mezayen A (2006) Aquaculture in Egypt, A fragile Colossus? . International Conference and Exhibition, Firenze Florence, Italy.

17. Sayyad Ruma (2008) 700,000 tonnes: The weight of Nigeria's slump.

18. Ngazy ZM (2004) Appraising Aquaculture: "The ZALA Park Fish Cultivation and Makoba Integrated Mariculture Pond System". State University of Zanzibar, Tanzania.

15

Effect of Temperature and Seasonality Principal Epizootiological Risk Factor on Vibriosis and Photobacteriosis Outbreaks for European Sea Bass in Greece (1998-2013)

Georgios Bellos[1]*, Panagiotis Angelidis[2] and Helen Miliou[1]

[1]*Department of Applied Hydrobiology, Faculty of Animal Science and Aquaculture, Agricultural University of Athens, GR-11855 Athens, Greece*
[2]*Laboratory of Ichthyology, Faculty of Veterinary Medicine, Aristotle University of Thessaloniki, GR-54124 Thessaloniki, Greece*

Abstract

Our epizootiological survey was focused on bacterial diseases of European sea bass, *Dicentrarchus labrax* L., an important Mediterranean species for aquaculture, in Greece. Vibriosis and Photobacteriosis were the most severe bacterial diseases, located in the majority of Greek mariculture areas (Argolikos gulf, North Evoic gulf Maliakos gulf and Thesprotia Sagiada coast, Amvrakikos gulf, Aitoloakarnania Mitikas coast, Ionian island coasts) during the period 1998-2013. A database of 152 cases was formulated, from which 134 cases concerned vibriosis and photobacteriosis, while the rest outbreaks were motile aeromonas septicemia and tenacibaculosis. PCA pointed out three principal components with the following ranking order: a) temperature and seasonality, b) group of mariculture areas and average body weight, and c) case year. The results of logistic analysis showed the temperature - seasonality as the first, in ranking, and the only statistically significant epizootiolological risk factor. Contrast Test (low vs. high temperature values) also proved the significant effect of temperature (p<0.05). Vibriosis pathogen Listonella (Vibrio) anguillarum was recorded in a wide temperature range (12-26°C) in the most of Greek rearing locations. In contrast, Vibriosis from rest vibria was found to a narrow temperature range, especially in Argolikos gulf, North Evoic gulf and Ionian Island coasts. Specifically, *V. harveyi* cases emerged at high temperatures (19-22°C), while those of *V. alginolyticus* and *V. splendidus* II at low temperatures (15-17°C). Photobacterium damselae subspecies piscicida in most of Greek rearing areas and Photobacterium damselae subspecies damselae in Argolikos gulf, North Evoic gulf and Ionian Island coasts appeared in a relatively wide range (19-25°C). However, they showed higher frequencies in warm period. The results will support an evolutionary epizootiological survey and will reinforce a preventive biosecurity program in Greek mariculture taking into consideration the temperature - seasonality factor along with the classical sanitary approach.

Keywords: Epizootiological survey; Risk factors; Vibrial diseases; Sea bass; Aquaculture; Greece; Preventive biosecurity

Short Communication

European sea basis [1] commonly known as "sea bass" in Mediterranean and European Union countries, is one of the most commercially important, mainly brackish and seawater fish species. It is characterized by the second higher global production amount of 48,000 tons in Greece in comparison with the Turkish highest production amount of 51,600 tons, in 2013 [2]. In Greece, as in other Mediterranean countries, mariculture is a dynamic aquaculture sector with an also intensive development of hatchery tanks and cage culture in Spain, Italy and France [3]. The hatchery production establish Greece as the most significant provider with a production amount of 192,000 thousand sea bass juveniles, not only nationally, but also for other Mediterranean countries, in 2013 [2]. European sea bass was the target fish species of our research with emphasis on causal pathogen agents, geographic distribution of their incidences through this epizootiological survey, for a long period 1998-2013. It has focused on clear case results, for the assessment of the epizootiological risk factors in most important coastline rearing areas in Greece. Similarly, epizootiological surveys have been performed in other Mediterranean or European countries, especially in France and Spain [4,5]. The Greek coastline most important mariculture locations of this survey consist of four Mediterranean fish culture area groups with approval from Greek Government Authorities have as follow: a) Thesprotia Sagiada and Mitikas gulf – Ionian Island coasts (Ionian Sea), b) West – Central – East Amvrakikos gulf (Ionian Sea), c) Maliacosgulf –– Biotia – Evoia (North Evoic gulf) (Central Aegean Sea) d) Argolicos gulf – (South Aegean Sea). A database of 152 cases of severe bacterial diseases, for

European sea bass, was formulated. The 134 cases concerned vibriosis and photobacteriosis outbreaks caused from the pathogen bacteria shown (Figure 1). The rest 18 cases concerned motile aeromonas septicemia and tenacibaculosis (former marine flexibacteriosis) caused from *Aeromonas hydrophila*, A. sobria, *Aeromonas spp.* and *Tenacibaculum maritimum* (former Flexibacter maritimus). The 152 cases were classified into groups according to: a) temperature as low (11-19°C) and high (20- 28°C) values, b) seasonality as winter (December – February), spring (March – May), summer (June - August) and autumn (September – November), c) the above area groups of Greek coastline, d) average body weight as 0.1-5 g (immature innate non-specific immunity and almost absent specific acquired immunity class of larvae and juveniles), 6-80 g (mature innate and immature acquired immunity class of elder juveniles), 81-150 g (mature innate non-specific and mature acquired specific immunity class of young fish), 151-400 g (mature innate non-specific and mature acquired specific immunity of adult fish), and e) case-year from the

---

*Corresponding author: Georgios Bellos, Department of Applied Hydrobiology, Faculty of Animal Science and Aquaculture, Agricultural University of Athens, GR-11855 Athens, Greece, E-mail: elenmi@aua.gr

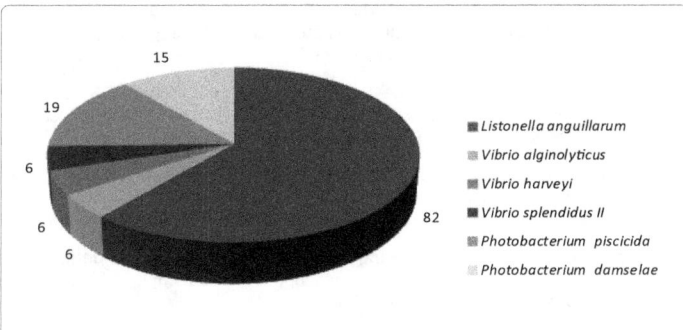

Figure 1: The Vibriosis and photobacteriosis cases pie for sea bass (1998-2013).

period (1998-2013). PCA pointed out three principal components with the following ranking order: a) temperature and seasonality, b) group area and average body weight, and c) case year. In a following Logistic Procedure, the two first principal components were considered. The results of logistic analysis showed the temperature - seasonality as the first, in ranking, and the only statistically significant epizootiological risk factor ($p < 0.05$). Contrast Test [6] with temperature as the only variable (low vs. high temperature values), also proved the significant effect of this risk factor ($p < 0.05$). Statistical analysis was performed with SAS [7]. In this survey, the sea bass classical vibriosis outbreaks from *Listonella anguillarum* appeared in a wide temperature value range (12-26°C). However, the acute classical vibriosis affected larvae and juveniles mainly at high temperatures, while the chronic or asymptomatic form appeared in young and adult fish at low temperatures. The chronic form skin ulcerative and visceral or intestinal of classical vibriosis has been reported only for adult fish at low temperatures in Greek coastline areas [8,9]. Larsen [10] reported that *L. anguillarum* can survive, not only in winter but also in summer, reaching a maximum population number. It has been suggested that vibriosis appears when the water temperature exceeds 10-12°C and occurs more often in areas where overcrowding exists and environmental equilibrium is fragile [11]. Outbreaks from Vibrio harveyi were recorded for the acute vibriosis form in juveniles at 19°C in May with high losses and for the sub-acute intestinal and visceral vibriosis in adult fish, at 22°C in August. Incidences from *V. harveyi* have also been reported for sea bass, at similar temperature values, in North Ionian Sea (South Italy) [12]. Contrary to the above, vibriosis from *Vibrio alginolyticus* and *V. splendidus* sero-type II outbreaks were recorded at low temperatures 15-17°C (November – February) in both juveniles and adults, causing higher mortalities in the former. Similarly, cases from *V. alginolyticus* have been reported for sea bass at low temperature values in Adriatic Sea (North Italy) and North Ionian Sea (South Italy) [12]. The photobacteriosis outbreaks from *Photobacterium damselae* subspecies *pisci-cida* (former Pasteurella piscicida) emerged at temperature values 19-25°C (April –September). At this temperature range, cases from sea bass juveniles and young fish with acute septicemia epizootics in Amvrakikos and Argolikos gulf have been reported [9,13-15]. In addition, chronic visceral or asymptomatic photobacteriosis cases have been referred to adult fish in Mediterranean coasts of Turkey [16,17] Egypt [18] Italy [19] and Spain [2]. Other researchers [3,20] have identified photobacteriosis epizootics mainly in summer, after heavy rain falls in brackish water coast areas, with high losses. The motile *Photobacterium damselae subspecies damselae*, in our survey, was found in sea bass at a temperature value range 19-24, 5°C (May – August). Specifically, chronic or asymptomatic photobacteriosis cases appeared in adults at 19°C, while acute septicemic photobacteriosis outbreaks with high losses in juvenile and young fish at 21-24.5°C,

particularly under stress conditions caused by cages transportation. This newer motile subspecies has been detected in Italy and Spain coasts from sea bass at high temperature values [21-24]. All the above data for the principal epizootiological risk factor, temperature and seasonality, effect on vibriosis and photobacteriosis, for European sea bass, may be utilized to an innovative marine biosecurity program. It can be applied especially for the stenothermal vibria like *V. harveyi* and *V. alginolyticus* and *V. splendidus* serotype II. For the eurythermal vibrio L. anguillarum a preventive medicine procedure has recently been applied in Greece for sea bass, including probiotics and immunostimulants administration to larvae (0.1 g), bath vaccinations of *L. anguillarum* serotypes O1 and O2 to 2/5 g and intraperitoneal vaccination to 25g juveniles [10,25,26]. A bivalent bacterin vaccine for the protection from both classical vibriosis and photobacteriosis has also been applied, particularly in areas with higher water temperatures [27]. A further research is needed for vaccination procedures concerning vibriosis from the rest vibria. In these vaccination programs, the role of temperature seasonality should be taken into consideration. In addition, the rearing temperature adjustment at adverse levels for each stenothermal Vibrio sp. can be applied where this is feasible, such as in recirculating aquaculture systems. Specifically, the up-to-date data point out optimum temperatures low (e.g. <19°C) for V. harveyi and high (e.g. >17°C) for V. alginolyticus and *V. splendidus* serotype II. It seems that rearing of sea bass at a temperature about 18°C prevents main health problems related to Vibrio pathogens. Therefore, under such preventive veterinary medicine biosecurity conditions, we can reinforce the sustainability and profitability through continual epizootiological survey and strict surveillance.

Acknowledgments

We are grateful to the Associate Professor Dr. Antonios Kominakis (Department of Animal Husbandry, Faculty of Animal Science and Aquaculture, Agricultural University of Athens) and the Lecturer Dr. Nikolaos Demiris (Department of Statistics, Athens University of Economics and Business) for their contribution to statistical analysis.

References

1. Athanassopoulou F, Billinis C, Psychas V, Karipoglou K (2003) Viral encephalo-pathy and retinopathy of Dicentrarchus labrax (L.) farmed in fresh water in Greece. J Fish Dis 26: 361-365.

2. Jung TS, Thompson KD, Morris DJ, Adams A, Sneddon K (2001) The produc-tion and characterization of monoclonal antibodies against Photobacterium damselae ssp. piscicida and initial observations using immunohistochemistry. J Fish Dis 24: 64-77.

3. Toranzo AE, Magariños B, Romalde JL (2005) A review of the bacterial fish diseas-es in mariculture systems. Aquaculture 246: 37-61.

4. Davies IM, Barg U, Black E (2004) GESAMP Initiative on Environmental Risk Analysis for Coastal Aquaculture.

5. De Blas N (2005) Searching for Evidence of Pathogen Exchange in Aquatic Environ-ments: Limits of Epidemiological Tools.

6. Abdi H, Williams LJ (2010) Contrast analysis.

7. SAS (2010) SAS/STAT© 9.22 User's Guide. SAS Institute Inc, Campus Drive, Cary, North Carolina, USA.

8. Noussias H (2007) Vibriosis (Vibrio spp.) in Fish Farms.

9. Alexopoulos A, Plessas S, Voidarou C, Noussias H, Stavropoulou E, et al. (2011) Microbial ecology of fish species ongrowing in Greek sea farms and their watery environment. Anaerobe 17: 264-266.

10. Larsen JL (1982) Vibrio anguillarum: prevalence in three carbohydrate loaded marine recipients and a control. Zentralbatt fur Bakteriologie und Hygiene 1, Abteilung Orig-inale C 3: 519-530.

11. Angelidis P (2014) Chapter 11: Vibrio anguillarum-associated vibriosis in the Mediter-ranean aquaculture. Blue Crab PC Publication, Chalastra, Greece.

12. Cavallo RA, Stabili L (2004) Culturable vibrios biodiversity in the Northern Ionian Sea (Italian coasts). Scientia Marina 68: 23-29.

13. Bakopoulos V, Hanif A, Poulos K, Galeotti M, Adams A, et al. (2003) The effect of in vivo growth on the cellular and extracellular components of the ma-rine bacterial pathogen Photobacterium damselae subsp. piscicida. J Fish Dis 27: 1-13.

14. Bakopoulos V, Pearson M, Volpatti D, Gusmani L, Adams A, et al. (2004) Investigation of media formulations promoting differential antigen expression by Photobacterium damselae ssp. piscicida and recognition by sea bass, Dicentrarchus labrax (L.), immune sera. J Fish Dis 26: 1-13.

15. Yiagnisis M, Athanassopoulou F (2011) Bacteria Isolated from Diseased Wild and Farmed Marine Fish in Greece.

16. Candan G, Ang Kucker M, Karatas S (1996) Pasteurellosis in cultured sea bass (Dicentrarchus labrax) in Turkey. Bulletin of the European Association of Fish Pathologists16: 150-153.

17. Korun J, Timur G (2005) The first Pasteurellosis case in cultured sea bass (Dicen-trarchuslabrax L.) at low marine water temperatures in Turkey. The Israeli Journal of Aquaculture – Bamidgeh 57: 197-206.

18. Almeida A, Cunha Â, Gomes N, Alves E, Costa L, et al. (2009) Phage therapy and photodynamic therapy low environmental impact approaches to inactivate microorganisms in fish farming plants. Marine Drugs 7: 268-313.

19. Zappulli V, Patarnello T, Patarnello P, Frassineti F, Franch R, et al. (2005) Direct identification of Photobacterium damselae subspecies piscicida by PCR-RFLP analysis. Dis Aquatic Organ 65: 53-61.

20. Magariños B, Couso N, Noya M, Merino P, Toranzo AE, et al. (2001) Effect of temperature on the development of pasteurellosis in carrier gilthead seabream (Spa-rus aurata). Aquaculture 195: 17-21.

21. Osorio C, Toranzo AE, Romalde JL, Barja JL (2000) Multiplex PCR assay for ureC and 16S rRNA genes clearly discriminates between both subspecies of Photo-bacterium damselae. Dis Aquatic Organ 40: 177-183.

22. Botella S, Pujalte MJ, Macián MC, Ferrús MA, Hernández J, et al. (2002) Amplified fragment length polymorphism (AFLP) and biochemical typing of Photo-bacterium damselae subsp. damselae. J App Microbiol 93: 681-688.

23. Rayan PR, Lin JHY, Ho MS, Yang HL (2003) Simple and rapid detection of Photobacterium damselae ssp. piscicidab y a PCR technique and plating method. J App Microbiol 95: 1375-1380.

24. Labella A, Vida M, Alonso MC, Infant C, Cardenas S, et al. (2006) First isolation of Photobacterium damselae ssp. damselaefrom cultured redbanded seabream, Pagrusauriga Valenciennes, in Spain. J Fish Dis 29: 175-179.

25. Costello MJ, Grant A, Davies IM, Cecchini S, Papoutsoglou S, et al. (2001) The control of chemicals used in aquaculture in Europe. J App Ichthyol 17: 173-180.

26. Angelidis P, Karagiannidis D, Crump EM (2006) Efficacy of Listonella anguilla-rum (syn. Vibrio anguillarum) vaccine for juvenile sea bass Dicentrarchus labrax. Diseases of Aquatic Organisms 71: 19-24.

27. Athanassopoulou F (2006) Diseases of Marine Fish. University of Thessaly, Volos, Greece.

16

Effects of Exogenous Melatonin in *Clarias Macrocephalus* Male Broodstock First Puberty Stage

Siti-Ariza Aripin[1,2]*, Orapint Jintasataporn[1] and Ruangvit Yoonpundh[1]

[1]*Faculty of Fisheries, Kasetsart University, 10900, Bangkok, Thailand*
[2]*School of Fisheries and Aquaculture Sciences, Universiti Malaysia Terengganu, 21030, Terengganu, Malaysia*

Abstract

The purpose of this study was to investigate the exogenous melatonin feeding administration to the first puberty stage in male broodstock of the Walking catfish, *Clarias macrocephalus*. The melatonin level of 0 (Control), 50 (Mt0.05) and 250 (Mt0.25) mg/kg in the diet mixed in isonitrogenous and isocaloric of 37% crude protein and 9.3% crude lipid was applied. The male maturation analysis for this study comprised of gonad histology, testosterone assay, gonadosomatic index, sperm abnormality, live sperm rate, sperm concentration, and sperm kinetic parameters. Significant differences were found in maturation analysis (P<0.05) compared with the control treatment. In addition, the histological analysis found that the mature spermatozoa cells were highest in the melatonin-treated male catfish. The present results showed that exogenous melatonin is able to enhance the reproductive system of male *C. macrocephalus*. The suitable exogenous melatonin level to enhance the *Clarias macrocephalus* male broodstock first puberty is Mt0.25 (50 mg/kg melatonin in the diet).

Keywords: Melatonin; Broodstock; First puberty; Male; *Clarias macrocephalus*

Introduction

Clarias *macrocephalus* (walking catfish) is one of the important fish species in Southeast Asia aquaculture industry. This species is now regarded in as near threatened by The International Union for Conservation of Nature (IUCN) Red List, because of disease susceptibility, hybrid catfish introduction, habitat fragmentation, low larval survival rate, low sperm quality and domestication problems [1,2]. Male catfish does not release semen by manual stripping and this has resulted in the sacrifice of male broodstock and removal of testis for artificial insemination of stripped eggs [3,4]. The circumstance of sufficient quality of C. *macrocephalus* semen will benefit the artificial propagation of walking catfish. Therefore, it is essential to enhance the first puberty stage and to increase the completion of spermatogenesis (spermiogenesis and spermiation) of walking catfish.

Photoperiod is one of the factors which regulate the animal rhythms including the reproductive cycle. The secretion of melatonin from the pineal gland is the effect from photoperiodic signals, and it affects the reproduction system [5]. It enhances the gonadotropin-releasing hormone (GnRH) from the hypothalamus, and it exercises its function mainly on the hypothalamus-pituitary-gonad axis [6]. Based on research by Almeida et al., pubertal testis maturation teleost fish are controlled by the endocrine system with environmental signals such as photoperiod to regulate the changes in Gnrh release.

Melatonin administration might be useful in enhancing and initiate the first puberty stage of the testicular event. Nevertheless, the existing information on the effect of melatonin to testicular event is based mostly on male mammalian studies, where melatonin is known to enhance the male reproductive system in human, ram, rat, and buck [7-10]. In addition, previous studies in female teleost show that induce melatonin are able to stimulate the maturation-inducing hormone, oocytes maturation and embryo development [11,12]. According to Tan et al. [13], melatonin has the ability to metabolize different kinds of reactive oxygen and nitrogen species. It also stimulates the ability to preserves cell membrane fluidity [14]. The studies proved that melatonin antioxidative effects on spermatozoa have significantly reduces the rate

of lipid peroxidation in sperm, and it can protect sperm mitochondria from the damage induced by reactive oxygen species (ROS) throughout its effective antioxidative effect [15,16]. To promote the understanding of the melatonin influence in male C. *macrocephalus*, this research investigated the effects of melatonin feeding administration in male broodstock of the C. *macrocephalus*.

Materials and Methods

Animals

The maiden C. *macrocephalus* male broodstock were obtained from the Fisheries Station of Kham Pheng Phet, Department of Fisheries, Ministry of Agriculture and Cooperative, Thailand. The experiment trials were conducted in the Laboratory of Nutrition and Aquafeed, Department of Aquaculture, Faculty of Fisheries, Kasetsart University, Bangkok, Thailand. The eighteen weeks old catfish were acclimatized in 500 L tanks at the density of 15ind/m²/fish/tank and fed with control feed for two weeks prior to the experiment. The source of water supply was provided with continuous aeration to maintain the oxygen supply in the experiment tank.

Experimental diets

Feeding trial used one control and two different levels of melatonin. The basal diet was formulated from practical ingredients containing 22% fishmeal, 35% soybean, 1% spirulina, 12% wheat flour, 11.8% tapioca, 5% ricebran, 2% fish oil, 3% soy oil, 1.2% mineral premix, 2% soy lecithin, 1.5% calcium phosphate, 1% attractant, 2% binder and

*Corresponding author: Siti-Ariza Aripin, School of Fisheries and Aquaculture Sciences, Universiti Malaysia Terengganu, 21030, Terengganu, Malaysia
E-mail: siti.ariza@umt.edu.my

0.5% vitamin premix. The diet also consisted of 37% crude protein and 9.3% crude lipid. Diets containing melatonin were prepared by adding graded levels of melatonin (Health Connection Labs Inc, USA) to the basal diet. These melatonin concentrations were 0 (control), 50 (Mt0.05) and 250 (Mt0.25) mg/kg in the diet [17].

Experimental condition

A total of 45 male of mean weight 57.43 ± 5.58 g (mean ± S.D.) were starved in tanks for two days prior to the experiment and all experiment population were subjected to normal photoperiod (12 hours daylight) and temperature (varied from 28°C to 30°C) prior and during the treatment. The fishes were fed twice daily with pellet feed at a level equivalent to 3% of their body weight. The fishes were randomly distributed in three treatments (control, Mt0.05 and Mt0.25) and with three replicates. Following acclimation, the fishes were exposed to melatonin feeding treatments for eight weeks.

Growth performance

Male broodstock was weighted before the final sampling to determine the growth performance by using the following formula:

Weight gain: WG (%)=[(Final body weight–Initial body weight)/ Initial body weight]×100

Histology

The sacrificed males from the treatment were dissected, and testes were fixed in 10% buffered formalin. Histological slides were made by using the standard paraffin technique where tissue samples were collected from the mid-part of the gonad. The 3-6 µm sections were stained with Haematoxylin-eosin in accordance to Drury and Wallington [18].

Immunohistochemistry

The 10% buffered formalin fixed brains were processed by following the standard method by Carvajal et al. [19]. All sections were pre-incubated for 30 minutes in normal goat serum (Sigma) at a dilution of 1:30 in 0.01 TBS buffer, pH 7.8, in order to prevent non-specific protein binding. The sections were incubated in goat anti-melatonin (Abcam, UK) at a dilution of 1:5000 in TBS buffer for 24 hours at 4°C.DAB Liquid Substrate Dropper System Kit (Sigma) was done for sections staining to localize peroxidase in tissue sections by following the method of Graham and Karnovsky [20].

Characteristics of semen and spermatozoa

Testes from the treatment fishes were surgically removed and were cut in small pieces to allow sperm release after gentle squeezing. Semen was stored with an extender solution in sterile Eppendorf tubes on ice before use. Composition of extenders was as follows: NaCl 8.760 g in 1000 ml distilled water [21].

Live sperm rate and sperm concentration: Eosin-nigrosin stained semen smears were evaluated with brightfield microscopy for the live cells by counting 400 cells per sample. Sperm concentration was determined in duplicate and expressed as the number of spermatozoa×10^6/ml. Number of spermatozoa was counted in improved Neuber haemacytometer under a light microscope.

Sperm kinetic characteristic: Kinetic characteristics were evaluated immediately after semen collection. Ten microlitres of the samples were diluted 1:1 with the extender solution and were laid over a pre-warmed Hamilton Thorne Biosciences chamber at 27.8°C. The analyses were done by Computer-Assisted Sperm Analysis (CASA). The recorded sperm parameters were [22,23]:

➢ motility (%)=the percentage of motile spermatozoa;

➢ progressive (%)=the percentage of spermatozoa with a progressive motility;

➢ path velocity (VAP; µm/s)=is a smoothed path constructed by averaging several neighbouring positions on the track (five points) and joining the averaged positions, which reduced the effect of lateral head displacement;

➢ prog. velocity (VSL; µm/s)=is the straightline distance between the first and last tracked points, divided by the acquisition time;

➢ track speed or curvilinear velocity (VCL; µm•s^{-1})=is the total distance between adjacent points, divided by the time elapsed;

➢ lateral amplitude (ALH; µm)=the mean width of the head oscillation as the sperm cells swim;

➢ beat frequency (BCF; Hz)=frequency of sperm head crossing the average path in either direction;

➢ straightness (STR; %)=is an index of the departure of the sperm path from a straight line;

➢ linearity (LIN; %)=is an index of the straightness of the path

Sperm abnormality: Fresh semen in extender solution was stained within 10 minutes of post collection. Eosin-nigrosin stained semen smears were evaluated by bright field microscopy at x 1000 magnification under oil immersion, and the percentages of abnormal and normal sperm cells were determined.

Gonadosomatic index

After feeding treatments for eight weeks, male C. *macrocephalus* were randomly selected from each treatment to measure body weight and testes weight. Then, the GSI (Mean ± S.D.%) of each individual was calculated (King, 1995).

The Gonadosomatic Index was determined as:

GSI=100 (Gm/Tm)

where;

Gm=Mass of Gonad, Tm=Total mass of fish

Testosterone analysis

The serum from treatments was used for testosterone analysis. The procedure was in accordance to the IMMULITE® Testosterone by Siemens Medical Solution Diagnostic standard procedure with the detection range from 0.2-15 ng/ml of testosterone level. The testosterone enzymes conjugate competes with 20 µl of serum samples, and the chemiluminescent substrate was added and the signal was generated according to the bound enzyme.

Statistical analysis

All data were analyzed by one-way ANOVA (analysis of variance), followed by the Tukey's honest significance test to analyze the significant between the treatment means. Duncan test was done when Tukey's honest significant test failed to analyze the significant between treatment means. All means comparisons significance was tested at P<0.05 using SPSS software.

Results

Melatonin trial did not significantly affect the weight gain between treatments where the weight gain was 10.94% (control), 15.55% (Mt0.05) and 16.98% (Mt0.25) with the value at p=0.1. Meanwhile, histological analysis is a useful tool for description of the testis after hormonal treatment where histological changes in testes during the melatonin trial were observed in C. macrocephalus (Figure 1). It was established in the present study that the control group demonstrates the most abundance in spermatogonium which was 33.9% (control), followed by 23% (Mt0.05) and 9.2% (Mt0.25) (Figure 1A and Table 1). On the other hand, spermatids were prominent in Mt0.05 group (28.4%), followed by control group (27.3%) and Mt0.25 group (5.4%) (Figure 1B and Table 1). Spermatozoa were found mainly in Mt0.25 group (85.4%), followed by Mt0.05 group (48.6%) and control group (38.8%) (Figure 1C and Table 1).

Treatment with melatonin increased the gonadosomatic index, live sperm rate and lowered the sperm abnormality rate. The significant level of melatonin on male gonadosomatic index, sperm abnormality and live sperm rate were p=0.001, p=0.001 and p=0.001, respectively (Figures 2A-2C, 3 and Table 1). However, exogenous melatonin

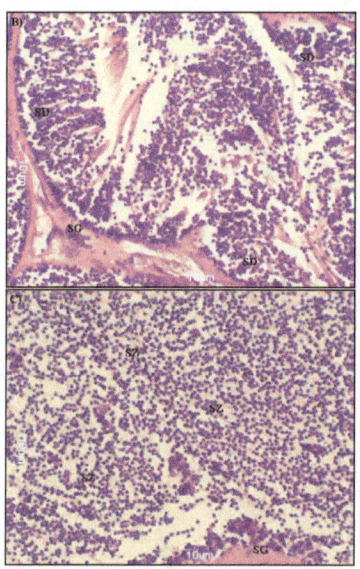

Figure 1: Effect of melatonin treatment on testis histology of the *C. macrocephalus*. Cross section of testis treated melatonin and control: A) Control, B) Mt0.05 and C) Mt0.25. Spermatogonia (SG), spermatids (SD), and spermatozoa (SZ) scale bar: 100 μm.

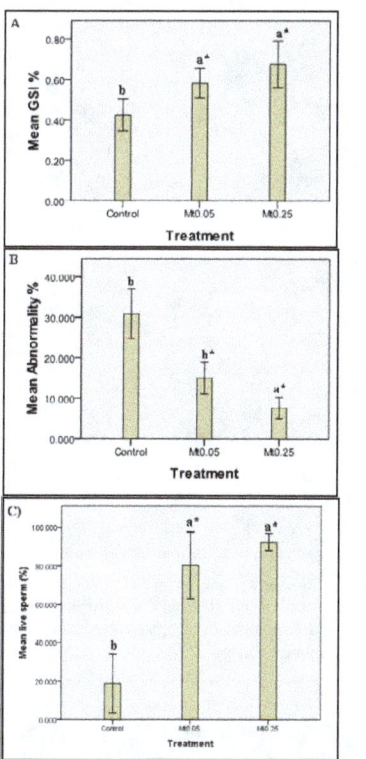

Figure 2: Mean gonadosomatic index (A), mean sperm abnormality (B), and live sperm rate (C) of male *C. macrocephalus* after eight weeks of melatonin treatment. Values are expressed as mean ± SEM p<0.05.

treatments did not alter the sperm concentration with the mean ranged from 127.3-220 3 6/ml; p=0.5. The investigation demonstrated that Mt0.25 was the optimum melatonin dosage because of the highest GSI, live sperm rate and lowest abnormality percentage.

Testosterone levels after eight weeks of treatment in male C. macrocephalus were not calculated for ANOVA analysis because the level of testosterone was beyond the chemiluminescent assay detection range. The detection range of testosterone level for chemiluminescent assay is from 0.2-15 ng/ml. In this current study, testosterone level for control was 13.45 ng/ml, while the testosterone level for melatonin treated catfish was 15>ng/ml (Mt0.05 and Mt0.25) (Table 1).

Melatonin trial was performed on the C. macrocephalus male broodstock to evaluate the effects of melatonin administration on brain immunohistochemistry. There were differences after eight weeks of treatment in melatonin intensity for immunohistochemistry slide sections between the treatments. The intense melatonin-like immunoreactivity in the brain section was found in melatonin treatment groups compared to the control treatment (Figure 4). The response was distinct in forebrain zone, pineal gland immediate zone and the brain wall of male broodstock.

After the experiment trial, there was a significant increment in sperm motility (Figure 5A) and sperm progressive (Figure 5B) among male fish in the presence of melatonin treatment after eight weeks of treatment where the mean ranged from 4.0-12.5%; p=0.004 and 1-2%; p=0.048, respectively (Table 2). Dietary melatonin treatments did not alter the kinetic parameters such as sperm path velocity, prog. velocity, track speed, lateral amplitude, beat frequency, straightness, and

Parameters	Control	Mt0.05	Mt0.25	P value
Weight gain (%)	10.94 ± 5.8	15.55 ± 7.6	16.98 ± 12.5	0.1
Histology spermatogonium cell	33.90%	23%	9.20%	-
Histology spermatids cell	27.30%	28.40%	5.40%	-
Histology spermatozoa cell	38.80%	48.60%	85.40%	-
Gonadosomatic index Mean (%)	0.425[b] ± 0.07	0.582[a] ± 0.07	0.677[a] ± 0.11	0.001
Sperm abnormality Mean (%)	30.8[b] ± 7.5	15[b] ± 4.8	7.6[a] ± 3.3	0.001
Live sperm rate (%)	18.7[b] ± 18.7	80.3[a] ± 21.2	92.5[a] ± 5.4	0.001
Sperm concentration (10⁶/ml)	220.3 ± 269	127.3 ± 104	133.3 ± 56	0.5
Testosterone (ng/ml)	13.45 ± 2.7	15>	15>	-

[a,b]Values with different superscripts in a row differ significantly (*P*<0.05).

Table 1: Maturation analysis in male *C. macrocephalus* with different levels of melatonin (Mean ± S.E.)

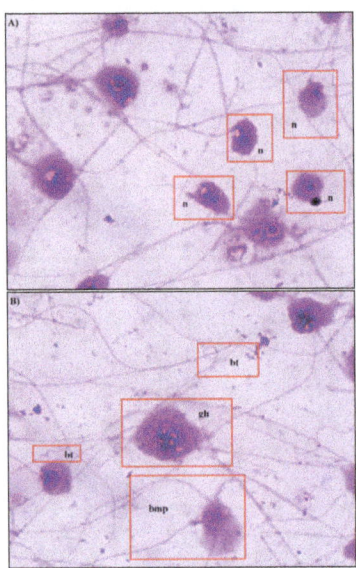

Figure 3: Some of the sperm abnormality traits in melatonin treatment of the male *C. macrocephalus*. A) Normal *C. macrocephalus* spermatozoa. B) Abnormal *C. macrocephalus* spermatozoa. Normal (n), bend tail (bt), giant head (gh), bowed midpiece (bmp).

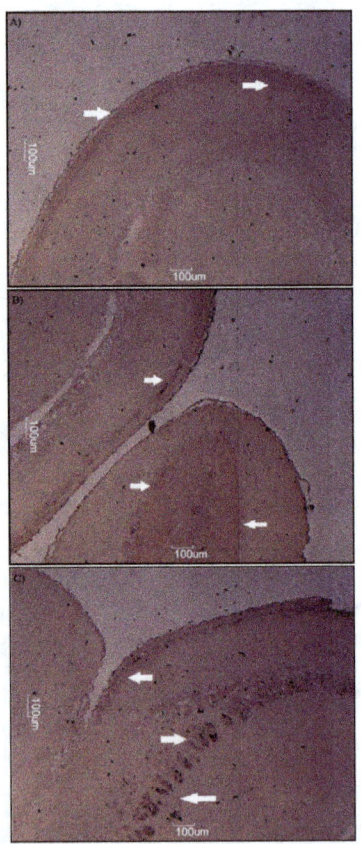

Figure 4: Light micrograph of the *C. macrocephalus* brain section incubated with anti-melatonin (dilution 1/5000) and DAB staining. Cross section of fish treated brains with control (A), Mt0.05 (B) and Mt0.25 (C). Note arrow for the intense reaction at the brain section. Scale bar=100 μm.

linearity with the mean ranged from 42.8-48.9 μm/s; p=0.3, 34.8-39.4 μm/s; p=0.4, 67.1-71.7 μm/s; p=0.8, 5.6-7.3 μm; p=0.5, 29.5-32.0 Hz;

p=0.7, 79.2-80.3%; p=0.9, and 52.7-57.2%; p=0.5, respectively (Table 2).

Discussion

It is understood that photoperiod manipulation is associated with melatonin treatment where photoperiod modulates the melatonin secretion by the pineal gland. It has been supported in the present study that the levels of GSI from both melatonin groups demonstrated a significant influence during the experiment trial compared to the control group. Similar findings have been reported in human, rams, rats, carp [24]. In recent histological study, the spermatogonia cells are abundance in the control group indicating that the testicular event in the group is immature. Spermatids and spermatozoa are more prominent in the male that were exposed to both doses of melatonin which indicates that both treatment enhanced the maturity of male *C. macrocephalus* and the results were consistent with the increase of GSI percentage. There were evidences from the previous studies that melatonin acts as an important part in the regulation of testicular events [25]. Although melatonin has an effect on spermatogenesis, more information is still needed. Some previous works have indicated that melatonin act in the reproductive axis through direct action on the Gonadotropin-releasing hormone (GnRH) gene expression and its regulation with the melatonin receptors on the GnRH neurons

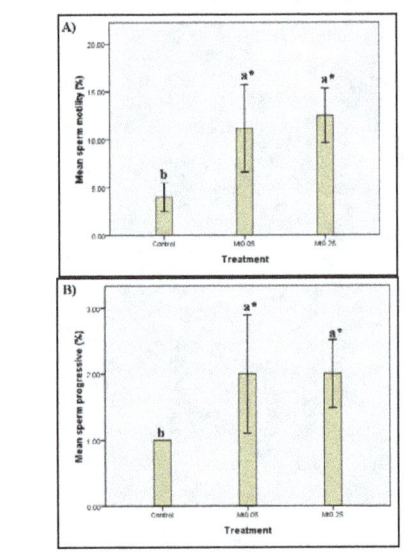

Figure 5: Mean sperm motility (A) and mean sperm progressive (B) of male *C. macrocephalus* after eight weeks of melatonin treatment. Values are expressed as mean ± SEM. p<0.05.

Parameters	Control	Mt0.05	Mt0.25	*P* value
Motility (%)	4.0[b] ± 1.8	11.2[a] ± 5.6	12.5[a] ±3.5	0.004
Progressive (%)	1[b] ± 0	2[a] ± 1.1	2[a] ± 0.6	0.048
Path velocity (VAP) (μm/s)	42.8 ± 4.3	48.7 ± 8.1	48.9 ± 11.6	0.3
Prog. velocity (VSL) (μm/s)	34.8 ± 3.6	38.3 ± 4.4	39.4 ± 10.1	0.4
Track speed (VCL) (μm/s)	67.1 ± 9.8	68.4 ± 12.4	71.7 ± 19	0.8
Lateral amplitude (ALH) (μm)	5.6 ± 2.4	6.2 ± 2.4	7.3 ± 2.9	0.5
Beat frequency (BCF) (Hz)	29.5 ± 7.5	32.0 ± 4.8	29.9 ± 4.2	0.7
Straightness (STR) (%)	79.2 ± 14	79.5 ± 7.4	80.3 ± 5.4	0.9
Linearity (LIN) (%)	52.7 ± 6.8	54.0 ± 9.6	57.2 ± 4.6	0.5

[a,b]Values with different superscripts in a row differ significantly (*P*<0.05).

Table 2: Sperm kinetic parameters measured by CASA for male *C. macrocephalus* with different levels of melatonin (*P*<0.05) (Mean ± S.E.)

of the hypothalamus [26]. This would have increased the luteinizing hormone (GTH-II) and follicle-stimulating hormone (GTH-I); with the confirmation from previous study that in vitro melatonin study stimulated GTH-II and GTH-I release from pituitary cells [27]. This indicates that melatonin may significantly elevate plasma GTH-II and GTH-I levels where it mainly regulates gonadal maturation and spermiation process. Schulz et al. [28] reported that GTH-II and GTH-I are the most significant pituitary hormones that regulate testicular physiology. In mammals, GTH-II regulates Leydig cell sex steroid production and GTH-I regulate the Sertoli cell activities [29].

For immunohistochemistry analysis, the brain sections from female broodstock that were treated with melatonin displayed more intense colouration area compared to the control treatment. The similar display was found in mammals. The high intensity of melatonin in treated broodstock occurs due to melatonin absorption from the treated fed. The melatonin treated fed is absorbed in the digestive system via transcellular pathway [30]. According to Zeuthen [31], transcellular pathway is a type of molecules transport that flows across both basal and apical membranes, which occurs in response to the osmotic stimuli created by salt transport, by entering the bloodstream and goes to the target organ such as brain and peripheral tissues.

According to Langford et al. [32], melatonin stimulates spermatogenic activity by increasing the sensitivity of Leydig cells to GTH-II. Besides enhancing the maturation via pituitary-hypothalamus-gonad axis, melatonin also acts as directly on the testes through Leydig cells [33]. Therefore, GTH-II regulates Leydig cell sex steroid production such as testosterone. The stimulation of live sperm rate, sperm motility, enhanced some of the CASA parameters and lowered the sperm abnormality by melatonin treatment might be caused by the enhancement of the production of testosterone. Testosterone in the testis is essential for spermatogenesis which regulates the spermatogenesis, maintaining the blood-testis barrier and releasing the mature sperm [34]. Additionally, testosterone could increase sperm quality by inhibiting the spermatocyte and spermatide apoptosis [35].

Conclusion

For the first time, this study shows that melatonin feeding treatment to male C. macrocephalus has significantly improved the first puberty event by enhancing the maturation of testes and sperm. The study recommended that the Mt0.25 is the optimum level to enhance the Clarias macrocephalus male broodstock first puberty. The results obtained from the present study provide information on melatonin enhancement to male reproductive system that will lead to more efficient gamete management and may increase yield of catfish in the aquaculture industry.

Acknowledgement

The authors would like to thank all members of Laboratory of Nutrition and Aquafeed, Department of Aquaculture, Faculty of Fisheries, Kasetsart University, Bangkok for their kind assistance during this study. This study was funded by Ministry of Education, Malaysia.

References

1. Petkam R, Moodie GEE (2001) Food particle size, feeding frequency, and the use of prepared food to culture larval walking catfish (Clarias macrocephalus). Aquaculture 194: 349-362.

2. Vuthiphandchai V, Thadsri I, Nimrat S (2009) Chilled storage of walking catfish (Clarias macrocephalus) semen. Aquaculture 296: 58-64.

3. Mansour N, Lahnsteiner F, Berger B (2004) Characterization of the testicular semen of the African catfish, Clarias gariepinus (Burchell, 1822), and its short-term storage. Aquac Res 35: 232-244.

4. Viveiros ATM, Fessehaye Y, ter Veld M, Schulz RW, Komen J (2002) Handstripping of semen and semen quality after maturational hormone treatments, in African catfish Clarias gariepinus. Aquaculture 213: 373-386.

5. Falcon J, Migaud H, Muñoz-Cueto JA, Carrillo M (2010) Current knowledge on the melatonin system in teleost fish. Gen Comp Endocrinol 165: 469-482.

6. Falcon J, Besseau L, Sauzet S, Boeuf G (2007) Melatonin effects on the hypothalamo-pituitary axis in fish. Trends Endocrinol Metab 18: 81-88.

7. Awad H, Halawa F, Mostafa T, Atta H (2006) Melatonin hormone profile in infertile males. Int J Androl 29: 409-413.

8. Casao A, Mendoza N, Perez-Pe R, Grasa P, Abecia J, et al. (2010) Melatonin prevents capacitation and apoptotic-like changes of ram spermatozoa and increases fertility rate. J Pineal Res 48: 39-46.

9. Drago F, Busa L (2000) Acute low doses of melatonin restore full sexual activity in impotent male rats. Brain Res. 878: 98-104.

10. Daramola JO, Adeloye AA, Fayeye TR, Fatoba TA, Soladoye AO (2006) Influence of photoperiods with or without melatonin on spermiograms in West African dwarf bucks. World J Zool 1: 86-90.

11. Chattoraj A, Bhattacharyya S, Basu D, Bhattacharya S, Bhattacharya S, et al (2005) Melatonin accelerates maturation inducing hormone (MIH): induced oocyte maturation in carps. Gen Comp Endocrinol 140: 145-155.

12. Danilova N, Krupnik VE, Sugden D, Zhdanova IV (2004) Melatonin stimulates cell proliferation in zebrafish embryo and accelerates its development. FASEB J 18: 751-753.

13. Tan DX, Reiter RJ, Manchester LC, Yan MT, El-Sawi M, et al. (2002) Chemical and physical properties and potential mechanisms: melatonin as a broad spectrum antioxidant and free radical scavenger. Curr Top Med Chem 2: 181-197.

14. Ashrafi I, Kohram H, Ardabili FF (2013) Antioxidative effects of melatonin on kinetics, microscopic and oxidative parameters of cryopreserved bull spermatozoa. Anim Reprod Sci 139: 25-30.

15. Gavella M, Lipovac V (2000) Antioxidative effect of melatonin on human spermatozoa. Arch Androl 44: 23-27.

16. Shang X, Huang Y, Ye Z, Yu X, Gu W (2004) Protection of melatonin against damage of sperm mitochondrial function induced by reactive oxygen species. Zhonghua Nan Ke Xue 10: 604-607.

17. Rubio VC, Sanchez-Vazquez FJ, Madrid JA (2004) Oral administration of melatonin reduces food intake and modifies macronutrient selection in European sea bass (Dicentrarchus labrax, L.). J Pineal Res 37: 42-47.

18. Drury RA, Wallington EA (1980) Carleton's Histological Technique. Oxford University Press, USA.

19. Carvajal JC, Esteban MBG, Carbajo S, Munoz-Barragan L (2004) Melatonin-like immunoreactivity in the pineal gland of the cow: an immunohistochemical study. Histol Histopathol 19: 1187-1192.

20. Graham RC, Karnovsky MJ (1966) The early stages of absorption of injection horseradish peroxidases in the proximal tubules of mouse kidney: Ultrastructural cytochemistry by a new technique. J. Histochem Cytochem 14: 291-302.

21. Sprecher DJ, Coe PH, Walker RD (1999) Relationships among seminal culture, seminal white blood cells, and the percentage of primary sperm abnormalities in bulls evaluated prior to the breeding season. Theriogenology 51: 1197-1206.

22. Katebi M, Movahedin M, Abdolvahabi MA, Akbari M, Abolhassani F, et al. (2005) Changes in motility parameters of mouse spermatozoa in response to different doses of progesterone during course of hyperactivation. Iran Biomed J 9: 73-79.

23. Klimowicz MD, Nizanski W, Batkowski F, Savic MA (2008) The comparison of assessment of pigeon semen motility and sperm concentration by conventional methods and the CASA system (HTM IVOS). Theriogenology 70: 77-82.

24. Lombardo F, Gioacchini G, Fabbrocini A, Candelma M, D'Adamo R, et al. (2014) Melatonin-mediated effects on killifish reproductive axis. Comp Biochem Physiol 172: 31-38.

25. Bhattacharya S, Chattoraj A, Maitra SK (2007) Melatonin in the regulation of annual testicular events in carp Catla catla: evidence from the studies on the effects of exogenous melatonin, continuous light, and continuous darkness. Chronobiol Int 24: 629-650.

26. Roy D, Angelini NL, Fujieda H, Brown GM, Belsham DD (2001) Cyclical

regulation of GnRH gene expression in GT1-7 GnRH-secreting neurons by melatonin. Endocrinology 142: 4711-4720.

27. Aizen J, Kowalsman N, Niv MY, Levavi-Sivan B (2014) Characterization of tilapia (*Oreochromis niloticus*) gonadotropins by modeling and immunoneutralization. Gen Comp Endocrinol 207: 28-33.

28. Schulz RW, Franca LR, Lareyre J, LeGac F, Chiarini-Garcia H, et al. (2010) Spermatogenesis in fish. Gen Comp Endocrinol 165: 390-411.

29. Huhtaniemi IT, Themmen AP (2005) Mutations in human gonadotropin and gonadotropin receptor genes. Endocrine 26: 207-217.

30. Tran HTT, Tran PHL, Lee B (2009) New findings on melatonin absorption and alterations by pharmaceutical excipients using the Ussing chamber technique with mounted rat gastrointestinal segments. Int J Pharm 378: 9-16.

31. Zeuthen T (2002) General models for water transport across leaky epithelia. Int Rev Cytol 215: 285-317.

32. Langford GA, Ainsworth L, Marcus G, Shrestha JNB (1987) Photoperiod entrainment of testosterone, luteinizing hormone, follicle stimulating hormone and prolactin cycles in rams in relation to testis size and semen quality. Biol Reprod 37: 489-499.

33. Shiu S, Yu Z, Chow P, Pang S (1996) Putative melatonin receptors in the male reproductive tissues. In: Tang P, Pang S, Reiter R, (eds) Melatonin a universal photoperiodic signal with diverse actions. AG, Hong Kong Karger 21: 90-100

34. Walker WH (2009) Molecular mechanisms of testosterone action in spermatogenesis. Steroids 74: 602-607.

35. Ruwanpura SM, McLachlan RI, Meachem SJ (2010) Hormonal regulation of male germ cell development. J Endocrinol 205: 117-131.

17

Genetic Variation among Cat Fish (*Mystus vittatus*) Population Assessed by Randomly Amplified Polymorphic (RAPD) Markers from Assam, India

Innifa Hasan[1]* and Mrigendra Mohan Goswami[2]

[1]*Department of Zoology, Handique Girls college Guwahati, Assam, India*
[2]*Department of Zoology, Gauhati University, Assam, India*

Abstract

Mystus vittatus is a small indigenous fish species having higher nutritional value in terms of protein, micronutrients, vitamins and minerals. But the catfish aquaculture including Mystus sp has not been developed extensively for its aquaculture potential even though the demand of catfishes in the Indian domestic markets are very high. Therefore for good aquacultural practices and to maintain a healthy gene pool, detailed knowledge on the population structure of Mystus sp. is needed.

In the present study molecular and morphological analysis of a population of *Mystus vittatus* caught from four different freshwater bodies of Assam about 100-400 km away from each other was done using RAPD markers. Total 412 RAPD fragments were generated using nine decamer primers of arbitrary nucleotide sequences. In the experiment 322 polymorphic bands and 90 monomorphic bands were produced which shows 78.15% of polymorphism and 21.84% of monomorphism. UPGMA dendrogram constructed on the basis of genetic distance formed three distinct clusters indicating comparatively higher level of genetic variations in the studied M. vittatus populations in Assam. Once the population structure is known, scientific management for optimal harvest and conservation of the catfish fishery resource can be undertaken. Therefore, the present study may serve as a reference for future examinations of genetic variations within the populations of fishes which are commercially important and the possible use of DNA markers in future may create new avenues for cat fish molecular biological research in this part of world.

Keywords: Genetic; Diversity; RAPD; Markers; Cat fish

Introduction

The striped dwarf catfish *Mystus vittatus* is an economically important and favourite food fish in the South East Asian countries. This important species for aquaculture is naturally distributed in India, Bangladesh, Pakistan, Sri Lanka, Nepal, Myanmar and Thailand [1]. Even though the catfishes are in great demand in the Indian domestic markets, the catfish aquaculture including *Mystus* sp has not yet been developed for its aquaculture potential [2]. The entire demand for this fish in the domestic market is met through capture from river bodies and hence the effective management of wild stocks is critical. For the development of effective management strategies information on population structure is essential to conserve the biodiversity associated with different species, sub-species, stocks and races [3]. Therefore for good aquacultural practices and to maintain a healthy gene pool, detailed knowledge on the population structure of *Mystus sp.* is needed.

All organisms are subject to mutations because of normal cellular operations or interactions with the environment, leading to genetic variation (polymorphism). Genetic variation in a species enhances the capability of organism to adapt to changing environment and is necessary for survival of the species [4]. Genetic variation arises between individuals leading to differentiation at the level of population, species and higher order taxonomic groups apart from other evolutionary forces like selection and genetic drift. Molecular markers along with the development of new statistical tools has revolutionized the analytical power necessary to explore the genetic diversity, both in native populations and in captive lots [5]. Nowadays, a wide range of new molecular techniques have been explored and reported for fishes [6,7]. Williams *et al.* first introduced the Random amplified polymorphic DNA (RAPD) technique [8]. RAPD technique is the one of the most frequently used molecular methods for taxonomic and systematic analyses of various organisms and has provided important applications in catfish [9]. RAPD has also been used to estimate genetic diversity

and variations needed to study fish management and conservation practices, even with endangered species [10,11]

It is based on the PCR amplification of discrete regions of genome with short oligonucleotide primers of arbitrary sequence [12]. The characters assessed through RAPD are useful for genetic studies because they provide various types of data-taxonomic population, inheritance pattern of various organisms including fishes [13].

The information on morphometric measurements of fishes and the study of statistical relationship among them are essential for taxonomic work [14]. Moreover, to know the origin of stock, separation of stocks or identification of commercially important species of fishes, morphometric characters are frequently used [15,16]. These type of study are important for understanding the interactive effect of environment, selection and heredity on the body shapes and sizes within a species [17]. Several studies on the comparative morphometrics of different fish populations have been conducted [18]. The study of genetic diversity of catfishes of Assam is very much limited, so in the present study, this technique was applied to analyze the genetic relationship among *Mystus vittatus* populations. The objectives of this study are focused on morphometric identification and

*Corresponding author: Innifa Hasan, Department of Zoology, Handique Girls College Guwahati, Assam, India
E-mail: innifa_hasan@rediffmail.com

detection of RAPD pattern for determination of the genetic variation of a population of *Mystus vittatus* from Assam, India.

Materials and Methods

Fish sampling sites and morphometric measurements of fishes

Geographically, populations of *Mystus vittatus* were caught from freshwater bodies of Assam about 100-400 km away from each other, that is, Kolong river at Morigaon (Morigaon District); Deepor beel at Guwahati (Kamrup District); Kani beel (Dibrugarh District); Dhemaji local fish market (Dhemaji District) in the month of June, 2013. A total of 60 fish specimens were collected from all the locations with the help of local fishermen and 20 fish specimens were randomly selected for morphometric measurements and estimating genetic variations. All the fish specimens were kept in the iceboxes and brought to the laboratory for further study. For the morphometric measurements, total 24 parameters were considered. Fish specimens were morphologically identified with taxonomic keys [19,20]. The muscle tissues were isolated from freshly caught fishes and preserved at -20°C for further use.

Isolation of genomic DNA from fish tissue

For the isolation of total genomic DNA, a modified protocol was followed using Sambrook and Russel Molecular Cloning-A Lab. Manual [21]. UV-VIS spectrophotometer was used to check quality as well as quantity of isolated DNA.Optical densities of the DNA samples were measured at 260 nm and 280nm and the concentration of extracted DNA was adjusted to 50 ng/μl for PCR amplification.

PCR primers

In the present study, 30 commercially available RAPD primers (10 to 20 base long) made by Xcelris Genomics, India were used to initiate PCR amplifications. Primers were randomly selected on the basis of GC content and annealing temperature for RAPD-PCR amplifications.

PCR amplification

The reaction mixture (10 μl) for PCR was composed of 1 μl of 10 X Taq polymerase buffer, 1 μl of 2.5 mM dNTPs, 1 μl of RAPD primer, 0.15 μl Taq DNA polymerase (2 U/μl), 5.55 μl PCR grade water, 0.3 μl of 50 mM $MgCl_2$ and 1 μl template DNA. A negative control, without template DNA was also included in each round of reactions. After preheating for 5 mins at 94°C, PCR was run for 35 cycles. It consisted of a 94°C denaturation step (1min), 37°C annealing step (1 min) and 72°C elongation step (2 min) in a thermal cycler (Biorad). At the end of the run, a final extension period was appended (72°C, 10 min) and then stored at 4°C until the PCR products were analyzed.

Agarose gel electrophoresis

The amplified DNA fragments were separated on 1.8% agarose gel and stained with Ethidium bromide. A low range DNA marker of 100 bp from Bangalore Genei, Bangalore, India was run with each gel. The amplified pattern was visualized on an UV transilluminator and photographed by gel documentation system (BIORAD, USA).

Statistical analysis

The RAPD fragments were scored for the presence and absence of fragments on the gel photographs and RAPD fragments were compared among the *M. vittatus* populations. RAPD banding patterns were recorded on spreadsheets, which were used to determine gene diversity, gene flow, number of polymorphic loci and genetic distance

through a construct by an un-weighted pair group method of arithmetic mean or UPGMA [22] using GGT 2.0 software (http://www.dpw.wau.nl/pv/pub/ggt/)

The similarity index (SI) values between the RAPD profiles of any 2 individuals on the same gel were calculated using following formula:

Similarity Index (SI)=2 $N_{AB}/(N_A+N_B)$

Where,

N_{AB}=total number of RAPD bands shared in common between individuals A and B

N_A=total number of bands scored for individual A

N_B=total number of bands scored for individual B [23],

Cluster analysis was carried out using GGT 2.0 version software. Dendograms were constructed by employing UPGMA (Unweighted Pair Group Method with Arithmetic Mean) based on Sneath and Sokal to study the genetic variability within the species.Similarly the same method was followed to construct the dendogram to study the phylogenetic relationship among the genotypes of *Mystus* vittatus [24].

Results and Discussion

The morphometric variation among the different individuals of *Mystus vittatus* was found to be very low (Table 1).

RAPD polymorphisms

Among the 30 primers initially tested, only nine R-4, R-5,R-6, R-11, R-12, R-13, R-20, R-21 and R-22 were selected that yielded relatively large number of good quality bands. All the primers produced different RAPD patterns, and the number of fragments amplified per primer varied. The nine primers yielded a total of 412 reproducible and consistently scorable RAPD bands of which 322 were found to be polymorphic and 90 were monomorphic. The number of bands per primer ranged from 5 to 13. Among the primers, R-4 and R-21 gave DNA profile with highest number of bands while R-13 gave the least (Table 2). The RAPD profile of the bands obtained in the population of *M.vittatus* with primer R-4,R-5,R-6 and R-11 is shown in the Figure 1 as representative photographs. The UPGMA dendrogam was prepared based on genetic distance by the GGT 2.0 software. The unweighted dendrogam divided all the genotypes in three clusters.

Little morphological differences were revealed among the *M.vittatus* populations by morphometric and meristic studies that were sampled from different rivers and localities in Assam. Most of the morphometric

Characters	Minimum	Maximum	Mean ± SD
Forked Furcal length:TL	6.2	7.9	7.16 ± 0.644
Eye Diameter:TL	0.2	0.5	0.35 ± 0.102
Caudal peduncle:TL	0.7	1.0	0.93 ± 0.11
Dorsal fin Height:TL	1.0	1.2	1.04 ± 0.066
Dorsal fin Length:TL	0.8	1.1	0.91 ± 0.094
Pectoral fin length:TL	0.8	1.8	1.24 ± 0.335
Ventral fin height:TL	0.1	0.3	0.8 ± 0.19
Ventral fin length:TL	0.1	0.3	0.21 ± 0.083
Anal fin height:TL	0.7	1.2	0.92 ± 0.178
Caudal fin length:TL	1.2	2.0	1.6 ± 0.293
2nd Dorsal fin length:TL	0.8	1.2	0.96 ± 0.128
Body width:TL	0.6	1.5	1.08 ± 0.346

Table 1: Morphometric measurements of *Mystus vittatus* (in proportion to Total length).

Characters	Minimum	Maximum	Mean ± SD
Body Weight,gm	2.35	5.15	3.59 ± 1.087
Total length,cm	7.1	9.5	8.17 ± 0.88
Standard length,cm	5.3	7.5	6.74 ± 0.9
Head length:SL	26.09	31.08	1.89 ± 0.281
Pre Dorsal length:SL	35.71	40.74	2.6 ± 0.335
Post Dorsal length:SL	55.71	64.81	3.96 ± 0.4
Pre Orbital length:SL	6.67	9.26	0.55 ± 0.081
Post Orbital length:SL	12	16.36	0.95 ± 0.067
Head Width:SL	7.55	13.51	0.77 ± 0.205
Body Depth:SL	6.67	20.29	0.91 ± 0.394
Head length excluding snout:SL	13.33	20.27	1.15 ± 0.216
Snout length:SL	3.77	6.76	0.36 ± 0.88
Anal Fin length:SL	7.55	11.59	0.64 ± 0.128

Table 2: Morphometric measurements of *Mystus vittatus* (in proportion to standard length).

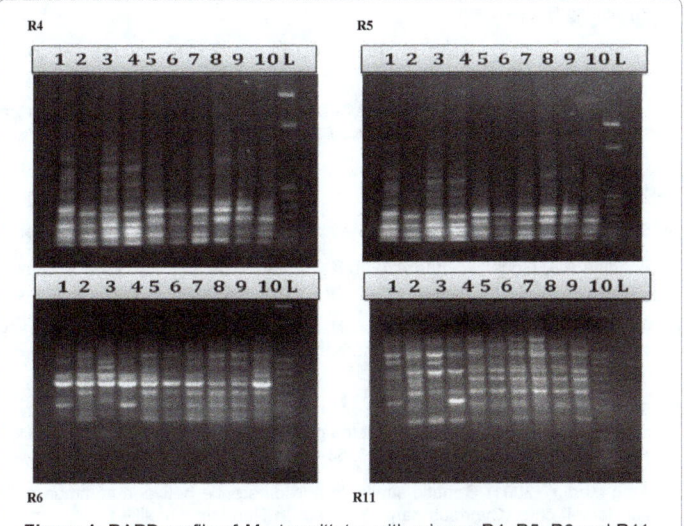

Figure 1: RAPD profile of *Mystus vittatus* with primers R4, R5, R6 and R11.

Sl no	Polymorphism	*Mystus vittatus*								
		R-4	R-5	R-6	R-11	R-12	R-13	R20	R21	R22
1	Total no of bands	58	36	38	48	37	35	53	58	49
2	Total no of polymorphic bands	48	26	28	38	27	25	43	48	39
3	Total no of monomorphic bands	10	10	10	10	10	10	10	10	10
5	Polymorphism (%)	82.75	78.8	73.6	79.16	72.9	71.4	81.1	82.7	79.5
6	Monomorphism (%)	18.18	27.7	26.3	27.7	27.1	28.5	18.8	17.2	20.5

Table 3: Pattern of polymorphism (primer wise) in 10 individuals of *Mystus vittatus*.

characteristics of the fishes in the present study were similar and there is overlap in the range of each of traditional morphometric measurements taken (Tables 2 and 3). The morphometric data may not be enough to support the established genetic structure of the population that often leads to taxonomic uncertainty in many occasions because of the considerable geographical and ecological variability in form [25,26]. In *Hilsa* sp., significant differences in allele frequencies and morphological variations were observed from nine different sites within Bangladesh by which may be due to the local environmental conditions [27].

DNA fingerprinting method has tremendous potential in aquaculture and in fisheries as a tool for identification of individuals

[28] and population genetics studies [29-31]. RAPD-PCR is a useful tool for estimating the genetic variability and degree of similarity among fish species as has been reported by other workers [32]. Using a RAPD analysis, the intrapopulation variation was detected with different primers in tilapia [33]. Chong et.al could identify and characterize *Mystus nemurus* populations by RAPD analysis in Malaysia [34]. RAPD technique has also been used to determine genetic variation within and among three populations of *Mystus vittatus* by Tamanna et.al in Bangladesh in the year 2012 [5].

In the present study among the 30 single decamer random primers, nine primers generated a total of 412 bands in the population which were found to be both polymorphic and monomorphic. In the experiment 322 polymorphic bands and 90 monomorphic bands were produced which shows 78.15% of polymorphism and 21.84% of monomorphism (Figure 1). Polymorphism for genetic similarity among the different individuals of *M. vittatus* which was analyzed using GGT software is expressed in Figure 2. The cluster analysis and dendogram showing genetic relationship between 10 genotypes of *M.vittatus* showed formation of 3 clusters (Figure 3). Cluster I include genotype 4, 7, 9 and 8; Cluster II include genotype 10,3 and 1; Cluster III include genotype 6, 5 and 2.

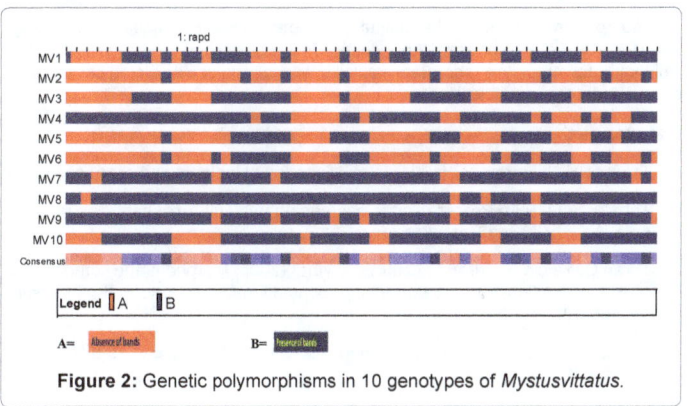

Figure 2: Genetic polymorphisms in 10 genotypes of *Mystusvittatus*.

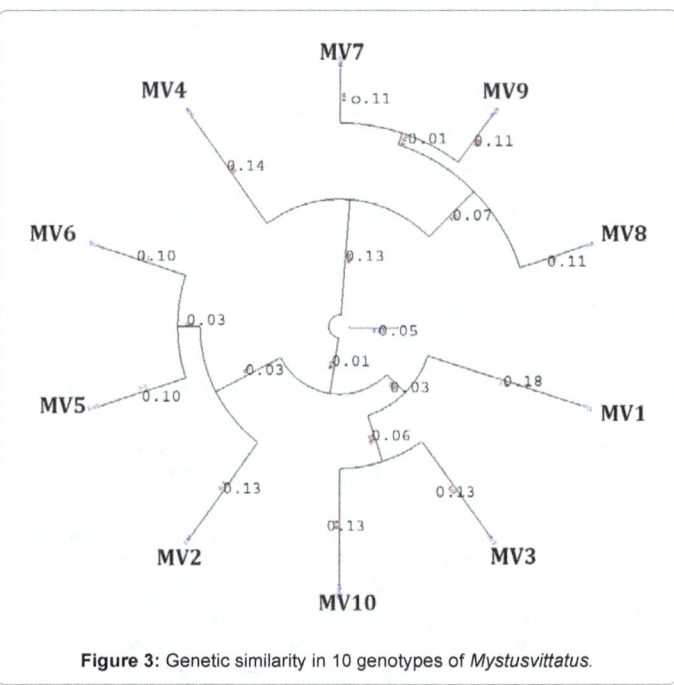

Figure 3: Genetic similarity in 10 genotypes of *Mystusvittatus*.

The present study indicates that comparatively higher level of genetic variation exists in the studied *M. vittatus* populations in Assam. Garg et al. could discriminate between the different populations of *M.vittatus* in reservoirs of Madhya Pradesh by RAPD analysis [35]. According to them the intraspecific genetic similarity between the individuals of the population was due to geological variations or changes in aquatic environment. Our statistical analysis showed considerable genetic variation among the genotypes of *M.vittatus* collected from different locations of Assam. This population genetic differentiation may be due to ecological, geographical and evolutionary factors. The genetic diversity data has varied applications in research on evolution, conservation and management of natural resources and genetic improvement programmes, etc. RAPD analysis for genetic diversity study provides a basis to obtain genetic variation within and among populations. Once the population structure is known, scientific management for optimal harvest and conservation of the catfish fishery resource can be undertaken. Therefore, the present study may serve as a reference point for future examinations of genetic variations within the populations of fishes which are commercially important and the possible use of DNA markers in future may create new avenues for fish molecular biological research.

Acknowledgement

Authors are thankful to the Institutional Biotech Hub, Handique Girls' College and UGC-SAP (DRF) phase II Aquaculture Project, Zoology Department, Gauhati University for providing laboratory facilities with all molecular biological tools for executing this research work.

References

1. Talwar PK, Jhingran AG (1991) Inland fisheries of India and adjacent countries.

2. Tripathi SD (1996) Present status of breeding and culture of catfishes in South Asia. Aquatic Living Resources 9: 219-228.

3. Turan C, Yalcin S, Turan F, Okur E, Akyurt I (2005) Morphometric comparisons of African catfish, Clarias gariepinus, populations in Turkey. Folia Zoologica 54: 165-172.

4. Fisher RA (1930) The Genetical Theory of Natural Selection. Oxford University Press, UK.

5. Tamanna FM, Rashid J, Alam MS (2012) High levels of genetic variation revealed in wild and hatchery populations of the stripped dwarf catfish Mystus vittatus (Bloch) (Bagridae: Siluriformes) in Bangladesh by Random Amplified Polymorphic DNA techniques. Int J Adv Biol Res 2: 322-327.

6. Lehmann J, Stohr T, Feldon J (2000) Long-term effects of prenatalstress experiences and postnatal maternal separation on emotionality and attentional processes. Beh. Brain Res 107: 133-144.

7. Jayasankar P (2004) Random amplified polymorphic DNA (RAPD) fingerprinting resolves species ambiguity of domesticated clown fish (genus: Amphiprion, family: Pomacentridae) from India. Aquacult Res 35: 1006-1009.

8. Williams JGK, Kubelik AR, Livak KJ, Rafalski JA, Tingey SV (1990) DNA polymorphisms amplified by arbitrary primers are useful as genetic markers. Nucleic Acids Res 18: 6531-6535.

9. Bartish IV, Gorkava LP, Rumpunen K, Nybom H (2000) Phylogenetic relationship and differentiation among and within populations of Chaenomeles Lindl. (Rosaceae) estimated with RAPD's and isozymes. Theor Appl Genet 101: 554-563.

10. Lopera-Barrero (2006) Genetic diversity in piracanjuba populations (Brycon orbignyanus) with the RAPD (Random Amplified Polymorphic DNA) markers. Journal Animal Science 84: 170.

11. Shair OHM, Al- Ssum RM, Bahkali AH (2011) Genetic variation investigation of tilapia grown under Saudi Arabian controlled environment. Am J Biochem Mol Biol 1: 89-94.

12. Peinado MA, Malkhosyan S, Velazquez A, Perucho M (1992) Isolation and characterization of allelic losses and gains in colorectal tumors by arbitrarily primed polymerase chain reaction. Proc Natl Acad Sci USA 89: 10065-10069.

13. David CJ, Pandian TJ (2006) GFP reporter gene confirms paternity in the androgenote Buenos Aires tetra, Hemigrammus caudovittatus. J Exp Zool A Comp Exp Biol 305: 83-95.

14. McCONNEL RHL (1978) Identification of fresh water fishes. In: Methods for assessment of fish production in fresh water. Blackwell Scientific Publications, London.

15. Royce WF (1963) A morphometric study of yellow fin tuna Thunnus albacore (Bonnaterre) U.S. Fish & Wildlife Services., Fishery Bull 63: 395-443.

16. Kramholz C, Cavanah F (1968) Comparative morphometry of fresh water drum from two mid-western localities. Trans Am Fish Soc 97: 429-441.

17. Cadrin SX (2000) Advances in morphometric identification of fishery stocks. Reviews in Fish Biology and Fisheries 10: 91-112.

18. Ibanez-Aguirre AL, Cabral-Solis E, Gallardo-Cabello M, Espino-Barr E (2006) Comparative morphometrics of two populations of Mugil curema (Pisces: Mugilidae) on the Atlantic and Mexican Pacific coasts. Scientia Marina 70: 139-45.

19. Srivastava CBL (2000) A Text book of fishery science and Indian Fisheries Kitab Mahal.

20. Jayaram KC (2002) The Fresh Water Fishes of the Indian Region. Narendra Publishing House, Delhi.

21. Sambrook J, Russell DW (1989) Molecular cloning: A laboratory manual. Harbor Laboratory Press, New York.

22. Nei M (1978) Estimation of average heterozygosity and genetic distances from small number of individuals. Genetics 89: 583-590.

23. Lynch M (1990) The similarity index and DNA fingerprinting. Mol Biol Evol 7: 478-484.

24. Sneath PHA, Sokal RR (1973) Numerical taxonomy. Freeman, San Francisco.

25. Ponniah AG, Gopalakrishan A (2000) Cultivable, ornamental, sport and food fishes endemic to Peninsular India with special references to Western Ghats.

26. Garg RK, Silawat N, Sairkar P, Vijay N, Mehrotra NN (2009b) Genetic diversity between two populations of Heteropneustes fossilis (Bloch) using RAPD profile. Int J Zool Res 5: 171-177.

27. Salini JP, Milton DA, Rahman MJ, Hussain MG (2004) Allozyme and morphological variation throughout the geographical range of the tropical Hilsa shad Tenualosa ilisha. Fish Res 66: 53-69.

28. Jong-Man Y (2001) Genetic similarity and difference between common carp and Israeli carp (Cyprinus carpio) based on Random Amplified Polymorphic DNA analyses. Kor J Biol Sci 5: 333-339.

29. Bielawski JP, Pumo DE (1997) Randomly amplified polymorphic DNA (RAPD) analysis of Atlantic Coast striped bass. Heredity (Edinb) 78: 32-40.

30. Smith PJ, Benson PG, Mcveagh SM (1997) Comparision of three genetic methods used for stock discrimination of orange Hoplostethys atlanticus: Alloenzyme mitochondrial DNA and Random Amplified Polymorphic DNA. Fish Bull 95: 800-811.

31. Mamuris Z, Apostolitis AP, Theodorou AJ, Triantaphyllidis C (1998) Application of random amplified polymorphic DNA (RAPD) markers to evaluate intraspecific genetic variation in red mullet (Mullus barbatus). Mar Biol 132: 171-178

32. Barman HK, Barat A, Yadav BM, Banerjee S, Meher PK, et al. (2003).Genetic variation between four species of Indian carp as revealed by random amplified polymorphic DNA assay. Aquaculture 217: 115-123.

33. Bardakci F, Skibinski DOF (1994) Applications of the RAPD technique in tilapia fish: species and subspecies identification. Heredity 73: 117-133.

34. Chong LK, Tan SG, Yusoff K, Siraj SS (2000) Identification and characterization of Malaysian river catfish, Mystus nemurus (C&V): RAPD and AFLP analysis. Biochem Genet 38: 63-76.

35. Garg RK, Silawat N, Sairkar P, Vijay N, Mehrotra NN (2009a) RAPD analysis for genetic diversity of two populations of Mystus vittatus (Bloch) of Madhya Pradesh, India. Afric J Biotechnol 8: 4032-4038.

18

Embryonic and Larval Development of Yellow Tail Catfish, *Pangasius pangasius*

Ferosekhan S*, Sahoo SK, Giri SS, Saha A and Paramanik M

Central Institute of Freshwater Aquaculture, Kausalyaganga, Bhubaneswar-751 002, India

Abstract

The yellow tail catfish, *Pangasius pangasius* embryonic and larval development study was carried out. The eggs were adhesive and transparent in colour with equal perivitelline space. First cleavage appeared at 00:49 ± 00:02 h resulting two equal blastomers. The eight cells, thirty two cell and morula stage appeared at 01:30 ± 00:06, 02:04 ± 00:10 and 03:43 ± 00:33 h respectively. The blastomeres looked overlapped during these multi-cell stages and the size got reduced during morula stage onwards. The fertilized eggs took 09:29 ± 01:24 and 25:27 ± 01:28 h respectively for attaining "C" shape embryo and hatching. The transparent larvae were 3-4 mm in length with compact oval shape yolk sac of 1.4-1.6 mm in length at hatching. Heart beat was detectable (2-3 times per minute) in the newly hatched larvae, whereas the mouth, barbells or elementary canal were not visible. The mouth was clearly visible in one day old larvae, which remained opened and complete closing of mouth with jaw movement was noticed at the age of 11-12 dph (days post hatch). The fins were not seen during their early life due to the encircling of a uniform membrane from behind the dorsal side to the posterior part of yolk sac. This continuous membrane started disintegrating during 5-10 dph, within which the caudal, pelvic, pectoral and dorsal fin started appearing. The 11 dph larva possessed dorsal, pectoral; pelvic and caudal fin possessed 6-7, 6-7, 5-6 and 19-20 fin rays, respectively. The larvae were resembled just like adult fish at 12 dph.

Keywords: Catfish; Embryonic development; *Pangasius pangasius*; Fertilization; Larval development

Introduction

Members under Pangasidae family are in demand for aquaculture activities due to their growth potentiality. Their contribution to aquaculture production is exemplary in south-east Asian countries [1,2]. Many species are also under research to explore their growth potentiality with an aim to supply fish protein to the growing population. *Pangasius pangasius* is one such species, which is native to Indian sub-continent and is found in major river system [3]. The species is gradually declined in the natural water bodies due to over exploitation to satisfy the market demand as well as degradation of its native ground due to anthropogenic causes, which pushed it to an endangered species [4]. The information on the biology of this species has been compiled [5]. It is also considered as an excellent species among other Pangasids for mollusc control in the aquaculture ponds. In spite of its potentiality as a candidate species, enough attempts have not been made for its captive production. Recently few literatures have come up on its induced breeding as a sign of beginning for the domestication [6-8]. Our Institution is also in command to develop its package of practice through several projects. It is vital that potential endemic species be researched with the information on the embryonic and larval development. The changes of features during larval development and to understand the organogenesis are of crucial important, which are essential during the development of management and rearing technology of any new species for seed production. The information on the embryonic and larval development is lacking in this catfish species. Considering the importance of *P. pangasius*, attempts were undertaken to study the detailed embryonic and larval history in the controlled condition.

Materials and Methods

Breeding protocol

The juveniles collected from river were maintained in the earthen pond till they attain maturity. The broods were selected for breeding operation during the month of July. The suitable females were selected on the basis of bulging abdomen and pinkish vent. Many occasion intra-ovarian oocytes were examined through catheter to see the uniformity of the oocyte size. Simultaneously the suitability of males was judged by seeing free oozing of milt. Ovaprim (sGnRHa + Domperidone) at the dosage level of 1.0 ml/kg body weight was injected to females for successful ovulation, whereas the males did not receive any hormone injection. The male and female were released separately in ferro-cement tanks (1.5 m diameter) with continuous aeration and showering. The tanks were well covered to avoid fish jumping and an outlet was provided for drainage of excess water. The females were found ready for stripping after 13-14 h of post injection. The eggs stripped from the females were mixed with the stripped milt of male for fertilization.

Observation of embryonic development

Few fertilized egg samples were collected in petridish to see the embryonic development under a dissecting microscope. The eggs containing uniform and round yolk sphere and smooth perivitelline space were considered for embryonic study. One of such egg was separated from each breeding attempt to study the embryonic development. The egg was kept in petridish and was put under water tap for continuous water exchange with an aim to provide oxygen to the developing embryo. Complete aeration was provided to all the tubs. Water temperature, pH, dissolved oxygen, nitrite-N and alkalinity were analyzed following standard methods [9]. Water quality parameters

*Corresponding author: Ferosekhan S, Aquaculture Production and Environment Division, Central Institute of Freshwater Aquaculture, Kausalyaganga, Bhubaneswar-751002, India, E-mail: feroseaqua@hotmail.com

recorded during the egg development were as follows: dissolved oxygen between 5.8-6.7 mg L^{-1}, temperature 27.5 to 28.5°C, pH 6.9-7.8, nitrite 0.002-0.004 mg L^{-1} and alkalinity 122-130 mg L^{-1}. The egg is checked at regular interval to record the timing of embryonic development for each stage. The developmental stage of the egg was also captured under microscope with photographic attachment (CKS41, Olympus). A total of two observations were made from the eggs collected from two breeding operations in different times in the season to record the maximum variability of developmental timing. Compound microscope fitted with ocular micrometer was used to record the size of fertilized egg.

Observation of larval development

The total length and yolk sac biometry of newly hatched larvae was recorded using ocular micrometer fitted to a compound microscope and the weight was taken with the help of electronic balance (XS 105, Mettler Toledo). The morphological changes of larvae in each day were recorded through compound microscope.

Statistical analysis

All the data related to timing of embryonic development were expressed in hour: minute (mean ± SD). The mean and standard deviation values of timing of embryonic development were calculated by using Microsoft Excel 2013.

Results

Embryonic development and Egg development

The events in embryonic development and their respective time of observation in *P. pangasius* are presented in Table 1. The ovulated or fertilized eggs were round, adhesive in nature and looked transparent. The diameter of eggs just after stripping and after fertilization was 1.09-1.28 and 1.2-1.45 mm, respectively. The egg membrane was fully separated by a thin space from the egg called perivitelline space (0.07-0.09 mm), which was equal all around and filled with yolk free clear fluid (Figure 1a).

Cleavage stages (Single cell to Sixty four cell stage)

The fertilized eggs of *P. pangasius* showed meroblastic cleavage. An outgrowth appeared at the animal pole with the accumulation of yolk free cytoplasm called as blastodisc or germinal disc stage at 00:21

± 00:01 h (Figure 1b). With the pass of time, the blastodisc became thick and a vertical line appeared over the blastodisc. The line moved down dividing the blastodisc into two blastomeres, representing two cell stages at 00:49 ± 00:02 h (Figure 1c). Further division of blastomere took place with the advancement of time to reach four (Figure 1d) and eight-cell (Figure 1e) stage at 01:07 ± 00:04 and 01:30 ± 00:06 h respectively. Sixteen (Figure 1f), thirty two (Figure 1g) and sixty four (Figure 1h) cell stage appeared at 01:51 ± 00:08, 02:04 ± 00:10 and 02:32 ± 00:20 h respectively. The blastomeres at these stages were unequal and reduced in size. The blastomeres were also overlapped and their boundaries were faintly visible.

Morula stage

This stage of development appeared at 03:43 ± 00:33 h. The blastomeres at this stage were divided into many cells. The small blastomeres resulted from repeated division of blastomers lost their boundaries and appeared compact. It looked just like a marigold flower at the animal pole (Figure 1i).

Blastula and gastrula stage

Blastula stage appeared at 05:12 ± 01:08 h. The flowery look of blastomeres at morula stage was compressed at this stage and the blastoderm occupied larger area. The blastomeres were completely lost their identity and moved in both the side of animal pole occupying 30-40% area over the yolk sphere (Figure 1j). Gastrula stage appeared at 07:27 ± 01:16 h (Figure 1k). The blastoderm over the vitelline sphere further spreaded in both the side compared to blastula stage, which covered about 60-70% area. The blastomeric cells over the yolk sphere were further compressed, giving an appearance of a thread line called as germ ring. One end of the germ ring observed to be little broad compared to the other end, indicating the future development of cephalic region. The "C" shape embryonic stage appeared at 09:29 ± 01:24 h (Figure 1l), where head and tail end was recognized. After another 6-7 h from this stage the differentiation of cephalic region, optic vesicle, dorsal fin fold and tail region were well marked. At this stage embryo appeared just like a miniature of fish larvae encircled over the yolk sphere.

Twitching and hatching

Twitching movement was observed at 21:30 ± 01:10 h of post fertilization. The embryo was fully formed with the clear differentiation of head and tail. The head portion was attached to yolk sac, whereas tail

Figure No.	Developmental stage	Time of appearance (h)	Characters
1a	Fertilization	00:00	Round, transparent and adhesive in nature
1b	Blastodisc	00:21 ± 00:01	An out growth over the vitelline sphere developed at animal pole
1c	Two cell	00:49 ± 00:02	First cleavage, where two cells over the yolk sphere was clearly visible.
1d	Four cell	01:07 ± 00:04	Second cleavage, where blastomeres were clearly visible and counted.
1e	Eight cell	01:30 ± 00:06	Third cleavage, where little overlapping of blastomeres was observed.
1f	Sixteen cell	01:51 ± 00:08	Fourth cleavage, where overlapping of blastomeres was observed and were placed in two rows.
1g	Thirty two cell	02:04 ± 00:10	Fifth cleavage where the blastomeres were visible in 2-3 layers.
1h	Sixty four cell	02:32 ± 00:20	Sixth cleavage, where overlapping of blastomeres was observed and were placed in 2-3 layers.
1i	Morula	03:43 ± 00:33	Blastodermal cells were very small due to repeated division and gave a flowery look at the animal pole.
1j	Blastula	05:12 ± 01:08	Difficult to recognize the blastomers. The blastoderm was compressed and occupied more than half area over the yolk sphere.
1k	Gastrula	07:27 ± 01:16	Thick layer of blastoderm or germinal ring occupied 3/4th area over the yolk sphere. One end of it was observed broad, which became the future cephalic part of the embryo.
1l	"C" shape embryo	09:29 ± 01:24	Embryo looked kidney shape over the yolk sphere with clear distinction of head and tail.
	Twitching	21:30 ± 01:10	Tail became free from yolk sphere and beat in quick succession to rupture the egg membrane.
1m	Hatching	25:27 ± 01:28	The larvae were transparent with straight body and having free swimming capability.

Time shown in the table for each developmental stage is the average value of two observations (Mean ± SD). Time is expressed as h:min ± h:min

Table 1: Different events of embryonic development of fertilized egg and their respective Time in *P. pangasius*.

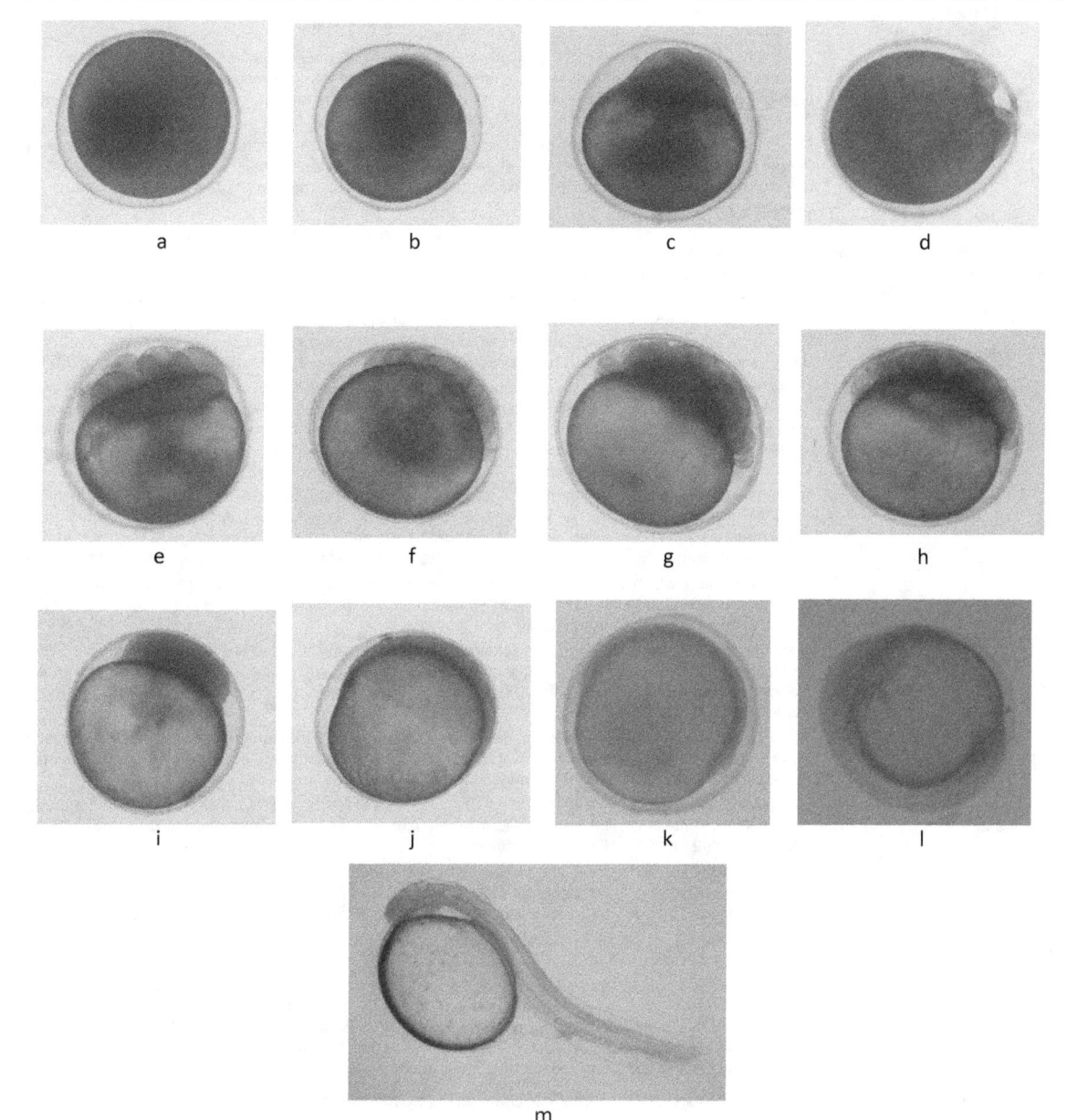

Figure 1: Embryonic development of *Pangasius pangasius* (a) fertilized egg (b) blastodisc (c) two cell (d) four cell (e) eight cell (f) sixteen cell (g) thirty two cell (h) sixty four cell (i) morula (j) blastula (k) gastrula (l) "C" shape embryo (m) hatchling.

became free. The beating of tail by the embryo just prior to hatching was very fast and continued for 7-10 second with a pause of 3-7.5 second. The larvae started hatching at 25:27 ± 01:28 h. The larvae remained dormant for few second just after hatching as if the embryo rested on the yolksac after which the activeness of larvae increased with the increase of tail movement. The tiny larvae were straight with the clear differentiation of head, trunk and tail (Figure 1m).

Larval development

The detail changes of morphometry in *P. pangasius* from the day of hatching to twelfth day of life are presented in Table 2. The larvae looked transparent with a straight body. The larvae were 3.0-4.5 mm in length bearing a round and compact yellowish yolksac, which were 1.4-1.6 mm in length and 1.2-1.4 mm in height. The mouth was not visible just

after hatching, but the mouth cleft was clearly visible after 12-13 h of hatching. Gradual reduction of yolksac was observed with the increase of age, which got completely absorbed at the end of third day. The alimentary canal also was visible at this age and the larvae accepted live mixed zooplankton as feed during their rearing in hatchery condition. The heart beat was observed to be 2-3 times per second in the newly hatched larvae. The mouth remained opened in one day hatchling and the gap between the jaws was gradually reduced as the larvae became older. Complete closing of both the jaws was observed during 9-11 days of life. Development of pigments were observed on the body surface in this larvae, which was first appeared on head surface followed by over the belly as well as just below the dorsal side and finally at the base of pectoral spine. These black pigments gradually increase in size and merged, giving a deep grey coloration to the head surface and dorsal

Age of larvae	Length (mm)/ Weight (mg)	Description
0 dph	3.0-4.5/0.8-1.1 (Yolksac Length: 1.4-1.6 mm; Yolksac height: 1.2-1.4 mm)	Larvae look silvery. The eyes are very small and appear like black dot on the anterior part of head. Head is attached to yolksac, hence looks little bent at the anterior portion. Yolksac is round and compact. Mouth or mouth cleft is not distinguishable. Small cleft is visible at the posterior part of yolksac, indicating future anus. A continuous membrane encircled from just behind the yolksac and continues in ventral portion to dorsal side encircling tail region.
1 dph	3.5-4.8/0.9-1.2 (Yolksac Length: 1.2-1.5 mm; Yolksac height: 1.0-1.2 mm)	Mouth remains opened and lower jaw shows occasional vibration. Vent is clear. No change of continuous membrane from ventral to dorsal side is marked as also seen in the 0 day old larvae. Barbells are not appeared clearly, but a rudimentary bulging or elevation is marked below the lower jaw. Heart looks transparent, but beats are clear, which were 2-3 times per second.
2 dph	4-5/1.2-1.6 (Yolksac Length: 0.9-1.1 mm; Yolksac height: 0.6-1.0 mm)	Mouth opening increases. Small bristle like structure is visible in lower and upper jaw indicating its teeth. Sometimes lower jaw closes for a while. Barbells of lower jaw extend up to yolksac. Tail fin is broad like a sark tail and appearance of rays noticed. But the continuous membrane over the body persists.
3 dph	6-7/2.4-3.0 (Yolksac Length: 0.4-0.6 mm; Yolksac height: 0.2-0.3 mm)	The yolksac does not show any round shape. The yolk content in the yolksac reduced drastically. The continuous membrane on the dorsal side reduced in size and showed dissolving sign with a gap in the dorsal side of peduncle. Tail looks single without any lobe as found in two days larvae. Fin rays are visible but not prominent. Barbells of upper jaw start developing. The visibility of heart beat is reduced. Dorsal, pectoral and pelvic fins are not originated.
4 dph	7-8/2.7-3.2	Lower jaw is movable but still mouth remains opened. Upper jaw barbell is clearly visible. Degeneration of continuous membrane on the dorsal side noticed and an elevation of fin is observed on the dorsal side nearer to neck indicating the initiation of dorsal fin formation. But the membrane at the ventral portion remains continuous as before. No yolk material is seen and elementary canal is visible. Heart beat is not seen due to increase of thickness of body. Bifurcation of caudal fin is initiated and fin rays appeared.
5 dph	7.5-8.5/3.5-4.2	Lower jaw barbells are longer than the upper jaw barbells. Dorsal side looks uniform without any membrane. Elevation of fin still persists as observed on the dorsal side nearer to the neck. The continuous membrane beyond anus to caudal region persists but a depression is marked just before the tail region on the ventral side of caudal peduncle. Faint rays in ventral fin appeared. Bifurcation of caudal fin is clear with 17 rays. Few pigments appear on the upper surface of head.
6 dph	8-9/4.6-5.1	Eye is prominent, which is 0.2 mm in diameter. The pigments on the head are increased in number. The barbells are 2.8-4.7 mm in length. Caudal fin deeply forked with 17 rays, where the upper lobe is longer than the lower one. Rays appear in ventral fin, which are 15 numbers approximately.
7 dph	12-13/12.8-13.4	Snout looks round, but mouth remains opened. Elementary canal is coiled and extends up to vent. Pigments on the head surface expand in size. Pigments in less numbers also appear just below the dorsal side and over the belly. Ventral fin clearly demarcated with 17-19 rays. A membrane is visible in both side of operculum indicating future pectoral fin. Adipose tissue clearly visible.
8 dph	10-15/16-21	The visibility of elementary canal in the belly drastically reduced. The pigments on the above portion of belly increased in number. Dorsal fin grows bigger with faint rays, which are not countable. Pectoral fin is clear with 3-4 fin rays. Pelvic fin is not at all clear but two bulged membranous structures are found just before the anus.
9 dph	14-18/22-28	Complete closing of mouth is not found, but lower jaw movement is frequent. The teeth in lower and upper jaw are clearly visible. The pigments on the head fused in many places. Dorsal fin is clear with 5-6 fin rays. The ventral and tail fin possesses 25-27 and 17-18 rays respectively.
10 dph	15-19/26-32	Black patch on the head surface and above the belly is observed due to merge of pigments. Few dots are also seen below the base of pectoral fin. Heart beats 50-53 times per minute. Pelvic fin is clear with 3-4 rays. Pectoral fin well developed with 4-5 fin rays. The ventral fin well developed with 23-25 fin rays. Tail fin also possesses 18-20 rays.
11 dph	18-22/26-44	The ventral portion of larvae look silvery and the dorsal portion look deep grey. The dorsal, pectoral, pelvic, ventral and caudal fin possesses 6-7(1+5-6), 6-7, 5-6, 24-25 and 19-20 fin rays.
12 dph	21-28/37-46	The color of the larvae is same as seen at the age of eleven days. Mouth is completely closed with lower jaw movement. Few pigments still persist at the base of pectoral spine. Different morphological parts like fins, adipose tissue, barbells etc are prominent. The dorsal, pectoral, pelvic, ventral and caudal fin possesses 6-7(1+5-6), 6-7 (1+5-6), 6-7, 25-26 and 19-20 fin rays, respectively. It resembles with that of an adult *P. pangasius*. The larvae at this stage swim actively and accept plankton and compound feed during their rearing in the hatchery.

Table 2: Brief description and summary of morphometrical changes of *P. pangasius* during its larval development.

side of fish at the age of eleven days. A continuous membrane was found in the newly hatched larvae from its dorsal side to the posterior part of yolksac. This membrane lost its originality by self-dissolution with the pass of time and different fins were developed synchronously. The well identified caudal fin developed first followed by ventral fin, pectoral fin, dorsal fin and pelvic fin. These fins respectively possessed 19-20, 25-26, 6-7(1+5-6), 6-7(1+5-6) and 6-7 fin rays at the age of twelve days old larvae. By this time the larvae resembled an adult fish of *P. pangasius*.

Discussion

Embryonic development

The ovulated eggs (1.09-1.28 mm) of *P. pangasius* increased to 1.2-1.45 mm in size after incubation of fertilized eggs in hatchery, which might be due to hydration of the eggs. Swelling of egg from 1.0-1.3 to 1.3-1.6 mm in *Rita rita* [10] and 1.1-1.2 to 1.3-1.5 mm in *H. fossilis* [11] has also been documented. The fertilized eggs were strongly adhesive and found in clutch among the eggs during egg incubation

in the hatchery. Many teleost under siluriformes show adhesive nature of the eggs [11-14]. This nature of eggs in *P. pangasius* could be due to sticky jelly like covering with radiating ridges on the egg surface as also reported in *Pseudobagrus ichikawani* [15]. The egg membrane got separated giving birth to the uniform perivitelline space. The yolk sphere pushed towards the vegetal pole as the embryonic development proceeded. This could be due to providing more space for the divisional activities of blastomeres at the animal pole. The clarity of blastomeres as in 2-4 cell stage was gradually reduced as the cleavage proceeded for 64 cell stage onwards. The identity of blastomeres was completely lost at morula and blastula stage. This loss of boundaries between the blastomeres might be due to repeated division and overlapping of cell resulting small and compact blastomeres at the animal pole as well as increased cellularity. This gradual change is a usual happening in the embryonic development of fertilized eggs in teleosts [12,16,17]. The sixty four cell stage and morula stage in the present study appeared at 2.53 h and 3.44 h respectively. In *P. sutchi* the 64 cell stage appeared earlier (1.35 h), whereas the morula stage appeared (2.1- 4 h) within the similar time range as seen in *P. pangasius*. The time variation or similarity at the same stage of development between these two close species is acceptable as also observed in the developmental stages between *O. bimaculatus* and *O. pabo* [14,18]. Even though the gastrula stage appeared at 7.27 h, further events in the germinal ring continued for another 8-9 h resulting clear visibility of cephalic region, optic vesicle, dorsal fin fold and tail region. This stage appeared at 19 h in *P. pardalis* [19] at 9-10 h in *O. pabo* [14] and at 11 h in *P. sutchi* [20]. Beating of tail by the embryo (twitching) was approximately 50-60 times per minute at 22-23 h post fertilization. Hatching occurred at 24-26 h (25:27 ± 01:28 h) in *P. pangasius*, which is greatly varying with other siluriformes: 18 h in *P. corruscans* [21] 22 h in *R. rita* [10] and 26 h in *C. batrachus* [22]. These variations are mostly related to species variability and due to water temperature. Islam [20] reported the inverse relation of hatching in Thai pangas with temperature fluctuation. The eggs took another 2-3 h more for complete hatching in the present study.

Larval development

The newly hatched transparent larvae have straight body posture and anterior part of head looks bent due to attachment to the yolksac. This type of morphometry is also observed in newly hatched catfish larvae of *P. sutchi* and *Mystus cavasius* [12,20,22]. The larvae were reported to be 3.0-4.5 mm length compared to 4-5 mm in *O. pabo* and 7.8 ± 0.12 mm in *P. pardalis*. This variation is related to the egg size between the species. The egg size reported for *O. pabo* and *P. pardalis* were 1.0-1.3 and 2-3 mm, respectively [14,19,23] were also agreed with the positive correlation of egg size with the hatchling. The round and compact yolksac got reduced as the hatchling grows in age and complete absorption took place at the end of third day of life when the larvae were nourished with external feed. In agreement to our observation, Islam [20] in *P. sutchi* and Jumawan et al. [19] in sucker mouth catfish reported gradual reduction of yolksac till 4-7 days after which larvae were ready for external feed. The development of mouth or barbell was not found in the just hatched out larvae. The mouth was clear at first day of hatching with an opening between the two jaws. The ambiguity of mouth formation was also reported in *Hemibagrus nemurus* just after hatching [24] which also varies 8-10 h in *C. striatus* [13] and 3-4 h in *Heterobranchus longifilis* [25]. The alimentary canal was also visible at day four post-hatch, just after the day of yolk absorption. This indicated the exogenous feeding to the larvae during their rearing. This period varies in the fish species like 48 h for *H. longifilis* [25] and 72 h for *Clarias batrachus* [26]. The pigments are also seen on the body of hatchlings in the present study like few other catfish species [11,19]. The pigments merged during latter part of life (11-12 days) to give deep grey colour to the dorsal part of the body. The change of body color to orange in *M. montanus* [27] and purple red in *C. striatus* [28] has also been documented. The continuous membrane present just behind the posterior part of yolk sac to the dorsal side started disintegrating during the age of 5-10 days of life, within which the caudal fin, ventral fin, pectoral fin and dorsal fin started appearing. Similar continuous membrane over the body is also found in *P. sutchi* and *H. fossilis* [11,20]. But the complete appearance of fins in *H. fossilis* is within 2-6 days post hatch, which is much less compared to *P. pangasius*. The fry of 11-12 days resembles like an adult fish.

Conclusion

The present study summarized the embryonic and larval development of *P. pangasius*. These findings can also contribute for a better understanding of the embryonic development in other Pangasid catfishes. The observations on its larval development may provide a basis for further studies on its ontogeny. The information on mouth development, day at first feeding and free swimming behaviour may provide knowledge to develop key management during its hatchery production.

Acknowledgement

We are very much grateful to the Director, Central Institute of Freshwater Aquaculture, Bhubaneswar for the kind support during this study.

References

1. De Silva S, Phuong NT (2011) Stripped catfish farming in the Mekong Delta, Vietnam: a tumultuous path to a global success. Reviews in Aquaculture 3: 45-73.

2. Jayasankar P, Giri BS (2013) Stripped catfish (Pangasinodon hypophthalmus) Gift from Vietnam to Indian aquaculture industry. Fishing Chimes 33: 36-40.

3. Talwar PK, Jhingran AG (1991) Inland fishes. Oxford and IBH Publishing Co. Pvt. Ltd. New Delhi

4. C.A.M.P. (1998) Conservation Assessment and Management Plan for Freshwater Fishes of India. Zoo Outreach Organization and National Bureau of Fish Genetic Resources, Lucknow, India.

5. Chondar SL (1999) Biology of finfish and shellfish. SCSC Publisher, Howrah, India.

6. Gupta SD, Reddy PVGK, Rath SC, Das Gupta S, Sahoo SK, et al. (1998) A note on artificial propagation of a teleosts, Pangasius pangasius (Ham). J Aqua 6: 23-26.

7. Khan MHK, Mollah MFA (2004) further trials on Induced breeding of Pangasius pangasius (Hamilton) in Bangladesh. Asian Fisheries Science 17: 135-14.

8. Sahoo SK, Giri SS, Chandra S (2008) Rearing performance of Clarias batrachus larvae: Effect of age at stocking on growth and survival during fingerling production. Aquaculture 280: 158-160.

9. APHA (1998) Standard methods for the estimation of water and waste water. American Public Health Association, American Water Works As-sociation, Water Environment Federation, Washington DC.

10. Mollah MFA, Talima K, Rashid H, Hossain Z, Sarowar MN, et al. (2011) Embryonic and larval development of critically endangered riverine catfish Rita rita. EurAsian Journal of BioSciences 5: 110-118.

11. Puvaneswari S, Marimuthu K, Karuppasamy R, Haniffa MA (2009) Early embryonic and larval development of Indian catfish, Heteropneustes fossilis. Eur Asian Journal of BioSciences 3: 84-96.

12. Rahman MR, Rahman MA, Khan MN, Hussain M (2004) Observation on the embryonic and larval development of Silurid catfish, Gulsha (Mystus cavasius Ham.). Pakistan Journal of Biological Science 7: 1070-1075.

13. Marimuthu K, Haniffa MA (2007) Embryonic and larval development of the stripped snakehead Channa striatus. Taiwania 52: 84-92.

14. Sarma D, Das J, Dutta A, Goswami UC (2012) Early embryonic and larval development of Ompok pabo with notes on its nursery rearing. Euro J Exp Bio 2: 253-260.

15. Watanabe K (1994) Mating behavior and larval development of Pseudobagrus ichikawai (siluriformes: Bagridae). Japan J Ichthyology 41: 243-251.

16. Thakur NK (1980) Notes on the embryonic and larval development of an air breathing catfish Clarias batrachus (Linn.). Journal of Inland Fisheries Society of India 12: 30-43.

17. Gonzalez-Doncel M, Okihir MS, Villalobos SA, Hinton DE, Tarazona V (2005) A quick reference guide to the normal development of Oryzias latipes (Teleostei, Adrianichthyidae). J Applied Ichthyology 21: 39-52.

18. Chakrabarty NM, Chakrabarty PP, Mondal SC, Sarangi N (2008) Embryonic development of pabda (Ompok pabda) with notes on its farming. Fishing Chimes 28: 55-59.

19. Jumawan JC, Herrera AA, Vallejo B (2014) Embryonic and larval development of the sucker mouth sailfin catfish Pterygoplichthys pardalis from Marikina river, Phillippines. EurAsian J Bioscience 8: 38-50.

20. Islam A (2005) Embryonic and larval development of Thai Pangas (Pangasius sutchi Fowler, 1937). Development Growth and Differentiation 47: 1-6.

21. Marques C, Nakaghi LSO, Faustino F, Ganeco LN, Senhorini JA (2008) Observation of the embryonic development in Pseudoplatystoma coruscans (Siluriformes: Pimelodidae) under light and scanning electron microscopy. Zygote 16: 333-342.

22. Das SK (2002) Seed production of magur (Clarias batrachus) using a rural model portable hatchery in Assam, India-A farmer proven technology. Aquaculture Asia 7: 19-21.

23. Bagarinao TU, Chua TE (1986) Egg size and larval size among teleosts: implications to survival potential.

24. Adebiyi FA, Siraj SS, Harmin SA, Christianus A (2013) Embryonic and larval development of river catfish Hemibagrus nemurus (Valenciennes, 1840). Asian J Ani Vet Adv 8: 237-246.

25. Ogunji JO, Rahe RE (1999) Larval development of the African catfish Heterobranchus longifilis Val, 1840 (Teleostei; Claridae) and its larval behavior. Journal of Aquaculture in the Tropics 14: 11-25.

26. Sahoo SK, Sahu AK, Giri SS, Chandra S (2006) Artificial propagation and rearing of Pangasius pangasius.

27. Arockiaraj AJ, Haniffa MA, Seetharaman S, Singh SP (2003) Early development of a threatened freshwater catfish Mystus montanus (Jerdon). Acta Zoologica Taiwanica 14: 23-32.

28. Yackob WAA, Ali AB (1992) Simple method for backyard production of snakehead Channa striatus (Bloch) fry. Naga 15: 22-23.

19

Evaluation of Seaweeds *Ulva rigida* and *Pterocladia capillaceaas* Dietary Supplements in Nile Tilapia Fingerlings

Malik M Khalafalla[1]* and Abd-elaziz M A El-Hais[2]

[1]*Department of Aquaculture, Faculty of Aquatic and Fisheries Sciences, Kafrelsheikh University, 33516–Kafr El-sheikh, Egypt*
[2]*Department of Animal Production, Faculty of Agriculture, Tanta University, Egypt*

Abstract

The present study was carried out to investigate the effect of green algae *Ulva lactuca* and red algae *Pterocladia capillacea* at 0.0, 2.5 and 5% on growth performance, feed utilization, carcass composition and blood indices of Nile tilapia, *Oreochromis niloticus* fingerlings. Fish (18.47 ± 1.25 gm) were randomly divided into fifteen aquaria in triplicates and fed diets contained about 29.51% total protein and 4.53 kcal/g gross energy. All the growth performance parameters and feed utilization values of experimental fish were increased significantly (P ≤ 0.05) by both of algae supplementation. Diet supplemented with 5% of *Ulva lactuca* had acceptable growth parameters compared with other diets. Fish fed supplemented diets had slight increases and decreases for carcass protein and lipids without significant differences (P ≥ 0.05). Also, no significant differences (P>0.05) were obtained for serum total protein, albumin and globulin and liver activity. It may be summarized that, algae supplementation especially at 5% of *Ulva lactuca* level may improves growth parameters and carcass composition without adverse effects on blood metabolites and liver activity.

Keywords: Algae; Growth Performance; Nile Tilapia

Introduction

The demand for feeds which are safer and effective than the traditional animal products especially plant origin ingredients mainly terrestrial and sometimes aquatic has recently renewed and increasing for aquafeeds preparation. Edible seaweeds or algae are a renewable natural resource existing in large quantities all along the Pacific coast.

In general, seaweeds are not considered a good source of PUNSFA [1], and the total lipid content has always been found to be greater than 4% [2], but their polyunsaturated fatty acids content can be equal to land plants content [3]. Besides that, seaweeds are not a main source of energy but they had a high nutritional value which related with acceptable vitamin, protein and mineral contents [4,5]. It is said that 100 g of seaweed provides more than the daily requirement of vitamin A, B2 and B12 and two thirds of the vitamin C requirement [6]. In addition to, seaweeds are an important source of dietary fiber, mainly soluble fiber [7], which is considered an important component for preventing constipation, colon cancer, cardiovascular disease and obesity, among others. The potential of algae as an alternate protein or feed supplement ingredient in aquatic feeds is currently being examined in many regions of the world [8], because of their high protein content and fast growth rate [9].

The green alga *Ulva lactuca* (Chlorophyta) and red alga *Pterocladia capillacea* (Rhodophyta) are among the dominant macroalgae along the Egyptian Mediterranean coast all the year around. They grow near the water level, in large amounts, and can easily be harvested by hand, from natural populations. In recent years, *Ulva* species have become important macroalgae, which have been investigated as a dietary ingredient for a wide range of fish species. Low-level dietary incorporation of *Ulva* meal has resulted in improved growth, feed utilization, physiological activity, disease resistance, carcass quality, and reduced stress response [10,11]. *Ulva* as one of seaweeds have a good vitamin and mineral profile and are especially rich in ascorbic acid [12,13]. Moreover, recent studies on *Ulva* meal addition in Nile tilapia [14] and rainbow trout diets [15] have been accepted to be good supplements for fish feeding. Also, many researchers have studied the

composition and the properties of polysaccharides from the red algae *Pterocladia capillacea*. *Pterocladia* as a red algae genus is used for industrial production of gelling galactans and is commonly distributed in the seas of Lebanon, Egypt, Brazil, Italy and other countries [16,17].

The Nile tilapia *Oreochromis niloticus* still the most widely cultured species of tilapia in Africa. Tilapia had positive aquacultural characteristics, adaption with their tolerance to poor water quality and resistance to viral, bacterial and parasitic diseases compared to other cultured fish, especially in Egypt at optimum temperatures for growth. Moreover, tilapia ingests a wide variety of natural food organisms. So, tilapia is a suitable fish to maximize production and meet human nutrition requirements under different culture conditions.

The purpose of the present study was to evaluate the nutritional value of the *Ulva lactuca* and *Pterocladia capillacea* species and their effects on feeding efficiency, growth performance, body composition and blood components of Nile tilapia fingerlings.

Material and Methods

This work was carried out at the Wet Fish Lab., Department of Animal Production, Faculty of Agriculture, Kafr elsheikh University, Egypt, during summer season, 2014.

Experimental fish

Nile tilapia fingerlings were bought from the private fish farm

---

*Corresponding author: Malik M. Khalafalla, Department of Aquaculture, Faculty of Aquatic and Fisheries Sciences, Kafrelsheikh University, 33516–Kafr El-sheikh, Egypt, E-mail: malikkhalafalla@yahoo.com

at Tolompate 7, Kafr El-Sheikh governorate. Prior to the start of the experiment, fingerlings were placed in a fiberglass tank and randomly distributed into glass aquaria to be adapted to the experimental condition until starting the experiment. All fish were fed the control diet during the first 7 days after stocking to adapt them to feeding and handling practices. After that, the fish were fed the experimental diets.

Experimental design of rearing fish

Fish (n=150; 18.47 ± 1.25 gm) were randomly divided into fifteen (70 L) aquaria in triplicates (10 fish per replication). One third of water aquaria were replaced daily and totally once every week after removing the wastes. Fresh tap water was stored in fiberglass tanks for 24 h under aeration for chlorination. Fifteen air stones were used for aerating the aquaria water. All aquaria were maintained under a constant photoperiod (12 h dark: 12 h light) created by fluorescent lamps. Fish feces and feed residue were removed daily by siphoning.

Experimental diets and feeding regime

Prior to the start of the experiment, the fish were adapted to a basal commercial diet [control diet (D1)] containing 29.55% crude protein (Table 2) for two weeks. Five experimental commercial diets were formulated to contain 0, 2.5 and 5% of green algae *Ulva lactuca* and red algae *Pterocladia capillacea*.

Algae were obtained from the Egyptian Mediterranean coast of Alexandria. The chemical analysis of *Ulva lactuca* and *Pterocladia capillacea* is presented in Table 1. A basal diet was formulated using the commercial ingredients. The dry ingredients were finely grounded and mixed by a dough mixer for 20 minutes for homogeneity. Oil was gradually added while mixing. After homogenous mixture, 40 ml water per 100 gm diet. Diet was slowly added to the mixture according to Shimeino et al. [18].

The diets were cooked on the water evaporator for 20 minutes and pelleted (3 mm) through fodder machine and the manufacture pellets were dried on oven at 70°C for 48 hours. The diets collected, tagged and stored in refrigerator at 4°C. Fish in all treatment were daily fed the experimental diets at a level of 5, 4 and 3% of the body weight daily for the first, the second and the third four weeks, respectively.

Feed amount was given three times daily (900, 1200 and 1500) in equal portions, six days a week for 12 weeks. Fish were weighed

Chemical composition	*Ulva lactuca*	*Petrocladia capillacea*
DM	80.98	79.01
CP	20.33	18.92
CF	9.87	12.02
EE	3.21	2.74
NFE	48.34	44.99
ASH	17.98	20.95
NDF	38.85	40.43
ADF	24.67	26.03
ADL	7.67	8.01
Hemicellulose	14.01	14.36
Cellulose	17.13	17.78
Calories, kcal	303.54	282.24
Minerals composition		
Sodium	195.8	205.1
Potassium	96.9	94.9
Calcium	69.8	68.7
Magnesium	207.0	196.5

Table 1: Chemical composition of *Ulva lactuca* and *Pterocladia capillacea*.

Ingredients	Diet[6] No.				
	D1	D2	D3	D4	D5
Fish meal	100	100	100	100	100
Soybean meal	380	375	380	375	380
Yellow corn	200	200	200	200	200
Wheat bran	120	100	100	100	100
Rice bran	150	150	120	150	120
Oil mixture[1]	30	30	30	30	30
Vitamin and mineral premix[2]	20	20	20	20	20
Ulva lactuca	0	25	50	0	0
Pterocladia capillacea	0	0	0	25	50
Total	1000	1000	1000	1000	1000
Proximate composition and energy content (% dry matter basis)					
Dry matter	90.35	90.54	90.41	90.45	90.38
Crude protein	29.55	29.48	29.62	29.42	29.50
Crude lipid	8.97	8.87	8.76	8.66	8.57
Crude fiber	5.43	5.64	5.78	5.86	6.05
Crude ash	7.54	7.64	7.87	8.12	8.20
Nitrogen free extract	48.51	48.37	47.97	47.94	47.68
GE (kcal/kg diet)[3]	456.4	454.5	452.6	450.5	448.9
DE (kcal/kg diet)[4]	342.3	340.9	339.5	337.9	336.7
P/E ratio (mg CP/kcal DE)[5]	86.4	86.5	87.2	87.1	87.6

[1]Mixture of sunflower oil and linseed oil with a ratio of 1:1.

[2]Commercial vitamin (Super Vit, Arab Veterinary Industrial Co., Jordan). 15,000 IU vitamin A, 0.7 g vitamin C (Stay C®, 35% active), 15,000 IU vitamin D3, 2 mg vitamin E, 2.5 mg vitamin B2, 2 mg vitamin K3, 10 mg nicotineamide, 3 mg vitamin B6, 5 mg vitamin B12, 2 mg vitamin B1, 2 mg folic acid, 5.5 mg Ca-D-pantothenate and mineral premix (Eco Vit, Egyptian Veterinary Produced and Feed a Additives Co., Demyatta, Egypt). 200 g calcium, 90 g phosphate, 40 g sodium, 2.5 g copper, 48 g magnesium, 3.6 g manganese, 23.5 g zinc, 8 g iron, 450 mg cobalt, 200 mg iodine and 20 mg selenium.

[3]GE (Gross energy) was calculated according to NRC [19] by using factors of 5.65, 9.45 and 4.22 K cal per gram of protein, lipid and carbohydrate, respectively.

[4]DE (Digestible energy) was calculated by applying the coefficient of 0.75 to convert gross energy to digestible energy according to Hepher et al. [20].

[5]P/E (protein energy ratio)=crude protein×1000/digestible energy/100 g, according to Hepher et al. [20].

[6]Diets: D1 (control): without supplements, D2: 2.5% *Ulvalactuca*, D3:5% *Ulvalactuca* and D4: 2.5% *Pterocladiacapillacea* and D5 5% *Pterocladiacapillacea*.

**Values of diets content were within the range suggested for tilapia by Jauncey and Ross [21] and NRC [19].

Table 2: Formulation and chemical proximate composition of the experimental diets (g/kg dry weight basis).

biweekly and feed amounts were adjusted on the basis of the new weight. Mortality was monitored daily and recorded.

Analytical procedures

Moisture, crude protein (%N x 6.25), crude lipid, crude fiber and ash contents of diet ingredients and a sample of fish at the beginning and end of the experiment were determined in triplicate according to A.O.A.C [19-22]. Gross energy (GE) contents of the experimental diets and fish samples were calculated by using factors of 5.65, 9.45 and 4.22 kcal/g of protein, lipid and carbohydrates, respectively [19].

Measurements of water parameters

Water samples were taken each two days for ammonia and pH analysis. Analytical methods were done according to the American Public Health Association (APHA) [23]. Water temperature and oxygen level were measured daily at 8 o'clock by (Oxygen meter model 9070). In all treatments water quality parameters for water temperature ranged between 27.00 to 28.00°C, pH ranged from 7.01 to 7.25; dissolved oxygen ranged from 5.75 to 6.12 mg/L and water ammonia ranged from 0.06 to 0.09 mg/L. In general, all the water quality parameters were within the acceptable ranges for fish growth.

Measurements of growth and feed utilization parameters

Body weight of fish in each aquarium was measured at start and every two weeks during the experimental period (12 weeks). Diet performance was evaluated as follows:

Average weight gain AWG (g/fish)=Wt – W0.

Average daily weight gain ADG (g/fish/day)=Wt – W0/t.

Specific growth rate % day SGR (%/day)=100×(In Wt – InW0)/t

**Where Wt is weight of fish at time t, W0 is weight of fish at time 0, and t is the experimental period in days.

Feed conversion ratio, FCR=dry feed fed/wet weight gain.

Protein efficiency ratio, PER=wet weight gain/ Protein fed.

Protein productive value, PPV (%)=100×(protein gain/protein fed).

Survival rate, SR=100(Total No. of fish at the end of the experimental/Total No. of fish at the start of the experiment].

ER (%)=(%Energy in fish carcass (kcal) at the end-Energy in fish carcass (kcal) at the start)×100/Energy intake (kcal)

Blood parameters determination

At the end of the experiment fish in each aquaria were weighed and three blood samples were taken randomly from the caudal vein for blood analysis and differential leukocyte count, Anti coagulated blood samples were prepared immediately for counting red and white blood cells etc. Red blood cells count (RBCs×10^6/mm^3) and white blood cells count (WBCs×10^3/mm^3): were measured on an A bright-line Haemocytometer model (Neubauer improved, Precicolor HBG, Germany) by using a commercial kits (Ranox company, Germany) according to the method described by Stoskopf [24]. Hemoglobin concentration (Hbgm/dl): was determined according to the method of Zinkl [25]. Packed cell volume (PCV %): was estimated by the microhaematocrite method as described by Dacie and Lewis [26]. A total protein (TP) was measured according to the method of Henry [27] using reagent kits obtained from Diamond Diagnostic Company (Egypt). ALT (U|L) and AST (U|L): Alanine Aminotransferase (ALT) and Aspartate Aminotransferase (AST) activities were assayed according to the method of Reitman and Frankel [28] using reagent kits purchased from Randox Company (UK).

Statistical analysis

The obtained numerical data were statistically analyzed using SPSS [29] for one-way analysis of variance. When F-test was significant, least significant difference was calculated according to Duncan [30].

Results and Discussion

Growth performance parameters of Nile tilapia (O. niloticus) fed on the experimental diets is presented in Table 3.

The data of the present study showed that, final body weight (FBW), average total gain (ATG), average daily gain (ADG), and specific growth rate (SGR) of experimental fish were influenced significantly (P ≤ 0.05) by different levels of Ulva lactuca and Pterocladia capillacea algae. Fish fed on diet supplemented with 5% Ulva lactuca (D3) had higher increases of growth parameters as FBW, ATG, ADG and

Items	Diets No (On DM basis, %)					SE*
	D_1, Control	D_2	D_3	D_4	D_5	
Initial weight, g/fish	18.57	18.24	18.67	18.35	18.55	1.25
Final body weight, g/fish	47.52^c	55.86^{ab}	60.45^a	53.54^b	55.28^{ab}	2.31
Average total gain[1], g/fish	28.95^c	37.62^{ab}	41.78^a	35.19^b	36.73^{ab}	1.54
Average daily gain[2], g/fish/day	0.34^c	0.45^{ab}	0.50^a	0.42^b	0.44^{ab}	0.01
Specific growth rate[3] (SGR %/day)	1.12^c	1.33^b	1.39^a	1.27^{bc}	1.30^b	0.30
Survival rate[4], %	97	97	100	97	100	3.12
Feed intake (FI), DM g/fish	55.76	56.98	57.24	58.50	57.36	2.41
Feed conversion ratio[5] (FCR)	1.74^a	1.37^c	1.24^d	1.50^b	1.41^c	0.31
Protein efficiency ratio[6] (PER)	1.94^d	2.47^b	2.73^a	2.26^c	2.40^b	0.45
Protein productive value[7] (PPV, %)	13.10^c	15.32^b	17.87^a	12.78^c	15.70^b	1.14

*Means in the same rows having different superscript letters were significantly different at 0.05 levels

Table 3: Growth performance parameters of Nile tilapia (O. niloticus) fed on the experimental diets.

SGR compared to other experimental fish groups. Also, fish fed diets supplemented with both algae at 5% had the best survival rate (SR) percentage (100%).

The results of feed utilization appeared that, no significant differences were detected (P>0.05) among supplemented fish groups for feed intake (FI). On the other hand, FCR of experimental fish groups was affected significantly (P ≤ 0.05) by algae supplementation. The lower FCR ratio was obtained by D3 (1.24), while the higher FCR was detected by control diet (D1, 1.74). Moreover, PER and PPV were influenced significantly (P ≤ 0.05) by using different levels of Ulva lactuca and Pterocladia capillacea. The preferable values of PER and PPV were observed with Ulva lactuca supplementation at 5 and 2.5%, respectively. Our data are in agreement with those reported for Nile tilapia (Oreochromis niloticus) using 5% U. rigida dietary supplementation for better growth and nutrient utilization [13]. Diler et al. [31] found that the inclusion of Ulva meal at 5 to 15% replacing wheat meal in carp diets improved the growth parameters and could be acceptable for common carp. Moreover, Guroy et al. [32] found that higher values were obtained for weight gain of Nile tilapia fed on diets supplemented with various levels of Ulva meal (5 to 10%). Besides, feeding trial with previous Ulva meal for mullet (Mugil cephalus) was successful and related with higher growth performance, feed intake and protein utilization [33]. These findings confirm the positive effects reported on promoted growth performance and survival rate of gilthead seabream fry (S. aurata) with the addition of 5% Ulva meal (UM) or 10% pterocladia meal (PM) to their diets which probably due to the relatively high protein and mineral contents and good essential amino acid profile of both algae meals [10]. On the other hand, Kissil et al. [34] observed no significant effect (P>0.05) after the inclusion of Ulva meal in the diets of grow-out gilthead seabream on growth rate but found a slightly better protein utilization at the highest tested level of 8%. Wassef et al. [35] indicated that feeding seabass only at low level (5%) of Ulva or Ptercladia meal produced the best growth, feed utilization, nutrient retention, and survival rates among all the dietary groups. Elmorshedy [36] reported that, there were positive trends between protein efficiency ratio (PER) and seaweeds inclusion levels in the diet up to the level of 14%.

Body composition of Nile tilapia

Effect of Ulva lactuca and Pterocladia capillacea algae

supplementation on Nile tilapia body composition is shown in Table 4. Data of this experiment explained that, the addition of algae had no significant influence (P>0.05) on DM, CP, lipids and ash content of fish body composition. Fish fed with the algae diets recorded relatively higher dry matter, protein and ash but lower lipids content than those of the respective control diet. Fish fed diet supplemented with *Ulva lactuca* at 5% had a higher content of CP (59.62%) and lower lipid content (8.55%) of fish body composition compared to other experimental fish groups. In agreement with our data, Olvera-Novoa et al. [37] investigated that the body composition of Nile tilapia was not clearly affected by the inclusion of *spirulina* from 20 to 100% in experimental diets. Besides, the partial replacement of fish meal by *Gracilaria bursa-pastoris* and *Ulva rigida* up to 10%, and by *Gracilaria cornea* at 5% did not affect the whole-body composition of sea bass, whereas fish fed 10% *Gracilaria cornea* exhibited higher ash content [10]. Carcass protein content in the present study increased and lipid content decreased gradually with increasing *Ulva* and *pterocladia* algae level in the diet to 5% without significant differences. This agrees with the results obtained by Azaza et al. [38] who observed that increasing supplemental levels of *Ulva rigida* meal decrease carcass lipid content of the Nile tilapia. In contrast, the feeding of algae was reported to enhance the body lipid in red sea bream [39]. *Ulva* species have a good vitamin and mineral content and are especially rich in ascorbic acid or vitamin C [11,12]. Vitamin C is considered as a promoter for lipid metabolism, which may effects on body metabolism and composition. The pervious reasons may result in the alteration of body nutrients deposition in fish, and therefore decrease carcass lipid and save a protein nutrient for tissue development [40,41].

Biochemical blood parameters

Table 5 explained the hematological and biochemical parameters of Nile tilapia fingerlings fed on the experimental diets is presented in.

Items	Initial Fish	Diets					SE*
		D$_1$, Control	D$_2$	D$_3$	D$_4$	D$_5$	
Dry matter	24.50	26.54	26.98	27.24	26.87	27.10	0.25
Crude protein	55.12	58.21	58.67	59.62	57.65	58.68	2.36
Lipids	9.21	8.59	8.67	8.55	8.96	8.76	0.54
Ash	13.25	14.21	13.68	14.02	14.25	14.26	0.20

Table 4: Effect of experimental diets on Nile tilapia body composition (%, on DM basis).

Items	Treatments					SE*
	Control, D1	D2	D3	D4	D5	
Blood picture						
T. W. BC's (10^3/mm^3)	22.35	21.54	23.54	22.01	22.78	2.15
T. R. BC^{2}s (10^6/mm^3)	3.54	3.65	3.47	3.21	3.84	0.10
Hb^3(g/dl)	6.70	7.21	7.24	6.89	6.84	0.44
Biochemical parameters Protein						
Total Protein (g/dl)	3.21	2.89	3.14	3.22	3.34	0.34
Albumin (g/dl)	1.34	1.45	1.39	1.28	1.54	0.08
Globulin (g/dl)	1.87	1.44	1.75	1.94	1.80	0.12
Liver function						
AST^5 (U/L)	54.21	55.32	53.45	54.25	56.31	3.14
ALT^6 (U/L)	22.54	23.10	24.53	22.45	21.03	0.18

*Means in the same columns having different superscript letters were significantly different at 0.05 levels.

**TWBC^1=Total White Blood Cell, TRBC^2=Total Red Blood Cells, Hb^3=Hemoglobin Concentrations; ALT^5=Alanine Aminotransferase and AST^6=Aspartat Amino Transferase.

Table 5: Hematological and biochemical parameters of Nile tilapia fed on the experimental diets. Aminotransferase.

Algae supplementation had no significant (P>0.05) on blood serum WBC, RBC and Hb concentrations with different levels and strain. There were slight increases with supplemented *Ulva* for the pervious parameters compared to other experimental diets. Also, blood protein fractions as albumin, globulin and total protein were not affected significantly (P>0.05) by both of algae addition. The higher blood concentration of total protein and albumin were found with D5 (3.34 and 1.54 g/dl, respectively) while, a higher blood concentration was obtained by D4 (1.94 g/dl).The results explained that, the function of liver through the blood ALT and AST concentrations was insignificantly influenced (P>0.05) by *Ulva* and *pterocladia* addition. Our results agree with Promya and Chitmant [42] who reported that fingerlings which received 5% *A. platensis* diets had higher values for red and white blood cell counts and immunity stimulating capacity. Jongkon and Chanagun [43] found that catfish fed with 5% *Spirulina platensis* (SD) exhibited higher red and white blood cell counts and a higher immunity stimulating capacity (measure by a lysozyme activity assay). Also, Fingerlings, which received feed 5% SD had higher values for red and white blood cell counts. Increased in WBC and RBC refer that *Ulva* and *pterocladia* in the diet may benefit the immune system in Nile tilapia as well. *Ulva* spp contains carotenoids, which affect the health of fish, specifically improving the ability to resistant the infections via the reduction of stress [44]. Also, this results attributed to the algae contain a different minerals and vitamins that improve the immunity of the fish.

Conclusions

The present work suggests that using red alga *P. capillacae* or green alga *U. lactuca*, at the optimum inclusion levels, in Nile tilapia diets is feasible which will lead to improved growth performance, feed utilization, survival rate and fish body composition. Moreover, the dietary supplements can impart beneficial effects on immunity and liver activity.

Acknowledgments

The authors wish to thank the team works of the faculty of agriculture, Tanta University, Egypt and Faculty of Aquatic and Fisheries Sciences, Kafrelsheikh University, Egypt for their technical assistance during this study.

References

1. Ratana-arporn P, Chirapart A (2006) Nutritional evaluation of tropical green seaweeds Caulerpa lentillifera and Ulva reticulate. Kasetsart J Nat Sci 40: 75-83.

2. Herbreteau FL, Coiffard JM, Derrien A, De RoeckHoltazahuer Y (1997) The fatty acid composition of five species of macro algae. Bot Mar 40: 1-6.

3. Sanchez-Machado DI, Lopez-Cervantes J, Lopez-Hern'andez J, Paseiro-Losada P (2004) Fatty acids, total lipid, protein and ash contents of processed edible seaweeds. Food Chem 85: 439-444.

4. Chan JCC, Cheung PCK, Ang PO (1997) Comparative studies on the effect of three drying methods on the nutritional composition of seaweeds Sargassum hemiphyllum (Turn) C. Ag. J Agri Food Chem 45: 3056-3059.

5. Norziah MH, Ching ChY (2000) Nutritional composition of edible seaweed Gracilariachanggi. Food Chem 68: 69-76.

6. Chapman VJ, Chapman DJ (1980) Seaweeds and their uses.

7. Lahaye M (1991) Marine algae as sources of fibers: determination of soluble and insoluble dietary fiber contents in some sea vegetables. J Sci Food Agri 54: 587-594.

8. Shields RJ, Lupatsch I (2012) Algae for aquaculture and animal feeds. Technikfolgenabschätzung-Theorie und Praxis 21: 23-37.

9. Mustafa MG, Nakagawa H (1995) A review: dietary benefits of algae as an additive in fish feed. Israel J Aquacult 47: 155-162.

10. Wassef EA, El-Sayed AFM, Kandeel KM, Sakr EM (2005) Evaluation of

Pterocladia and Ulva meals as additives to gilthead seabream Sparus aurata diets. Egypt J Aquat Res 31: 321-332.

11. Valente LMP, Gouveia A, Rema P, Matos J, Gomes EF, et al. (2006) Evaluation of three seaweeds Gracilaria bursa-pastoris, Ulva rigida and Gracilaria cornea as dietary ingredients in European sea bass Dicentra rchuslabrax juveniles. Aquaculture 252: 85-91.

12. Ortiz J, Romero N, Robert P, Araya J, Lopez- Herna´ndez J, et al. (2006) Dietary fiber, amino acid, fatty acid and tocopherol contents of the edible seaweeds Ulva lactuca and Durvillaea antarctica. Food Chem 9: 98-104.

13. García-Casal MN, Pereira AC, Leets I, Ramírez J, Quiroga MF (2007) High iron content and bioavailability in humans from four species of marine algae. J Nutr 137: 2691-2695.

14. Ergün S, Soyutürk M, Güroy B, Güroy D (2009) Merrifield, Influence of Ulva meal on growth, feed utilization and body composition of juvenile Nile tilapia (Oreochromis niloticus) at two levels of dietary lipid. AquacultInt 17: 355-361.

15. Güroy D, Güroy B, Merrifield DL, Ergün S, Tekinay AA, et al. (2011) Effect of dietary Ulva and Spirulina on weight loss and body composition of rainbow trout, Oncorhynchus mykiss (Walbaum), during a starvation period. J Anim Physiol Anim Nutr (Berl) 95: 320-327.

16. Wassef EA, El Sayed AM, Kandeel KM, Mansour HA, Sakr EM (2002) Effect of feeding Pterocladia and Ulva meals in diets for gilthead bream Sparus aurata. Cahiers Options Mediterranean's.

17. Bottalico A, Foglie CID, Fanelli M (2008) Growth and reproductive phenology of Pterocla diellacapillacea (Rhodophyta: Gelidiales) from the southern Adriatic Sea. Botanica Marina 51: 124-131.

18. Shimeino S, Masumoto T, Hujita T (1993) Alternative protein sources for fish meal diets of young yellow tail. Nippon Suisan Gakkaishi 59: 137-143.

19. NRC (1993) Nutrient requirements of fish. National Academy Press, Washington DC, USA.

20. Hepher B, Liao IC, Cheng SH, Haseih CS (1983) Food utilization by red tilapia- Effect of diet composition, feeding level and temperature on utilization efficiency for maintenance and growth. Aquaculture 32: 255-272.

21. Jauncey K, Ross B (1982) A guide to tilapia feeds and feeding Ins. Aquaculture Univ, Scotland, U.K.

22. AOAC (2000) Association of Official Analytical Chemists. Official Methods of Analysis.AOAC, Arlington, Virginia, USA.

23. APHA American Public Health Association (1985) Standard methods for the examination of water and waste. 12th addition, Inc. New York.

24. Stoskopf MK (1993) Fish Medicine, WB Saunders Company, Harcourt.

25. Zinkl JG (1986) Veterinary Hematology, Philadelphia, Paihea and Fibiger, 256-260.

26. Dacie SIV, Lewis SM (2006) Practical Haematology. 10th ed., Churchill Livingstone, London.

27. Henry RJ (1964) Colorimetric determination of total protein. In: Clinical Chemistry. Harper and Row Publ., New York, USA.

28. Reitman S, Frankel S (1957) A colorimetric method for the determination of serum glutamic oxalacetic and glutamic pyruvic transaminases. Am J Clin Pathol 28: 56-63.

29. SPSS (1997) Statistical package for the social sciences, Versions 6, SPSS in Ch, Chi-USA.

30. Duncan MB (1955) Multiple ranges and multiple F-tests. Biometrics 11:1-42.

31. Diler IA, Tekinay A, Guroy D, Guroy BK, Soyuturk M (2007) Effects of Ulva rigida on the growth, Feed intake and body composition of Common Carp, Cyprinus carpio L. J Biol Sci 7: 305-308.

32. Guroy BK, Cirik, Guroy S, Sanver D, Tekiny AA (2007) Effects of Ulva rigida and Cystoseira barbata meals as a feed additive on growth performance, feed utilization and body composition of Nile tilapia, Oreochromis niloticus. Turkey J Vet Anim Sci 31: 91-97.

33. El Masry MH, Eissa MA, Mikhail FR (2001) Evaluation of five supplementary feeds for grey mullet Mugil cephalus fry. Egypt J Nutrition and Feeds 4: 731-741.

34. Kissil GW, Lupatsch I, Neori A (1992) Approaches to fish feed in Israeli mariculture as a result of environmental constraints.

35. Wassef EA, El-Sayed AM, Eman M, Sakr (2013) Pterocladia (Rhodophyta) and Ulva (Chlorophyta) as feed supplements for European seabass, Dicentrarchuslabrax L., fry. J Appl Phycol 25: 1369-1376.

36. Elmorshedy I (2010) Using of algae and seaweeds in the diets of marine fish larvae.

37. Olvera-Novoa MA, Dominguez-Cen IJ, Olivera Castillo L, Martinez-Palacios CA (1998) Effects of the use of the microalga Spirulina maxima as fishmeal replacements in diets for tilapia, Oreochromis mossambicus, Peters, fry. Aquac. Res. 29: 709-715.

38. Azaza MS, Mensi F, Ksouri J, Dhraief MN, Brini B, et al. (2008) Growth of Nile tilapia Oreochromis niloticus L fed with diets containing graded levels of green algae Ulva meal (Ulva rigida) reared in geothermal waters of southern Tunisia, Journal of Applied Ichthyology 24: 202-207.

39. Mustafa MG, Wakamatsu S, Takeda T, Umino T, Nakagawa H (1995) Effects of algae meal as feed additive on growth, feed efficiency and body composition in red sea bream. Fishery Science 61: 25-28.

40. Miyasaki T, Sato M, Yoshinaka R, Sakaguchi M (1995) Effect of vitamin C on lipid and carnitine metabolism in rainbow trout. Fish Sci 61: 501-506.

41. Om HJA, Yoshimatsu T, Hayashi M, Umino T, Nakagawa H, et al. (2003) Effect of dietary vitamins C and E fortification on lipid metabolism in red sea bream Pagrus major and black sea bream Acanthopagrusschlegeli. Fish Sci 69: 1001-1009.

42. Promya J, Chitmant C (2011) The effects of Spirulina platensis and Cladophora algae on the growth performance, meat quality and immunity stimulating capacity of the African Sharptooth Catfish (Clarias gariepinus). Int J Agri Bio 13: 77-82.

43. Jongkon P, Chanagun C (2011) The effects of Spirulina platensis and Cladophora. Algae on the Growth Performance, Meat Quality and Immunity Stimulating Capacity of the African Sharptooth Catfish (Clarias gariepinus). Int J Agric Biol 13: 77-82.

44. Nakono T, Yamaguchi T, Sato M, Iwama GK (2003) Biological Effects of Carotenoids in Fish, International Seminar "Effective Utilization of Marine Food Resource", Songkhla, Thailand.

20

Gene2Path: A Data Analysis Tool to Study Fish Gene Pathways by Automatic Search of Orthologous Genes

Natalia Ballesteros[1], Néstor Aguirre[2], Julio Coll[3], Sara I Pérez-Prieto[1] and Sylvia Rodríguez Saint-Jean[1]*

[1]*Centro de Investigaciones Biológicas (CSIC), C/ Ramiro de Maeztu 9, 28040 Madrid, Spain*
[2]*Instituto de Física Fundamental (CSIC), C/ Serrano 123, 28006 Madrid (Spain)*
[3]*Instituto Nacional de Investigaciones Agrarias (INIA), Crta La Coruña km7, Madrid 28040, Spain*

Abstract

Most of the gene regulation pathways data from biochemical and molecular experiments are drawn from humans or from species commonly used as experimental animal models. Accordingly, the software packages to analyse these data on the basis of specific gene identification codes (IDs) or accession numbers (AN) are not easy to apply to other organisms that are less characterized at the genomic level. Here, we have developed the Gene2Path programme which automatically searches pathway databases to analyse microarray data in an independent, species-specific way. We have illustrated the method with data obtained from an immune targeted rainbow trout microarray to search for orthologous pathways defined for other well known biological species, such as zebrafish, although the software can be applied to any other case or species of interest. The scripts and programme are available and free at the "GENE2PATH" web site http://gene2path.no-ip.org/cgi-bin/gene2path/index.cgi. A user guide and examples are provided with the package. The Gene2Path software allows the automated searching of NCBI databases and the straightforward visualization of the data retrieved based on a graphic network environment.

Keywords: Pathway analysis tool; Microarrays; Species-independent pathway analysis; Orthologous genes

Introduction

The use of high-density microarrays had a significant impact on studies of gene expression, attracting much interest among biologists. Microarray technology have been used to test the expression of thousands of genes in a single experiment, exploiting the ability of messenger RNA (mRNA) to bind specifically to the DNA template from which it was derived. Microarray gene expression screening can identify the genes involved in a given process, as well as predict interactions among thousands of genes by studying genome transcription. Many fields have benefited considerably from DNA microarray technology, such as drug discovery and toxicological research [1,2], as well as human disease diagnosis. However, studies in the field of veterinary sciences have been more restricted due to the lack of their genome sequences. In fish, for instance, most microarray and/or RNAseq studies have been carried out on zebrafish (*Danio rerio*), which is a well characterized species with large amount of sequenced genome for which commercial arrays are available. However, other economically relevant cultured fish are still far from having their complete genome sequenced, such as turbot (*Scophthalmus maximus*), sea bass (*Dicentrarchus labrax*), sea bream (*Sparus aurata*). Although some commercial microarrays have recently been made available for some of these species such Atlantic salmon (*Salmo salar*) and rainbow trout (*Oncorynchus mykiss* Walbaum); their genome coverage is still far from that of the zebrafish [3].

Several studies have highlighted the importance of presenting microarray data in the framework of documented biological pathways [4,5]. Typically, microarray gene expression experiments produce long lists of genes that are differentially expressed in two different circumstances. The information regarding these pathways is difficult to apply to species that are not well characterized, mainly because most of the actual software packages use species-specific gene identification data (IDs) from a few biological species that cannot handle genomic data for other less well known species.

To circumvent this problem it has been proposed that "....most of the genetics and physiology of the less well-represented species will be similar or comparable with the data of human and laboratory animal species stored in the database" [6]. Thus, ranking pathways in terms of their relevance to a particular phenotype or metabolic route can help researchers focus on a few sets of genes and such an approach may be very useful to answer some biological hypotheses.

However most biologists and veterinarians who are not familiar with simultaneous upload of thousands of data may have some difficulty because both, the need to configure on a local computer and the excessively long computing times required for analyzing several genes at once, are prohibitive.

In this study, we present an open access web server called Gene2Path for analyzing microarrays results and automatic searches of orthologous genes to be associated in pathways.

Therefore, the main objective of the present study was to develop and validate a tool that extrapolates information associated with different pathways across different species. The programme provides a software tool that uses species-independent gene IDs and streamlines that process searching for information regarding pathways in online public databases. The software uses lists of genes in microarrays (IDs), combining pathway information with this microarray data. Other researchers working with less well studied species, such as the chicken, have also faced the same problem when analyzing the pathways associated with the data derived from microarrays. A tool to study a

*Corresponding author: Sylvia Rodríguez Saint-Jean, Centro de Investigaciones Biológicas (CSIC), C/ Ramiro de Maeztu 9, 28040 Madrid, Spain
E-mail: sylvia@cib.csic.es

chicken-specific reaction to bacterial infection was developed by Pas et al. [6] based on a set of PERL scripts to extract data from databases via internet and using the names of the genes in the microarray. The gene short names and synonyms were extracted from the GO and KEGG databases, and the results were visualized using colour codes when they combined the pathway data with the microarray data.

In our case example rainbow trout-specific pathways are not available. Therefore, we used pathway schemes from other species (such as zebrafish), to obtain orthologous pathways. There are some free programs for the analysis of data from microarrays in an specific pathway mammal, but they mostly work for human, mouse or other models species not for commercial and marine fish. Our Gene2Path program has the so called "Orthology step", which offers the possibility to detect homology between the non annotated genes or sequences of the uncommon specie with other eukaryote genomes and annotated gene IDs.

We are interested in teleost fish virology and immunology, and accordingly we validated this tool by identifying immune and infection-realted signalling pathways for fish species and chose for that the KEGG database (http://www.genome.ad.jp/kegg/).

Infectious pathogens are a serious problem in aquaculture, and salmonid fish viruses are responsible for important losses in rainbow trout and salmon farming, reflected in the intense research into these viruses within the field of aquaculture [7-9]. Infectious pancreatic necrosis virus (IPNV) for example is the aetiological agent of a well characterized acute disease that produces systemic infection and relevant mortality in farmed rainbow trout (*Oncorhynchus mykiss*) and other salmonid species. The mortality of this virus may be as high as 70% in young fish. The virus establishes an asymptomatic carrier state in survivors [10], both in different species of salmonids and in other species of farmed fish such as turbot and Atlantic cod (*Gadus morua*). Nevertheless, the production of vaccines against this virus is an area that has been little investigated.

In a previous study, we assayed oral DNA-based immunotherapy against IPNV [11] and the immune specific host reaction to a VP2-IPNV vaccine). The transcriptional changes produced by infection were determined in a rainbow trout 15k microarray designed by including annotated genes selected by key-words in the GenBank. However, since difficulties arose when trying to analize the results in terms of pathways, we tried to solve those problems by designing a user friendly and amenable programme to study this data, which is described below. We have illustrated the method with data obtained from this rainbow trout microarray to search for orthologous pathways in zebrafish, although the software can be applied to any other case or species of interest. Finally, the present work reports the search of some orthologous genes or proteins involved in several pathways from three teleost fish species (*Dicentrarchus labrax, Salmo salar and Oncorhynchus mykiss)* from KEGG database.

Materials and Methods

Database searches

We have used the following databases to design the programme's algorithm: (See 132 Supplementary File1) Nucleotide and Expressed Survey Sequence (EST): http://www.ncbi.nlm.nih.gov/nuccore/ and http://www.ncbi.nlm.nih.gov/nucest/): The Nucleotide Genome Survey Sequence (GSS) and Expressed Sequence Tag (EST) databases contain nucleic acid sequences typically uncharacterized such as short genomic (GSS) or cDNA (EST) sequences.

HomoloGene: (http://www.ncbi.nlm.nih.gov/homologene). It is a program that makes use of amino acid sequence searches (blastp) to find more distant relationships, although the procedure still refers to the DNA sequence to perform some of the statistics. Moreover, HomoloGene entries now include paralogues in addition to orthologues. Nevertheless, data for all species is still not available. For example, fish are only represented by zebrafish (*Danio rerio*).

Gene (http://www.ncbi.nlm.nih.gov/gene/): This database contains information on gene specificity, structure, function, homology between species and citations. The database supplies gene-specific connections in the nexus of a map, sequence, expression, structure, function, citation and homology data.

UniGene (http://www.ncbi.nlm.nih.gov/unigene/): This database groups transcript sequences from different loci based on genomic sequences. The availability of a genomic sequence is helpful to identify sets of transcript sequences that correspond to distinct transcript loci or to annotated genes.

Blast2GO (http://www.blast2go.com/b2ghome) is a programme to get homologous amino-acid sequences from nucleotide sequences. This research tool uses BLASTx to find the most similar sequence between several input sequences in a FASTA format.

KEGG (http://www.genome.ad.jp/kegg/) is a collection of manually drawn pathway maps, representing the molecular interaction and reaction networks for a number of cellular processes and genetic events. The database contains gene names and information on biological species-specific pathways. While searching the KEGG database with known pathways, we found that genes may be represented with several synonyms not all of which were linked to a pathway. Therefore, when the species of interest corresponding ID was not available in the KEGG database, it was first necessary to find the gene ID of the corresponding orthologous gene. Our programme provides this tool.

Microarray data

We previously studied the transcriptional changes induced by an oral DNA vaccine against the IPNV by using a rainbow trout microarray. Pooled RNA from the fish was hybridized on an Agilent rainbow trout microarray (8x15K format custom microarrays -ID032303-) containing 6442 60-mer oligo sequences. The annotation file of the microarrays was designed by us and might be provided by the supplier in ".txt" format (Ballesteros et al., 2012 for further details of the microarray used, the hybridization conditions and the first analysis). For the corresponding raw data see NCBI (http://www.ncbi.nlm.nih.gov/geo/), accession Number GSE31591. We used this microarray data as an example for the analysis and interpretation of the results from the Gene2Path programme.

Results

Software package for the analysis and interpretation of DNA microarray data

An scheme of the steps to follow the Gene2Path programme is shown in Figure 1, briefly the next steps are shown in Figures 2 and 3 (Supplementary File 2):

Dictionary, gene symbol and download of the FASTA data file

The dictionary section of the programme Gene2Path converts an input as accession numbers IDs into other code types available in the NCBI database (such as gene identification -GI- in GenBank).

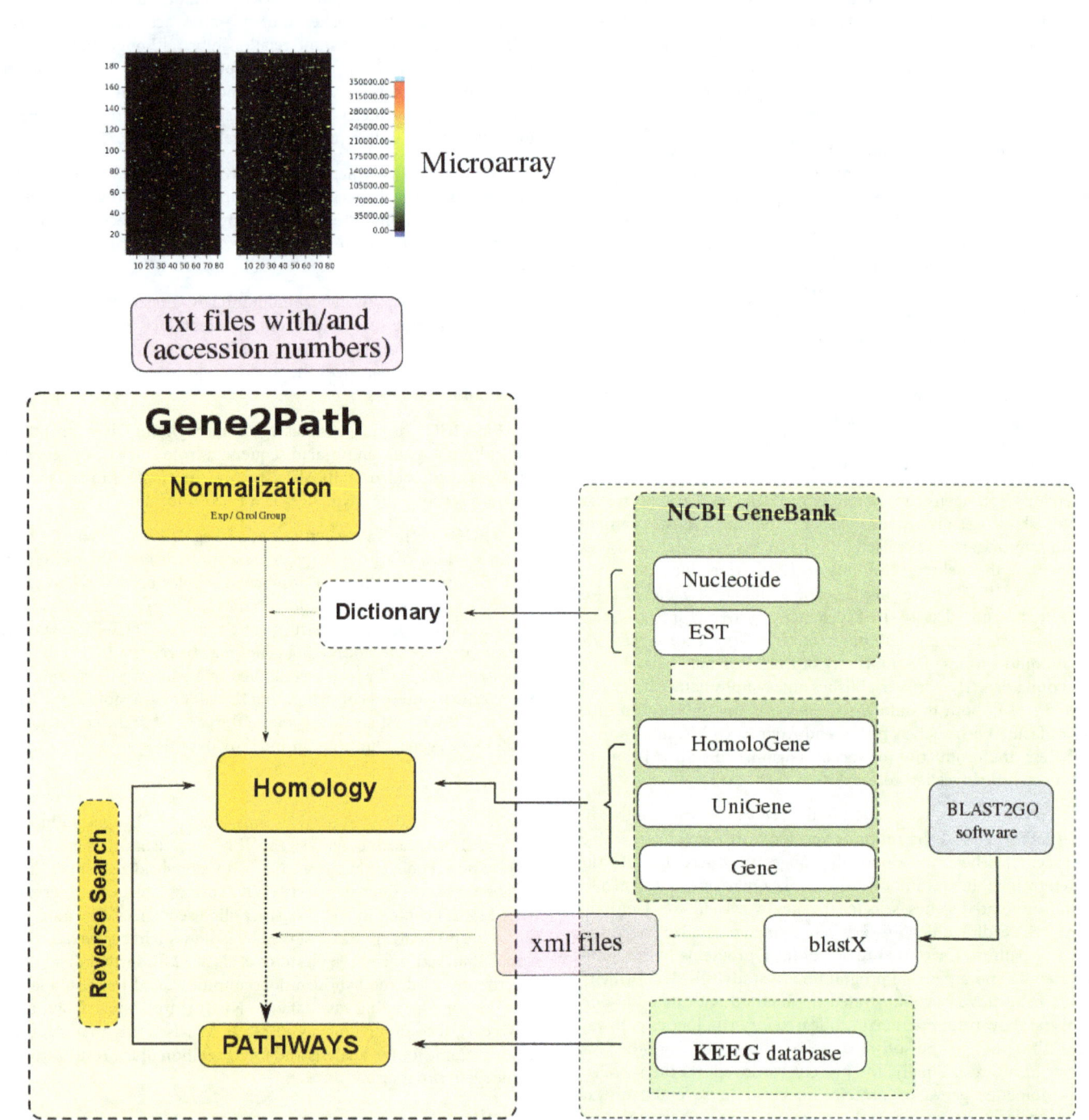

Figure 1: Scheme of the pathway analysis program and its search into different databases such as NCBI and KEGG. The diagram of Gene2Path is represented in the yellow cells, showing the automated steps of the program. The cells in green represent the databases or external programs used by Gene2Path. Arrows indicate the databases needed to follow up each one of the program steps. Into the rose cells are mentioned the archives extensions used by Gene2Path: txt are the gProccesedSignal of each of the genes from the microarray "input data"; the xml archives were obtained from the external program BLAST2GO and are used by the Gene2Path program to process and organize the blastX information. The Dictionary section provided, if needed, the gene identification codes (ID), obtained from the NCBI GeneBank for each one of the genes.

The scripts and programme are available at the "GENE2PATH" web site and they include the following features: (1) The addition of synonyms for gene identification by searching the NCBI database; (2) the analysis of the intensity of the gProccesedSignal reading of the synonym names microarray input files (extension txt); (3) Automated gene homology searches through the Homologene, Gene and Unigene NCBI databases; (4) An automatic filter was developed to find sequences similar by BLAST and to visualize the xml output file and to deal with several files at the same time; (5) Search for pathway information in the KEGG database using Gene IDs.

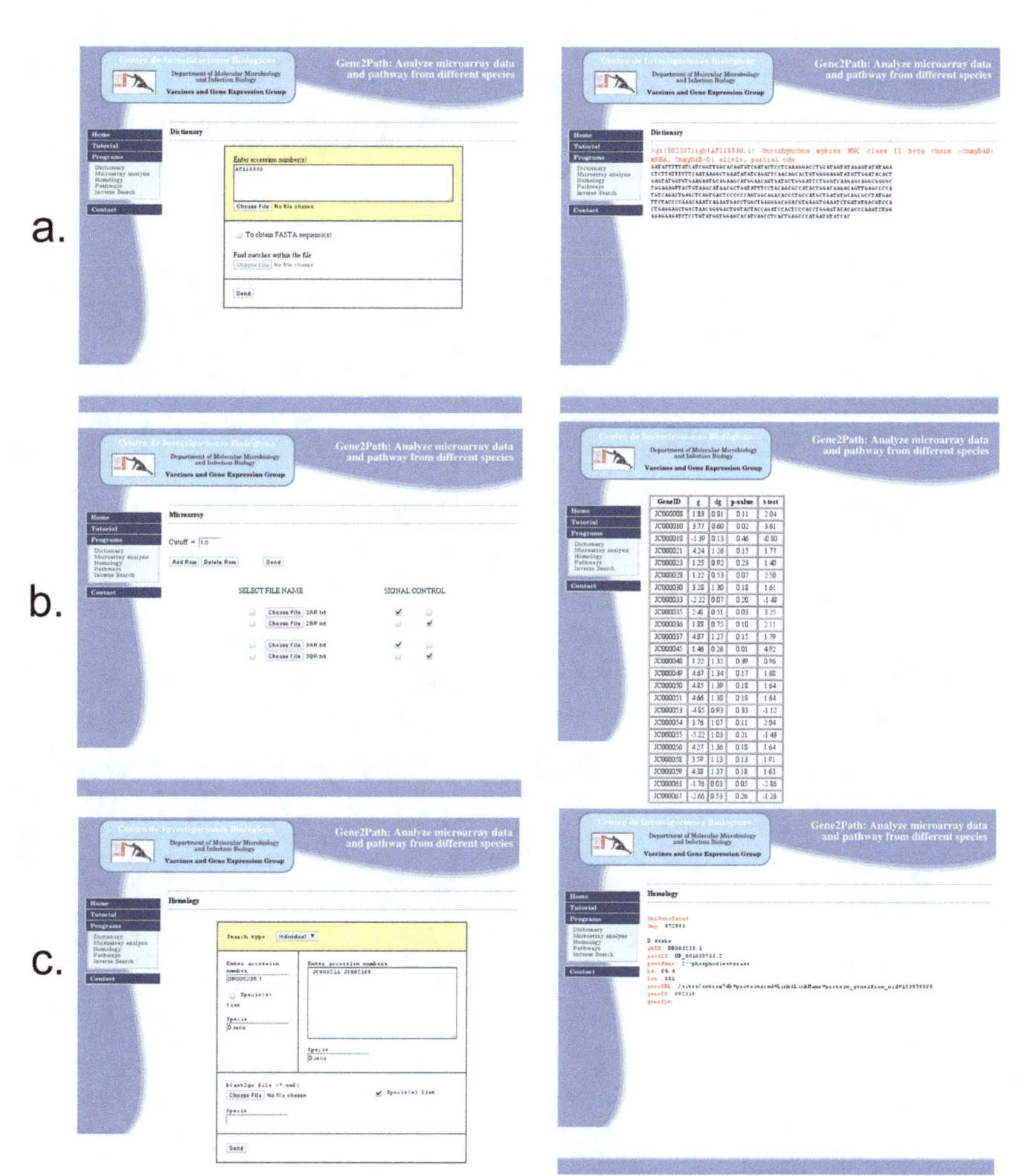

Figure 2: Illustrates the web tutorial of Gene2Path program. Successive images of the screens are shown **(a)** Dictionary step. **(b)** Normalization step. Values from the "g" signal process file in the ".txt" format were used. Data files correspond to control or experimental groups, and the gene expression fold with their t-test paired p-value. **(c)** Orthology step: Searching through NCBI.

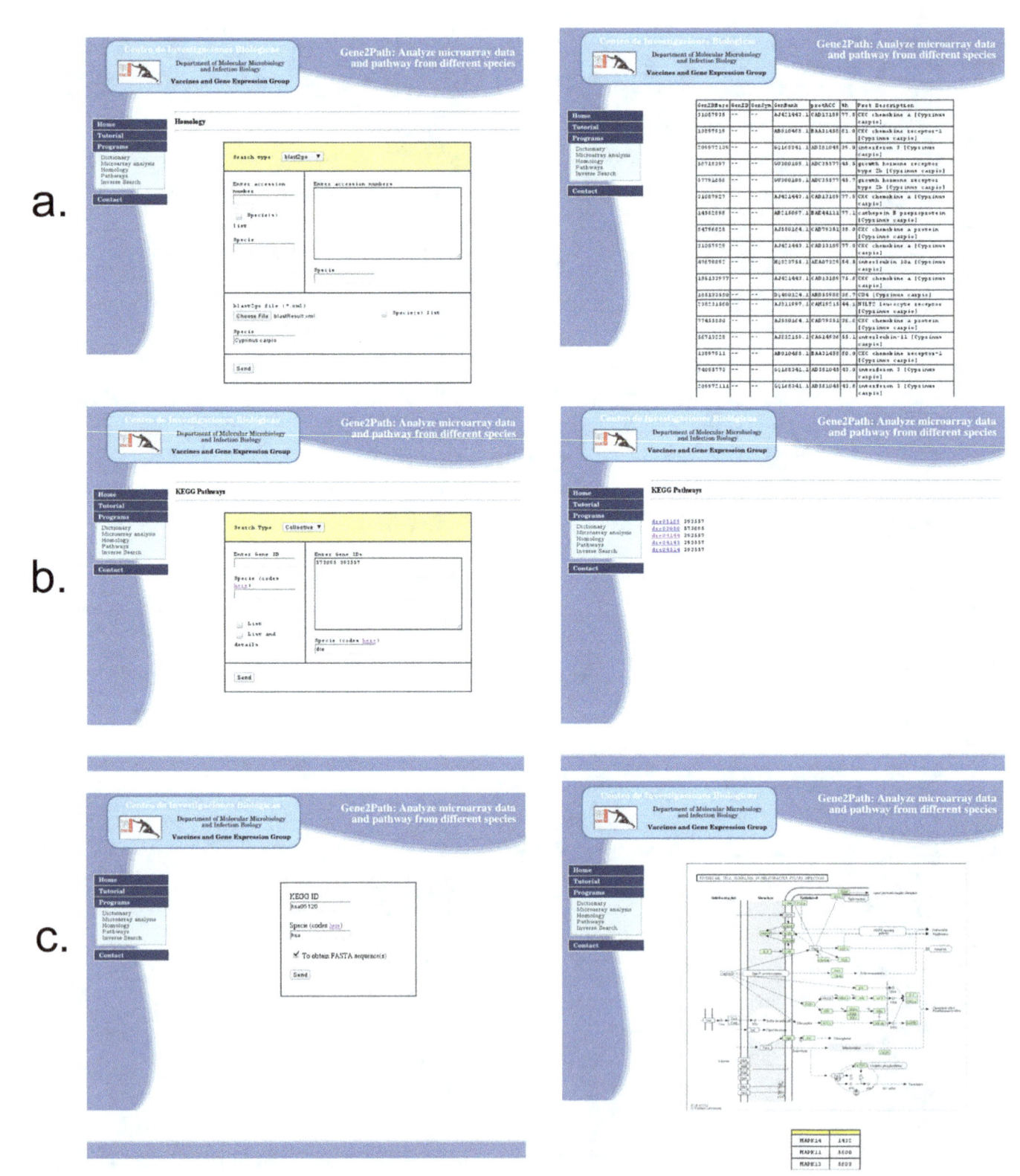

Figure 3: Illustrates the web tutorial of Gene2Path program. Successive images of the screens are shown **(a)** Orthology step: Searching through Blast2GO program. **(b)** Identifying pathways through KEGG. Gene IDs are necessary to search genes in the KEGG pathways involved. **(c)** Reverse Search of pathway. The web page shows the boxes or components separated by horizontal broken lines, which permits the genes to be assigned to a particular box.

Moreover, it is possible to obtain the corresponding gene sequences in FASTA format and to visualize the gene symbol (short name) for each of the genes in the microarray. The IDs given by the user are run through the Nucleotide or EST sequence database, named "nuccore" or "nucest", respectively (Figure 2a) and a final output list is provided.

Normalization and the level of gene expression from DNA microarray data

The normalization section of programme Gene2Path automates the analysis of the data generated by the microarray scanner. In our case example, values from the "g" signal process file in ".txt" format were used and most simple calculations were chosen (Supplementary File 3). Among the programme's options, the user can choose whether the data files correspond to control or experimental groups, and the level of gene expression to be visualized on the web site http://gene2path.no-ip.org/cgi-bin/index.cgi can be established. In our example, values ≥ 2 fold were selected. Finally, the program facilitates the process of data with t-test to obtain the p-value for statistics (Figure 2b).

Orthology

Identifying gene orthologous through NCBI

The programme allows the search of similarity of a deduced amino acid sequence from translated nucleotide to be determined between two species selected by the user. The procedure involves first searching for gene IDs in the GenBank web using Unigen, Gene and HomoloGene to detect orthology between the annotated genes or sequences of entire eukaryote genomes (Figure 2c). The results are visualized on the web and the percentage of homology is shown.

Identifying orthologous genes through BLAST2GO

Gene2Path filters and organizes the results from the Blast2GO program (".xml" files). Blast2GO uses the deduced amino acid sequence in order to find orthologous proteins in other species (blastx). As FASTA sequences from each of the genes are needed to run the Blast2GO programme, we can use data obtained in the Dictionary section (Figure 3a). The programme produces a table with information such as: The gene ID from the original or input species, gene ID in the orthologous selected by the user, gene symbol (short name) of the orthologous gene, GenBank entry of the gene sequence from the orthologous sequence, protein accession ID of the orthogous protein, percentage of homology and a short description of the gene (Figure 3a, right side).

Identifying KEGG pathways for the orthologous genes

The Gene2Path programme finds routes defined by the genes (pathways) in the KEGG database. Gene IDs are necessary to search for KEGG pathways containing those genes and gene IDs can be obtained from the orthologous genes, as described above. Each pathway is identified by its own ID (KEGG alias). To automate the search and the retrieval of the gene data from the KEGG database, a BASH script was written using the KEGG API. Direct links in each pathway were added to the file for each of the genes. This software can be used freely http://gene2path.no-ip.org/cgi-bin/index.cgi, with no need to register (Figure 3b).

Reverse Search of Pathway

The programme provides a tool to find genes involved in specific pathways by using the KEGG ID and recovering each of the genes shown in the pathway box through their gene IDs. The web page shows the boxes or components separated by horizontal broken lines,

which permits the genes to be assigned to a particular box (Figure 3c). The user may obtain the sequences for each of the genes in FASTA format for the procedure. An example of a selected pathway and table, with the names of the boxes or KEGG components, as well as their corresponding Gene IDs is shown. This tool can be used to search for gene or protein orthologous involved in pathways because not all the pathways are available for a particular species in the KEGG data base. In this study, we selected potentially-relevant zebra fish and human (*Homo sapiens*, hsa and *Danio rerio*, dre respectively) pathways from the Kyoto Encyclopedia of Genes and Genomes (KEGG) database (http://www.genome.ad.jp/kegg/) for *Dicentrarchus labrax*, *Oncorhynchus mykiss* and *Salmo salar*.

The human ("hsa") and zebra fish ("dre") pathways were selected because they are the most complete and phylogenetically close, respectively, to our species interest. The KEGG pathways selected for study were: "Mapk signaling–dre04010; Apoptosis– dre04210; TGF-BETA signaling–dre04350; Toll-like receptor-dre04620; NOD-like receptor-dre04321; Cytosolic DNA-sensing-dre04623; Jak-stat signaling-dre04630; Herpes simple Infection-dre05168, Chemokine signaling-hsa04062; B-cell receptor-hsa04662; Fc-epsilon RI signaling-hsa04664; Bacterial invasión-hsa05100; Hepatitis C-hsa05160; Measles virus-hsa05162; Influenza A-hsa05164; HTLV-1-hsa05166 and NK-cell mediated-hsa04650". Some of these mammalian pathways have unknown fish equivalents. On the other hand, four of seventeen pathways were not found to *D. labrax* (see Supplementary File 4). The Mapk signalling pathway-dre04010 for all species is showed in Figure 4.

In summary, given one gene ID the user of Gene2Path program would be able to obtain which KEGG pathway is in. And given a pathway the user would got how many of the genes are in it.

Gene2Path analysis of the microarray gene expression results of rainbow trout genes following the administration of an oral fish DNA vaccine. A case study

To demonstrate the use of the Gene2Path, we used a dataset derived from an experiment previously reported. Thus, an oral alginate-microencapsulated DNA vaccine against VP2- IPNV protected in rainbow trout *Oncorhynchus mykiss* [11] against IPNV. The kidney and pyloric ceca from vaccinated and control fish, obtained 7 days post vaccination, were assayed using a microarray enriched in rainbow trout immune-related genes from the GeneBank and selected genes from a previous design. Our rainbow trout microarray (8x 15K), called minitrout 12.8 (Agilent ID032303), contains 6,442 unique 60-mer oligo sequences, each in duplicates arranged randomly in the microarray [11,12].

Data analysis procedure

Step 1: Obtaining the rainbow trout gene IDs. Gene2Path was used to convert the rainbow trout mRNA accession number used for the microarray design into Gene IDs.

Step 2: Searching for orthologous genes in zebrafish. (Orthology step). The gene IDs or the accession numbers were used to search homoloGene, Gene and Unigene sections of the NCBI database, and BLASTx using BLAST2GO software for zebrafish (*Danio rerio*) orthologous. This step was obligatory to subsequently search for pathways in the KEGG database, as they are only available for some species. In this example from an input of 6442 rainbow trout accession numbers 1,282 orthologous IDs for zebrafish genes were found in the NCBI database.

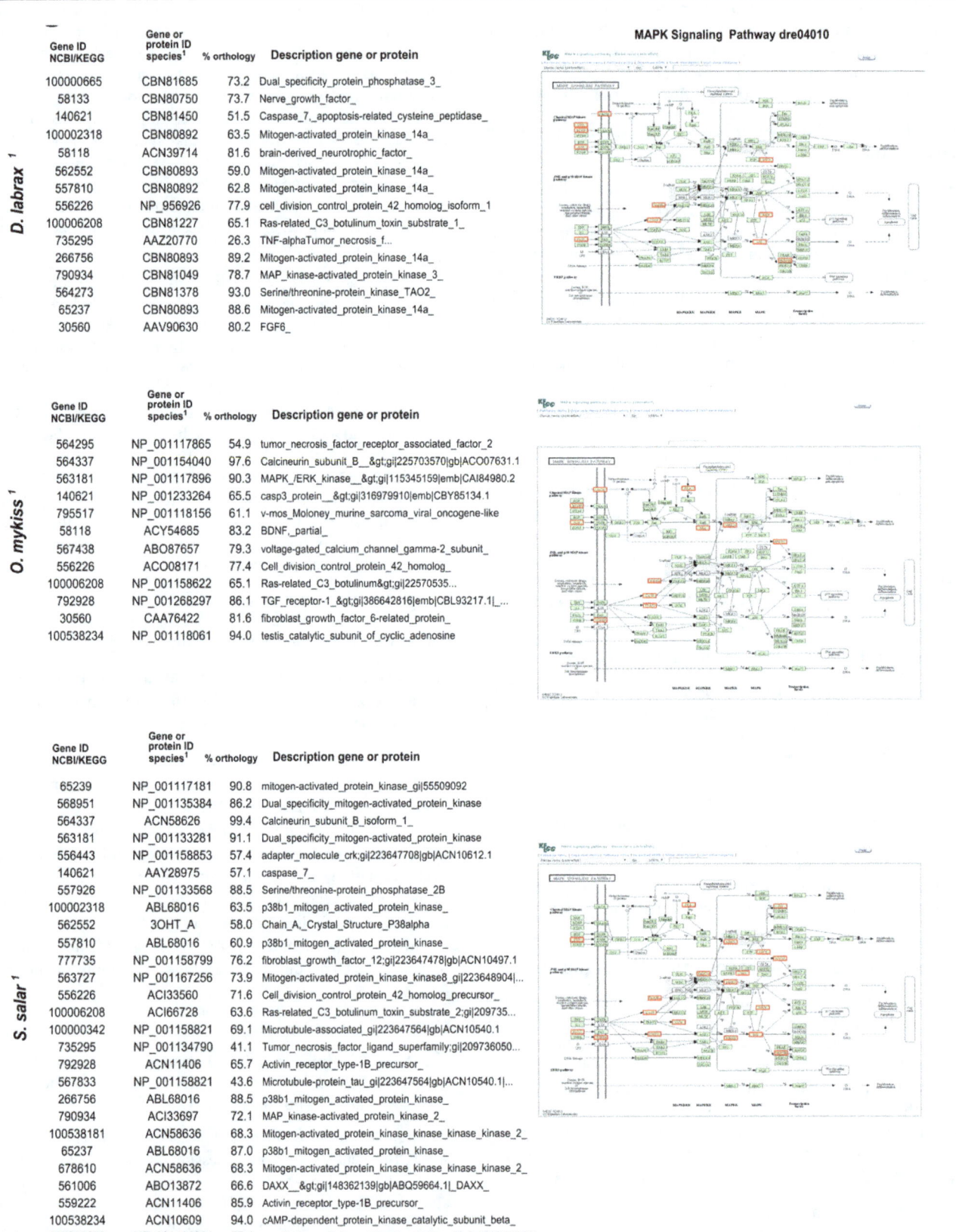

MAPK Signaling Pathway dre04010

D. labrax [1]

Gene ID NCBI/KEGG	Gene or protein ID species[1]	% orthology	Description gene or protein
100000665	CBN81685	73.2	Dual_specificity_protein_phosphatase_3_
58133	CBN80750	73.7	Nerve_growth_factor_
140621	CBN81450	51.5	Caspase_7,_apoptosis-related_cysteine_peptidase_
100002318	CBN80892	63.5	Mitogen-activated_protein_kinase_14a_
58118	ACN39714	81.6	brain-derived_neurotrophic_factor_
562552	CBN80893	59.0	Mitogen-activated_protein_kinase_14a_
557810	CBN80892	62.8	Mitogen-activated_protein_kinase_14a_
556226	NP_956926	77.9	cell_division_control_protein_42_homolog_isoform_1
100006208	CBN81227	65.1	Ras-related_C3_botulinum_toxin_substrate_1_
735295	AAZ20770	26.3	TNF-alphaTumor_necrosis_f...
266756	CBN80893	89.2	Mitogen-activated_protein_kinase_14a_
790934	CBN81049	78.7	MAP_kinase-activated_protein_kinase_3_
564273	CBN81378	93.0	Serine/threonine-protein_kinase_TAO2_
65237	CBN80893	88.6	Mitogen-activated_protein_kinase_14a_
30560	AAV90630	80.2	FGF6_

O. mykiss [1]

Gene ID NCBI/KEGG	Gene or protein ID species[1]	% orthology	Description gene or protein
564295	NP_001117865	54.9	tumor_necrosis_factor_receptor_associated_factor_2
564337	NP_001154040	97.6	Calcineurin_subunit_B__>gi|225703570|gb|ACO07631.1
563181	NP_001117896	90.3	MAPK_/ERK_kinase__>gi|115345159|emb|CAI84980.2
140621	NP_001233264	65.5	casp3_protein__>gi|316979910|emb|CBY85134.1
795517	NP_001118156	61.1	v-mos_Moloney_murine_sarcoma_viral_oncogene-like
58118	ACY54685	83.2	BDNF,_partial_
567438	ABO87657	79.3	voltage-gated_calcium_channel_gamma-2_subunit_
556226	ACO08171	77.4	Cell_division_control_protein_42_homolog_
100006208	NP_001158622	65.1	Ras-related_C3_botulinum>gi|22570535...
792928	NP_001268297	86.1	TGF_receptor-1_>gi|386642816|emb|CBL93217.1|...
30560	CAA76422	81.6	fibroblast_growth_factor_6-related_protein_
100538234	NP_001118061	94.0	testis_catalytic_subunit_of_cyclic_adenosine

S. salar [1]

Gene ID NCBI/KEGG	Gene or protein ID species[1]	% orthology	Description gene or protein
65239	NP_001117181	90.8	mitogen-activated_protein_kinase_gi|55509092
568951	NP_001135384	86.2	Dual_specificity_mitogen-activated_protein_kinase
564337	ACN58626	99.4	Calcineurin_subunit_B_isoform_1_
563181	NP_001133281	91.1	Dual_specificity_mitogen-activated_protein_kinase
556443	NP_001158853	57.4	adapter_molecule_crk;gi|223647708|gb|ACN10612.1
140621	AAY28975	57.1	caspase_7_
557926	NP_001133568	88.5	Serine/threonine-protein_phosphatase_2B
100002318	ABL68016	63.5	p38b1_mitogen_activated_protein_kinase_
562552	3OHT_A	58.0	Chain_A,_Crystal_Structure_P38alpha
557810	ABL68016	60.9	p38b1_mitogen_activated_protein_kinase_
777735	NP_001158799	76.2	fibroblast_growth_factor_12;gi|223647478|gb|ACN10497.1
563727	NP_001167256	73.9	Mitogen-activated_protein_kinase_kinase8_gi|223648904|...
556226	ACI33560	71.6	Cell_division_control_protein_42_homolog_precursor_
100006208	ACI66728	63.6	Ras-related_C3_botulinum_toxin_substrate_2;gi|209735...
100000342	NP_001158821	69.1	Microtubule-associated_gi|223647564|gb|ACN10540.1
735295	NP_001134790	41.1	Tumor_necrosis_factor_ligand_superfamily;gi|209736050...
792928	ACN11406	65.7	Activin_receptor_type-1B_precursor_
567833	NP_001158821	43.6	Microtubule-protein_tau_gi|223647564|gb|ACN10540.1|...
266756	ABL68016	88.5	p38b1_mitogen_activated_protein_kinase_
790934	ACI33697	72.1	MAP_kinase-activated_protein_kinase_2_
100538181	ACN58636	68.3	Mitogen-activated_protein_kinase_kinase_kinase_kinase_2_
65237	ABL68016	87.0	p38b1_mitogen_activated_protein_kinase_
678610	ACN58636	68.3	Mitogen-activated_protein_kinase_kinase_kinase_kinase_2_
561006	ABO13872	66.6	DAXX__>gi|148362139|gb|ABQ59664.1|_DAXX_
559222	ACN11406	85.9	Activin_receptor_type-1B_precursor_
100538234	ACN10609	94.0	cAMP-dependent_protein_kinase_catalytic_subunit_beta_
100002272	NP_001133901	57.4	Dual_specificity_protein_phosphatase_22-Agi|209155756|

Figure 4: Results obtained after running the Reverse Search step of Gene2Path program in the MAPK signalling pathway dre04010 of three fish species. Headings are: the gene ID from *D. rerio* (first column), the gene ID of the fish species under study (second column), the percentage orthology (third column) and a short description of the gene or protein of the species selected (fourth column). The figure (right side) illustrates the situation of orthologes genes into the pathway.

Step 3: Search zebrafish pathways with orthologous genes: It is interesting to note that only 12 genes into 10 different pathways were detected in KEGGS by using the ID gene symbols provided by the microarray. This search was run without the orthology step (original data https://earray.chem.agilent.com/earray/search.do?search1/4arrayDesign); however, after running the Gene2Path program (accession data 2013-01-15), the numbers increased to 1169 genes and 179 pathways (Figure 5). The identification of 169 additional pathways with the Gene2Path programme demonstrates its efficacy. Although genes may be active in more than one pathway. The microarray used was designed to determine transcriptional changes in selected immune genes induced by a DNA vaccine and hence, identification of immune related pathways is to be expected. Nevertheless several other pathways were also detected, such as those involved in "apoptosis and the cell cycle" (23 different KEGG pathways), "regulation of energy metabolism" (42 different KEGG pathways), and several pathways that could be grouped without direct network associations. Figure 6 illustrates the data of the intestinal immune network for IgA production (KEGG alias: dre04672).

Only two genes were obtained manually from the KEGG database, without using the "orthology step" are represented in Figure 6a. The same pathway is shown after running the "orthology step" on the Gene2Path software (Figure 6b); from the 55 genes of the pathway, 18 genes were detected, which represents a 33% increase. IgA production pathway has not been yet identified in fish, but functional similarities with the IgT genes implicated in the *Danio rerio* orthologous could exist, and pathways showed similar trends than most of those pathways described.

Moreover if a user needs to know the specific genes involved in a pathway, or if there is interest in focusing on a specific section of the pathway, it is possible to obtain these genes through the "Reverse Search of pathways" application of Gene2Path. The results of the pathway analysis of the microarray data provide insight into differential organ-specific biological processes that may explain the differences in host response to the VP2-IPNV vaccine.

Discussion

In this article, we describe a pathway-based approach to analyze microarray data from uncommon biological species using newly designed Gene2Path software. To achieve this, we relied on the orthology of the genes identified in a microarray with those from other species included in the KEGG database. The programme

automatically searches pathway databases to analyse microarray data in an independent, species-specific way. (For installation, instructions, examples and source code see Supplementary File 5). We have illustrated the method with data obtained from a rainbow trout microarray to search for orthologous pathways in other well known biological species, such as zebrafish, although the software can be applied to any other case or species of interest. Large scale gene expression studies represent an important advance in experimental molecular biology. Microarrays have become an important tool in functional genomics studies and they are often used to address a variety of biological situations but in some cases they are structured to well know biological species [3]. Such studies rapidly generate large quantities of gene expression data, the handling of which represents a major challenge for biologists. Indeed, the importance of presenting microarray data in the framework of documented biological pathways has often been noted [4,5]. In biology, pathway is a set of interactions or functional relationships between the physical and genetic components of a cell that operate in concert to fulfil a biological requirement. The databases that capture information on these functional interactions of molecular species are numerous, and the lack of uniformity of models and the methods to access this data makes integrating pathway data extremely difficult for uncommon biological species. Thus most of the software packages use species-specific gene IDs and they cannot handle gene data from other species. Yet it is necessary to make such pathway information systems more flexible and efficient, since while data for humans and common laboratory animals such as mice are widely available in databases through the internet, this is not so for other species such as those economically important fish species. The zebrafish (*Danio rerio*) is a model organism for genomic studies and a variety of functional pathways from this species can be found in the databases. The same is true for human databases, the most studied species at the genomic level. These pathways can be used as models to integrate and visualize data from microarray experiments from other species (rainbow trout in our case).

The microarray pathway analysis tool described here can be applied to a typical experiment in which two conditions are compared to identify genes whose differential expression changes significantly with respect to the reference condition. We used a microarray to analyse the differential expression of immune-related genes induced by the administration of an oral DNA vaccine in two rainbow trout organs, a species that is much less well represented than zebrafish in the pathway databases. The principal motivation for building pathway databases is to make tools available that help answer specific biological questions. The majority of genes in most genomes have no known function and examining genes in the context of a particular pathway may help to elucidate their role. For example in our case, a gene of unknown function connected to a set of genes involved in early immunity is likely to also act in this process. However, the power of many pathway analysis techniques is proportional to the amount of input data. Rainbow trout-specific pathways are not available and therefore, we have used pathway data from other species (such as humans or zebrafish), for comparision. Pathway analysis software tools, such as STARNET 2 [13], Reactome database [14] and CYTOSCAPE [15], are available, although again they are only applicable to humans and some experimental animals (mouse and rat). On the other hand, Babelomic (http://babelomics.bioinfo.cipf.es), GEPAS (http://www.gepas.org), are a set of free programs for the analysis of data from microarrays [16]. These softwares are very comprehensive and useful; however, it is necessary to work with human, mouse or other species, because the programs lack an orthologous step beyond the most usual

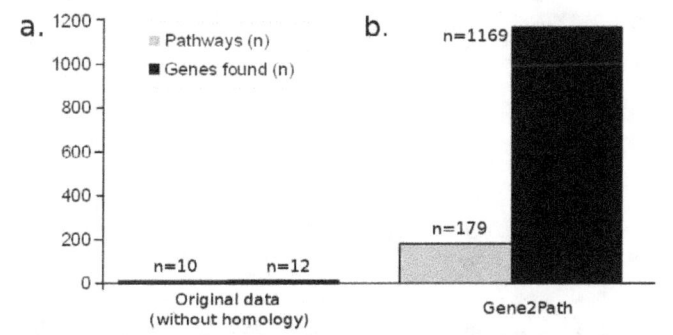

Figure 5: Number of pathways and genes found by using the gene IDs from a rainbow trout microarray (original data of the microarray species) and numbers obtained after running the Gene2Path software. The reference orthologous species was *Danio rerio*. The program provides the gene IDs in this species for searching pathways into KEGG database.

a.

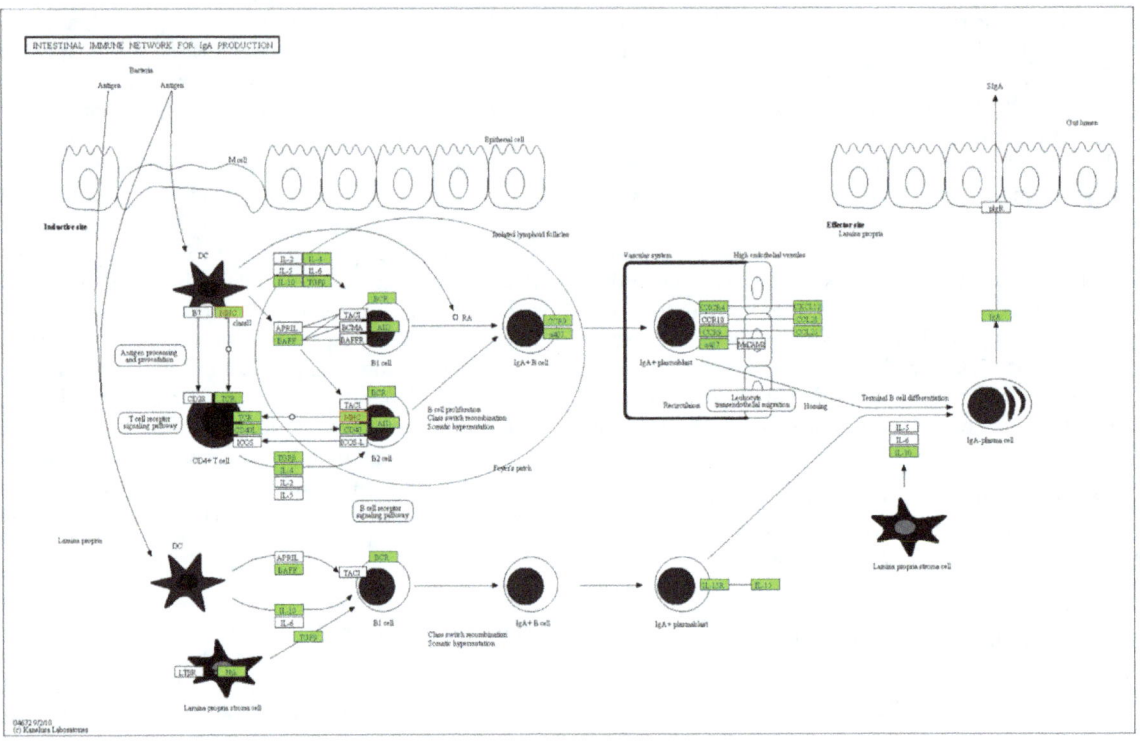

b.

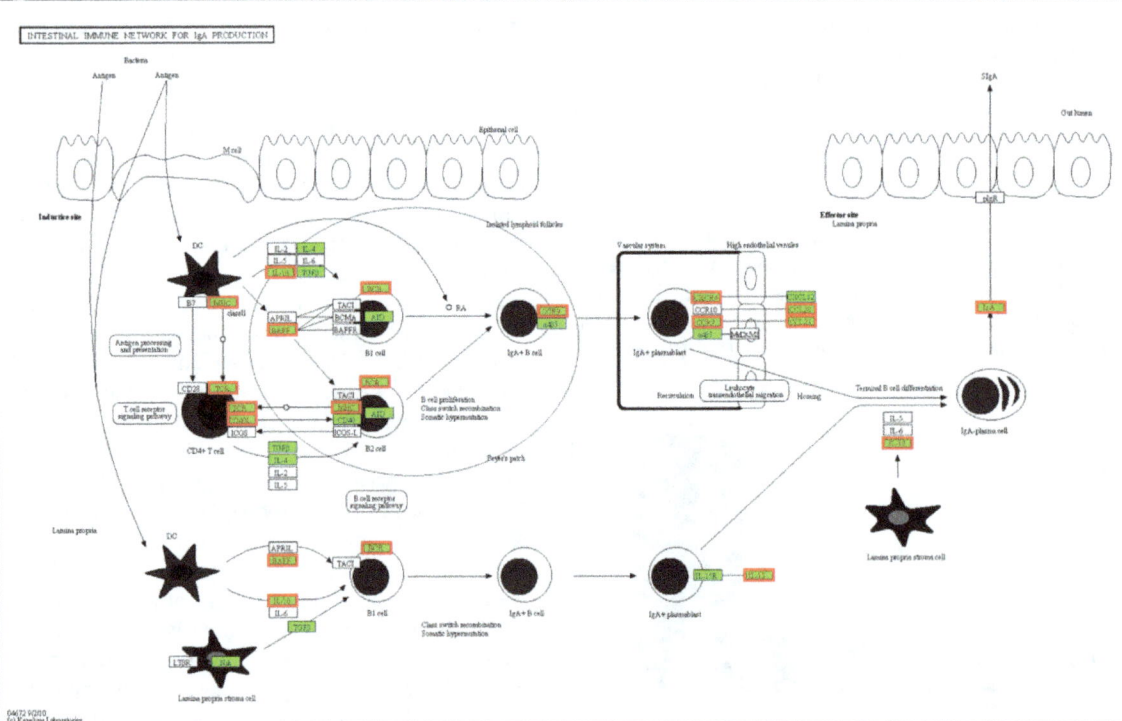

Figure 6: An example showing the results obtained in a pathway when processed or not by Gene2Path: Illustration of the "Intestinal immune network for IgA Production" using the data from a rainbow trout microarray. (A) Genes (red boxes) found in the *Danio rerio* pathway (KEGG alias: dre04672) obtained manually from the databases (n=2/55 genes), without the orthology step. (B) Genes found (red boxes) in the same pathway after running the orthology step of the program (n=20/55 genes). IgA have not been identified in fish, but functional similarities with the IgT genes implicated in the *Danio rerio* orthologous could exist and pathways showed similar trends than most of those pathways described.

biological species. Our program offers the possibility to perform orthology searches in other biological species such as fish. The software described here, and the application to one example, show how results from microarray experiments can be integrated into pathways and visualized by using one "gene orthology" step even with uncommon biological species. This enables to drawn pathways in species which are not supported in the KEGG database. The issue here was to derive knowledge of biological relevant patterns in genetic profiling data that were related to the teleost's immune defences. Accordingly, the role of several genes revealed by the pathway comparison could be defined. Another advantage of the automated procedure used by the Gene2Path software is that no direct supervision is needed and once the process has been initiated, the user can leave the programme running and visualize the results later. In the Agronomic, soils and environmental sciences department have been developed some user friendly software that can be used in industrial companies related with healthy, safety, environment (HSE). Some examples of that software are: Environmental flow diagram (EFD) [17], Soil Heat Calculator Program (SHCP) [18], and Optimize the infiltration parameters in Furrow irrigation using Visual Basic and genetic algorithm [19]. In summary the Gene2Path programme performs an automated search of several databases over 5 steps. (1) The addition of synonyms to identify genes by searching the NCBI database by their IDs. (2) The analysis of the intensity of the gProccesedSignal reading in the one-channel microarray output files (extension ".txt"); (3) An automated search of gene orthology through the homologene, Gene and Unigene NCBI databases, whereby the tool compares nucleotide sequences and comparative 3D models of proteins (constructing an atomic-resolution model of the "target" protein from its amino acid sequence, and producing an experimental 3D structure of a related orthologous protein). (4) The identification of sequences similar to the query set in NCBI nr and EST databases using Xblast. Since other programmes now exist for this step, such as BLAST2GO, we developed an automatic filter to readily visualize the output file (".xml") that enables several files to be analyzed at the same time; (5) A search of KEGG database pathway information using orthologous genes (Gene IDs).

Conclusion

All the software steps were applied to the microarray data we had obtained previously from vaccinated fish: The software proved to be very efficient in terms of automation and data processing. For instance, running a search of different NCBI databases renders 2/3 genes per second (depending on the internet connection speed). The analysis of the data from the vaccinated fish in our example rendered 179 targeted pathways. The Gene2Path software allows the automated searching of NCBI databases and the straightforward visualization of the data retrieved based on a graphic network environment.

Acknowledgement

We thank Mario García-Lacoba for his advice and helpful comments. This work was funded by Consejo Superior de Investigaciones Científicas (project 2010-20E084) and by Ministerio de Economia y Competitividad, (MINECO) project AGL2010- 18454 of Spain. N. Ballesteros wants to thank the MINECO for a PhD student fellowship.

References

1. Afshari CA, Hamadeh HK, Bushel PR (2011) The Evolution of Bioinformatics in Toxicology: Advancing Toxicogenomics. Toxicol Sci 120: S225-S237.

2. Haab BB (2003) Methods and applications of antibody microarrays in cancer research. Proteomics 3: 2116-2122.

3. Salem M, Kenney PB, Rexroad CE, Yao J (2008) Development of a 37 k high-density oligonucleotide microarray: a new tool for functional genome research in rainbow trout. J Fish Biol 72: 2187-2206.

4. Huang D, Pan W (2006) Incorporating biological knowledge into distance-based clustering analysis of microarray gene expression data. Bioinformatics 22: 1259-1268.

5. Cary MP, Bader GD, Sander C (2005) Pathway information for systems biology. FEBS Lett. 579: 1815-1820.

6. Pas MFW (2008) A pathway analysis tool for analyzing microarray data for species with low physiological information. Adv Bioinform. 2008: 1-7.

7. Mortensen SH (1993) The Relevance of Infectious Pancreatic Necrosis Virus (Ipnv) in Farmed Norwegian Turbot (Scophthalmus-Maximus). Aquaculture 115: 243-252.

8. Rodriguez Saint-Jean S, Vilas Minondo MP, Perez Prieto S (1993) A viral diagnostic survey of Spanish rainbow trout farms: I Sensitivity of four cell lines to wild IPNV isolates. Bull Eur Assoc Fish Pathol 13: 119-122.

9. Rodriguez Saint-Jean SP, Borrego JJ, Perez-Prieto SI (2003) Infectious Pancreatic Necrosis Virus: Biology, Pathogenesis, and Diagnostic Methods. Adv Virus Res Volume 62: 113-165.

10. Murray AG (2006) Persistence of infectious pancreatic necrosis virus (IPNV) in Scottish salmon (Salmo salar L.) farms. Prev Vet Med 76: 97-108.

11. Ballesteros NA, Saint-Jean SSR, Encinas PA, Perez-Prieto SI, Coll JM (2012) Oral immunization of rainbow trout to infectious pancreatic necrosis virus (Ipnv) induces different immune gene expression profiles in head kidney and pyloric ceca. Fish Shellfish Immunol 33: 174-185.

12. Ballesteros NA (2012) Trout oral VP2 DNA vaccination mimics transcriptional responses occurring after infection with infectious pancreatic necrosis virus (IPNV). Fish Shellfish Immunol 33: 1249-1257.

13. Jupiter D, Chen H, VanBuren V (2009) STARNET 2: a web-based tool for accelerating discovery of gene regulatory networks using microarray co-expression data. BMC Bioinformatics 10: 332.

14. Haw R, Stein L (2012) Using the reactome database. Curr. Protoc. Bioinformatics Chapter 8, Unit8.7.

15. Shannon P (2003) Cytoscape: A software environment for integrated models of biomolecular interaction networks. Genome Res 13: 2498-2504.

16. Al-Shahrour F (2008) Babelomics: advanced functional profiling of transcriptomics, proteomics and genomics experiments. Nucleic Acids Res 36: W341-W346.

17. Valipour M, Morteza Mousavi S, Valipour R, Rezaei E (2012) Air, Water, and Soil Pollution Study in Industrial Units Using Environmental Flow Diagram. J Basic Appl Sci Res 2: 12365-12372.

18. Valipour M, Morteza Mousavi S, Valipour R, Rezaei E (2012) SHCP: Soil Heat Calculator Program. IOSR J Appl Phys 2: 44-50.

19. Valipour, Mohammad Montazar Asghar A (2012) Optimize of all Effective Infiltration Parameters in Furrow Irrigation Using Visual Basic and Genetic Algorithm Programming. Aust J Basic Appl Sci 6: 132.

21

Effects of Stocking Density on Growth, Body Composition, Yield and Economic Returns of Monosex Tilapia (*Oreochromis niloticus* L.) under Cage Culture System in Kaptai Lake of Bangladesh

Mohammad Moniruzzaman[1,2]*, Kazi Belal Uddin[1], Sanjib Basak[1], Yahia Mahmud[1], Muhammad Zaher[1] and Sungchul C Bai[2]

[1]*Riverine Sub-Station (Lake Fisheries), Bangladesh Fisheries Research Institute (BFRI), Rangamati Hill District, Bangladesh*
[2]*Department of Marine Bio-Materials and Aquaculture, Feeds & Foods Nutrition Research Center, Pukyong National University, Busan, 608-737, Republic of Korea*

Abstract

A 120-day research was conducted to evaluate the effects of different stocking densities on growth, body composition, survival, yield and economic returns of monosex male Nile tilapia, *Oreochromis niloticus* in net cages in Kaptai Lake of Bangladesh. Juvenile monosex tilapia with an average weight of 15.20 ± 0.15 g (mean ± SD) were randomly stocked in 12 floating net cages (3 m × 3 m × 2 m) at densities of 50 fish/m^3 (T_{50}), 75 fish/m^3 (T_{75}), 100 fish/m^3 (T_{100}) and 125 fish/m^3 (T_{125}) in triplicate groups. Fish were fed with a commercial pelleted floating feed (29% protein) at 3-5% of body weight, twice daily in all the treatments. The physico-chemical parameters of lake water were within suitable ranges for fish cultured in cages. After 120 days of trial, growth in terms of body final length, final weight, weight gain, percent weight gain, daily weight gain and specific growth rate of fish from T_{50} were significantly higher than those of fish from T_{75}, T_{100} and T_{125}. Feed conversion ratio was significantly lower in T_{50} followed by T_{75}, T_{100} and T_{125} consecutively. Survival rate was not significantly different in T_{50}, T_{75} and T_{100} while lowest survival was found in T_{125}. Significantly lower amount of body lipid and carbohydrate contents were found in T_{125} than those of fish from T_{50}, T_{75} and T_{100}. Gross and net production levels from T_{100} were significantly higher than those from T_{50}, T_{75} and T_{125}. However, the benefit cost ratio from T_{50} was better than those from T_{75}, T_{100} and T_{125}. The results demonstrated that on the basis of growth and economic return 50 fish/m^3 was the best stocking density for monosex tilapia culture in cages which might be technically feasible and economically viable.

Keywords: Cage culture; Stocking density; Growth; Economic return; Monosex tilapia; Kaptai lake

Introduction

Cage aquaculture is an important technology to increase fish production. A widespread and profitable culture of fish and prawns in cages has already been developed successfully in Asia, Europe and America [1,2]. This technique in South-East Asia first started from late 1800s, since then, many countries in this area were practicing cage culture in freshwater and marine environments, including open sea, estuaries, lakes, reservoirs, ponds and river [3,4]. However, in Bangladesh, aquaculture activities are still concentrated largely in pond culture system. Cage aquaculture in open water bodies could provide an opportunity for mitigating protein demand in the nation.

Kaptai Lake in Bangladesh (latitude 22°22'-23°18' N, longitude 92°00'-92°26'E), is the largest manmade lake of South-East Asia (comprised about 68,800 hectare water body) [5]. The present annual fish production of Kaptai Lake reached to 8,537MT in 2012 [6] which plays a significant role in the national economy of Bangladesh. In addition, it may offer a tremendous scope for cage aquaculture in terms of animal protein intake.

Tilapia is considered as the 'aquatic chickens' of warm-water aquaculture [7]. It is the second most important farmed fish in the world, after carps [8]. Tilapia culture is being practiced in most of the tropical, subtropical and temperate regions. In Bangladesh, tilapia production reached to 136,000 MT in 2012 [6] which is expected to increase up to 160,000 MT by 2015 [9]. The drawback of mixed sex tilapia culture is female tilapia shows precocious reproduction which results poor growth performance, overcrowding and feed competition [10,11]. Nowadays, great attention has been paid to sex-reversed male Nile tilapia, *Oreochromis niloticus* culture due to their sexual size dimorphism, males grow significantly faster, larger and are more

uniform in size than females [12-14]. The sex-reversion process can be done by manual sexing, interspecific hybridization, hormonal sex reversal or YY male production [15]. In Bangladesh, hormonal sex reversal technique in genetically improved farmed tilapia (GIFT) has been disseminated by Bangladesh Fisheries Research Institute (BFRI) to public and private hatchery operators [9]. Presently, more than 400 monosex tilapia hatcheries are actively operating around the country and they are producing over 4 billion monosex male fry every year [9]. Experiences suggested that, this fish species can grow rapidly, require minimum oxygen, can tolerate wide range of temperature, are resistant to disease and has high yielding performance [16-18]. It can be successfully cultured in brackish water environment with shrimp in coastal ponds [19]. Dan and Little [20] reported that monosex tilapia did not exhibit better growth performance over mixed sex tilapia in cage culture system. In another study, Kamruzzaman [11] proposed that monosex all male tilapia cage culture has no advantage over mixed sex tilapia in terms of growth performance.

In cage aquaculture, fish stocking density has great impact on growth, survival, health, water quality and production [21]. For

*Corresponding author: Mohammad Moniruzzaman, Department of Marine Bio-Materials and Aquaculture, Feeds and Foods Nutrition Research Center, Pukyong National University, Busan, 608-737, Republic of Korea, E-mail: mzaman_bfri@yahoo.com

the maximization of monosex tilapia production, profitability and sustainability in cage culture system, it is essential to determine its optimum stocking density. So far our knowledge, very little work has been done on the stocking density of monosex tilapia in cages. The aim of the research work was to investigate monosex male tilapia, *Oreochromis niloticus* as a potential culture species and to identify its suitable stocking density and yield in net cage culture system in Kaptai Lake of Bangladesh.

Materials and methods

Experimental site

The present experiment was conducted for a period of four months (120 days) from March to June 2012 in Kaptai Lake near to Riverine Sub-Station (Lake Fisheries) of Bangladesh Fisheries Research Institute (BFRI) in Rangamati Hill District of Bangladesh.

Cage construction and installation

In this experiment, 12 floating net cage each having an area of 3 m × 3 m × 2 m (L × W × H, 18 m³) made of synthetic nylon net (mesh size 1.1 cm) has been installed in the lake. Each net cage was tied and hanged with bamboo pole frame and covered at the top with another piece of plastic net (mesh size 2.5 cm) to prevent escape of fish by jumping and bird predation. The bamboo poles were covered with long pieces of wooden raft for easy movement, feeding and sampling of the experimental fish on the cage structure. Empty vacuum plastic drums of 250 liters size were used as cage float for buoyancy of cage structure. The outer side of each experimental net was also covered with a fine meshed net (20 cm depth) according to surface of the lake water level to inhibit floating feeds to go out from the net and to reduce the entrance of non-caged fishes from wild sources [22]. The whole cage structure was tied with anchors at both shore sides by nylon rope to facilitate easy floating and moving of whole cage structure with 12 individual cages depending on water level.

Fish collection and stocking

In the present study, we tested four different stocking densities of monosex tilapia like 50 fish/m³, 75 fish/m³, 100 fish/m³ and 125 fish/m³ designated as T_{50}, T_{75}, T_{100} and T_{125}, respectively, in triplicates for each treatment group. In brief, hatchery produced and hormonally sex-reversed juvenile monosex male tilapia, *Oreochromis niloticus* (GIFT strain) averaging 15.20 ± 0.15 g (mean ± SD) were transported to the experimental site by oxygenated polyethylene bags and they were kept in three net hapas for 24 hours for acclimation with environment and then initial length and weight of fish were recorded individually in 'cm' and 'g' with the help of a measuring scale and a digital electronic balance (OHAUS, Model CT 1200-S, USA), respectively. Finally, the cages were randomly stocked with monosex tilapia and the number of fish stocked in each cages (18 m³) were recorded simultaneously.

Feeding and rearing

A commercial floating pellet feed was used in the experimental period. The diet's pellet size was 1.8-2.2 mm, and the feed contained 28.76% crude protein, 5.95% crude lipid, 14.23% crude ash and 9.43% moisture (Table 1). Feeding was done at 3-5% of the initial body weight. The total amount of feed was divided into two equal rations for feeding at 8 a.m. and 4 p.m. according to Chapmen [23] and Cruz [24]. Feeding was done by hand very slowly and carefully to ensure ingestion of feed completely. Around 20% of fish in each treatment were sampled fortnightly in order to determine fish weight (TANITA digital kitchen

Component	Composition (%)
Moisture	9.43
Crude protein	28.76
Crude lipid	5.95
Crude fiber	6.47
Crude ash	14.23
NFE[1]	35.16
Diet size (mm)	1.8-2.2

[1]NFE: Nitrogen Free Extract [100%-(protein+lipid+ash+fiber+moisture)], a measure of soluble carbohydrates [54].

Table 1: Proximate composition of the experimental diet (% dry basis) for monosex tilapia.

scale, model KD-160, Japan, ± 1 g) and the feed amounts were adjusted as per changes of body weight in each trial. A record of supplied feed was maintained to determine the food conversion ratio (FCR).

During the study period, dead fish were recorded and removed quickly. The cages were lifted from water at every 15 days interval to check the net and for cleaning purposes. Cages were cleaned with soft brush to remove algae, sponges and other organisms. Loose twine, net mesh torn by predators, anchors and sinkers were checked routinely and immediately mended or replaced as needed.

Sampling and data analysis

Physico-chemical parameters of water such as temperature (°C), transparency (cm), pH, dissolved oxygen, DO (mg/L), hardness (mg/L) and total alkalinity (mg/L) were monitored weekly in the morning between 7 and 8 a.m. during the whole culture period [25]. Water temperature was recorded with a glass Celsius thermometer, water transparency was recorded with a Secchi disc, pH and DO were measured using a digital pocket pH meter (model-HI 98107 pHep® HANNA Instruments, Carrollton, TX, USA) and DO meter (DO-5509, Lutron Electronic Enterprise Co. Ltd., Taipei, Taiwan), respectively. Other chemical parameters were measured using a HACH kit box (model FF-2, No. 243001, Loveland, CO, USA).

After 120 days of trial, the whole cage structure moored to shore of the lake and all fish were harvested by repeated scoop netting and then fish were counted, measured and weighed for each cage. To determine the growth response, yield and survivability of experimental fish, the following parameters were calculated:

Weight gain (WG)=final fish weight (g)-initial fish weight (g)

Weight gain (%)=(final weight-initial weight) × 100/initial weight

Average daily weight gain (ADWG)=(final fish weight-initial fish weight)/days

Feed Conversation Ratio (FCR)=weight of feed given (g)/fish weight gain (g)

Specific Growth Rate (SGR%)=100 × (ln final wt-ln initial wt)/days.

Gross yield=fish production (kg)/cage (18 m³)

Net yield=fish production (kg)/m³ of cage

Survival Rate (SR%)=100 × (number of fish survived/number of fish stocked)

Proximate composition analyses

Proximate composition analyses of experimental diet and whole fish body were performed by the standard methods of Association of Official Analytical Chemists [26]. After the completion of experiment, three fishes from each treatment were collected for further analyses of body carcass composition. For determining moisture content, samples of diets and wet fishes were dried at 135°C for 2 h. Ash content was determined by using a muffle furnace (600°C for 4 h). Crude lipid content was determined by the soxhlet apparatus using Soxtec system 1046 (Foss, Hoganas, Sweden) and crude protein content by Kjeldahl method (N × 6.25) after acid digestion, distillation and titration of samples. The samples were analyzed in the Fish Nutrition Lab, Department of Aquaculture, Faculty of Fisheries, Bangladesh Agricultural University, Mymensingh, Bangladesh.

Statistical analysis

The mean values for water quality parameters, growth, survival and production of different treatments were subjected to one-way ANOVA followed by Duncan's New Multiple Range test [27]. All statistical analyses were performed using SAS software version 9.1 (SAS Institute Inc, Cary, NC, USA). Standard deviation of each parameter and treatment was determined and expressed as mean ± SD. Treatment effects were considered with the significant level at $P<0.05$.

Economic analysis

After the termination of experiment, an economic analysis was performed to estimate the net return and benefit–cost ratio on the basis of different stocking densities of monosex tilapia. The following simple equation was used according to Asaduzzaman et al. [28]:

$R=I-(FC+VC+I_i)$

Where, R=net return, I=income from monosex tilapia sale,

FC=fixed/common costs, VC=variable costs and I_i=interest on inputs

The benefit-cost ratio was determined as:

Benefit cost ratio (BCR)=Total net return/Total input cost

At the end of the experiment, all fish were sold and the prices of fish were attributed to the Rangamati local fish market price in July 2012 and expressed in Bangladeshi taka (1US\$=80 BDT). The local price per kg of monosex tilapia was 250 BDT for near or more than 250 g fish (T_{50}), 200 BDT for 200-250 g fish (T_{75} and T_{100}) and 180 BDT for less than 200 g fish (T_{125}). In the study, fish selling price was higher than normal price, because every year from April/May to June/July, Kaptai Lake is closed by the government authority for capture fishing.

Results

Mean values (mean ± SD) and ranges of water quality parameters measured in the cages over the experimental period are summarized in Table 2. The average temperature, pH, transparency, DO, free CO_2, total hardness and total alkalinity did not change significantly among the treatments (P>0.05).

During the experimental period, increments of growth at fortnightly intervals are shown in Figure 1. The highest final weight was observed in T_{50} followed by T_{75}, T_{100} and T_{125}. At the end of 120 days of fish rearing in cages, biological performances and production of monosex tilapia at different stocking densities were presented in Table 3. All growth parameters in terms of final weight, final length, weight gain, percent weight gain, average daily gain and SGR were significantly decreases from lower to higher stocking densities (P<0.05). Significantly lower survival rate was found in T_{125} than those in T_{50}, T_{75} and T_{100}. However, no significant differences were found in survival rate of T_{50}, T_{75} and T_{100}. Feed conversion ratio (FCR) was significantly increased with increasing

Parameters	Stocking densities (fish/m³)			
	T_{50}	T_{75}	T_{100}	T_{125}
Water temperature (°C)	30.2 ± 1.1ᵃ (26.3-31.8)	29.8 ± 1.2ᵃ (26.4-31.6)	30.3 ± 1.6ᵃ (26.8-31.3)	30.7 ± 1.8ᵃ (26.6-31.8)
DO (mg/L)	6.8 ± 1.2ᵃ (6.5-7.8)	6.6 ± 1.4ᵃ (6.4-7.5)	6.6 ± 1.2ᵃ (6.4-7.2)	6.5 ± 1.5ᵃ (6.3-7.1)
Free CO_2 (mg/L)	2.1 ± 0.3ᵃ (1.8-2.6)	2.3 ± 0.3ᵃ (2.2-2.8)	2.8 ± 0.6ᵃ (2.4-3.2)	3.2 ± 0.8ᵃ (2.7-3.5)
pH	6.8 ± 0.6ᵃ (6.4-7.4)	6.8 ± 0.7ᵃ (6.3-7.2)	6.7 ± 0.6ᵃ (6.3-7.1)	6.6 ± 0.5ᵃ (6.2-6.9)
Transparency (cm)	54.1 ± 1.2ᵃ (51.4-58.4)	53.8 ± 1.5ᵃ (50.8-57.8)	53.5 ± 1.9ᵃ (50.6-57.6)	53.1 ± 1.6ᵃ (50.5-57.2)
Total Hardness (mg/L)	44.6 ± 2.3ᵃ (39.5-53.2)	43.8 ± 2.8ᵃ (38.7-50.2)	44.4 ± 3.1ᵃ (38.5-52.8)	43.5 ± 3.6ᵃ (37.6-51.7)
Total Alkalinity (mg/L)	58.3 ± 3.2ᵃ (48.4-67.4)	57.8 ± 3.8ᵃ (47.9-66.2)	59.3 ± 3.3ᵃ (48.3-67.3)	59.6 ± 4.1ᵃ (48.3-67.1)

Values in each row having the same superscripts are not significantly different (P>0.05).

Table 2: Mean values (± SD) and ranges (parentheses) of water quality parameters in cages over 120 days.

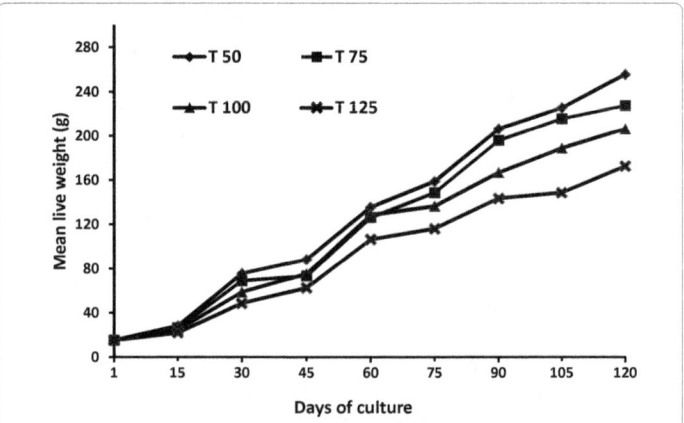

Figure 1: Growth increment of monosex tilapia at different stocking densities on each sampling day over 120 days of cage culture.

fish stocking density. The best FCR was found for the lowest stocking density of T_{50} while highest in T_{125} (P<0.05). Significantly higher gross and net fish production were found in T_{100} followed by T_{125}, T_{75} and T_{50} (P<0.05).

In the present study, we did not find any significant differences in whole-body moisture, protein and ash content in T_{50}, T_{75} and T_{100} (P>0.05) (Table 4). However, significantly lower amount of body lipid and carbohydrate contents were found in T_{125} than those of the other treatments (P<0.05) (Table 4).

In consideration of economic analysis, total cost of inputs in T_{50} (BDT/cage) was lower than that in T_{75}, T_{100} and T_{125} (Table 5). A highest total net return was also obtained in T_{50} followed by T_{100}, T_{75} and T_{125}. In overall, the highest benefit cost ratio (BCR) was found in T_{50} and a lowest in T_{125}.

Discussion

In the present study, different physico-chemical parameters of water in the cages were within the suitable and safe ranges for monosex tilapia growth throughout the experimental period. The range of water temperature (26.3-31.8°C) in the experimental cages is in agreement with Boyd [29], who proposed that the range of water temperature from 26.06 to 31.97°C is suitable for tropical fish culture. Similarly, Dan and Little [20] suggested that tilapia is suitable for raising between April and December when temperatures range from 25°C to 35°C. Oxygen is

Parameters	Stocking densities (fish/m³)			
	T$_{50}$	T$_{75}$	T$_{100}$	T$_{125}$
Initial length (cm)	9.15 ± 0.05[a]	9.16 ± 0.06[a]	9.17 ± 0.05[a]	9.16 ± 0.04[a]
Final length (cm)	23.10 ± 6.17[a]	21.84 ± 5.81[b]	20.72 ± 6.52[b]	18.48 ± 7.96[c]
Initial weight (g)	15.20 ± 0.15[a]	15.21 ± 0.14[a]	15.21 ± 0.13[a]	15.20 ± 0.14[a]
Final weight (g)	255.53 ± 16.36[a]	227.24 ± 15.05[b]	206.36 ± 14.06[c]	172.62 ± 14.89[d]
Weight gain (g)	240.33 ± 5.86[a]	212.03 ± 4.73[b]	191.15 ± 3.68[c]	157.42 ± 3.92[d]
Weight gain (%)	1581.12 ± 6.32[a]	1394.02 ± 7.16[b]	1256.74 ± 7.55[c]	1035.70 ± 9.23[d]
Average daily weight gain (g)	2.01[a]	1.76[b]	1.59[c]	1.31[d]
Specific growth rate (%)	2.35[a]	2.25[b]	2.17[c]	2.02[d]
Feed conversion ratio (FCR)	1.81[c]	1.92[b]	1.97[b]	2.05[a]
Survival rate (%)	96.8[a]	92.6[a]	91.2[a]	83.1[b]
Gross fish yield (kg/cage/120 days)	223[a]	284[b]	339[c]	323[c]
Net fish yield (kg/m³/120 days)	12.4[a]	15.8[b]	18.8[c]	17.9[c]

Values are means from triplicate groups (n=3) of fish where the values in each row with different superscripts are significantly different ($P<0.05$)

Table 3: Growth performance, survival, feed utilization and yield of monosex tilapia in cages after 120 days of experimental period..

Stocking density (fish/m³)	Moisture (%)	Protein (%)	Lipid (%)	Ash (%)	Carbohydrate (%)
T$_{50}$	69.02 ± 0.26[a]	15.05 ± 0.56[a]	10.38 ± 0.19[a]	4.02 ± 0.28[a]	1.53 ± 0.08[a]
T$_{75}$	68.33 ± 0.42[a]	16.35 ± 0.72[a]	9.32 ± 0.12[a]	4.07 ± 0.18[a]	1.93 ± 0.16[a]
T$_{100}$	71.08 ± 0.58[a]	14.51 ± 0.63[a]	8.33 ± 0.18[a]	4.00 ± 0.13[a]	2.08 ± 0.11[a]
T$_{125}$	73.57 ± 0.32[a]	15.82 ± 0.52[a]	6.04 ± 0.27[b]	3.79 ± 0.22[a]	0.78 ± 0.21[b]

Values are means from triplicate groups (n=3) of fish where the values in each row with different superscripts are significantly different ($P<0.05$).

Table 4: Proximate composition (% wet basis) of monosex tilapia after 120 days of cage culture.

Variables	Price rate (BDT)*	In Bangladeshi Taka (BDT)			
		T$_{50}$	T$_{75}$	T$_{100}$	T$_{125}$
Fixed/common cost					
Net cage (18 m³)	4000/unit	4000	4000	4000	4000
Plastic drum (250 L)	500/unit	2000	2000	2000	2000
Subtotal		6000	6000	6000	6000
Variable costs					
Juvenile monosex tilapia	3/juvenile⁻¹	2700	4050	5400	6750
Feed cost	42/kg	16921	22886	27990	27664
Auxiliary materials (bamboo, anchors etc) for supporting cage		5000	5000	5000	5000
Subtotal		24621	31936	38390	39414
Total		30621	37936	44390	45414
Interest on inputs (4 months)	10% annually	1021	1264	1480	1514
Total inputs		31642	39200	45870	46928
Financial returns					
Monosex tilapia sale as total returns	250/kg† 200/kg‡ 180/kg¶	55655	56820	67752	58095
Total net returns		24013	17620	21882	11167
Benefit cost ratio (BCR)		0.759	0.449	0.477	0.238

*1 US $=80 Bangladeshi Taka (BDT)
†price of monosex tilapia of T$_{50}$ group
‡price of monosex tilapia of T$_{75}$ and T$_{100}$ groups
¶price of monosex tilapia of T$_{125}$ group

Table 5: Cost and return analysis of monosex tilapia in different stocking densities from single unit of cage (18 m³) after 120 days.

the most important stress factor which has direct impact on health and survival of caged fishes [30]. For optimal fish growth, dissolved oxygen (DO) levels should be above 5 ppm for warm water fish species [29]. In our experiment, concentration of DO in all the cages ranged between 6.3-7.8 mg/L. The pH (6.3-7.4) and transparency or Secchi disc depth (50.5-58.4) were within the favorable ranges for cage culture of tilapia [30]. The value of free CO_2 ranged from 1.8 to 3.5 mg/L which supports the results of Rahman [31]. The alkalinity denotes the concentration of calcium or magnesium carbonate or bicarbonate in water. In this study,

growth in terms of final length, final weight, weight gain, percent weight gain, average daily weight gain and specific growth rate of monosex tilapia were significantly higher in lower stocking density, T$_{50}$ (50 fish/m³) compared to higher stocking densities followed by T$_{75}$ (75 fish/m³), T$_{100}$ (100 fish/m³) and T$_{125}$ (125 fish/m³). Albeit a same feed was applied at an equal ratio, the growth performance differed significantly in all the stocking groups. The reason behind this might be due to lower density fishes got more spaces and there was less competition for feeds compared to higher density treatments [35]. This result is in agreement with those reported by Chakraborty [36] for monosex tilapia, Ouattara [37], Gibtan [30] and Osofero [38] for mixed sex tilapia, Hengsawat [39] for African catfish, Haque [40] for Thai silver barb and Ahmed [41] for common carp in cage culture system. In contrast, increased growth, survival and feed conversion ratio were also observed in high density culture of juvenile silver perch, *Bidyanus bidyanus* and tilapia, *Oreochromis spirulus* [42,43].

In the present study, the overall growth performance of monosex tilapia was higher than Gibtan [30] and Osofero [38] at T_{50} and T_{100}. Similarly, Chakraborty [36] found lower performance than our study at T_{50}, T_{75} and T_{100}. Mridha [44] found much lower growth performance of all male tilapia in rain fed rice-fish ecosystem after 4 months culture period than the present study. Contrary to our findings, Dan and Little [20] reported better final weight and daily weight gain in overwintered GIFT monosex tilapia at the density of 30 fish/m³. The performances were lower than our results in case of new-season (first production of fish after spawning in the season) with same strain of tilapia. Moreover, Dan and Little [20] confirmed that overwintered caged tilapia have better growth performance than that of earthen ponds, whereas in case of new season tilapia the phenomena were opposite. In comparison with our study, Ahmed [18] found lower final weight and average daily weight gain of monosex tilapia in earthen ponds. Shofiquzzoha [19] researched on polyculture of shrimp along with three strains of tilapia- hormonally sex-reversed Bangla FISHGEN (all male) and GenoMar Supreme Tilapia™ (all male), and BFRI-GIFT (mixed sex) tilapia in earthen brackish water ponds. The authors found best final weight in monosex GenoMar Supreme Tilapia™ (291 g) after 120 days of polyculture which was higher than the present study (255.53 g). However, other strains like BanglaFISHGEN (225.29 g) and BFRI-GIFT (193.10 g) showed lower final weight than our findings.

Weight gain and survival rate are the most important issues for a successful cage aquaculture because they determine the production performance and profit of the system. Yi and Lin [45] reported that higher biomass of caged tilapia had a significant negative effect on final body weight. In this experiment, lowest growth performance and survival rate (83.1%) was found in T125 (125 fish/m³). The result probably due to overcrowding effect for limited living spaces, oxygen depletion, limited surfaces for proper feeding which might cause serious feed competition and nutritional deficits, more energy expenditure and finally increased stress, stunted growth and overwhelming of fishes in net cages for stocking huge number of monosex tilapia [35,37,46,47]. The SR reported by some researchers, Liti [48] and Abou [49] found 70% and 75%, respectively in caged tilapia and Ahmed [18] found 75.55% in monosex tilapia in earthen ponds which were lower than our result. In the present study, treatments T_{50}, T_{75} and T_{100} showed more than 90% survivability which is in close agreement with Dan and Little [20], Osofero [35] and Gibtan [27]. Mridha [44] reported that the survival of monosex tilapia varied from 79-88% which was also very near to our study.

Feeds and feeding are the major costs of production in cage culture [47]. In the present study, we used 29% crude protein contained floating pelleted feed which is in close accordance with Dan and Little [20] for monosex tilapia. The experimental feed was well accepted by the fish among all the treatment groups. FCR values of T_{50} were significantly lower than those in T_{75}, T_{100} and T_{125}. This result indicates probably low density stocked fishes might have high efficiency to convert given feed to flesh than fish stocked with high density in terms of growth [48,50]. The FCR of this study are very similar to the values reported by Osofero [38], Ahmed [18] and Saha [51] but much lower than Gibtan [30] and higher than Shofiquzzoha [19].

In the present study, there was an increasing trend of gross and net fish yield with increasing stocking density. However, the total yield was increased from 50 fish/m³ (T_{50}) up to 100 fish/m³ (T_{100}) then declined at highest stocking density, 125 fish/m³ (T_{125}). Significantly highest biomass was found in T_{100} which agree with that reported for caged tilapia at the same density [52]. The negative correlation between stocking density and fish production at T_{125} might be postulated for density-dependent mortality and poor growth performance. In line with our study, Hengsawat [39] reported that final harvest and production values were directly related to stocking density and there must be a limit where mortality will be severe and growth and production will be reduced. In contrast, Ridha [8] found better growth and production performance in mixed sex GIFT tilapia at T_{125} than the present study.

In this experiment, after the analyses of growth performance, we also analyzed the whole body proximate composition of monosex tilapia at each stocking density to confirm their nutritional status. The results demonstrated that T_{125} group fishes contained significantly low amount of lipid and carbohydrate contents compared to other groups. The possible reason might be due to over expenditure of body energy for maintaining normal metabolic activity by the fish (125 fish/m³) during the experimental period. At final harvest, we also noticed that apparently in T_{50}, T_{75} and T_{100} fish were more robust than in T_{125}.

In the present study, production economics was affected by stocking densities of monosex tilapia. Benefit-cost ratio (BCR) of each treatment was determined on the basis of input costs of fish, feeds and cage materials and it returns from total fish sale. In this experiment, very cheap bamboo poles were used for cage structure. Moreover, durable net cages and plastic drums could be use up to 5-6 harvests. So, in initiating year or during the first harvest, the profit index will be lower than in the following years because of the fixed costs. In this study, we used little bit bigger size of cages (18 m³ each) to reduce input costs. McGinty and Rakocy [47] proposed that as cage size increases, cost per unit volume also decreases. In the present study, near about 50-60% input cost came from feed which is usual in any aquaculture operation [8]. Through the economic analysis we found that even though T_{100} (100 fish/m³) attributed best production performance, T_{50} (50 fish/m³) showed the highest economic return due to preferred market size and selling price of fish which is in agreement with Zonneveld and Fadholi [53]. In addition, the lowest economic return was found in T_{125} (125 fish/m³) possibly due to smallest size, lowest total production and lowest selling price of monosex tilapia which is in agreement with Mridha [54,55].

Conclusions

In conclusion, it can be corroborated that on the basis of growth performance and economic return, 50 fish/m³ exhibited the highest performance to all stocking densities. Therefore, stocking density of 50 fish/m³ could be recommended for the successful cage culture of monosex tilapia. This study has implications of sustainable and cost-effective cage culture practices in lake environment. However, further research could be addressed on the fish stocking size and cage size.

Acknowledgments

This research work was funded by Fish production, Conservation and Strengthning Management at Kaptai Lake (Component - C, BFRI Part) Project (Code no. 05-4405-5470) under the Ministry of Fisheries & Livestock (MoFL) of the government of Bangladesh.

References

1. Bardach E, Ryther H, McLarney O (1972) Aquaculture: The Farming and Husbandry of Freshwater and Marine Organisms. John Wiley and Sons Ltd., USA.

2. Beveridge MCM (1987) Cage aquaculture. Fishing News Books Ltd. Farnhan, Surrey, England.

3. Balcazar J, Aguirre A, Gomez G, Paredes W (2006) Culture of Hybrid Red Tilapia (Oreochromis mossambicus × Oreochromis niloticus) in Marine Cages: Effects of Stocking Density on Survival and Growth.

4. Eng CT, Tech E (2002) Introduction and history of cage culture.

5. Fernando CH (1980) The fishery potential of man-made lakes in Southeast Asia and some strategies for its optimization.

6. Department of Fisheries (2013) National Fish week compendium 2013.

7. Little D (1998) Options in the Development of the Aquatic Chicken.

8. Ridha MT (2006) Comparative study of growth performance of three strains of Nile tilapia, Oreochromis niloticus, L. at two stocking densities. Aquac Res 37: 172-179.

9. Hussain MG, Barman BK, Karim M, Keus EKJ (2013) Progress and the future for Tilapia farming and seed production in Bangladesh.

10. Lèveque C (2002) Out of Africa: the success story of tilapias. Environ Biol Fish 64: 461-464.

11. Kamaruzzaman N, Nguyen NH, Hamzah A, Ponzoni RW (2009) Growth performance of mixed sex, hormonally sex reversed and progeny of YY male tilapia of the GIFT strain, Oreochromis niloticus. Aquac Res 40: 720-728.

12. Ponzoni RW, Hamzah A, Saadiah T, Kamaruzzaman N (2005) Genetic parameters and response to selection for live weight in the GIFT strain of Nile tilapia (Oreochromis niloticus). Aquaculture 247: 203-210.

13. Bwanika GN, Murie DJ, Chapman LJ (2007) Comparative age and growth of Nile tilapia (Oreochromis niloticus L.) in lakes Nabugabo and Wamala, Uganda. Hydrobiologia 589: 287-301.

14. Nguyen NH, Khaw HL, Ponzoni RW, Hamzah A, Kamaruzzaman N (2007) Can sexual dimorphism and body shape be altered in Nile tilapia by genetic means?. Aquaculture 272: S36-S48.

15. Beardmore JA, Mair GC, Lewis RI (2001) Monosex male production in finfish as exemplified by tilapia: applications, problems, and prospects. Aquaculture 197: 283-301.

16. Hussain MG, Rahman MA, Akhteruzzaman M, Kohinoor AHM (1989) A study on the production of Oreochromis niloticus (L.) under semi-intensive system. Bangladesh J Fish 12: 59-65.

17. Shamsuddin M, Hossain MB, Rahman MM, Asadujjaman M, Ali MY (2012) Performance of monosex fry production of two Nile tilapia strains: GIFT and NEW GIPU. World J Fish and Mar Sci 4: 68-72.

18. Ahmed GU, Sultana N, Shamsuddin M, Hossain MB (2013) Growth and production performance of monosex tilapia (Oreochromis niloticus) fed with homemade feed in earthen mini ponds. Pak J Biol Sci 16: 1781-1785.

19. Shofiquzzoha AFM, Alam MJ, Moniruzzaman M (2009) Species diversification in coastal aquaculture: Production potentials of shrimp (Penaeus monodon) with mono and mixed sex tilapia. Bangladesh J Fish Res 13: 179-184.

20. Dan NC, Little DC (2000) The culture performance of monosex and mixed-sex new-season and overwintered fry in three strains of Nile tilapia (Oreochromis niloticus) in northern Vietnam. Aquaculture 184: 221-231.

21. Costa C, Menesatti P, Rambaldi E, Argenti L, Bianchini ML (2013) Preliminary evidence of colour differences in European sea bass reared under organic protocols. Aqua Eng 57: 82-88.

22. McGinty A (1991) Tilapia production in cages: effects of cage size and number of non-caged fish. Prog Fish Cult 53: 246-249.

23. Chapman F (2006) Culture of Hybrid Tilapia: A Reference Profile. University of Florida, Institution of Food and Agriculture Science, Gainesville, FL, USA.

24. Cruz P (1997) Aquaculture feed and fertilizer resource Atlas of the Philippines. FAO Fisheries Technical Paper, Rome, Italy.

25. APHA (1992) Standards Methods for the Examination of the Water and Wastewater.

26. AOAC (1995) Association of official analytical chemists.

27. Duncan DB (1955) Multiple range and multiple F tests. Biometrics 11: 1-42.

28. Asaduzzaman M, Wahab MA, Verdegem MCJ, Adhikary RK, Rahman SMS, et al. (2010) Effects of carbohydrate source for maintaining a high C:N ratio and fish driven re-suspension on pond ecology and production in periphyton-based freshwater prawn culture systems. Aquaculture 301: 37-46.

29. Boyd CE (1982) Water Quality Management for Pond Fish Culture.

30. Masser PM (1997) Cage culture: site selection and water quality.

31. Gibtan A, Getahun A, Mengistu S (2008) Effect of stocking density on the growth performance and yield of Nile tilapia (Oreochromis niloticus) in a cage culture system in Lake Kuriftu, Ethiopia. Aquac Res 39: 1450-1460.

32. Rahman MM, Bashar MA, Farhana Z, Hossain MY (2014) Temporal Variation of Physicochemical Parameters in Kaptai Lake, Bangladesh. World J Fish and Mar Sci 6: 475-478.

33. ARG (Aquatic Research Group-University of Chittagong, Bangladesh) (1986) Hydrobiology of the Kaptai reservoir. FAO, Rome, Italy.

34. Chowdhury SH, Mazumder A (1981) Limnology of Lake Kaptai-I: Physicochemical features. Bangladesh J Zool 9: 59-72.

35. Beveridge MCM (1996) Cage Aquaculture.

36. Rahman MM, Verdegem MCJ (2010) Effects of intra- and interspecific competition on diet, growth and behaviour of Labeo calbasu (Hamilton) and Cirrhinus cirrhosus (Bloch). Appl Anim Beh Sci 128: 103-108.

37. Chakraborty SB, Mazumdar D, Banerjee S (2010) Determination of ideal stocking density for cage culture of monosex Nile Tilapia (Oreochromis niloticus) in India. Proc Zool Soc 63: 53-59.

38. Ouattara NI, Teugels GG, Douba VN, Philippart JC (2003) Aquaculture potential of the black-chinned tilapia, Sarotherodon melanotheron (Cichlidae). Comparative study of the effect of stocking density on growth performance of landlocked and natural populations under cage culture conditions in Lake Ayame (Cote d'Ivoire). Aquac Res 34: 1223-1229.

39. Osofero SA, Otubusin SO, Daramola JA (2009) Effect of stocking density on tilapia (Oreochromis niloticus Linnaeus 1757) growth and survival in bamboo-net cages trial. African J Biotech 8: 1322-1325.

40. Hengsawat K, Ward FJ, Jaruratjamorn P (1997) The effect of stocking density on yield, growth and mortality of African catfish (Clarias gariepinus Burchell 1822) cultured in cages. Aquaculture 152: 67-76.

41. Haque MKI, Ahmed KK, Saha SB (1997) Effects of Stocking Density on the Growth of Puntius gonionotus Cultured in Floating Net Cages at Kaptai Lake. Bangladesh J Agri 22: 55-62.

42. Ahmed KK, Haque MKI, Paul SK, Saha SB (2002) Effect of Stocking Density on the Production of Common Carp (Cyprinus carpio Lin.) in Cages at Kaptai Lake, Bangladesh. Bangladesh J Fish Res 6:135-140.

43. Rowland SJ, Mifsud C, Nixon M, Boyd P (2006) Effects of stocking density on the performance of the Australian freshwater silver perch (Bidyanus bidyanus). Aquaculture 253: 301-308.

44. Cruz EM, Ridha M (1991) Production of the tilapia Oreochromis spirulus (Gunther) stocked at different densities in sea cages. Aquaculture 99: 95-103.

45. Mridha MAR, Hossain MA, Azad Shah AKM, Uddin MS, Nahiduzzaman M (2014) Effects of stocking density on production and economics of all-male Tilapia (Oreochromis niloticus) culture in a rain fed rice-fish ecosystem. J Appl Aqua 26: 60-70.

46. Yi Y, Lin CK (2001) Effects of biomass of caged Nile tilapia (Oreochromis niloticus) and aeration on the growth and yields in an integrated cage-cum-pond system. Aquaculture 195: 253-267.

47. Diana JS, Yi Y, Lin CK (2004) Stocking densities and fertilization regimes for Nile tilapia (Oreochromis niloticus) production in ponds with supplemental feeding.

48. McGinty AS, Rakocy JE (1995) Cage culture of tilapia.

49. Liti D, Fulanda B, Munguti J, Straif M, Waidbacher H, et al. (2005) Effect of open-pond density and caged biomass of Nile Tilapia (Oreochromis niloticus L.) on growth, feed utilization, economic returns and water quality in fertilized ponds. Aquac Res 36: 1535-1543.

50. Abou Y, Fiogbe E, Micha J (2007) Effect of stocking density on growth, yield and profitability Nile tilapia, Oreochromis niloticus L., fed Azolla diet, in earthen ponds. Aquac Res 38: 595-604.

51. Bhijkajee M, Gobin P (1997) Effect of temperature on the feeding rate and growth of a red tilapia hybrid.

52. Saha SB, Moniruzzaman M, Alam MJ (2009) Optimization of stocking density of black tiger shrimp (Penaeus monodon) for concurrent culture with genetically improved farmed tilapia (GIFT) in brackish water ponds. Bangladesh J Zool 37: 174-185.

53. Daungsawasdi S, Chomchei C, Yamorbsin R, Kertkomut B (1986) Net cage culture of tilapia and puntius in Klong Praew irrigation tank.

54. Zonneveld N, Fadholi R (1991) Feed intake and growth of red tilapia at different stocking densities in ponds in Indonesia. Aquaculture 99: 83-94.

55. Castell JD, Tiews K (1980) Report of the EIFAC, IUNS and ICES Working Group on the Standardization of Methodology in Fish Nutrition Research.

22

Microbiological Evaluation of Water and Fillets in the Production Chain of Nile Tilapia (*Oreochromis niloticus*)

Gabriel Marcos Domingues de Souza[1], Lucienne Garcia Pretto-Giordano[2], Gislayne Trindade Vilas-Bôas[1], Túlio Oliveira de Carvalho[3], Ângela Teresa Silva-Souza [4,5], Mauro Caetano Filho[5], Ronaldos Tamanini[2] and Laurival Antônio Vilas-Boas[1]

[1]*Departamento de Biologia Geral, Universidade Estadual de Londrina, CP 10.011, CEP 86057.970, Londrina/PR, Brazil*
[2]*Departamento de Medicina Veterinária Preventiva, Universidade Estadual de Londrina, CP 10.011, CEP 86057.970, Londrina/PR, Brazil*
[3]*Departamento de Matemática, Universidade Estadual de Londrina, CP 10.011, CEP 86057.970, Londrina/PR, Brazil*
[4]*Programa de Pós-graduação em Ciências Biológicas, Universidade Estadual de Londrina, CP 10.011, CEP 86057.970, Londrina/PR, Brazil*
[5]*Departamento de Biologia Animal e Vegetal, Universidade Estadual de Londrina, CP 10.011, CEP 86057.970, Londrina/PR, Brazil*

Abstract

This study evaluated the quantity of total aerobic bacteria, total coliform and *Escherichia coli* in samples of the Nile tilapia fillet (*Oreochromis niloticus*) and in water samples at different stages of a tilapia production chain in Brazil. Furthermore, the effect of temperature on the number of such bacterial groups in the fillet and water samples, and the impact of the type of tank, net cages or ponds, during the fattening stage of tilapia on bacterial quantification was ascertained. A generalized linear model with negative binomial distribution was used, aiming to assess the influence of predictor variables like climate season and type of tank, with interaction term, on bacterial counts. Analysis of the results showed that, for water samples, the highest scores of all bacterial groups evaluated were obtained at debugging stage. Additionally, the increase in temperature positively influenced the counting of bacteria, both in water and in fillet samples. Moreover, microbial quantification in fillet samples did not only result from the bacteria present in the animal growth water sites, but also from intrinsic factors involved in the slaughter and filleting processes. In relation to the type of tank, EC and AB counts were significantly lower in net cages than in ponds. In conclusion, the results of this study indicate the need for deployment of additional care, especially at debugging stage, aiming to obtain safer fillets for consumption. Furthermore, in the event of increased temperature, greater attention should be given to the slaughter and filleting processes.

Keywords: Water quality; Coliform and aerobic bacteria; Fillets quality aerobic bacteria; Total coliform; *Escherichia coli*

Introduction

Fish is considered an important source of protein for humans. Consequently, an increase in consumption, with the resulting increase in the demand for fishes has been occurring worldwide. Thus, fish farming has been increasing in several regions worldwide [1,2] including Brazil, where it has had an intense development in recent years due to various factors, such as climatic and hydrographic conditions favorable for this activity [3]. Brazil produces different species of fish. Among the most frequently cultivated is the Nile tilapia (*Oreochromis niloticus-Linnaeus,* 1758), which has a faster growth rate, tolerance to harsh environmental conditions and ease of culture techniques [4].

The fish quality is influenced by several factors. Water quality in the different production stages deserves special attention in intensive systems [5,6] as they have high population density where fish are unable to reach sites with better water quality. The monitoring of water quality can be done by measuring physical, chemical and biological parameters. However, special attention should be given to assessing the microbiological quality of the water, which should be monitored by both research and quantification of bacterial indicators of environmental contamination.

Several studies have been conducted to monitor the microbiological quality of water in fish cultures [7-9]. These studies have been performed using mainly Gram-negative bacteria belonging to the Enterobacteriaceae family, represented especially by facultative anaerobic bacteria, which are components of the intestinal microbiota of humans and warm-blooded animals. Some of these are associated with human diseases [10,11] especially the group of total coliforms (TC) and fecal coliforms (FC) [12].

These microbial groups have also been monitored in samples of processed fish, and it has often been shown that the microbiota present in fish is influenced by the microbiota of the aquatic environment [4,13,14]. However, to assess the microbiological quality of water and fish, most studies concentrate only on the animal fattening stage, not evaluating other important stages of the production chain, such as the production of fingerlings, juveniles, or debugging and slaughter stages. To fill in these gaps, this study was also conducted to evaluate the microbiological quality of the tilapia fillet, as well as of the water at the different stages of the tilapia production chain in the northern region of Paraná State, Brazil. Finally, this study evaluated the influence of the tilapia culture system in cages and ponds, as well as the potential effect of temperature on the amount of mesophilic aerobic bacteria (AB), total coliform (TC) and *Escherichia coli* (EC) in water samples and fillets.

*****Corresponding author:** Laurival Antônio Vilas-Boas, Departamento de Biologia Geral, CCB, Universidade Estadual de Londrina, CP 10.011, CEP 86057.970, Londrina/PR, Brazil, E-mail: lavboas@uel.br

Materials and Methods

Study area

Five fish farming that comprise the Nile tilapia (*O. niloticus*) production chain in northern Paraná State, Brazil, were assessed for evaluation of microbiological quality of both water and produced fillets. This chain was composed of a single property responsible for the production and supply of larvae and pre-juveniles (animals up to 5 g) to all other properties. Among those are five properties with tanks for fish at the juvenile (animals up to 10 g), intermediate (animals up to 100 g) and fattening stages (animals with approximately 500 g), which three properties had ponds, while two properties had floating net cages systems located on the Tibagi and Paranapanema basins, respectively. The debugging, slaughter and filleting stages were accomplished in exactly one of these properties.

Definition of sample collection sites

The collections of water and *O. niloticus* fillet samples occurred on four different occasions, corresponding to the four seasons of the year: i) April 2010 (autumn); ii) August 2010 (winter); iii) November 2010 (spring) and iv) March 2011 (summer).

The water sample collection points on each of the properties were selected according to the characteristics of each fish farm. On the property responsible for the production of fingerlings, water samples from the cages maintained in hothouses and from the stream supplying such tanks at two distinct points were also collected. The first took place before the supply to the cages, and the second at the outflow point of them. On all properties, collections from juvenile fattening tanks, as well as from the stream feeding these tanks were accomplished. On the property responsible for fish slaughtering, water samples were also collected from the debugging tanks and from the slaughterhouse. In this same fish farm, collections of fillet of *O. niloticus* directly from the slaughterhouse refrigerator were conducted.

In the fish culture stations with net cages, the cages were arranged in rows, and fish separated by size and stage of development. Water samples were collected from tanks containing fish in three different development stages: juvenile, intermediate and adult. Collections of water samples from two different locations of installation of net cages were also performed: i) between the riverbank and the cages and ii) between the riverbed and the cages.

Sampling collection

For the collection of water samples, 50 ml sterilized polypropylene bottles (121°C, 20 minutes) were kept sealed until the time of use. Each bottle was submerged until its upper part reached the depth of about five centimeters, when the bottle cap was removed. Each bottle was identified according to the collection point and the samples were immediately stored in isothermal boxes containing ice and protected from light. All tests were performed within a maximum of four hours after collection of the samples. Water temperature of the tanks was recorded at 30 cm depth.

For the study of the fillets in each season of the year, three samples, each consisting of three fillets, which were collected just after slaughter. These samples were stored in trays sealed with plastic film and transported to the laboratory in isothermal boxes at 10°C and protected from light. All samples were performed under the same conditions and the analyses not exceeding a maximum of four hours after collection.

Research of microorganisms

The assessment of indicator microorganisms was conducted both from water and tilapia fillets samples employing a ready-to-use method, commercially available (Petrifilm, 3M Brazil Ltda), following the instructions of the manufacturer regarding the procedures of inoculation, incubation of the samples and reading of the results. For this purpose, the volume of 1 ml of each sample of sterile water was used for the preparation of serial decimal dilutions. The fillet samples were analyzed using 25 g diluted in 225 mL of 1% peptone water, homogenized in Stomacher (ITR, Brazil) for 3 min and followed by serial decimal dilutions thereafter. Further, both dilutions of water and fillet samples were plated on Petrifilm AC plates for the enumeration of mesophilic aerobes (AB) and on Petrifilm EC plates for the enumeration of total coliforms (TC) and *Escherichia coli* (EC). The results of the counts were corrected according to the dilutions accomplished and expressed in CFU/ml for AB and CFU/ml 100 for TC and *E. coli*.

Statistical approach

The models for the AB, TC or EC response variables comply with the formulation

$$\log (C) \sim \text{Weather Station} * \text{Tank}$$

where "C" denotes the observed count (AB, TC or EC) in each water sample; the weather station has numeric values: 1 for winter, 2 for spring, 3 for fall and 4 for summer; and "tank" refers to the type of tank, a nominal variable with two possible values: ponds or net cages, the first being the reference value (baseline).

To analyze the influence of predictor variables (weather station and type of tank) on bacterial counts (AB, TC and EC) and the possibility of interaction between two predictors, the R software, version 2.14.0, and in particular the MASS package was used for adjusting generalized linear models with negative binomial distribution, due to the high dispersion of data. In this model, the coefficient of each predictor variable was evaluated, as well as the coefficient of the interaction between the predictor variables and their significances. The 5% descriptive level is used as cut-off value to establish the significance of each coefficient.

A linear regression between the temperature data from the water samples collected in the growing tanks for the juvenile, pre-juvenile, intermediate, fattening and purification stages in the four seasons (autumn, winter, spring and summer) was elaborated to justify considering season as a numerical discrete variable with values from 1 to 4. Thus, the factor underlying the influence of the season that effectively denotes the significance of the model coefficients is in fact the effect of temperature on bacterial counts.

Results

The variables used were "weather station", "source", "stage", "tank" and bacterial counts. The characteristics of all variables have already been explained above, except for the variable "source", which indicates the source of water for breeding fish, comprising a total of 18 diverse categories, including stream, tap, bank and riverbed. Unquestionably due to such diversity, this variable could not be used as a predictor, but the data from these sources were used to help interpret the results of the other variables. "Stage" denotes the development phase of the fish, a categorical variable with the following values: larval, pre-juvenile, juvenile, intermediate, fattening, debugging and filleting. At the larval stage, there is control of temperature, and therefore the variable weather station has NA (not available) value at this stage of the production chain.

For the variable tank, there were 24 observations in the ponds and 72 in net cages, besides 32 observations categorized as NA, because they cannot be categorized in one way or another. In the variable weather station, there were 12 cases listed as NA because they were in the larval stage. The analysis was performed on 128 samples collected for each of the three types of bacteria, in the four seasons of the year, in five fish farmers, with 16 missing data for AB, 18 missing data for TC and 44 missing data for EC.

In the Figure 1 is showed the quantification of AB, TC and EC bacteria in water samples collected from tanks representative of the different stages (larval, pre-juvenile, juvenile, intermediate, fattening, debugging and filleting) along the production chain of *O. niloticus* in northern Paraná State, Brazil. The quantification of AB bacteria presented a median value of 10^4 CFU/ml at all stages of the production chain. As for the TC bacteria, samples from most of the production chain stages showed a median of 10^3 CFU/100 ml, except for the debugging and filleting stages, which presented a median of 10^4 CFU/100 ml. The quantification of EC at the different stages of the production chain showed greater variation in median values than the quantification of AB and TC bacteria. The stages of larval, juvenile, intermediate and fattening showed a median of 1 CFU/100 ml and the detection of these bacteria in a few isolated samples in these steps (as evidenced by the isolated points on the graphs), while the other stages of development (slaughter, pre-juvenile and debugging) showed medians of 200 CFU/100 mL^{-1}, 800 CFU/100 mL^{-1} and 10^4 CFU/100 mL^{-1}, respectively.

Figure 2 displays the quantification of AB, TC and EC bacteria in the water samples of *O. niloticus* growing tanks, as well as the comparison between the two tank types analyzed, i.e., net cages or ponds. In this Figure, only the pre-juvenile, juvenile, intermediate and fattening stages were included. The quantification of AB bacteria generated a median of about 10^4 CFU/100 mL^{-1} for both types of tank, while the number of TC bacteria generated a median of 10^3 CFU/100 mL^{-1} for both types of tank, and the EC showed a median of 1 CFU/100 mL^{-1} for both types of tank. However, the samples of ponds presented a greater degree of data dispersion (as evidenced by the height of the column on the graph) than the samples from the netcages.

Figure 3 shows the quantification of AB, TC and EC bacteria in the water samples of *O. niloticus* growing tanks and also the results obtained in the four weather seasons of the year, ranging from April 2010 to March 2011, throughout the entire production chain, with the exception of the larval rearing stage. The quantification of AB and TC bacteria generated medians around 10^4 CFU/m L^{-1} and 10^3 CFU/100 m L^{-1}, respectively, for the four sampling periods. The quantification of EC in the winter and spring seasons generated a median of 1 CFU/100 m L^{-1}, while in the autumn, the median was around 400 CFU/100 m L^{-1}. No observations were made for EC in the summer.

In the Figure 4 are showed the temperature data from water samples collected from the tanks for the juvenile, pre-juvenile, intermediate, fattening and debugging stages in the four climate seasons evaluated. The average winter temperature was 20°C, followed by spring and fall, with 24°C, and summer with 28°C. The linear regression analysis, T ~ Station, presented a coefficient of 1.77 and p-value of 3.2×10^{-5},

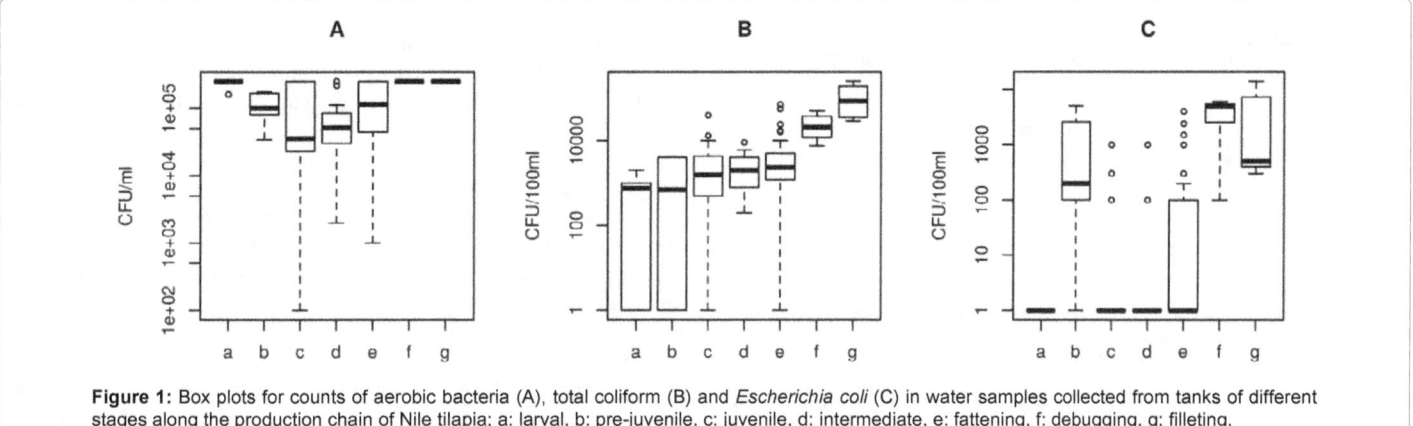

Figure 1: Box plots for counts of aerobic bacteria (A), total coliform (B) and *Escherichia coli* (C) in water samples collected from tanks of different stages along the production chain of Nile tilapia; a: larval, b: pre-juvenile, c: juvenile, d: intermediate, e: fattening, f: debugging, g: filleting.

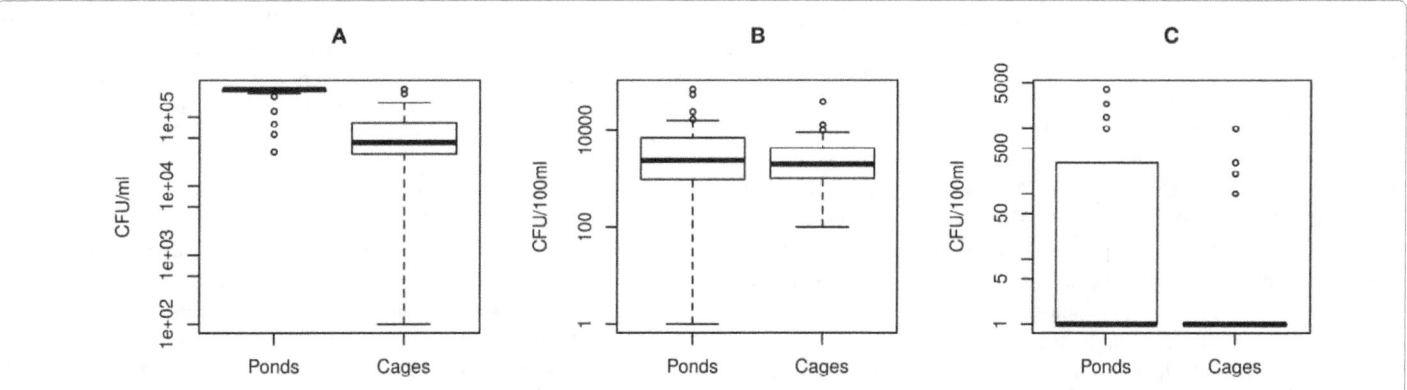

Figure 2: Box plots for counts of aerobic bacteria (A), total coliform (B) and *Escherichia coli* (C) in water samples collected from different tank types (net cages or ponds) of the production chain of Nile tilapia.

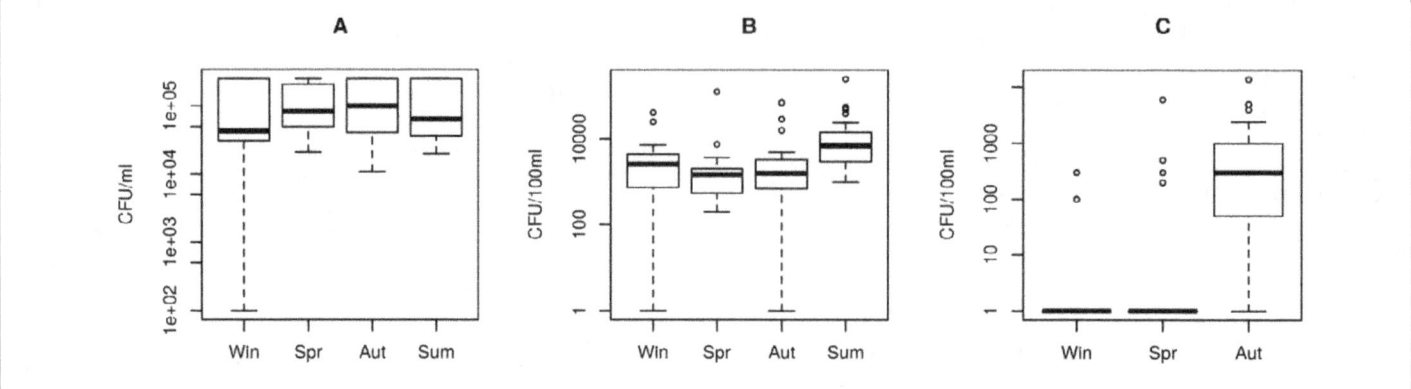

Figure 3: Box plots for counts of aerobic bacteria (A), total coliform (B) and *Escherichia coli* (C) in water samples collected from tanks (net cages or ponds) of the production chain of Nile tilapia along seasons between April 2010 and March 2011.

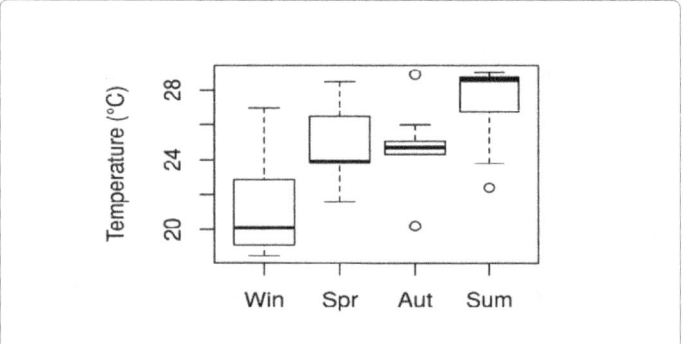

Figure 4: Box plots of water temperatures (30 cm depth) from production chain of Nile tilapia collected along seasons between April 2010 and March 2011.

Variable	AB		TC		EC	
	Coefficient	P-value	Coefficiente	P-value	Coefficiente	P-value
Intercept	1.83e5	<2e-16	5.92e2	<2e-16	0.34	0.15
Season	1.1	0.66	2.48	5.8e-6	8.88	1.6e-15
Tank	0.29	0.011	2.39	0.17	0.19	0.065
Season*tank	1.0	0.87	0.56	0.012	0.91	0.78

Table 1: Coefficients and significance level of the significant variables applying the negative binomial model for the quantification analysis of AB, TC and EC bacteria throughout the *O. niloticus* production chain.

evidencing significant differences between the recorded temperatures in winter and summer, and between the temperatures recorded in spring and summer. However, the temperatures recorded in fall showed no significant differences from the temperatures recorded in spring and summer. This analysis justifies the use of the variable Station as a discrete variable.

Table 1 shows a summary of the models adjusted for counting bacteria in the water samples analyzed during the period from April 2010 to March 2011 throughout the tilapia production chain. For each bacterial category analyzed (AB, TC or EC) coefficients and their respective significance levels (p-values) are presented. The significant coefficients with values greater than 1.0 (significance with value equal to or lower than 0.05 or 5%) substantially influence the increase in bacterial count in relation to the base line, whereas significant coefficients with values lower than 1.0 influence the reduction in the bacterial count, also in relation to the baseline.

The results presented in Table 1 show that the weather season

positively and significantly influenced the TC and EC bacterial count, which presented coefficients of 2.48 and 8.88, respectively, and were significant (5.8×10^{-6} to 1.6×10^{-15}, respectively). However, this variable did not exert a significant influence on the count of the AB bacteria. The TC bacteria exhibit a tendency to increasing counts according to the weather station, and were ordered according to the average temperatures observed, with the lowest counts observed in samples collected in the winter, and the highest counts in samples collected in the summer. For EC, the ordering of the seasons was similar, with the lowest counts obtained in samples collected in the winter, followed by samples collected in the spring and fall. No samples were collected in the summer.

When the data were analyzed in relation to the type of tank used for raising tilapia, i.e., net cages or ponds, one can see that the type of tank is a factor that influences the count of the three categories of bacteria evaluated (AB, TC and EC). The data in Table 1 show that the count of AB bacteria is significantly lower in the net cages than in the ponds, presenting a coefficient of 0.29 and p-value of 1.1%. This behavior repeats in the counting of EC, which presented a coefficient of 0.19, but with marginal significance (6.5%). The interaction term was not significant for both AB and EC counts. For TC, the direct influence of the tank was not significant (p-value 0.17). Nevertheless, the significance of the interaction term indicated that the growth effect throughout the seasons on this bacterial count is attenuated in net cages by a factor of 0.56, i.e., the growth is 44% lower compared to the variation in temperature throughout the seasons. The interaction effect shows significance of 1.2%.

The determination of the presence of AB, TC and EC bacteria in fillets of tilapia collected from the slaughterhouse was done immediately after slaughter of the animals. The data were analyzed aiming to verify the influence of the climate season, since the tank variable has no meaning at the slaughter stage. There were 12 repetitions, three for each season of the year. The data in Figure 5 show that the AB bacteria count ranged from 10^3-10^5 CFU/g^{-1} of fillet, while the TC bacteria count ranged from 500 to 10^4 CFU/g^{-1}, according to the season evaluated. Additionally, the count of EC was always less than 10 CFU/g^{-1}. Data analysis applying the negative binomial model allowed the classification of the weather station as a predictor variable with a significant influence on the AB bacterial count in fillets, with coefficient of 1.8 and significance of 3%. Likewise, the weather station influenced the TC bacteria count in fillets, showing a coefficient of 4.4 and significance lower than 0.1% (p-value$<2\times10^{-16}$).

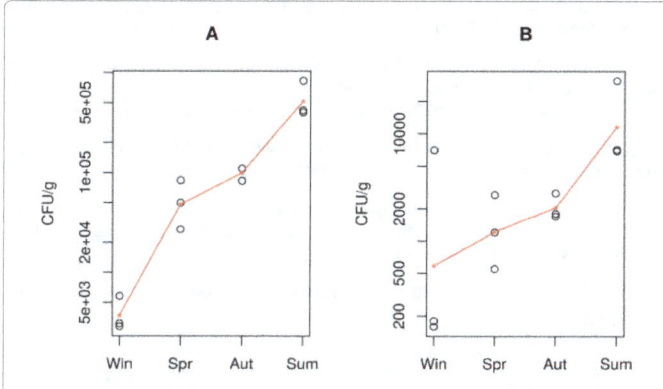

Figure 5: Count of aerobic bacteria (A) and total coliform (B) in fillets of Nile tilapia collected along seasons between April 2010 and March 2011.

Discussion

The water quality is one of the main factors that directly influence the productivity and health of farmed fish in net cages or ponds. There are several water parameters that must be taken into consideration in the growth of fish, including physical parameters such as temperature, color and turbidity; chemical parameters, such as pH, dissolved oxygen, amount of ammonia and nitrite; and microbiological parameters, which include the quantification of different bacterial groups. Therefore, one must consider water as a potential source of contamination of fish and derived products, especially microbiological contamination [14].

The count of AB bacteria was assessed at all stages of the Nile tilapia production chain of Figure 1A. These bacteria act on the decomposition of nutrients, producing part of the food consumed by other small organisms that live in the waters of lakes and rivers, being the base of the food chain. In Brazil, there is no legislation determining the minimum or maximum amount of bacteria in fish farming waters, thus, our results were compared to those obtained by other authors, and were lower than the scores obtained by [3,7] and higher than those reported by [15] in tilapia culture ponds.

Regarding counts of TC and EC in the water samples, the highest values were obtained in the debugging and filleting stages. In Brazil, there are no specific regulations concerning the presence of bacteria in the waters at these stages. However, the presence of these bacteria indicates the contamination of the environment, since they are not part of the fish microbiota, but of warm-blooded animals, including humans [16]. Previous studies [4,5] showed the importance of monitoring the presence of coliforms in the fish farming waters, including the culture of Nile tilapia, since these authors detected the presence of these contaminants in the organs of fish grown in waters with the presence of these bacteria [17] experimentally inoculated *E. coli* in the water tank and recovered the bacteria from the fish muscle.

The comparison between the two types of tank, as shown in Table 1, indicates that the bacteria counts in net cages were significantly smaller than those obtained from the ponds. Similar results were also found for EC, but with marginal significance (6.5%), indicating that further studies should be conducted to evaluate the importance of the type of tank in the count of EC. Regarding TC bacteria, the estimates showed in Table 1 permits the identification of the climate station as a variable that significantly influenced the increase in the count of these bacterial groups in both types of tanks, since the analysis of the data demonstrated a tendency to increase in TC bacterial count according to the average temperatures observed in water samples collected in the

different climate seasons. It was also observed that this increase is more pronounced in ponds.

The results in Table 1 show that net cages had lower AB and EC bacterial counts throughout the year compared to ponds and, therefore, respond differently to environmental alterations resulting from changes in the climate seasons. Among the factors that may contribute to the increase in bacterial counts of ponds, is the dilution factor, i.e., after the occurrence of the summer rainy season, which may carry contaminants from the neighborhood to the water bodies of both types of tank, the incidence of the dilution factor is smaller in ponds compared to net cages.

During the assessment of samples of tilapia fillet, increased AB and TC bacterial counts were found in combination with the increase in temperature. Considering that the count of these bacterial groups in water samples responded to climate changes differently from the count observed in fillets, intrinsic factors of the slaughter and filleting processes are responsible for this increase. The counts of EC in fillets showed a low contamination level of same, regardless of the climate station, being always below the limits recommended by FAO [18]. Thus, in the event of increased temperature, greater attention should be given to the slaughter and filleting processes, aiming to reduce the risk of contamination of fillets.

The joint analysis of the results shows that the highest counts of all bacterial groups assessed were obtained in samples of water at the debugging stage, which precedes the slaughter stage to obtain filet. The application of preventive approaches based on the target of reducing the risk of contamination of fish may curtail the risk of food borne diseases. Thus, the results of this study suggest that additional care must be deployed at this stage, which may result in obtaining fillets with low bacterial counts and, consequently, safer for consumption and with greater shelf life.

Acknowledgments

This work was supported by Conselho Nacional de Desenvolvimento Científico e Tecnológico (CNPq), Brazil (process 472648/2009-8). We thank to the fish farmers for their participation in this study.

References

1. Naylor RL, Goldburg RJ, Primavera JH, Kautsky N, Beveridge MCM (2000) Effect of aquaculture on world fish supplies. Nature 405: 1017-1024.

2. Naylor RL, Hardy RW, Bureau DP, Chiu A, Elliott M, et al. (2009) Feeding aquaculture in an era of finite resources. Proceedings of the National Academy of Sciences 106: 15103-15110.

3. Gorlach-Lira K, Pacheco C, Carvalho LC, Júnior, HNM, Crispim MC (2013) The influence of fish culture in floating net cages on microbial indicators of water quality. Braz J Biol 73: 457-463.

4. Mandal, SC, Hasan MS, Rahman, Manik ZH, Mandal M, et al. (2009) Coliform Bacteria in Nile Tilapia, *Oreochromis niloticus* of Shrimp-Gher, Pond and Fish Market. World Journal of Fish and Marine Sciences 1: 160-166.

5. Pal D, Gupta D, Chanchal (1992) Microbial pollution in water and its effect on fish. Journal of Aquatic Animal Health 4: 32-39.

6. Azevedo, PA, Podemski CL, Hesslein RH, Kasian SEM, Findlay DL, et al. (2011) Estimation of waste outputs by a rainbow trout cage farm using a nutritional approach and monitoring of lake water quality. Aquaculture 311: 175-186.

7. Ntengwe FW, Edema MO (2008) Physico-chemical and microbiological characteristics of water for fish production using small ponds. Physics and Chemistry of the Earth 33: 701-707.

8. Degefu F, Mengistu S, Schagerl M (2011) Influence of fish cage farming on water quality and plankton in fish ponds: A case study in the Rift Valley and North Shoa reservoirs, Ethiopia. Aquaculture 316: 129-135.

9. Schenone NF, Vackova L, Cirelli AF (2011) Fish-farming water quality and

environmental concerns in Argentina: a regional approach. Aquacult Int 19: 855-863.

10. Plumb JA (1997) Infectious diseases of tilapia.

11. Shoemaker CA, Evans JJ, Klesius PH (2000) Density and dose: factors affecting mortality of Streptococcus iniae infected tilapia (*Oreochromis niloticus*). Aquaculture 188, 229-235.

12. Godfree AF, Kay D, Wyer MD (2003) Fecal streptococci as indicators of fecal contamination in water. Journal Applied of Microbiology 83: 110-119.

13. Al-Harbi AH, Uddin N (2005) Bacterial diversity of tilapia (Oreochromis niloticus) cultured in brackish water in Saudi Arabia. Aquaculture 250: 566-572.

14. Surendraraj A, Farvin KHS, Yathavamoorthi R, Thampuran N (2009) Enteric

bacteria associated with farmed freshwater fish and its culture environment in Kerala, India. Research Journal of Microbiology 4: 334-344.

15. Al-Harbi AH, Uddin N (2003) Quantitative and qualitative studies on bacterial flora of hybrid tilapia (Oreochromis niloticus × Oreochromis aureus) cultured in earthen ponds in Saudi Arabia. Aquaculture Research 34: 43-48.

16. Cohen J, Shuval HI (1973) Coliform, fecal coliform and fecal streptococci as indicators of water Pollution. Water Soil Pollution 2: 85-95.

17. Guzmán MC, Bistoni ML, Tamagnini LM, González RD (2004) Recovery of Escherichia coli in fresh water fish, Jensynsia multidentata and Bryconamericus iheringi. Water Research 38: 2367-2374.

18. FAO (1979) Manuals of food quality control. FAO Food and Nutrition paper 14/4.

23

Evaluation of 3 Tagging Methods in Marking Sea Urchin, Paracentrotuslividus, Populations under Both Laboratory and Field Conditions

Cipriano A*, Burnell G, Culloty S and Long S

Aquaculture and Fisheries Development Centre School of Biological, Earth and Environmental Sciences University College Cork Distillery Fields North Mall Campus Cork, Ireland

Abstract

The purple sea urchin, "Paracentrotuslividus" is an Atlanto-Mediterranean species that is of commercial interest for its gonads (or roe) in Europe and Pacific/Asian countries. Individual identification of sea urchins is difficult due to the presence of spines and the structure of the skeletal-like test. However, a successful tagging technique is important for monitoring growth rate and survival of marked individuals in the laboratory and in the field. In addition, tagging can denote ownership, help in brood stock management, and allow for the tracking of animals in the market chain and laboratory experiments. In this study, smaller than previously reported passive integrated.

Transponder (PIT) tags and two external methods (fingernail polish and beads glued to the spines) were tested on "*P. lividus*" individuals to assess tagging capability, survival, and host response (e.g. lysozyme activity, nitric oxide levels, and cell viability). Additionally, PIT tagged individuals were released in an intertidal rock pool and monitored in order to test field application. Of the three different tagging methodologies, PIT tags were found to be the most successful in both studies carried out in the laboratory in regards to survival and tag retention. In the field, PIT tagged individuals were released and recaptured successfully. Furthermore, host response to individual tagging showed that the individuals were challenged by the sampling methodology which caused increased mortality.

Keywords: Fisheries; Paracentrotuslividus; Lysozyme; Aquaculture industry

Introduction

Demand for high quality seafood is increasing; resulting in the rapid growth of the aquaculture industry, which in turn reduces pressure on capture fisheries. Maximising the potential of the aquaculture industry though, requires innovation to refine existing techniques and apply new technologies [1]. Individual identification in holding conditions is important to monitor growth, behavior, genetics, and population dynamics [2]. The mechanism used to identify individuals must be easily distinguishable, be retained for long periods of time, and have minimal impact on

Growth and behavior if it is to have practical applications [3,4]. Individual identification also has further applications in brood stock management, denoting ownership, tracking animals in the market chain, and ecological studies [5] by serving as a marker within the population. However, many commonly farmed aquatic organisms, such as shrimp, sea urchins, and other marine invertebrates, have proven difficult to tag despite recent advances in tagging technology [5-7].

The purple sea urchin, Paracentrotuslividus, is an Atlanto-Mediterranean species that is of commercial interest for its gonads (or roe) in Europe and Pacific/Asian countries [8,9]. This commercial demand has placed pressure on wild sea urchin populations worldwide and has led to an increased need for aquaculture and hatcheries. In 2010, marine aquaculture produced an estimated 384,300 tons of echinoderms for consumption [10,11] necessitating the establishment of more hatcheries. Individual tagging or identification of sea urchins is difficult due to the presence of spines and the structure of the skeletal-like test. Previous studies have focused on external markings to the spines and test [12-14], drilling holes in the test [15-18], fingernail polish plus dental adhesive [14] internal markings, such as tetracycline injections [19-22] and passive integrated transponders (PITs) [6,23-

26]. However, these techniques are often invasive and result in altered behavior and high mortality rates [27,28].

The invasiveness of the tags challenges the individual and could lead to a compromised immune system and possible mortality. Factors affecting the immune system include diseases, condition, and diet [29]. Any factor that challenges an individual can elicit a host response. A tag, whether attached to the spine or test, or inserted into the coelomic cavity, may be treated by the sea urchins immune system as potential invaders. The sea urchins immune system is defined by its immune effectors 35 which have the capacity to respond to injuries, host invasion, and cytotoxic agents [29]. Using immune parameters, there are two means of evaluating host response: 1) humoral components such as nitric oxide and lysozyme activity assays and 2) cellular components such cell differentiation counts and cell viability assay. The humoral responses use antimicrobial compounds as a first response to invaders. Nitric oxide, a nitrogen radical produced from L-arginine during phagocytosis, serves as a mechanism of fighting off invasive pathogens [30]. Additionally, lysozyme levels demonstrate defense capabilities through the enzymatic break down of pathogenic cell membranes [31]. Cellular responses directly involve coelomocytes, circulating immune

*Corresponding author: Ashlie Cipriano, Aquaculture & Fisheries Development Centre School of Biological, Earth, & Environmental Sciences University College Cork Distillery Fields North Mall Campus Cork, Ireland
E-mail: acipriano@ucc.ie

cells, Located within the coelomic cavity. Therefore, cell viability and immune cell differentiation are important immune parameters which allow insight into the effects of tagging on *P. lividus*.

In this study, two laboratory trials looked at internal implanted passive integrated transponder (PIT) tags and two external methods (fingernail polish and beads glued to the spines) when tested on *P. lividus* individuals over a two or four month time period (February–May 2013) in order to assess individual tagging viability and host response. Additionally, PIT tagged individuals were released and detected in the field in West Cork, Ireland using a portable universal microchip reader (RealTrace® RT100) in a water proof scuba bag. The overall aim of the study was to identify a tag that was the most suitable for identifying an individual based on (a) tag retention, (b) host response to tags, and (c) survival of P.lividus in the laboratory and in the field.

Materials and Methods

In this study, three trials were conducted on sea urchins. The first trial (Trial 1) looked at tag retention and survival for 8 weeks in the laboratory. The second trial (Trial 2) looked at tag retention and host response for 4 weeks in the laboratory. The final trial (Trial 3) assessed PIT tag viability in the field.

All P. lividus individuals were sourced from Dunmannus 68 Sea foods sea urchin hatchery in West Cork, Ireland. Both laboratory trials were carried out at ambient temperature (14.0 ± 1.0°C; maintained with PSA Aquaclim 10 reversible heatpump/chiller) with continuous water circulation (1000 L sump filled with fresh sea water every 3 days) in four 400 L black plastic circular tanks. pH and oxygen saturation (DO) was monitored throughout the experiment to ensure water quality (pH: 8.0 ± 0.05 and DO: 8.0 mg/L ± 0.4 mg/L). Each tank (both trials) contained 4-5 plastic mesh baskets which each held 10 P. lividus individuals and underwent a different tag treatment. In order to acclimate the sea urchins, they were held for 7 days prior to trial commencement. No sea urchin mortalities were recorded during the acclimation period. Animals were fed ablibitum with Laminariasp.

Trial 1: Assessment of different tag options in sea urchins

General set-up: In each of the tanks, 1 basket held controls, 1 held sea urchins tagged with fingernail polish (40 ± 5 mm individuals), 1 basket held specimens tagged with beads (40 ± 5 mm individuals), and 2 baskets each held a different specimen size class (20 ± 5 mm or 40 ± 5 mm individuals) tagged with PIT tags. The control consisted of 10 un-tagged sea urchins (40 ± 5 mm).

External tagging: Two external tags were used (Figure 1b-1c). The first tag type was fingernail polish (Boot's Natural Collection®) applied to the top of an individual sea urchin's spines (approx. 20 spines) after drying spines with cotton. The second tag type was 2mm craft beads glued to 5 spines per sea urchin with a BISON non-toxic, non-drip formula super glue gel.

Internal tagging: 1.4 mm×8 mm PIT tags (Trovan®) (Chips4Fish, Zoo Chip, UK) programmed with a unique 12-digit identification number [32] were inserted through the peristome membrane via syringe application [33] (Figure 1a).

Monitoring: Animals were monitored daily for 8 weeks (119 days). External tags (fingernail polish and beads) on each individual sea urchin were counted. Internal tags (PIT Tags) were scanned using a portable universal microchip reader (RealTrace® RT100). Any Dead Sea urchins or sea urchins that had lost their tags (expelled PIT tag or

Figure 1: Physical representation of different tagging methodologies.

lost all beads/fingernail polish) were removed from the experimental system.

Trial 2: Host response from tagging of sea urchins in the laboratory

Tagging: Please refer to Trial 1 for general set-up, tagging, and monitoring with the exception that only smaller individuals (20 mm ± 5 mm) were used for this trial.

Host response measurements:

Coelomic fluid collection: Due to the small size of each individual urchin (20 ± 5 mm), samples were pooled from the 10 individuals per treatment per tank. An initial baseline sample was taken from 10 individuals prior to tagging. Sampling took place at 2 hrs (T2), 24 hrs (T48), 48 hrs (T48), and then occurred once a week for 4 weeks after tagging to allow for coelomic fluid levels to return to normal and for animals to de-stress between sampling episodes. Each week, 30 μl of coelomic fluid was taken from each individual and placed into a 2 ml eppendorf tube for pooling. Host response was monitored using lysozyme activity, nitric oxide measurements, and cell viability assays. In total, 40 individuals were analyzed per treatment.

Lysozyme activity assay: 200 μl of coelomic fluid, pooled from 10 sea urchins per treatment was immediately placed in a 2 ml eppendorf and placed on ice to prevent degradation of the samples. The samples were centrifuged at 3,000 rpm for 10 min to separate the cells from the serum. The supernatant was removed without disturbing the pellet formed at the bottom of the tube, placed into a clean 2ml eppendorf and stored at -20°C until analysis. The corresponding Pellet was also frozen at -20°C. The lysozyme activity assay was carried out as described by Cronin et al. [31], according to Caraballal et al. [34], a modification of Shugar [35]. Lysozyme activity was measured using a 96 well plate reader at a wavelength (λ) of 450 nm which calculated the mean decrease in absorbance at T0, 1 min, 2 min, 3 min and 4 min. Duplicate lysozyme standard solutions (30 µl) made from hen egg white lysozyme (SIGMA) were serially diluted, were included on each plate and consisted of seven concentrations 5.0 µg/ml, 2.5 µg/ml, 1.25 µg/ml, 0.625 µg/ml, 0.3125 µg/ml etc. A corresponding number of blank wells, consisting of 200 µl phosphate buffer (0.1M; pH 7.5), were included on each plate. 30 µl of the supernatant of each sea urchin sample (in triplicate) was added to the wells of each plate. 170 µl of *M.* lysodeikticussuspension (pH 6.4) was added to the wells containing the standard solutions and the sample solutions on each plate to make up to a total volume of 200 µl per well.

Nitric Oxide production, Griess reaction: 100 µl of coelomic fluid, pooled from 10 sea urchins per treatment, was incubated in a 96 well-plate for 30 min at room temperature (in triplicate). The same volume of filtered sea water (FSW) was added to the controls and left to incubate for 2 hours. 50 µl of supernatant was removed and transferred to a new plate. 50 µl of each of the Sodium Nitrite standards: 0.1, 0.5, 1, 5, 10, 50, 100 µM, was added to new wells. 100 µl of Solution A (1% sulphanil amide in 2.5% phosphoric acid) then 100 µl of Solution B (0.1% N157 naphthyl-ethylenediamine in 2.5% phosphoric acid) was then added to all wells (samples, standards and blank) and incubated for 5 min at room temperature. The 96 well plates were placed in a spectrophotometer reader (Elx808 Ultra Microplate Reader, BIO-TEK instruments, INC.) and read at 540 nm.

Cell viability: 100 µl of coelomic fluid, pooled from 10 sea urchins per treatment, was incubated in a 96 well-plate for 30 min at room temperature (in triplicate). The supernatant was removed by overturning the plate. 100 µl of working neutral red solution (1/50 of stock solution: 0.02 g in 5 ml of filtered sea water (FSW), filter and maintain in dark) was added to each well and incubated for 2 hours. The supernant was discarded by overturning. Samples were washed with FSW and discarded again. 100 µl Lysis169 Solution (1% Acetic acid and 50% Ethanol in distilled H$_2$O) was then added to each well. The 96 well plates were placed in a spectrophotometer reader (540 nm) after being shaken for 1 min.

Trial 3: Individual tagging of sea urchins in the field

The remaining 44 PIT tagged sea urchins from Trial 1 and 2 (sizes ranging from 20 ± 5 mm to 40 ± 5 mm) were ranched in shallow rocky tide pools near the Dunmannus Sea foods sea urchin hatchery in West Cork, Ireland from August to October 2013 for a total of 6 weeks. The urchins were simply released into the rock pool and monitored fortnightly at spring tides using a portable universal microchip reader (Real Trace® RT100).

Data analysis

For both laboratory experiments (Trial 1 and 2), a chi-squared (χ^2) 183 test was used to indicate the significance of a particular chosen tag type on sea urchin mortality. In Trial 2, post hoc analyses were conducted given the statistically significant ANOVA (p<0.05) for the cell viability and nitric oxide assay results on day 14 (last day where all treatments were still measured). Specifically, Tukey HSD tests were conducted on all possible pair wise contrasts. For the lysozyme activity assay results, individual Kruskal Wallis tests were used to test

for significance on day 14 due to the missing data from later sampling points (Table 1).

Results

Trial 1: Assessment of different tag options in sea urchins

The controls (urchin size: 40 mm ± 5 mm) had a 25% mortality rate over the four month study period. Within 24hrs, the individuals with fingernail polish painted on their spines (urchin size: 40 mm ± 5 mm) started to lift the entire epidermal layer off their tests and drop their spines. By day 29, 100% mortality was observed in this group. Although the bead methodology had 100% survival, it was only successful in the short-term as sea urchins survived with the beads up to 29 days before all the beads fell off or the sea urchins dropped the spines holding the beads. Lastly, two size classes of sea urchins contained the PIT tags; small (20 ± 5 mm) and large (40 ± 5 mm). The large sea 203 urchins showed a 52.5% mortality rate and the small urchins had a 22.5% mortality rate over the 8 week study period. All PIT tags remained operational throughout the experiment. A chi-square test indicated that the Choice of a particular tag type was associated with the survival of the sea urchin (χ2206; p<6x10 -5207).

Trial 2: Host response from tagging of sea urchins in the laboratory

For *Trial 2*, only 20 mm (± 5 mm) individuals were used. The controls showed a 20% mortality rate. Within the first week, the individuals with fingernail polish started to lift the entire epidermal layer off their tests and dropped their spines with 90% mortality. By day 21,100% mortality was observed. As in *Trial 1*, the bead methodology was more successful, but still observed a 90% loss of tags by the end of the 28 day experiment. Lastly, urchins that

Contained the PIT tags had 60% mortality. The individuals that retained their tags survived and were healthy until the end of the experiment. A chi-square test indicated that there was a significant difference between the tag retention by the sea urchins based on the tagging option employed (χ^2; p<5.2×10-3219). Cell viability, nitric oxide levels, and lysozyme activity were measured to evaluate host response to tagging. The tag types were statistically compared on day 14 as it was the last sampling point when all tag types were still viable.

The cell viability measurements (Figure 2) indicated an overall decrease for all treatments and the control. Two hours after tagging, 0.36 OD540 and stabilized at 0.12 OD540 throughout the remainder of the trial. The different tags types followed a similar pattern. PIT tagged individuals after hours measured 0.36 OD540 and stabilized at 0.12 OD540, while beaded individuals, after two hours, were 0.58 OD540 and stabilized at 0.10 OD540. Fingernail polished individuals after two hours was measured at 0.32 OD540 and stabilized at 0.11 OD540. The fingernail polish tag type was significantly different (p<0.05) from other tag types on day 14. There was not enough sea urchin coelomocyte left over to sample until day 28 due to mortality. Nitric oxide

Tag Option	Trial 1.1		Trial 1.2 (Mortality)
	Mortality	**Tag Loss**	
Control	25%		20%
Bead		100%	90%
PIT Tag			
20 mm Individuals	22.5%		60%
40 mm Individuals		52.5%	
Nail Polish	100%		100%

Table 1: Tag retention and survival in *P. lividus*.

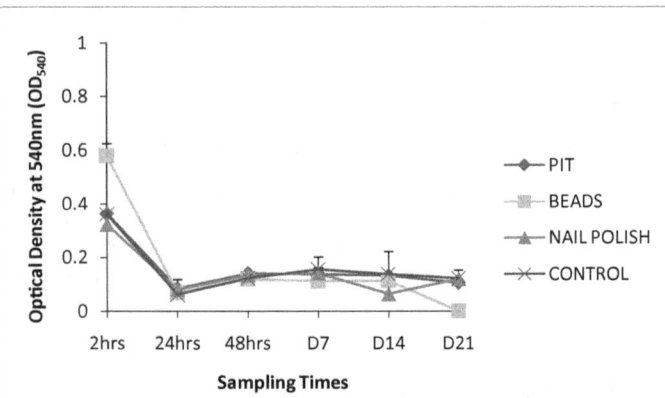

Figure 2: Cell viability, measured in optical density, for pooled *P. lividus* individuals tagged with different tagging options over a 28 day period.

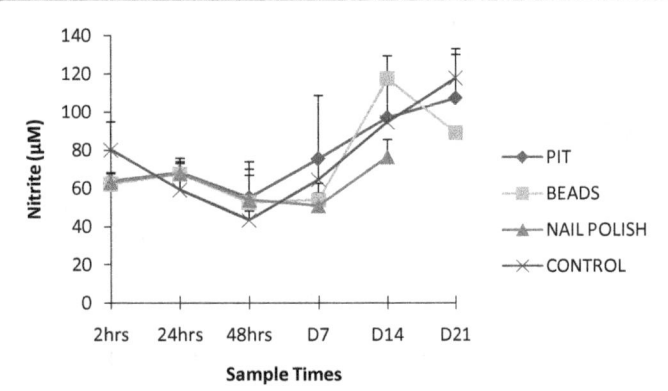

Figure 3: Nitric oxide levels for pooled *P. lividus* individuals tagged with different tagging options over a 28 day period.

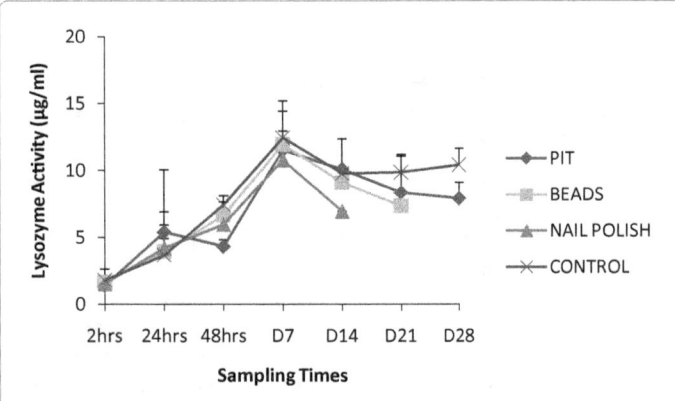

Figure 4: Lysozyme activity data pooled *P. lividus* individuals tagged with different tagging options over a 28 day period.

levels (Figure 3) showed an initial 48hr decrease after tagging followed by general increases in all tag types and controls until the end of the experiment. The controls at T2 measured 80.04 µM which decreased to 43.3 µM237 at T48 and then increased to 117.78 µM on day 21. The different tag types followed a similar pattern. PIT tagged individuals at T2 measured 63.8 µM which decreased to 55.01 µM at T48 and then increased to 107.28 µM on day 21. Beaded individuals at T2 were 62.35 µM which decreased to 52.7 µM at T48 and then increased to 107.28 µM at T21. Fingernail polished individuals at T2 were 63.26 µM which decreased to 54 µM at T48 and then increased until they died off on day

14. The bead tag type was significantly different (p<0.001) from other tag types on day 14. There was not enough sea urchin coelomocyte left over to sample until day 28 due to mortality. Lysozyme activity (Figure 4) showed a general increase in all tag types and the control until day 7. The controls at T2 measured 1.79 µg/ml, peaked at 12.42 µg/ml and measured 10.41 µg/ml at TDay28. The different tags types followed a similar pattern. PIT tagged individuals at T2 measured 1.37 µg/ml, peaked at 11.51 µg/ml and measured 7.89 µg/ml at TDay28, beaded individuals at T2 were 1.63 µg/ml peaked at 11.92 µg/ml and wasn't viable at TDay28, and fingernail polished individuals at T2 were 1.52 µg/ml, peaked at 10.77 µg/ml and wasn't viable at TDay28. The fingernail polish tag type was significantly different (p<0.05) from other tag types on day 14 [36].

Trial 3: Field detection of PIT tags

In the field, individuals (n=44; 20 ± 5 mm and 40 ± 5 mm) were monitored four times over a two month period (Table 2). Tag identification numbers were recorded in order to observe individual occurrences and tag feasibility. One week after releasing the PIT tagged individuals into the rock pool, 2.3% of individuals were recaptured and identified. However, 3 weeks after the animals were replaced, 13.6% of individuals were recaptured. 11.4%, 20.5%, and 4.5% individuals were caught and recaptured on weeks 4, 5, and 6, respectively. Altogether 12 individuals (27%) were identified and captured over the 6 week period [37,38].

Discussion

Previous studies using PIT tagged aquatic invertebrates claim that PIT tagging does not adversely affect survival [6,39-42]. Therefore, the higher mortality from this study could be due to sensitivity of P. lividus, condition of the individuals at the time of the experiment or the host response to the tags. Three tag types were used 271 in this study: glued craft beads, nail polish and PIT tags. The glued crafts beads could be viable for short term experiments that last less than two weeks and don't involve host response measurements. It was observed that the sea urchins dropped the spines with the beads, possibly as a defense mechanism. The fingernail polish resulted in 100% mortality. This was could be due to the fact that a) the epithelial layer with the fingernail polish became detached within 24 hrs of application layer making the urchin susceptible to infection and b) some urchins were observed to have a swollen peristomal membrane possibly indicating the fingernail polish, when applied, covered the anus preventing waste expulsion (*personal observation*). The specific PIT tag used in this study was 4 mm smaller than tags used in previous studies [6,36-39]. In our initial tag retention trial, the mortality rate observed was lower in the smaller individuals than in the larger individuals possibly due to the life history and previous holding conditions of the urchins. The most successful marker in this study was the PIT tag with applications in denoting ownership and tracking individuals.

Trial 1 showed that survival in the smaller sized urchins with PIT tags was the same as in the control group. The poorer survival in the larger PIT tagged individuals may have been due to their previous life history in the hatchery. Upon dissection of PIT tagged individuals, it was observed that the tag was lodged in the membrane covering the inside of the test (personal observation). In a study by Parker and Ranken [39] on PIT tagging in Black Rockfish, it was suggested that the movement of tags could be the cause of observed mortalities. Christy [26] observed PIT tags lodged in the outer peritoneum of two frogs (Limnodynastesperonii). This may have contributed to the mortalities observed in PIT tagged urchins in this study as the PIT tag was inserted

	Weeks post releasing to the field (n=44)				
Individual PIT Codes	Week 1 Aug. 23 2013	Week 3 Sept. 04 2013	Week 4 Sept. 12 2013	Week 5 Sept. 17 2013	Week 6 Oct. 21 2013
3458321 (15.4 g)	✓				
3480739 (41.2 g)		✓	✓	✓	
3440163 (12.1 g)		✓		✓	
3483893 (6.6 g)		✓		✓	
3474164 (8.4 g)		✓	✓	✓	
3483706 (38.6 g)		✓		✓	
3475231 (26.5 g)		✓	✓	✓	
3466602 (25.8 g)			✓	✓	
3467376 (30.2 g)				✓	
3447800 (12.6 g)				✓	
3476882 (33.5 g)					✓
3784860 (10. 4 g)					✓
% of individuals recaptured	2.3	13.6	11.4	20.5	4.5

Table 2: Recapture rate for PIT tagged urchins in the rock pool over a 6 week period.

into the coelomic cavity where it could move around, conceivably causing internal damage. Additionally, upon the completion of the experiment, eight of the surviving PIT tagged animals were dissected in order to locate the PIT tag. It was observed that the tag was starting to be encased in the tissue lining of the test. This observation has not been reported before and warrants further investigation. Although individual tagging has been used for many years in different species (mainly vertebrates) [43], other studies using nPIT tags on aquatic invertebrates include prawns, Macrobrachiumrosenbergii [44], freshwater signal crawfish, Pacifastacusleniusculus [38], pot-bellied seahorses, Hippocampus abdominalis [41], green sea urchins, Strongylocentrotusdroebachiensis [6,45], the kina sea urchin, Evechinuschloroticus, sea cucumbers, Holothuriawhitmaeiand Actinopygamiliaris [4], and freshwater pearl mussels, Margaritiferamargaritifera [46], easternlampmussels, Lampsilisradiataradiata [42] 305 with varying success as viable markers. Other studies using similar tagging techniques to this study, such as Agatsuma et al. [47] reported successful tagging of *Strongylocentrotusnuduss* pines with different colors of fingernail polish and covering the polish with a quick drying dental adhesive but only for trials lasting for shorter than three days. No studies to our knowledge have used craft beads glued to the spines.

There have been few studies on the effects of tagging on the host, especially in invertebrates [46]. This is the first study looking at host response to tagging in P. lividus using immune parameters as indicators of host response. Trial 1 was designed to assess tagging viability and mortality, while Trial 2 was designed to test immune parameters of the tagged host. All animals in *Trial 2* showed host response, and higher mortality when compared to *Trial 1*, including the controls, due to handling and sampling of the coelomic fluid via insertion of a syringe into the peristomal membrane. The overall decrease in cell viability within the treatment groups, as well as the controls, may be due to tag effects. With less viable cells, phagocytosis decreases as well; therefore decreasing capability of an immune response. Upon introduction of a

stressor, the host will liberate oxygen and free radicals, such as nitric oxide which is a potent bactericidal [48-51]. Our nitric oxide (NO) measurements indicate that the sea urchin coelomocytes produced NO with increased production at 48hrs; however, whether it was due to the tags or to the sampling is unclear and would need further verification. Lysozyme is an enzyme that can hydrolyse components of bacterial walls; therefore, aiding in immune defense and digestion [52]. Lysozyme results from this study indicate an increase in host response until day seven followed by a decrease in lysozyme activity in all treatments excluding the controls. Again, the initiation of the host response, whether it was due to the tags or to the sampling methodology, is unclear and would need further verification. Behavior studies should be developed to determine a less invasive ways of measuring stress and host response because the methodology used in this study challenged all individuals including the controls. Therefore, it is necessary to develop an alternative way of measuring host response (i.e. the activity of the podia and the configuration of the spines). In the capture, monitor, and release study, surviving tagged individuals from the tag retention experiment were ranched in natural rock pools on the west coast of Ireland. The recaptured urchins had retained their PIT tags for at least five months (laboratory and field) and were easily scanned with a portable universal microchip reader (RealTrace® RT100) (standard in veterinary practices); however, there are two limitations 339 to using PIT tags in the field: 1) sea urchins preferably hide in the crevasses between rocks which limits accessibility and 2) the relatively short distance from which tags can be detected (also reported in Bubb et al. [38]). One way to address these limitations is to apply a technique suggested by Bubb et al. [38] and Roussel et al. [53], which calls for the use of a coil antenna or and 'open coil' antenna mounted on a pole to facilitate searching for tagged individuals in aquatic environments full of rocky crevasses. PIT tagging permits repeated non-destructive sampling of individuals. The claim that this technique has a theoretically indefinite life span, negligible tagging mortality, high tag retention, and no apparent long term effects on growth and survival of tagged individuals [38], needs further verification in *P. lividus*. The PIT tags used in this study was a useful mechanism for individual sea urchin identification in the laboratory and in the subsequent field study. Additionally, this method, provided tagged individuals are held in captivity for three months to test for tag retention, allows for a large number of animals to be marked and has the potential to address numerous questions relating to the behaviour, mobility, habitat use, brood stock management, and denotes ownership within the laboratory and in the field.

Fingernail polish was the least successful tagging technique and caused 100% mortality. The bead technique is a temporary tagging solution but is highly stressful when compared to the control and PIT tagged individuals. Urchins released into rock pools were detected up to 6 weeks after release indicating that the use of these smaller PIT tags are a viable option in sea urchin culture.

Acknowledgement

We would like to thank Luke Harman, Maria O'Mahoney, and Elaine Brennan for all their assistance in the laboratory and John Chamberlin.

References

1. Charles Y, Muki S, Thierry C, Shawn R (2010) Global Conference on Aquaculture.

2. Emery L, Wydoski RS (1987) Marking and tagging of aquatic animals: an indexed bibliography. US Dep Inter Fish Wildl Serv 57: 165.

3. Freilich JE (1989) A Method for Tagging Individual Benthic Macroinvertebrates. J North Am Benthol Soc 8: 351-354.

4. Purcell SW, Agudo NS, Gossuin, H (2008) Poor retention of passive induced transponder (PIT) tags for mark-recapture studies on tropical sea cucumbers. SPC Beche Mer Inf Bull 28: 53-55.

5. Reisser J, Proietti M, Kinas P, Sazima, I (2008) Photographic identification of sea turtles: method of description and validation, with an estimation of tag loss. Endang Species Res 5: 73-82.

6. Hagen NT (1996) Tagging sea urchins: a new technique for individual identification. Aquaculture 139: 271-284.

7. Houghton JDR, Doyle TK, Davenport J, Hays GC (2006) Developing a simple, rapid method for identifying and monitoring jellyfish aggregations from the air. Mar Ecol Prog Ser 314: 159-170.

8. Turon X, Giribet G, Lopez S, Palacin C (1995) Growth and population structure of Paracentrotus lividus (Echinodermata: Echinoidea) in two contrasting habitats. Mar Ecol Prog Ser 122: 193-204.

9. Boudouresque CF, Verlaque M (2002) Biological pollution in the Mediterranean Sea: invasive versus introduced macrophytes. Mar Pollut Bull 44: 32-38.

10. Pais A, Chessa LA, Serra S, Ruiu A, Meloni G, Donno Y (2007) The impact of commercial and recreational harvesting for Paracentrotus lividus on shallow rocky reef sea urchin communities in North-western Sardinia, Italy. Estuar. Coast. Shelf Sci 73: 589-597.

11. FAO (2012) The State of World Fisheries and Aquaculture 2012.

12. Moore HB (1935) A comparison of the biology of Echinus esculentus in different habitats. Part II J Mar Biol Assoc United Kingdom New Ser 20: 109-128.

13. Sinclair AN (1959) Observations on the behaviour of sea urchins. Aust Mus Mag 13: 3-8.

14. Agatsuma Y (2001) Ecology of Hemicentrotus pulcherrimus, Pseodocentrotus depressus and Anthocidaris crassispina in southern Japan. Developments in Aquaculture and Fisheries Science Elsevier 32: 363-374.

15. Fuji AR (1963) On the growth of the sea urchin, Hemicentrotus pulcherrimus (A.Agassiz). Bull Jpn Soc Sci Fish 29: 118-126.

16. Ebert TA (1965) A Technique for the Individual Marking of Sea Urchins. Ecology 46: 193-194.

17. Lees DC (1968) Tagging subtidal echinoderms.

18. Olsen M, Newton G (1979) A simple, rapid method for marking individual sea urchins. Calif. Fish Game 65: 58-62.

19. Ebert TA (1977) An experimental analysis of sea urchin dynamics and community interactions on a rock jetty. J Exp Mar Bio Ecol 27: 1-22.

20. Gage JD (1992) Natural growth bands and growth variability in the sea urchin Echinus esculentus: results from tetracycline tagging. Mar Biol 114: 607-616.

21. Kenner MC (1992) Population dynamics of the sea urchin Strongylocentrotus purpuratus in a Central California kelp forest: recruitment, mortality, growth, and diet. Mar Biol 112: 107-118.

22. Ebert TA, Russell MP (1993) Growth and mortality of subtidal red sea urchins (Strongylocentrotus franciscanus) at San Nicolas Island, California, USA: problems with models. Mar Biol 117: 79-89.

23. Prentice EF, Park DL, Sims CW (1984) A study to determine the biological feasibility of a new fish tagging system. Portland.

24. Prentice EF, Flagg TA, McCutcheon CS (1990) Feasibility of using implantable passive integrated transponder (PIT) tags in salmonids. Am Fish Soc Symp 7: 317-322.

25. Hagen NT (1991) A new technique for the individual tagging of sea urchins.

26. Galimberti F, Sanvito S, Boitani L (2000) Marking of southern elephant seals with passive integrated transponders. Mar Mammal Sci 16: 500-504.

27. Gauthier Clerc M, Le Maho Y (2001) Beyond Bird Marking With Rings. Ardea 89: 221-230.

28. Godfrey JD, Bryant DM, Williams MJ (2003) Radio-telemetry increases free-living energy costs in the endangered Takahe Porphyrio mantelli. Biol Conserv 114: 35-38.

29. Matranga V (Ed.) (2005) Echinodermata: Progress 440 in Molecular and Sub-molecular Biology.

30. Tafalla C, Gómez-León J, Novoa B, Figueras A (2003) Nitric oxide by carpet shell clam (Ruditapes decussatus) hemocytes. Development and Comparative Immunology 27: 197-205.

31. Cronin MA, Culloty SC, Mulcahy MF (2001) Lysozyme activity and protein concentration in the haemolymph of the flat oyster Ostrea edulis (L.). Fish Shellfish Immunol 11: 611-622.

32. Rogers LM, Hounsome TD, Cheeseman CL (2002) An evaluation of passive integrated transponders (PITs) as a means of permanently marking badgers (Meles meles). Mamm Rev 32: 63-65.

33. Woods CMC, James PJ (2005) Evaluation of passive integrated transponder tags for individually identifying the sea urchin Evechinus chloroticus (Valenciennes). Aquac Res 36: 730-732.

34. Carballal MJ, Lopez C, Azevedo C, Villalba A (1997) Enzymes Involved in Defense Functions of Hemocytes of Mussel Mytilus galloprovincialis. J Invertebr Pathol 70: 96-105.

35. Shugar D (1952) The measurement of lysozyme activity and the ultra-violet inactivation of lysozyme. Biochem Biophys Acta, 8: 302-309.

36. Dutton PH, McDonald D, MacDonald DL (1994) Use of PIT tags to identify adult leatherback. Mar Turt Newsl 67: 13-14.

37. Christy MT (2006) The efficacy of using Passive Integrated Transponder (PIT) tags without anaesthetic in free-living frogs. Aust Zool 30: 139-142.

38. Bubb DH, Lucas MC, Thom TJ, Rycroft P (2002) The potential use of PIT telemetry for identifying and tracking crayfish in their natural environment. Hydrobiologia 483: 225-230.

39. Parker SJ, Rankin PS (2003) Tag Location and Retention in Black Rockfish: Feasibility of Using PIT Tags in a Wild Marine Species. North Am J Fish Manag 23: 993-996.

40. Wiles PR, Guan RZ (1993) Studies on a new method for permanently tagging crayfish with microchip implants. Freshwater Crayfish 9: 419-425.

41. Woods CMC (2005) Evaluation of VI-alpha and PIT-tagging of the seahorse Hippocampus abdominalis. Aquac Int 13: 175-186.

42. Kurth J, Loftin C, Zydlewski J, Rhymer J (2007) PIT tags increase effectiveness of freshwater mussel recaptures. J North Am Benthol Soc 26: 253-260.

43. Jenkins WE, Smith TIJ (1990) Use of 474 PIT tags to individually identify striped bass and red drum brood stocks.

44. Caceci T, Smith SA, Toth TE, Duncan RB, Walker SC (1999) Identification of individual prawns with implanted microchip transponders. Aquaculture 180: 41-51

45. Duggan RE, Miller RJ (2001) External and internal tags for the green sea urchin. J Exp Mar Bio Ecol 258: 115-122.

46. Wilson CD, Arnott G, Reid N, Roberts D (2011) The pitfall with PIT tags: marking freshwater bivalves for translocation induces short-term behavioural costs. Anim Behav 81: 341-346.

47. Agatsuma YU, Nakata AK, Matsuyama KE (2000) Seasonal foraging activity of the sea urchin Strongylocentrotus nudus on coralline flats in Oshoro Bay in south-western Hokkaido, Japan. Fish Sci 66: 198-203.

48. Kumar V, Jindal SK, Ganguly NK (1995) Release of reactive oxygen and nitrogen intermediates from monocytes of patients with pulmonary tuberculosis. Scand J Clin Lab Invest 55: 163.

49. Pacelli R, Wink DA, Cook JA, Krishna MC, De Graff W, et al. (1995) Nitric oxide potentiates hydrogen-peroxide induced killing of Escherichia coli. J Exp Med 182: 1469-1479.

50. Wheeler MA, Smith SD, García-Cardenã G, Nathan CF, Weiss RM, et al. (1997) Bacterial infection induces nitric oxide synthase in human neutrophils. J Clin Invest 99: 110-116.

51. Fang FC (1997) Perspectives series: host/pathogen interactions. Mechanisms of nitric oxide-related antimicrobial activity. J Clin Invest 99: 2818-2825.

52. Cheng TC (1983) The role of lysozymes in molluscan inflammation. American Zoologist 23: 129-144.

53. Roussel JM, Haro A, Cunjak RA (2000) Field test of a new method for tracking small fishes in shallow rivers using passive integrated transponder (PIT) technology. Can J Fish Aquat Sci 57: 1326-1329.

24

Microbiological Quality of Catfish (*Clarias Gariepinus*) and Tilapia (*Tilapia Mossambica*) Obtained from Wet Markets and Ponds in Malaysia

Titik Budiati[1]*, Gulam Rusul[2], Wan Nadiah Wan-Abdullah[3], Rosma Ahmad[3] and Yahya Mat Arip[4]

[1]*Food Technology Department, State Polytechnic of Jember, 68121, Jember, Indonesia*
[2]*Food Technology Division, School of Industrial Technology, University of Science, Malaysia, 11800, Penang, Malaysia*
[3]*Bioprocess Technology, School of Industrial Technology, University of Science, 11800, Penang, Malaysia*
[4]*School of Biological Science, University of Science, Malaysia, 11800, Penang Malaysia*

Abstract

The aim of this study was to determine the microbiological quality in catfish (*Clarias gariepinus*) and tilapia (*Tilapia mossambica*) obtained from wet markets and ponds in Malaysia. A total of 108 samples (32 catfish, 32 tilapia, and 44 water samples) were obtained from nine wet markets and eight ponds in Penang, Malaysia. The feed in fish ponds were chicken offal, spoiled eggs and commercial fish feed. Using standard procedures, aerobic plate counts (APC), coliform, fecal coliform including *E. coli* were performed. A total 31/32 of catfish and 31/32 of tilapia exceeded the recommended microbiological standard for the APC. *E. coli* was less than 3 MPN/g for all catfish and tilapia samples. Temperature and pH of water ponds ranged from 26 to 27.5°C and 6 to 6.8, respectively. Home-made feed using chicken offal and spoiled egg may contribute to the microbiological quality in fish. This highlights the importance of feed in aquaculture system.

Keywords: Catfish; Microbiological quality; Pond; Tilapia; Wet market

Introduction

Aquaculture, an important sector, has significant economic impact in many countries including Malaysia. Based on the recent data, freshwater fish is cultured using pond culture, ex-mining pool, freshwater cage, cement tank, canvas tank, and freshwater pen culture systems [1]. The highest freshwater fish production in Malaysia was catfish reared in pond culture (64.9%) followed by tilapia reared in ex-mining pools (18.2%) [1]. FAO [2] reported that aquaculture production in Malaysia is marketed locally for domestic consumption. Besides quantity of aquaculture production, the application of hygiene and food safety procedures in fish production should be taken into account [3]. The inappropriate aquaculture practice may become a concern for food safety issue [4]. Yet, intensification of fish production is still raising the risk of disease due to high density stocking of fish, antibiotics, poor water control and other factors [3]. Feed, an important diet for the growth of fish [5,6], may promoted the risk of disease for fish and ultimately for human [7]. In Asia-Pacific region, commercial feed and home-made feed have been used to feed the fish in freshwater aquaculture production. The home-made feed was chicken offals, spoiled eggs or waste food [8] which may affect to the quality of fish.

However, there is still limited data regarding the microbiological quality of fish distributed in Malaysia which were fed using home-made feed (chicken offal or spoiled eggs) and commercial feed. The microbiological quality of fish can be measured by using aerobic plate counts, coliform counts, fecal coliform counts and *E. coli* counts. Thus, this study was carried out to fill in this gap. The aim of this study was to determine the microbiological quality of catfish (*Clarias gariepinus*) and tilapia (*Tilapia mossambica*) obtained from wet markets and ponds (fed with chicken offals, spoiled eggs or commercial fish feed) in Malaysia.

Materials and Methods

Samples

Catfish and water samples were collected from local wet markets (A,B,C,D, and E), pond A1, Pond A2, Pond B1 and Pond B2. Tilapia and water samples were collected from local wet markets (E,F,G, and H), pond C1, Pond C2, Pond D1 and Pond D2. Alive catfish and tilapia was placed in sterile plastic bag that was filled with water. Further, the plastic bag was filled with oxygen and banned with rubber. Dead tilapia obtained from wet market was placed in sterile plastic bag and kept in the box with temperature approximately 4°C during transportation to the laboratory. Water samples were collected using sterile test tubes, and then transferred to the laboratory in ice box with temperature approximately 4°C. The samples were proceeded in the laboratory within 6 hours. Ponds A1 and Pond A2 used chicken offal to fed catfish. Pond C1 and C2 used spoiled egg to fed tilapia. Pond B1, B2, C1 and C2 used commercial fish feed.

Determination of aerobic plate count (APC)

Twenty five grams of catfish or tilapia was mixed with 225 mL of 0.1% Pepton Water (PW, Oxoid, Baringstoke, Hampshire, United Kingdom) and homogenized by using stomacher (Interscience, France) for 120 sec. The dilution was prepared by pipeting 1 mL of aliquot and mixed with 9 mL of 0.1% PW. The dilution was made from 10^{-1} to 10^{-6}. About 100 mL of aliquot was spread on Plate Count Agar (PCA, Merck KGaA, Darmstadt, Germany) and incubated at 37°C for 24-48 h. Total number of colonies were counted and calculated as BAM Manual Protocol [9]. Total aerobic count was expressed as log cfu g^{-1}. Twenty five millilitres of water sample was mixed with 225 mL of 0.1%

*Corresponding author: Titik Budiati, Food Technology Department, State Polytechnic of Jember, 68121, Jember, Indonesia
E-mail: titik.budiati@gmail.com

PW and homogenized by using stomacher for 120 sec. The dilution was prepared from 10^{-1} until 10^{-6} and plated on PCA as BAM Manual Protocol. Aerobic Plate Count (APC) was determined by using spread method and incubated at 37°C for 24-48 h [9]. This was expressed as log cfu mL^{-1}.

Determination fecal coliform count

Fecal coliform counts were determined using MPN method [10]. The equipment and solutions were sterilized in an autoclave before each use. The microbial quality of the catfish and tilapia were assessed by sampling 25 g of catfish and tilapia. One mL of each dilution (10^{-1} until 10^{-5}) was transferred into three tubes of Lauryl Sulfate Tryptose Broth (LST, Oxoid, Baringstoke, Hampshire, United Kingdom) and incubated at 37°C for 24 hours. Approximately, 10 µL of LST broth from positive tubes were transferred into 10 mL of Brilliant Green Lactose Bile Broth (BGLB, Merck KGaA, Darmstadt, Germany) and incubated at 37°C for 24 h. Turbid tubes with gas were considered as positive and coliform counts were expressed as MPN g^{-1} or MPN mL^{-1}.

Fecal coliform bacteria are used as the fecal indicator in guidelines for wastewater reuse in irrigation and aquaculture [11]. Therefore, the water in the fish ponds was subjected to microbiological investigation using fecal coliforms as indicators of fecal pollution. Fecal coliform count was determined by transferring 10 µL of LST broth from positive tubes into three tube of *Escherichia coli* Broth (EC, Merck KGaA, Darmstadt, Germany) and was incubated at 44-45°C for 24 h. Tubes showing gas and turbidity were considered positive for the presence of fecal coliform and these were expressed as MPN g^{-1} or MPN mL^{-1}. A loopful of broth in gasing tube was streaked on L-EMB agar (Merck, Germany) and incubated at 35°C for 18-24 h. Presumptive E. coli grew as distinctive metallic green sheen colony on L-EMB Agar. Biochemical tests such as gram staining, catalase, cytochrome oxidase, microscopic observation, indole production, Voges-Proskauer reaction, methyl read reaction, citrate production and re-inoculated back into LST to confirm gas production were carried out as the Bacteriological

Analytical manual [10]. *E. coli* culture (Food Microbiology Laboratory, School of Industrial Technology, USM) was used as control.

Temperature and pH of water analyses

Temperature and pH of water ponds were measured using Portable Digital pH Meter (Hanna, Model HI 8424, Romania). The sampling was measured every sampling day between 09.00 and 10.00 a.m.

Statistical Analysis

The statistical analysis of APC, coliform, faecal coliform and E. Coli in catfish obtained from different wet market and ponds were determined by using one-way ANOVA (SPSS software for Windows Version 13) at the significance level (P<0.05). The same statistical analysis was also applied for APC, coliform, faecal coliform and *E. coli* in tilapia obtained from different wet market and ponds. In similar, one way ANOVA was applied to determine APC, coliform, faecal coliform, *E. coli*, temperature and pH of water obtained from different wet market and ponds.

Results

Aerobic plate counts

Aerobic plate counts (APC) are a widely accepted measure of the general degree of microbial contamination [12]. In this study, the mean APC values were ranged from 5.30 to 6.84 log 10 cfu g^{-1} for catfish and 5.77 to 9.12 log 10 cfu g^{-1} for tilapia (Table 1). There was no significant different (P>0.05) among total aerobic bacteria in the catfish obtained from the ponds. Similarly, there was also found in tilapia obtained from all of the ponds (Table 1). However, there was significant different (P<0.05) between total bacteria in catfish obtained from wet market D (Gelugor wet market) and other markets. The significant different (P<0.05) was also observed between total bacteria in tilapia obtained from two markets (F,G) and other markets (E,H).

This present study found that 31/32 of catfish and 31/32 of tilapia

Location	Total aerobic counts (log CFU g^{-1} or log CFU mL^{-1})			Coliform counts (Log MPN g^{-1} or Log MPN mL^{-1})			Fecal coliform counts (Log MPN g^{-1} or Log MPN mL^{-1})		
	Catfish	Tilapia	Water	Catfish	Tilapia	Water	Catfish	Tilapia	Water
Wet market									
A (Bukit Mertajam)	6.84 ± 0.04^e	NA	7.12 ± 0.82	2.81 ± 1.09^{cd}	NA	3.7 ± 0.43^{de}	1.11 ± 0.44^c	NA	1.12 ± 0.48
B (Bagan Ajam)	6.76 ± 0.02^{de}	NA	7.07 ± 0.41	3.56 ± 0.34^{cd}	NA	3.96 ± 0.07^{de}	1.04 ± 0.49^c	NA	1.01 ± 0.37
C (Nibong Tebal)	5.86 ± 0.04^{cd}	NA	7.25 ± 0.39	3.75 ± 0.26^{cd}	NA	3.77 ± 0.23^{cde}	0.97 ± 0.33^c	NA	1 ± 0.43
D (Gelugor)	5.30 ± 0.02^c	NA	6.27 ± 0.05	2.52 ± 1.07^c	NA	3.17 ± 0.47^{cd}	0.95 ± 0.34^c	NA	1.38 ± 0.70
E (Bayan Baru)	6.55 ± 0.06^{de}	7.85 ± 0.55^d	^{b}6.35 ± 0.06	3.73 ± 0.27^{cd}	3.63 ± 0.30^{de}	^{b}3.86 ± 0.18^{cde}	0.94 ± 0.41^c	1.52 ± 0.19^d	^{b}1.41 ± 0.75
F (Hypermarket S1)	^{a}NA	8.49 ± 0.13^e	NA	NA	3.93 ± 0.09^e	NA	^{a}NA	1.44 ± 0.27^d	NA
G (Hypermarket S2)	NA	9.12 ± 0.46^e	NA	NA	3.96 ± 0.06^e	NA	NA	1.57 ± 0.19^d	NA
H (Chowrasta)	NA	8.95 ± 0.21^{de}	NA	NA	3.55 ± 0.32^{de}	NA	NA	1.35 ± 0.51^d	NA
Ponds									
A1	6.29 ± 0.49^{de}	NA	6.56 ± 0.53	4.1 ± 0.07^d	NA	3.8 ± 0.42^{cde}	0.99 ± 0.5^c	NA	1.63 ± 0.28
A2	6.48 ± 1.07^{de}	NA	6.93 ± 0.40	4.16 ± 0.18^d	NA	4.19 ± 0.16^e	1.2 ± 0.63^c	NA	1.37 ± 0.79
B1	6.23 ± 0.39^{de}	NA	6.55 ± 0.03	3.32 ± 0.48^{cd}	NA	3.09 ± 0.08^{cd}	0.79 ± 0.29^c	NA	0.67 ± 0.33
B2	6.69 ± 0.01^{de}	NA	6.45 ± 0.03	2.46 ± 0.09^c	NA	2.90 ± 0.84^c	0.90 ± 0.13^c	NA	0.96 ± 0.09
C1	NA	6.28 ± 0.45^c	5.78 ± 0.58	NA	3.38 ± 0.58^{de}	3.38 ± 0.58	NA	0.6 ± 0.02^c	0.6 ± 0.22
C2	NA	6.62 ± 0.25^c	6.35 ± 0.47	NA	3.84 ± 0.34^e	3.92 ± 0.22	NA	0.5 ± 0.05^c	0.58 ± 0.18
D1	NA	5.77 ± 0.39^c	5.80 ± 0.65	NA	3.09 ± 0.03^d	2.39 ± 0.60	NA	0.5 ± 0.05^c	0.5 ± 0.05
D2	NA	5.98 ± 0.58^c	5.76 ± 0.40	NA	1.35 ± 0.44^c	2.28 ± 0.27	NA	0.48 ± 0.0^c	0.48 ± 0.0

a=not available; b=water obtained from catfish tank; c,d,e =in the same column with different superscript letters are different (*P <0.05*); A1 and A2=catfish pond use chicken offal; B1 and B2=catfish pond use commercial fish feed; C1 and C2=tilapia pond use spoiled egg; D1 and D2=tilapia pond use commercial fish feed; Total aerobic and fecal coliform counts in water samples obtained from all sources were not significant different (P>0.05). *E Coli* was counted less than 3 MPN/gr (0.47 log MPN/gr) for all catfish and tilapia samples and less than 3 MPN/mL (0.47 log MPN/L) for all water samples.

Table 1: Total aerobic, coliform and fecal coliform counts in catfish, tilapia and water obtained from ponds and wet markets in Malaysia.

Ponds	Temperature (°C)	pH
A1	26.5 ± 0.5	6.2 ± 0.17
A2	26.83 ± 0.76	6.37 ± 0.38
B1	26.8 ± 0.72	6.4 ± 0.36
B2	26.93 ± 0.12	6.3 ± 0.21
C1	26.47 ± 0.06	6.3 ± 0.09
C2	26.85 ± 0.21	6.2 ± 0.06
D1	26.7 ± 0.34	6.42 ± 0.27
D2	26.67 ± 0.40	6.49 ± 0.28

A1 and A2=catfish pond use chicken offal; B1 and B2=catfish pond use commercial fish feed; C1 and C2=tilapia pond use spoiled egg; D1 and D2=tilapia pond use commercial fish feed.

Table 2: Temperature and pH of water obtained from ponds in Malaysia.

samples exceeded the recommended microbiological standard. International Commission of Microbiological Specific for Foods [13] stated that the total aerobic bacteria in fresh and frozen fish should be less than 5.7 log g^{-1} to meet Good Manufacturing Practise Criteria. However, 1/32 of catfish and 20/32 of tilapia samples were observed to be more than 7 log cfu g^{-1} which exceeded the safety or quality limit standard. International Commission of Microbiological Specific for Foods [13] revealed that the total aerobic bacteria in fresh and frozen fish should be less than 7 log g^{-1} to meet safety or quality standard.

Coliform, fecal coliform and *E. coli*

The coliform count ranged from 1.46 to 4.18 log MPN gr^{-1} for catfish, 1.6 to 4.04 log MPN gr^{-1} for tilapia, and 2.04 to 4.36 log MPN mL^{-1} for water. The coliform count was significant different (P<0.05) among catfish obtained from pond A1-A2 (chicken offals feed) and pond B2 (commercial fish feed). Similarly, there was the significant different (P<0.05) among tilapia obtained from pond C1-C2 (spoiled eggs feed) and pond D2 (commercial fish feed). Coliform count in water samples obtained from pond A2 (chicken offals feed) was also significant different (P<0.05) with those obtained from pond B1-B2 (commercial fish feed) (Table 1). However, coliform count in water samples obtained from pond C1-C2 (spoiled eggs feed) and pond D1-D2 (commercial fish feed) was not significant different (P>0.05). Similarly, the coliform count was not significant different (P>0.05) among water samples obtained from catfish tank at each wet market.

Fecal coliform ranged from 0.48 to 1.63 log MPN g^{-1} for catfish, 0.48. to 1.81 log MPN g^{-1} for tilapia, 0.48 to 1.97 log MPN mL^{-1} for water. There was not significant different (P>0.05) between fecal coliform in catfish obtained from wet market and pond. This was also not significant different between fecal coliform in tilapia obtained from wet market and pond. Similarly, fecal coliform was not significantly different (P>0.05) between water obtained from wet market and pond.

In this present study, *E. coli* was found for less than 3 MPN g^{-1} for catfish, tilapia and water. There was no significant different (P>0.05) between water obtained from wet market and pond. Similar statistics results were also observed in catfish and tilapia obtained from wet market and pond.

Temperature and pH

This present study observed temperature of water ponds ranged from 26 to 27.5°C (Table 2). Temperature of water in catfish tank obtained from different wet market was no significant difference (P>0.05). This was also observed in temperature of water ponds. There was no significant difference (P>0.05) between temperature of water ponds in Pond A1, A2, B1, and B2. This present study also observed

that pH of water ponds ranged 6 to 6.8 (Table 2). There is no significant different (P>0.05) between pH of water obtained from wet market A, B, C, D and E. The similar statistical analysis result was also observed in wet market E, F, G and H. There is no significant different (P>0.05) between pH of water obtained from Pond A1, A2, B1 and B2. This was also observed in pH water obtained from Pond C1, C2, D1 and D2. However, pH of water pond obtained from Pond A1, A2, C1, and C2 was relatively higher compared to those obtained from Pond B1, B2, D1, and D2.

Discussion

The results showed variation in total aerobic count in water form ponds, and catfish and tilapia from wet markets and ponds. The highest total aerobic count was observed in tilapia obtained from wet market H. This result similar to Shinkafi and Ukwaja [14] whom reported the high load of bacteria in tilapia sold in the central market of Sokoto Nigeria. The present study found that tilapia was not delivered as live fish but fresh or chilled fish. This condition might alter the growth of bacteria in the fish due to the spoiled process and temperature of storage. Keller [15] revealed that bacteria might increase due to the temperature and time of storage. Other study reported that spoilage was evident after 13 h at room temperature (26-29°C) in fresh common carp (Cyprinus carpio L) [16].

This present study indicated that type of feed did not affect to the bacterial load in the fish. The nutrient of different feed was shown to be similar each other for the growth of bacteria. The present study also found that high bacterial load in fish occurred in the high temperature of water. Catfish and tilapia are cold-blood animals and have the same temperature as their surroundings [17]. Al-harbi [18] reported that temperature of water correlated with the load of bacteria in fish. In the present study, temperature of water ranged from 26 to 27.5°C. Adam and Moss [19] revealed that mesophiles and psychrotrops bacteria grew in the ranges of 15 to 47°C and -5 to 35°C, respectively. Boyd and Tanner [20] reported that coliform in catfish ponds at Auburn. Alabama was greater in summer and spring compared to other seasons when the temperature was going to decrease.

This present study observed that pH of water ranged from 6 to 6.8. However, pH of earthen pond water (3/12) and pH of ex-mining pools water (12/12) was exceeded from the recommended pH.

Chapman [21] revealed that the recommended pH of water pond ranged from 6.5 to 9 for catfish farming. Ross [22] stated that the recommended pH of water pond for tilapia farming ranged from 7 to 9. Accumulation of waste feed and fish faecal material results in changes in the sediment, characterized by high content of organic material and accumulation of nitrogenous and phosphorous compounds which may induce the benthic communities [23] and affect the pH of water ponds [24]. Fish will be stressed and die if the pH reach below 5 or above 10 [24]. In the present study, the depth of ponds was ranging from 4 to 7 m for catfish ponds and 18 to 50 m for tilapia ponds. Tilapia was farmed in ex-mining pools. Mente et al. [23] revealed that the growth of benthic algal bloom occurred in the depth ranging from 20 to 50 m. Thus, this may induce the pH changes in water ponds and promote the stress of the fish.

This study found that coliform in catfish and tilapia fed with chicken offals or spoiled eggs were relatively higher compared to those fed with commercial fish feed. These were observed also in water samples. Boyd and Tanner [20] reported that the high organic matter input in feed could increase the coliform in the catfish ponds. Other study reported that chicken [25] and eggs [26] were potential agents for coliform. Thus, chicken offals and spoiled eggs introduced to the aquaculture system will increase coliform and reduce the hygiene level in the ponds.

This study indicated that the level of coliform correlated to the density of the ponds. The density of fish in the ponds ranged from 10-32 of catfish m^{-2} and 6-10 of tilapia m^{-2}. Coliform count was shown to be relatively higher in catfish compared to tilapia. These were also observed in water samples (Table 1). Previous study revealed that high density of fish related with coliform and other bacteria in water and fish [27].

Besides that, stream and hold water used in earthen ponds and ex-mining pools might be contaminated by coliform bacteria. Všetičková and Adámek [28] reported that the pattern of water quality changes after the flow through the pond was predominantly influenced by inlet water quality. Francy et al. [29] reported that total coliform in 136 stream and 143 ground water samples collected in five hydrology system of the United States were found in 99% and 20%, respectively. That study reported that the land use related to the density of coliform in stream water. Francy et al. [29] also reported that the presence of septic systems and well depth related with the density of coliform in ground water. Blogoslawski et al. [30] reported that pathogenic bacteria can introduce to the hatchery systems though a contaminated water source.

The coliform level was found relatively higher in fresh or chilled tilapia compared to live catfish in wet markets. The coliform level might promoted the deterioration in tilapia and increase the density of coliform. Gelman et al. [16] reported that spoilage was evident after 13 h at room temperature (26-29°C) in fresh common carp (Cyprinus carpio L).

This study found that the density of fecal coliform was observed to be significant different (P<0.05) in tilapia which were sold in wet market and reared in ponds. These might be occurred due to deterioration in tilapia. Keeping alive tilapia during distribution will make the high cost of transport, water and storage tanks. Thus, this fish was delivered as fresh or chilled tilapia in wet markets. Geldreich and Clarke [31] reported that the fate of fecal coliforms in the fish indicated that these organisms can probably survive and multiply when fish temperatures were 20°C or higher, but only when the organisms are retained in the gut for periods beyond 24 h. Moreover, the human health risk may rise when E. coli contaminate the fish. The present study found level of total aerobic bacteria, coliform and fecal coliform in catfish, tilapia and water increase and exceed the microbiological standard due to the use of chicken offal and spoiled egg as feed.

Conclusions

Chicken offals and spoiled eggs can be potential source for the bacterial contamination to water and fish. This evidence highlights the importance of feed quality in aquaculture system.

Acknowledgment

Financial assistance provided by MOSTI (305/PTEKIND/613512) is gratefully acknowledged.

References

1. Department of Fisheries Malaysia (2010) Annual fishery statistic 2010. Department of Fisheries Malaysia, Malaysia.

2. http://www.fao.org/fishery/countrysector/naso_malaysia/en Cited 5 November 2014.

3. Delgado CL, Wada N, Rosegrant MW, Meijer S, Ahmed M (2003) The future of fish issues and trends to 2020. International Food Policy Research Institute, Washington DC, USA.

4. Focardi S, Corsi I, Franchi E (2005) Safety issues and sustainable development of European aquaculture: new tools for environmentally sound aquaculture. Aquacul Inter 13: 3-17.

5. Workagegn KB, Ababbo ED, Tossa BT (2013) The Effect of Dietary Inclusion of Jatropha curcas Kernel Meal on Growth Performance, Feed Utilization Efficiency and Survival Rate of Juvenile Nile tilapia. J Aquacul Res Develop. 4: 193.

6. Workagegn KB, Ababboa ED, Yimer GT, Amare TA (2014) Growth Performance of the Nile Tilapia (Oreochromis niloticus L.) Fed Different Types of Diets Formulated From Varieties of Feed Ingredients. J Aquacul Res Develop 5: 235.

7. Lunestad BT, Nesse L, Lassen J, Svihus B, Nesbakken T, et al. (2007) Salmonella in fish feed; occurrence and implications for fish and human health in Norway. Aquaculture 265: 1-8.

8. http://www.fao.org/docrep/003/V4430E/V4430E00.HTM. Cited 10 Agustus 2012

9. http://www.fda.gov/Food/ScienceResearch/LaboratoryMethods/

10. http://www.fda.gov/Food/ScienceResearch/LaboratoryMethods/ BacteriologicalAnalyticalManualBAM/ucm064948.htm. Cited 1 September 2008

11. World Health Organization (1989) Health Guidelines for the Use of Wastewater in Agriculture and Aquaculture. World Health Organization, Geneva, Switzerland.

12. Department of Agriculture, Animal Health and Product (2004) Animal and animal origin foods.

13. International Commision of Microbiological Spesific for Foods (1986) Microorganisms in Foods 2. Sampling for Microbiological Analysis: Principles and Specific Applications. Blackwell Scientific Publication, Oxford, United Kingdom.

14. Shinkafi SA, Ukwaja VC (2010) Bacteria Associated with Fresh Tilapia Fish (Oreochromis niloticus) Sold At Sokoto Central Market in Sokoto, Nigeria. Nig J Bas App Sci 18: 217-221.

15. Keller JJ (2004) Employee Food Safety Handbook. Kelly and Associates Inc, Neenah, Wisconsin.

16. Gelman A, Pasteur R, Rave M (1990) Quality changes and storage life of common carp (Cyprinus carpio) at various storage temperatures. J Sci Food Agricul 52: 231-241.

17. Swann L, Morris JE, Selock, DE, Riepe J (1994) Cage Culture of Fish in the North Central Region Site selection. Iowa State University, Ames, Iowa.

18. Al-harbi AH (2003) Faecal coliforms in pond water, sediments and hybrid tilapia Oreochromis niloticus x Oreochromis aureus in Saudi Arabia. Aquacul Res 34: 517-524.

19. Adam MR, Moss MO (2004) Food microbiology. The Royal of Chemistry. Cambridge, United Kingdom.

20. Boyd CE, Tanner M (1998) Coliform organism in water of channel catfish ponds. J Wor Aquacul Soc 29: 74-78.

21. http://edis.ifas.ufl.edu/fa010.

22. Ross LG (2000) Environmental physiology and energetics. Kluwer Academic Publishers. Dordrecht, Netherlands.

23. Mente E, Pierce GJ, Santos MB, Neofitou C (2006) Effect of feed and feeding in the culture of salmonids on the marine aquatic environment: a synthesis for European aquaculture. Aquacul Intern 14: 499-522.

24. https://srac.tamu.edu/index.cfm/event/getFactSheet/whichfactsheet/112. Cited 29 Agustus 2012

25. Rodrigo S, Adesiyun A, Asgarali Z, Swanston W (2006) Occurrence of selected foodborne pathogens on poultry and poultry giblets from small retail processing operations in Trinidad. J Food Prot 69: 1096-1105.

26. Jones DR, Anderson KE, Guard JY (2012) Prevalence of coliforms, Salmonella, Listeria, and Campylobacter associated with egg and the environment of conventional cage and free-range egg production. Poult Sci 91: 1195-1202.

27. Mandal, SC, Hasan M, Rahman MS, Manik MH, Mahmud ZH, et al. (2009)

Coliform Bacteria in Nile Tilapia (*Oreochromis niloticus*) of Shrimp-Gher, Pond and Fish Market. Wor J Fish Mar Sci 1: 160-166.

28. Všetičková L, Adámek Z (2012) The impact of carp pond management upon macrozoobenthos assemblages in recipient pond canals. Aquacul Intern 21: 897-925.

29. Francy DS, Helsel DR, Nally RA (2000) Occurrence and distribution of microbiological indicators in groundwater and stream water. Water Environ Res 72: 152-161.

30. Blogoslawski WJ, Stewart ME, Rhodes EW (1978) Bacterial disinfection in shellfish hatchery disease control. Proc ann mee–Wor Maricul Soc 9: 587-602.

31. Geldreich, EE, Clarke NA (1966) Bacterial pollution indicators in the intestinal tract of freshwater fish. Appl Microbiol 14: 429-437.

25

Evaluation of Date Fiber as Feed Ingredient for Nile Tilapia *Oreochromis niloticus* Fingerlings

Belal IEH[1], El-Tarabily KA[2], Kassab AA[3], El-Sayed AFM[4] and Rasheed NM[5]

[1]*Ibrahim E H Belal Department of Aridland Agriculture, Faculty of Food and Agriculture, United Arab Emirates University, P. O Box 15551, Al-Ain, UAE*
[2]*Department of Biology, Faculty of Science, United Arab Emirates University, PO Box 15551, Al-Ain, UAE*
[3]*Department of Anatomy and Embryology, Faculty of Veterinary Medicine (Moshtohor), Benha University, Egypt*
[4]*Oceanography Department, Faculty of Science, Alexandria University, Alexandria, Egypt*
[5]*Aquaculturist, MAHY Khoory, Aquaculture Centre, P. O. Box 11944, Dubai, UAE*

Abstract

This study was carried out to evaluate the use of date fiber (DF) as a feed ingredient for tilapia fingerlings in terms of growth parameters, body composition, anatomical alterations of the intestinal villi. In addition to, pellet strength and bacterial type and population in the test diets. Four isonitrogenous isocaloric diets containing 0, 100, 200 and 300 g kg^{-1} DF as replacement of wheat bran were fed to triplicate groups of ten *O. niloticus* fingerlings (0.65 g) in a recirculating water system for 70 days. Fish fed diets contain up to 200 g kg^{-1} DF had similar growth parameters. Further increase in dietary DF to 300 g kg^{-1} resulted in significant retardation in all parameters. Body fat was reduced while protein, ash and moisture were increased by increasing DF level. Increasing dietary DF level caused changes in tilapia's intestinal villi, reduced dietary microbial activity and bacterial population of selected species, and produced stronger pellets.

Keywords: Date fiber; Fish; Growth, Bacteria; Feed; Electron microscopy

Introduction

Tilapia *Oreochromis niloticus* culture has been growing at an outstanding rate during the past decade in most of the tropical, subtropical and temperate regions. As a result, the production of farmed tilapia has jumped from 308,234 mt in 1988 to 2.8 million mt in 2008 [1]. In addition, tilapia culture has been gradually shifted from the traditional semi-intensive systems to the more intensive systems, which rely exclusively on artificial feeds. Therefore, formulating economic tilapia feeds has become a necessity.

Nutrition represents over 50% of total culture financial inputs in tilapia aquaculture [2]. In addition, the prices of major feed ingredients, including fish meal (FM), soybean meal (SBM), corn, bran and oils have been sharply increasing during the past few years [2]. This has been attributed mainly to one or more of the following reasons: 1) declining production, 2) increasing demands and competition among users, 3) increasing production cost, particularly fuel and fertilizer prices, and 4) conversion of some plant ingredients to biofuel (e.g. corn to ethanol) [2]. For example, the price of corn has jumped from US$ 95 in 2006 to US$ 230 in 2008. Similarly, the prices of soybean meal increased from about US$200 to $450 during the same period [2]. Therefore, the major challenge facing tilapia aquaculture industry is the production of cost effective and environmentally performing feeds for farmed tilapia, using inexpensive, locally available ingredients. Several studies have been conducted to evaluate the incorporation of different unconventional animal and plant proteins and energy sources for farmed tilapia with varying results [3].

Date palm tree is one of the most important cultivated trees in arid and semi-arid regions, especially in North Africa and Arabian Gulf countries. Date fruits play an important role in the economies of these countries, as a major source of nutrition. Egypt, Iraq, Iran, Saudi Arabia, UAE, Pakistan, Algeria, Sudan, Oman, Libya, China and Tunisia are the major date fruit producers. Over 6,700,000 mt of date fruits were produced in 2004 [4].

Date wastes include date pits and DF is produced annually. These by-products may have high potential as energy sources for farm animals and farmed fishes [5].

Date Fiber (DF), a by-product of date syrup production, is an insoluble, powder-like, connected with non-nutritive portion of the date flesh [5]. It is composed mainly of cellulose, hemicelluloses, lignin, ligno-cellulose, and insoluble proteins. This fiber is naturally broken down, by enzymes, during the ripening process, to more soluble compounds (glucose, sucrose, mannose and soluble pectin and galactomannan) to render the fruit more tender and soft. Date fiber represents 20-100 g kg^{-1} of the date flesh, depending on the type and quality of the dates. This means that a substantial amount of DF is produced annually, especially in tropical and subtropical regions, where dates are a major agricultural crop [5].

Only few studies have been carried out to investigate the use of this by-product as a feed ingredient in rats [6] and in human food fortification [7].They reported that patty formula replaced with up to 150 g kg^{-1} DF produced healthier and better quality beef patties by possessing hypolipidemic effects. Dietary fibers, particularly water soluble, might influence lipid metabolism in rats [6] as it was found to possess hypolipidemic in rats fed 2 g kg^{-1} cholesterol. Additionally, DF concentrates showed a high water and oil holding capacity [8].

The present study was conducted to investigate the use of DF as

*Corresponding author: Belal IEB, Department of Aridland Agriculture, Faculty of Food and Agriculture, United Arab Emirates University, P. O Box 15551, Al-Ain, UAE, E-mail: ibelal@uaeu.ac.ae

a replacement of dietary wheat bran in the diets of tilapia fingerlings growth and proximate body composition. Additionally, the effect of DF on pellet quality in terms of pellet strength, total and specific bacterial count, and anatomical alterations of the intestinal villi were investigated.

Methods

Culture condition

Nile Tilapia *O. niloticus* fingerlings (0.65 g average initial weight) were produced from tilapia brood stock kept in captivity at the Aquaculture Unit, College of Food and Agriculture, United Arab Emirates University, Al Ain, United Arab Emirates. Ten fish were stocked into 20 L fiberglass tanks in a closed, recirculating indoor system. The tanks were provided with central drainage pipes surrounded by outer pipes, perforated at the bottom, to facilitate self-cleaning and waste removal. The culture system was provided with a biological filter, aeration through an air blower, and heaters to maintain water temperature at 27°C. Approximately 10% of the water volume was replaced by new freshwater daily. Lighting in the culture unit was set at 12:12 L:D cycle. Water quality parameters, including dissolved oxygen (DO) (Oxygen meter, YSI, model 58), ammonia (NH4-N), Nitrates (NO3-N), and nitrites (NO2-N) (Orion Aquafast, Germany) and pH (pH meter, Jenway, UK) were monitored weekly.

Dietary formulations

Four isonitrogenous (320 g kg $^{-1}$ CP), isocaloric (18.84 kJ g -1) test diets with varying levels of DF as a replacement of wheat bran at 0, 100, 200 and 300 g kg $^{-1}$ were formulated. The diets were prepared as follows: all feed ingredients were ground in a commercial blender and then mixed in a kitchen mixer. Vitamin and mineral mixes were gradually added with continuous mixing. Distilled water (60°C) was slowly added while mixing until the mixture began to clump. Then, the diet passed through a kitchen meat grinder and was dried for 24 hours at 60°C in a vacuum drying oven. The dried diet was then chopped into pellets in a blender and then passed through laboratory test sieves (mesh 2 and 0.88 mm) to ensure homogenous particle size of sinking pellets and stored at -8°C until used. The amount of waste (powder form) as a result of the pelleting process for every test feed was calculated separately as a percentage of the total amount of every feed. This was used as an indicator of weak (high percentage) or strong (low percentage) pellets. The chemical composition of the DF and all test diets were determined according to [9] methods (Tables 1 & 2).

Each diet was fed to triplicate groups of 10 fish each (0.65 g ± 0.4) to satiation level, twice a day (09:00 and 16:00 h) for 70 days. Fish were weighed collectively at 10-day intervals, their average weights recorded.

Feed efficiency performance

Feed efficiency performance including fish Weight Gain (WG), Specific Growth Rate (SGR), Feed Conversion Ratio (FCR), Protein Efficiency Ratio (PER), were calculated with the following equations:

Nutrient	DF g kg^{-1}
Moisture	69
Crude protein	24
Crude fat	7
Crude fiber	515
Total ash	25
NFE	429

Table 1: Proximate analyses of DF on dry weight bases.

Ingredients (g kg^{-1})	0 g kg^{-1}	100 g kg^{-1}	200 g kg^{-1}	300 g kg^{-1}
Fish meal (700g kg^{-1} CP)	380	390	410	430
Wheat bran	530	420	290	160
DF	0	100	200	300
Sunflower oil	50	50	60	70
Vitamin and mineral mixes[1]	20	20	20	20
Binder (CMC)	20	20	20	20
Total	1000	1000	1000	1000
Proximate analysis				
Crude protein	327.8	324.3	328.9	320.2
Crude lipids	100.3	121.4	134.4	148.2
Total ash	120.1	122.1	123.2	121.1
Crude Fiber	54.2	96.7	120.1	174.5
NFE[2]	399.6	335.5	293.4	286
GE[3] (kJ g^{-1})	185.3	182.7	181.7	183.9

[1]Vitamins content are Thiamine 2.5 g kg^{-1}. Riboflavin I g kg^{-1}, Pyridoxine 2 g kg^{-1}, Pantothenic acid 5 g kg^{-1}. Inositol100 g kg^{-1}, Biotin 0.32.5 g kg^{-1}, Folic acid 0.75 g/kg^{-1}, Para aminobenzoic acid 2.5 g kg^{-1}, Choline 200 g kg^{-1}, Niacin I0 g kg^{-1}. Cyanocibalmin 0.005 g kg^{-1}. Retinolpalmitate100, WO III, ∞ tochpeml acetate 20.1 g kg^{-1}, ascorbic acid 50 g kg^{-1}, menadione 2 g kg^{-1}, cholecalciferol 500,000 IU. Minerals conent are "CaHP0,.2H20 727.775 g kg^{-1}, MgSO. 7H20 127.5 g kg^{-1}, NaCl60 g kg^{-1}, KCl 50 g kg^{-1}, FeSO, 7H20 2 g kg^{-1} 5, ZnS0, 4H20 5.5 g kg^{-1}, MnS04 .4H20 2.5375 g kg^{-1}, CuSO,.2H$_2$O 0.7850 g kg^{-1}, CoSO. 6H20 0.4775 g kg^{-1}, CalO. 6H20 0 295 g kg^{-1} CrCl, 6H20 0 127 g kg^{-1}. Similar to [18].

[2]Nitrogen-free extract was calculated by difference. [3]Gross energy, calculated based on 23.67, 17.17 and 39.79 kJ g^{-1}) for protein, carbohydrate and lipids, respectively

Table 2: Composition and proximate analyses of the test diets (g kg^{-1}dry matter). Values represent the means of three replicates. Means in each row followed by a different letter are significantly different (P>0.05).

WG= W_2-W_1, where WG is the mean of weight, W_2 is the Mean final Weight, W_1 is the Mean Initial Weight,

SGR = (ln W_2-In W_1)/time in days * 100

FCR = feed (dry) intake (g)/wet weight gain (g)

PER= average weight gain (g)/average weight of protein fed

Investigating intestinal wall under the scanning electron microscopy

Two fish from the control 0 g kg $^{-1}$ DF, 100 g kg $^{-1}$ DF, and 200 g kg $^{-1}$ DF treatments were used for this study. All fish were killed and the ventral body wall was opened. The entire gastrointestinal tract of the six fish was excised and fixed in 30 g kg $^{-1}$ glutaraldehyde in phosphate buffered saline. Cross sections of the gastrointestinal tract were performed at several levels and processed for scanning electron microscopy. Selected gut fragments were taken from different levels of the intestine, fixed, dehydrated to critical dried point, further dissected if necessary, mounted, sputter coated with gold and viewed on a JEOL JSM 5500 LV SEM. The scanning electron microscopy was done in the Central Laboratory Unit of United Arab Emirates University.

Microbial analyses: enumeration of microbial populations

The microbial populations of the test diet samples were estimated using the soil dilution plate method [10]. Three 10 g replicates, of each sample were dispensed into 100 mL of sterile 0.1% (w/v) agar (Gibco Brl, Paisley, Scotland) solution in deionized water containing 20 g glass beads (3 mm diameter). The suspension was shaken 50 times and then placed in an ultra-sonic cleaner at a frequency of 55,000 cycles sec^{-1} for 20 sec (Model: B- 221, 185 Warr, Branson Cleaning Equipment Company, USA). Ten-fold dilutions were made in sterile deionized water and 0.2 mL aliquots of what were considered appropriate dilutions were spread on the surface of the different media in sterile plastic Petri dishes (90 mm diameter) with a sterile glass rod. Nine plates were used per dilution. The plates were dried in a laminar flow cabinet for 1 h and then incubated at 25°C (± 2°C) and colony counts were carried out from day 2 onwards. The groups of organisms selected for enumeration and the media used were as follows: (i) total aerobic bacteria on 1/5 M32 medium [11], incubated for 2-4 days; (ii) fluorescent pseudomonads on 1/10 tryptic-soy agar (Difco laboratories, Michigan, USA) (TSA) containing ampicillin 50 mg mL^{-1}, (Sodium salt, Instituto Biochimico Italiano, Milano, Italy), cycloheximide 75 mg mL^{-1} (Sigma) and chloramphenicol 12.5 mg mL^{-1} (Sigma) (TSA + ACC), incubated for 2-4 days [12], (iii) Gram-negative bacteria on 1/10 TSA containing crystal violet (Sigma) at a concentrations of 2 μg mL^{-1} (TSA + CV), incubated for 2-4 days [13]; (iv) filamentous fungi and yeasts on Martin's medium containing rose bengal 33 μg mL^{-1} (Sigma) and streptomycin 30 μg mL^{-1} (Sigma) incubated for 4-6 days [14]. Bacterial and fungal colonies were counted from each medium and were expressed as log 10 colony forming units (cfu) g dry^{-1} sample.

Estimation of the total microbial activity

The microbial activity of all test diet samples were measured by fluorescein diacetate hydrolysis and by arginine ammonification. The hydrolysis of fluorescein diacetate (FDA) (Sigma Chemical Co., St Louis, Mo., USA) was measured by the method of [15]. Briefly, 5 g of each sample were added to 20 mL of sterile 60 mM potassium phosphate buffer (8.7 g K2HPO4 and 1.3 g KH2PO4 in 1 L distilled water, pH 7.6) in 250 mL flasks. FDA was dissolved in acetone and stored as a stock solution (2 mg mL^{-1}) at -20°C. The reaction was started by adding 0.2 mL of FDA (400 μg) from the stock solution to a buffer-sample mix. Each treatment consisted of eight replicates and one blank to which no FDA was added. The reaction flasks were shaken (90 rpm) at 25°C for 20 min on a rotary shaker (Model G76, New Brunswick Scientific, Edison, NJ, USA). The reaction was then stopped by adding 20 mL acetone to all samples. Sample residues were removed from the mixture by centrifugation at 500 rpm for 10 min and filtered through a No. 1 Whatman filter paper (Whatman, Maidstone, England). The filtrate was collected in a test tube, covered with parafilm and placed into an ice bath to reduce volatilisation of the acetone. The concentration of fluorescein was determined by reading the optical density at 490 nm, using a Shimadzu UV-2101/3101 PC scanning spectrophotometer (Shimadzu Corporation Analytical Instruments Division, Kyoto, Japan). This permitted the rapid handling of many samples, the concentrations of which were compared against a standard curve. The background absorbance was corrected for each treatment with the blank sample run under identical conditions but without the addition of FDA. Standard curves were prepared as described by [16].

The results were converted to μg hydrolysed FDA g dry^{-1} sample.

Body composition analysis

At the termination of the study, all fish in each tank were netted, weighed and frozen for body composition analyses at -20°C. Initial body analyses were performed on a sample of fish, which were weighed and frozen prior to the study. Proximate analyses of body water, protein, lipid, and ash were performed according to standard [9] methods.

Statistical analyses

Fish growth rates, feed utilization efficiency and body composition results were subjected to a one-way analysis of variance (ANOVA) to test the effects of DF inclusion level on fish performance. Orthogonal polynomial procedure [17] were used to compare means at P=0.05. Least significant difference (LSD) was used to test for the differences among treatment means when F-values from the ANOVA were significant.

Results

The average values of water quality parameters throughout the study were; DO = 6.4 ± 13 mg L^{-1}, NH$_4$-N=0.06 ± 0.002 mg L^{-1}, NO$_3$-N=8.4 ± 1.72 mg L^{-1}, NO$_2$=0.00 mg L^{-1} and pH=8.0 ± 0.09. Good binding properties were noted with increasing levels of DF in the experimental diet. The level of fines during the pelleting process decreased (277, 234, 183, 121 g kg^{-1}) for diets containing 0, 100, 200 and 300 g kg^{-1} DF, respectively, with very high correlation (r^2=0.97, P<0.05).

The proximate composition of DF is shown in Table 1. The proximate composition of the experimental diets (Table 2) showed little variation in nutrient levels of various diets and agreed with estimated values. The dietary DF significantly affected the growth performance of O. niloticus fingerlings (P<0.05).

The growth rates and feed conversion ratios of fish fed DF-based diets up to 200 g kg^{-1} inclusion level were similar to that of fish fed the control (date fiber-free) diet (Table 3). Further increase in dietary DF to 300 g kg^{-1} resulted in significant retardation in fish performance.

Proximate body composition, namely moisture, crude protein, and total ash, of O. niloticus fingerlings fed test diets with DF up to a level of 200 in g kg^{-1} were not affected (P<0.05) by replacing dietary wheat bran while body fat was reduced at the 200 in g kg^{-1}. As the level of the DF incorporation increased to 300 in g kg^{-1} DF in the test diets, body moisture, body protein and total ash were increased while body fat decreased significantly (Table 4).

The results of total microbial activity and microbial populations in the test diets (Table 5) showed that samples with 0 g kg^{-1} DF had

Test diet g^{-1}DF	IW^1	FW^2	WG^3	SGR^4	FCR^5	PER^6	Survival
0	0.65a	6.5 ± 0.17a	9.14 ± 0.42a	2.77 ± 0.06a	1.86 ± 0.02a	1.64 ± 0.06a	97 ± 0.77a
100	0.65a	6.0 ± 0.36a	8.82 ± 0.23a	2.65 ± 0.11a	1.84 ± 0.07a	1.78 ± 0.16a	98 ± 0.29a
200	0.67a	6.9 ± 0.59a	9.60 ± 0.53a	2.84 ± 0.08a	1.99 ± 0.21a	1.53 ± 0.11a	97 ± 0.85a
300	0.66a	4.5 ± 0.05b	4.16 ± 0.15b	2.27 ± 0.04b	3.32 ± 0.12b	0.97 ± 0.21a	97 ± 0.87a

1 Mean Initial Weight

2 Mean Final Weight

3 Weight Gain=FW-IW

4 SGR, Specific Growth Rate=(In FW-In IW)/time in days × 100

5 FCR, Food Conversion Ratios=feed (dry) intake (g)/wet weight gain (g)

6 PER, Protein Efficiency Ratio=average weight gain (g)/average weight of protein fed (g).

Table 3: Performance of O. niloticus fingerlings fed DF-based diets. Values represent the means of three replicates. Means ± SD in each column followed by a different letter are significantly different (P>0.05).

DF g kg⁻¹	Moisture g kg⁻¹	Crude protein g kg⁻¹	Lipid g kg⁻¹	Total Ash g kg⁻¹
0 (control)	700.6c	147.9a	73c	58a
100	716.5bc	146.1a	68.7b	58a
200	723ab	138.1a	55.3ab	74a
300	739a	138.5a	46.8a	71a

Table 4: Proximate body composition of *O. niloticus* fingerlings fed test diets with different percentages of DF. Values represent the means of three replicates. Means in each column followed by a different letter are significantly different (P>0.05).

DF g kg⁻¹	Total aerobic bacteria	Gram-negative bacteria	Fluorescent pseudomonads	Filamentous fungi and yeasts	Microbial activity
0 (control)	6.53 ± (0.12)a	4.54 ± (0.12)a	3.70 ± (0.11)a	2.23 ± (0.12)a	79.54 ± (2.56)a
100	5.81 ± (0.10)b	3.50 ± (0.15)b	2.63 ± (0.14)b	1.83 ± (0.09)b	48.82 ± (2.74)b
200	3.47 ± (0.14)c	2.30 ± (0.12)c	1.84 ± (0.09)c	1.57 ± (0.13)b	27.52 ± (1.80)c
300	2.26 ± (0.10)d	1.57 ± (0.13)d	1.16 ± (0.10)d	1.05 ± (0.11)c	17.71 ± (2.02)d

Table 5: Microbial population densities in log10 colony-forming units (cfu) g⁻¹ dry sample and total microbial activity (µg hydrolyzed FDA g⁻¹ dry sample of fish feed on different concentrations of date fibers (DF).

Values are means of eight replicates for each treatment and the values in brackets are the standard error of the mean. Values followed by the same letter within a column are not significantly different (P>0.05) according to Fisher's Protected LSD Test.

a significantly (P<0.05) highest total microbial activity as compared to all samples with 100, 200, 300 g kg⁻¹ DF. There was a significant (P<0.05) gradual reduction of total microbial activity and microbial populations as the level of DF increased in the test diet. The estimated total populations of aerobic bacteria, fluorescent pseudomonads, Gram-negative bacteria, filamentous fungi and yeasts were significantly (P<0.05) higher in the samples without DF than samples with DF. The population was gradually and significantly reduced as the level of DF was increased.

The scanning electron microscopy (SEM) of *O. reochomis niloticus* intestines is shown in Figures 1-3. It is important to indicate that the entire samples of fish intestine which were fed 100 g kg⁻¹ DF were lost during the analyses and could not recovered. The intestinal villi from fish fed 0 g kg⁻¹ DF (control group) were the smallest of all and the walls were the thinnest (Average width 793 nm) as compared to those from fed 200 and 300 g kg⁻¹ DF (average width 1.16 µm and 2.28 µm respectively). In other words there were gradual increases in size, height and thickness of the intestinal villi of *O. niloticus* fingerlings as the level of DF in the test diets fed increased (Figures 1-3). Unfortunately the intestinal samples of fish fed 100 g kg⁻¹ DF was lost, however, we can still get a clear idea of the effect of DF on the fish intestinal villi.

Discussion

Overall the closed recirculating culture system used in the experiment was capable of maintaining suitable water quality parameters for experimental fish [19]. DF inclusion in the test diets produced stronger pellets which were indicated by the reduction of powder after grinding. Up to our knowledge, there is no study on evaluating date fiber as a feed ingredient for fish. There is only one trial on feeding DF in starter ration for broiler with negative results which was due to inability of broiler to handle high fiber in their diet [20]. Few studies have been conducted on the use of dates and dates byproducts (date fiber not included) as feed ingredients in fish diets. For example, studies on *O. niloticus* [21-24]. They revealed that dates and date by-products could be used as a nutritional source for these fish. Similarly, it was found [25] that date pits can replace wheat bran-barley mixture in common carp feed at up to 750 g kg⁻¹ inclusion level, without any significant retardation in fish growth and feed utilization efficiency. The present study indicated that even though, nutrient content of wheat bran is better than DF, no significant differences (P<0.05) were

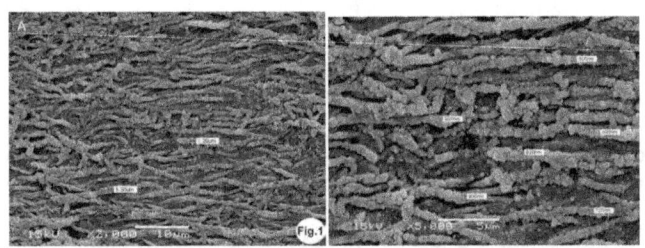

Figure 1: Scanning electron micrograph of tilapia intestinal villi of that fed control diet (0 g kg⁻¹ DF). R: low magnification and L: high magnification.

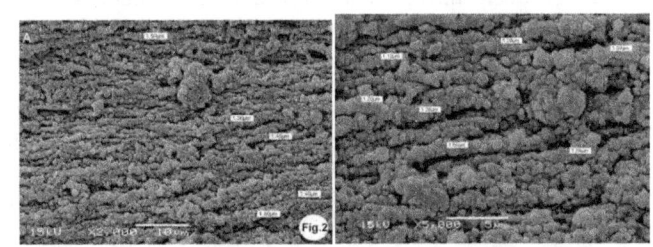

Figure 2: Scanning electron micrograph of tilapia intestine of that fed with 200 g kg⁻¹ DF. R: low magnification and L: high magnification.

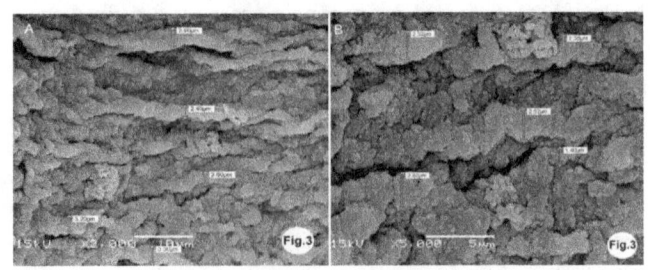

Figure 3: Scanning electron micrograph of the intestine of tilapia fed diet with 300 g kg⁻¹ DF. R: low magnification and L: high magnification.

detected in tilapia fed diets with DF at 0, 100 and 200 g kg⁻¹ in terms of growth parameters and feed utilization efficiency parameters (feed conversion ratios, specific growth rates, protein efficiency ratios). That was probably due to a combination of the following: first, the increase of digestible carbohydrates (oligo and monosaccharides). Tilapia gut microflora plays an important part in fiber digestion [26]. Date fiber contains simple sugars (glucose and fructose) and polysaccharides (glucan, xylan, galactan, mannan, arabinan, and acid soluble and insoluble lignin) [27]. Secondly, it may also be due to a free sugar such as mannose which is a part of partly digested mannan which worked as growth promoters. Mannose and oligomanan are good growth promoters for chicken, turkey [28] and fish [24]. Thirdly, the increase in intestinal villi in number, size, and thickness in fish fed test diet with DF 200 g kg^{-1} (Figures 1-3) could have improved nutrients absorption and make up for nutrients deficiency in DF composition. A researcher [29] has described that increased villus height suggests an increased surface area capable of greater absorption of available nutrients. Additionally, [30] showed that some fiber constituents (methoxylated pectin) causes changes in jejunal villus length and width and number in rates which villi function in digested feed absorption. It is understood that greater villus height and numerous cell mitoses in the intestine indicate that the function of the intestinal villi is activated [31-33]. Fourthly, the reducing effect of DF on microbial population, activity, total aerobic bacteria, fluorescent pseudomonads, filamentous fungi and yeasts may have played a role that cause a probiotic like effect to enhance tilapia growth (Table 5). On the other hand, the present study indicates that growth and growth parameters were negatively affected when the level of DF increased to a level of 300 g kg^{-1}. This may be due to the significant reduction in feed intake (Table 3). Feed intake reduction may have been due to the increased levels of fibers in DF in the feed while the ability of tilapia to utilize them is limited, as has been reported by [34]. Additionally feeds with high fiber intake increases the passage rat which reduces digestion and absorption [35-38] and increases fecal fat content in rats [39]. Approximate body composition of O. niloticus fingerlings fed the test diets with 0, 100 and 200 g kg^{-1}DF were similar. This indicates that fish digestive system was able to adapt itself with the DF at those levels as shown in the scanning electron microscope pictures in figures 1-3. However when DF level reached 300 g kg^{-1}DF, fish body moisture was significantly increased while body fat was decreased when compared to fish fed at lower levels of DF. This could be due to lower feed intake of fish test diet with 300 g kg^{-1}DF (Table 4) as compared to those fed diets with lower DF levels.

Conclusion

DF improved fish pellet quality; DF had significant effect on fish diet in reducing microbial counts of namely total aerobic bacteria, fluorescent pseudomonads, Gram-negative bacteria, filamentous fungi and yeasts. DF increased the intestinal mucosa surface area of tilapia which might play a role in dietary absorption. The present study suggests 200 g kg^{-1} of dietary wheat bran in tilapia feeds can be replaced with DF. This replacement can lead to a significant reduction in feed costs.

References

1. Food and Agriculture Organization of the United Nations (FAO), Globefish Databank 2011: FAO Rome, Italy.

2. Hardy R (2008) Utilization of plant proteins in fish diets: Effects of global demand and supplies of grains and oilseeds 16-18. Paper presented at aquaculture Europe 2008, Krakow, Poland.

3. El-Sayed AFM (2006) Tilapia Culture. CAB International Publishers, UK.

4. http://www.fao.org/es/ess/top/commodity.jsp?commodity=577&lang=EN&year=2004

5. Barreveld WH (1993) Date palm products FAO Agricultural Services Bulletin No. 101, FAO, Rome, Italy.

6. Kerkadi A (2006) Date fiber, a byproduct of date syrup (Debis) extraction influences serum lipid concentrations in rats fed 0.2% cholesterol. International Journal of Food, Agriculture and Environment 4:10-14.

7. Hashim IB, Khalil AH (2008) Quality characteristics of beef patties extended with DF Presented on Section 3B-9, 54th International Congress on Meat Science and Technology, Cape Town, South Africa.

8. Elleuch M, Besbes S, Roiseux O, Blecker C, Deroanne C, et al. (2008) Date flesh: Chemical composition and characteristics of the dietary fiber. Food Chemistry 111: 676-682.

9. AOAC Association of Official Analytical Chemists.

10. Johnson LF, Curl EA (1972) Methods for Research on the Ecology of Soil-Borne Plant Pathogens: Burgess Publishing Company, Minneapolis, MN, USA.

11. Sivasithamparam K, Parker CA, Edwards CS (1979) Rhizosphere microorganisms of seminal and nodal roots of Wheat grown in pots. Soil Biology and Biochemistry 11: 155-160.

12. Simon A, Ridge EH (1974) The use of ampicillin in a simplified medium for the isolation of fluorescent pseudomonads. Journal of Applied Bacteriology 37: 459- 460.

13. Bakerspigel A, Miller JJ (1953) Comparison of oxgall, crystal violet, streptomycin and penicillin on bacterial growth inhibitors. Soil Science 76: 123-126.

14. Martin JP (1950) Use of acid, rose bengal and streptomycin in the plate method for estimating soil fungi. Soil Science 69: 215- 232.

15. Schnurer J, Rosswall T (1982) Fluorescein diacetate hydrolysis as a measure of total microbial activity in soil and litter. Applied and Environmental Microbiology 6:1256-1261.

16. Chen W, Hoitink HAJ, Madden LV (1988) Microbial activity and biomass in container media for predicting suppressiveness to damping-off caused by Pythium ultimum. Phytopathology 78: 1447-1450.

17. Snedecor GW, Cochran WG (1981) Statistical Methods. The Iowa State University Press, Ames, IA Jauncey K (1989) A Guide to Tilapia Feeds and Feedings.

18. Jauncey, K, (1989) A Guide to Tilapia Feeds and Feedings. Institute of Aquaculture, University of Stirling, Scotland. pp. 49-56.

19. Wheaton FW, Hochheimer JN, Kaiser GE, Malone RF, Krones, et al. (1994) CC Nitrification filter design methods in Development of Aquaculture and Fisheries Science. Aquaculture Water Reuse Systems Engineering Design and Management 27: 149. Elsevier, Amsterdam.

20. Al-Marzooki W, Al-Halhali A, Al-Maqbaly R, Ritchie A, Annamalai K, et al. (2000) Date fiber as a constituent of broiler starter diets. Journal of Scientific Research in-Agricultural Sciences 5: 59-61.

21. Omar E, Nour A (1993) Utilization of Droppings of immature date fruits in feeding of Nile tilapia Oreochromis niloticus. Mars publishing House, Saudi Arabia.

22. Belal IEH, Al-Jasser MS (1997) Replacing dietary starch with pitted date fruit in Nile tilapia, Oreochromis niloticus (L.) feed. Aquaculture Research 28: 385-389.

23. Belal IEH, Al Owaifeir MA (2004) Incorporating Date Pits Phoenix dactylifera and their Sprouts in Semi-purified Diets for Nile Tilapia Oreochromis niloticus (L). Journal of World Aquaculture Society 35: 452-459.

24. Belal IEH (2008) Evaluating fungi-degraded date pits as a feed ingredient for Nile tilapia, Oreochromis niloticus L. Aquaculture Nutrition 14: 445-452.

25. Al-Asgah NA (1988) Date palm seeds as food for carp Cyprinus carpio L. Journal of College of Science, King Saud University 19: 59-64.

26. Saha S, Roy RN, Sen SK, Ray AK (2006) Characterization of cellulase-producing bacteria from the digestive tract of tilapia, Oreochromis mossambica (Peters) and grass carp, Ctenopharyngodon idella (Valenciennes). Aquaculture Research 37: 380-388.

27. Shafiei M, Karimi K, Taherzadeh MJ (2010) Palm Date Fibers: Analysis and Enzymatic Hydrolysis. International Journal of Molecular Sciences 11: 4285-4296.

28. Ferket PR, Parks CW, Grimes JL (2002) Benefits of dietary antibiotics and mananoligosaccharide supplementation for poultry Multi State Poultry Meeting, Department of Poultry Science, North Carolina State University, Raleigh.

29. Caspary WF (1992) Physiology and pathophysiology of intestinal absorption. American Journal of Clinical Nutrition 55: 299S-308S.

30. Sigleo S, Jackson MJ, Vahouny GV (1984) Effect of dietary fiber constituent on intestinal morphology and nutrient transport. American Journal of physiology 246: G34-G39.

31. Langhout DJ, Schutte JB, Van LP, Wiebenga J, Tamminga S, et.al. (1999) Effect of dietary high and low methylated citrus pectin on the activity of the ileal microflora and morphology of the small intestinal wall of broiler chicks. British Journal of Poultry Science 40: 340-347.

32. Yasar S, Forbes JM (1999) Performance and gastro-intestinal response of broiler chicks fed on cereal grain-based foods soaked in water. British Journal of Poultry Science 40: 65-76.

33. Shamoto K, Yamauchi K (2000) Recovery responses of chick intestinal villus morphology to different re-feeding procedures. Poultry Science 79: 718-723.

34. Shiau SY, Kwok CC (1988) Effects of cellulose, agar, carrageenan, guar gum and carboxymethylcellulose on tilapia growth Proceeding of Aquaculture International. 88: 93-94.

35. Maina JG, Beames RM, Higgs D, Mbugua P, Iwama NG, et al. (2007) Digestibility and feeding value of some feed ingredients fed to tilapia Oreochromis niloticus (L). Aquaculture Research 33: 853-862.

36. Hilton JW, JAtkinson JL, Slinger SJ (1983) Effects of increased dietary fibre on the Growth of rainbow trout Salmo gairdneri. Canadian Journal of Fish and Aquatic Sciences 40: 81-85.

37. Dioundick OB, Stom DI (1990) Effects of dietary-cellulose levels on the juvenile tilapia, Oreochromis mossambicus (Peters). Aquaculture 91: 311-315.

38. Shiau SY, Liang HS (1994) Nutrient digestibility and growth of hybrid tilapia, Oreochromis niloticus x O. aureus, as influenced by agar supplementation at two dietary protein levels. Aquaculture 127: 41-48.

39. Kritchevsky D, Tepper SA (2005) Influence of a fiber mixture on serum and liver lipids and on fecal fat excretion in rats. Nutrition Research 25: 485-489.

26

Effects of Grasshopper Meal in the Diet of *Clarias Gariepinus* Fingerlings

Olaleye Ibukun Grace*

Fisheries and Aquaculture unit, Institute of Oceanography, University of Calabar, Cross River State, Nigeria

Abstract

A study was conducted to assess the effects of grasshopper meal in the diet of *Clarias gariepinus* fingerlings. The aim was to substitute fishmeal with grasshopper meal in the formulation of *Clarias gariepinus* fingerlings feed. Feeds were formulated using different quantities of fishmeal and grasshopper meal and were used in feeding Clarias gariepinus fingerlings. Result shows that the best growth and feed utilization indices were recorded in the fingerlings fed 20% fishmeal and 10% grasshopper meal followed by those fed 15% fishmeal and 15% grasshopper meal. The least growth rate was recorded in fingerlings fed only 30% grasshopper meal. It could be concluded that Clarias fed with diet containing 10% grasshopper meal combined with 20% fishmeal produced the best growth rate.

Keywords: Growth; Grasshopper meal; *Clarias gariepinus;* Fingerling

Introduction

Fish is the major source of protein for most Nigerians. The increasing human population and the desire to obtain a nutritionally balanced level of protein intake is a major cause of the high fish demand in Nigeria. Aquaculture which is expected to bridge the gap between fish supply and demand is constrained generally by inappropriate technologies [1].

Fish feed is presently very expensive; both imported and locally produced ones. This is among the problems facing successful aquaculture in Nigeria coupled with good quality fish Falaye [2,3]. This is as a result of the competing need of the agricultural produce and by-products between man and livestock Salami et al. [4] and between livestock and fish in the formulation and production of the animal feed. Various protein sources have different amino acids, both essential and non-essential. A deficiency of one or more of these essential nutrients results in reduced growth rate, depressed diet, disease or even death NRC [5].

In the last decade, much effort has been made with the use of soybean meal as a good alternative to fishmeal in the diet of *Clarias gariepinus* [6-12]. Many researchers have attempted to use varied substitutes to fishmeal in *Clarias gariepinus* production with varying results. Faturoti and Oyelese [13] found yellow maize and sweet potato as a good energy source in the diet of *Clarias gariepinus* while, Eyo [14] obtained poor growth rate while feeding *Clarias anguillaris* with soybean diet. Ufodike and Ekokotu [15] confirm that excess levels of dietary protein might retard fish growth due to energy expenditure in deamination and excretion of excess protein. Ofojekwu and Ejike [16] also obtained poor results with cottonseed meal in Clarias food.

Edible Grasshoppers and locusts which include Nomadacris septemfasciata, Kraussaria sp., Katantop sp., Anacridium sp., Cataloipus sp., Hieroglyphycus sp., Gelestorhinus sp., and Locusta sp. are found to invade most of the North-eastern and Central States of Nigeria at a particular season of the year causing great consequences on crops Sharah [17]. These grasshoppers also serve as a delicacy to nation of North Eastern Nigeria during these invasions. These grasshoppers are as rich as the fishmeal in terms of its amino acid profile (Table 1). Encouraged by the similarity in the quality of the amino acid profile of fish and grasshopper meal, this research decided to replace fishmeal with grasshopper meal to ascertain if these qualities of the grasshopper can compare favorable in growth production of *Clarias gariepinus* as that obtained or fishmeal in the same species.

Materials and Methods

Preparation of grasshopper meal

Samples of edible grasshoppers and locusts were collected from the market located in Maiduguri irrespective of their sizes and species. The samples were dewinged, all appendages removed, sundried and crushed into powder with milling machine. Proximate analysis of the powdered samples was performed using standard methods AOAC [18]. Fibre content was assessed according to Cullison. The protein was measured by calorimetric method (Vanadomolybdale yellow method) (Table 2) with a varian 634UV visible spectrometer. Crude protein was calculated as total Kjeldahl N x 6.25.

Experimental diet

The feedstuffs used were obtained locally within Maiduguri town. The soybean was toasted for 15minutes according to Eyo [14]. Other ingredients such as groundnut cake, fishmeal, yellow maize, maize bran were obtained and ground into powder with the toasted soybean and grasshopper. A 45% cp feed was obtained from the combination of the feed ingredient in the diet and mixed with the premix.

Different diets (those containing only fishmeal and those containing grasshopper meal at various inclusion levels) were formulated using different treatments which include feed containing only fishmeal, feed containing only grasshopper meal and feed containing both fishmeal and grasshopper meal. The feed was pelleted using kitchen hand cranker. The pelleted feed was crushed into crumbles before administering them to the fish.

Experimental design and treatments

Fingerlings weighing between 15-20 g were obtained from the hatchery and conditioned in net hapa (1 m x 1 m x 1.2 m) installed in

***Corresponding author:** Olaleye I.G, Fisheries and Aquaculture unit, Institute of Oceanography, University of Calabar, Cross River State, Nigeria
E-mail: bknonair@gmail.com

Amino acid	Fish meal	Grasshopper meal
Lysine	7.85	5.87
Histidine	2.22	4.24
Arginine	5.82	7.62
Aspartic	9.35	9.32
Threonine	4.55	4.08
Serine	4.55	5.22
Glutamic	13.3	15.21
Proline	4.35	5.02
Glycine	5.90	4.78
Alanine	6.34	5.29
Cysteine	0.70	1.79
Valine	5.65	3.47
Methionine	2.84	1.96
Isoleucine	4.85	4.21
Leucine	7.35	5.30
Tyrosine	3.45	2.88
Phenylalanine	4.35	4.50

SOURCE: Okoye(2003).

Table 1: Comparative Amino acid profile of the proteins of fishmeal and grasshopper meal

FEEDSTUFFS	%INCLUSION LEVEL
Yellow maize	10.11
Groundnut cake	25.80
Soybean meal	25.80
Fishmeal/grasshopper meal	30
Cassava tuber starch	5
Premix (vitamin)	2
Salt	0.29
Bone	1
Total	100.00

Table 2: Product file for formulating 45% crude protein for *Clarias gariepinus*

11 m x 10 m x 1.2 m concrete tank for 48hours. The fish were stocked at 10 fish per meter square. Five different diets were tried with two replicates for each treatment for a period of 56 days. Below is a table showing the different treatments inclusion level (Table 3).

Fish in all treatments were fed 5% of their body weight daily split into two feeding frequency and the weight were recorded bi-weekly. Feeding rate was adjusted weekly based on body weight. Water quality parameter such as temperature and pH were monitored.

At the end of the research, weight gained (g), daily average growth rate (ADG), specific growth rate (SGR), food conversion rate (FCR) and protein efficiency ratio (PER) were calculated.

Data analysis

Data obtained from the trials were subjected to one way analysis of variance and statistical different between the means were separated using Turkey-HSD at 95% degree of confidence using SPSS 15.0 statistics package.

Results

Key

TFC=Total Feed Consumed, ADG=Average Daily Growth, SGR= Specific Growth Ratio, FCR=Food Conversion Ratio and PER=Protein Efficiency Ratio.

Means with the same superscripts along columns are not significantly different (p>0.05).

Discussion

The result of the nutrient composition shows that grasshopper meal has high crude protein of 64.51. This is a very high value that could completely replace fishmeal in fish feed. The value compares favorably with the result obtained by Njidda and Isadahomen [19] which was 64.32%cp. This value of grasshopper compares with that of fishmeal obtained by Okoye [20] from clupeid with 68.47%cp. The ether extract was 12.0 and closely related to that reported by Njidda and Isadahomen [19]. The value of the ether extract of grasshopper meal is greater than that obtained in fishmeal (Tables 4-6). This is good as it is being used as component of encapsulment of feed nutrient meant for fish to prevent loss of water soluble nutrients such as proteins and amino acids because of its insoluble property in water Lopez-Alverado et al. [21]. The crude fibre content was high due to the fact that grasshopper has an exoskeleton made of chitin Okoye and Nnaji [20]. The Nitrogen free extract was 5.49 which is the small amount of carbohydrates that can be digested easily because of its solubility Falayi [2,3]. The dry matter of grasshopper meal is very high 94.9 with low moisture content of 5.1. This implies quick drying of the feed compared to dry matter of fishmeal 90.0 and moisture content of 10% according to Eyo [22].

The calcium content is greater than those obtained from soybean meal and groundnut cake. It compares favorably with that of bloodmeal and less than that of fishmeal Haruna [23]. The phosphorus content is low due to low ash content. The sodium content compares favorably with that of soybean meal and yellow maize which has being used to replace fishmeal obtained by different researchers. The potassium content compares favorably with that of fishmeal obtained by Haruna [23]. The above nutrients composition of grasshopper meal and its quality makes it a good dietary supplement in fish feed production.

The result of the study shows that Treatment 2 (20% fishmeal and 10% grasshopper meal) has the highest weight gain, Average daily weight gain(ADG), Specific growth rate(SGR), Protein efficiency ratio(PER) and high Food conversion ratio(FCR) compared to other treatments despite the fact that they were of the same crude protein levels (45%cp). These might be attributed to good odour, colour and stability in water in line with Dupree and Haylor who reported that color and odour attract cultured organisms to pelleted feed.

High weight gain, Protein efficiency ratio, Food conversion ratio, Average daily weight gain and highest Specific growth rate was recorded from fish fed Treatment 3 (15% fishmeal and 15% grasshopper meal). This is related to Gbadamosi et al. [24] who recorded high weight gain, specific growth rate and food conversion ratio from *Clarias gariepinus* post juvenile fed with ration of 42% crude protein at 50% mixture level. Treatment 4 (10% fishmeal and 20% grasshopper meal) and Treatment 5 (30% grasshopper meal) have a lower weight gain, protein efficiency ratio, food conversion ratio, specific growth rate and average daily weight gain when compared with Treatment 1,2 and 3.

Okoye and Nnaji [20] reported that the inclusion of 10% grasshopper meal with 30% fishmeal gave a better growth performance than the diet with 40% fishmeal and no grasshopper meal. This is as a

Treatments	Fishmeal inclusion (%)	Grasshoppermeal Inclusion (%)
1	30	-
2	20	10
3	15	15
4	10	20
5	-	30

Table 3: Experimental Design with Grasshopper/ Fishmeal inclusion in the diets

Sample	%dry matter	%moisture content	%crude protein	%ether extract	%Ash	%Crude fibre	%NFE
Grasshopper meal	94.9	5.1	64.51	12.0	1.0	17.0	5.49

Table 4: Proximate analysis of Grasshopper meal

Sample	%calcium	%phosphorus	%sodium	%potassium
Grasshopper meal	0.55	0.12	0.1	0.73

Table 5: Essential mineral content of Grasshopper meal

Treatment	Initial weight(g)	Final weight(g)	Weight gain(g)	TFC(g)	ADG(g)	SGR(g)	FCR(g)	PER(g)	%Survival rate
1	15.60	60.50b	44.90b	85.12d	0.80b	0.024b	1.90d	0.99	80a
2	19.45	71.75a	52.30a	118.72a	0.93a	0.024b	2.26b	1.16a	70b
3	16.50	65.65b	49.15b	107.8b	0.88b	0.025a	2.14c	1.10b	75b
4	19.10	58.80c	39.75c	92.26c	0.71c	0.020d	2.32a	0.89d	65c
5	16.35	53.90c	37.55c	79.10c	0.67d	0.022c	2.10c	0.84d	80a

Table 6: Feed utilization and survival of *Clarias* fingerlings fed with five different diets for 56days

result of good quality essential amino acid present in both feedstuffs when combined.

Little mortality was recorded in all the treatments as a result of improper acclimatization and low temperature during the first 2weeks (which was between 21-23°C) of the study. This is in line with Falayi [2,3] who say that warm water fish grows best at temperatures between 25-32°C.

Conclusion

A lot of research had been carried out on suitable substitutes for fishmeal in fish diet. Grasshoppermeal has been shown to contain most of the essential amino acids in higher proportions than other protein feedstuff like bloodmeal, groundnut cake and soybean meal.

The growth performance of Clarias gariepinus fed with five different diets containing grasshoppermeal at varying inclusion level was monitored for 56 days in net hapas installed in concrete tank. The overall best performance was obtained in treatment 2 and 3 respectively. This is an indication of the potentials of grasshopper meal to substitute fishmeal for *Clarias gariepinus* to achieve optimal growth.

Based on the result obtained, more studies should be carried out on other conventional feedstuffs of least cost for growth performance of *Clarias gariepinus* and possibly other aquacultural fish.

References

1. Ajana AM (2002) Over view highlight and protein of fisheries in Nigerian aquaculture.

2. Falaye BA (2009a) Feed nutrients chemistry and importance in fish and livestock production.

3. Falaye BA (2009b) Tropical feedstuffs composition tables and biological catalogues in fish and livestock production.

4. Salami AA, Balogun OB, Fagbenro, Edibite L (1992) Utilization of non-pituitary extract in breeding of *Clarias gariepinus*.

5. NRC (1993) Nutritional requirements of warm water fish and shellfishes. National Academy Press. Washington DC, USA.

6. Balogun AM, Ologbobo AD (1989) Growth performance and nutrient utilization of fingerlings of *Clariasgariepinus* (Burchell) fed raw and cooked soybean diets. Aquaculture 76: 119-126.

7. Sadiku SOE, Jauncy K (1998a) Utilization of enriched soybean flour by *Clarias gariepinus*. J Aqua Tropics 13: 1-10.

8. Sadiku SOE, Jauncy K (1998b) Digestibility, apparent amino acid availability and waste generation potential of soybean flour-poultry meat meal blend diets for the sharp-toothed catfish fingerlings. J Applied Aquaculture 8: 69-75.

9. Fagbenro OA, Davies SJ (2002) Use of soybean flour (dehulled solvent extracted soybean) as fishmeal substitute in practical diet for African catfish, Clariasgariepinus (Burchell 1822) growth, feed utilization and digestibility. Journal of Applied Ichthyology 17: 64-69.

10. Fagbenro OA, Davies SJ (2003) Use of high percentages of soy protein concentrate as fishmeal substitute in practical diets for African catfish growth, feed utilization and digestibility. Journal of Aquaculture 16: (1).

11. Eyo AA (1994) Fish feed production techniques in agro products.

12. Davies SJ, Fagbenro OA, Abdel-Waritho, Diller I (1999) Use of soybean products as fishmeal substitute in African Catfish *Clariasgariepinus*, diets. Applied Tropical Agriculture 4: 10-19.

13. Faturoti EO, Oyelese I (1989) Digestibility and utilization of yellow maize and sweet potatoe based diets by Clariasgariepinus.

14. Eyo AA (1999) The effect of different method of soybean processing on the growth and food utilization of African mudfish *Clariasanguillaris* (L) fingerlings. J Biotech 10: 9-10.

15. Ufodike EBC, Ekotutu PA (1986) Protein digestibility and growth of African catfish fed blood meal and algae diets. Acta hydrobiologica 28: 237-243.

16. Ofojekwu PC, Ejike C (1984) Growth response and feed utilization in tropical Cichlid Oreochromisniloticus (Lin) fed on cottonseed based artificial diets. Aquaculture 42: 27-37.

17. Sharah HA (2012) The driving force behind increasing grasshopper frying business in Maiduguri: Profitability or Joblessness?. Int J Eco Dev Res Invest 3: 110-117.

18. Association of Official Analytical Chemists (A.O.A.C) (1995) Official methods of analysis of A.O.A.C. Washington DC, USA.

19. Njidda AA, Isidahomen CE (2010) Haematology, blood chemistry and carcass characteristics of growing rabbits fed grasshopper meal as a substitute for fishmeal. Pak Vet J 30: 7-12.

20. Okoye FC, Nnaji JC (2004) Effect of substituting fishmeal with grasshopper meal on the growth and food utilization of the Nile Tilapia, *Oreochromis niloticus* fingerlings.

21. Lopez-Alverado J, Langdon CJ, Teshima S, Kana- Sawa A (1994) Effect of coating and encapsulating of crystalline amino acids on leaching in larva feeds. Aquaculture 122: 335-345.

22. Eyo AA (2001b) Chemical composition and amino acid content of the commonly available feedstuffs used in fish feed in Nigeria.

23. Haruna BA (2003) Aquaculture in the tropics. Theory and practice. Al-Hassana Publishers Abuja, Kaduna, Kano- Nigeria.

24. Gbadamosi OK, Daramola JA, Osungbemiro (2007) Growth performance and nutrional utilization of vitamin c in diet of African catfish fingerlings. Acta Zoological science 58: 763-766.

27

Molecular Characteristic of Giant Grouper (*Epinephelus Lanceolatus*) Vitellogenin

Om AD[1]*, Sharif S[2], Jasmani S[2], Sung YY[3] and Bolong AA[3]

[1]*Fisheries Research Institute (FRI), Tanjong Demong, 22200 Besut, Terengganu, Malaysia*
[2]*Institute of Tropical Aquaculture (AQUATROP), University Malaysia Terengganu, 21030 Kuala Terengganu, Malaysia*
[3]*School of Fisheries and Aquaculture Science, University Malaysia Terengganu, 21030 Kuala Terengganu, Terengganu, Malaysia*

Abstract

The Vitellogenin (Vtg) gene sequence acts as an indicator to the fish reproduction, which can adapt to the environment factor or can influence the gonad development. The Vtg nucleotide sequence from Giant grouper was characterized using bioinformatics software. A homology search of the deduced amino acid sequence of the obtained Vtg DNAs (compare with 13 species) was carried out using the National Centre for Biotechnology Information website. Clustal W analysis was constructed a phylogenetic tree by using Molecular Evolutionary Genetic Analysis MEGA version 5.2.2. In order to verify the Vtg gene sequences obtain and elucidate structure-function relationship in Vtg, by using DELTA BLAST of 3-D structure of Giant grouper with others fishes.

Result of phylogenetic analysis using Maximum Likelihood (ML) and Neighbour Joining (NJ) showed tree analysis generated two separated tree topology. This similarity (0.015 distance matric viewer) was closely related in terms of their Vtg gene sequence although from different environmental and ecological conditions. In general, showed that *Epinephelus lanceolatus* Vtg is evolutionary more related to *Poecilia latipinna*. *Epinephelus lanceolatus* shows four main domains (Vitellogenin-N, DUF1943, DUF1944 and VMD), similarly found in *Dicentrachus labrax* but was different compare to *Clarias macrochepalus*, *Catla catla*, and *Danio rerio*. This indicates characteristic of Vtg domain for freshwater species is control by present of VMD in Vtg. The molecular approach can be done on Giant Grouper to understand the molecular respond towards fish growth and determine the individual of Giant grouper that has potential to increasing the Vtg production for increasing eggs quality.

Keywords: Nucleotide; Molecular; Environmental; Ecological; *Poecilia latipinna*

Introduction

The development of molecular tools has recently opened new direction and facilitated the discovery of the genes involved in these processes and their evolutionary functional significance. Fish oocyte development attracted specific interest in the last century. Morphological investigations were followed by biochemical, physiological, and endocrinological analyses that extended our knowledge of dynamic events that take place during oocyte development and egg formation.

Molecular characterization of vitellogenin (Vtg) gene is important because it indirectly leads to the understanding of the role-play in the molecular basis of gonad development in terms of their structure and function [1]. Basically, each gene has its own molecular characteristic that is specific to their action. This includes the Vtg genes, which has certain features that are fundamental and responsible for its actions. The Vtg gene sequence acts as an indicator to the fish reproduction, which can adapt to the environment factor or can influence the gonad development [2]. Study on the molecular levels could permit the understanding of gonad development, gene regulation, structural-function relationships, evolution and adaptation to environment.

In teleost fish, as in other oviparous, Vtg is specifically incorporated in the oocyte by receptor-mediated endocytosis through receptors belonging to the low density lipoprotein receptor (LDR) family, which have been named very low density lipoprotein receptors (VLDLRs), Vtg receptors (VtgRs), due to the presence of eight ligand-binding repeats [2]. The other members of the gene family bind various ligands and are involved in lipid metabolism in both vertebrates and invertebrates. Therefore, this function of Vtg component needs to clarify for better understanding in physiology process during oocyte development.

The objective of the present study was to characterize the Giant grouper Vtg gene, and to compare Vtg gene expression between other species as basic information to develop Vtg as biomarker indicator in sex identification of Giant grouper.

Material and Methods

Molecular phylogenetic analysis of giant grouper Vtg

The Vtg nucleotide sequence from Giant grouper was characterized using bioinformatics software. A homology search of the deduced amino acid sequence of the obtained Vtg DNAs was carried out using the National Center for Biotechnology Information website (http://www.ncbi.nlm.nih.gov/). The Vtg Giant grouper sequence (Figure 1) from previous study [3], was compared with 13 species such as lamprey (*Ichthyomyzon unicuspis*, GenBank; AAA49327.1), sailfin molly (*Poecilia latipinna*, GenBank; ACV65040.1), rohu (*Catla catla*; GenBank; ABP04034.2), tuna (*Thunnus thynnus*; GenBank;), catfish (*Clarias macrocephalus*; GenBank; ABW96364.1), carp (*Cyprinus carpio*; GenBank; AGZ80880.1), european seabass (*Dicentrarchus labrax*; GenBank; AFA26670.1), mummichog (*Fundulus heteroclitus*; GenBank; AAB17152.1), mangrove rivulus (*Kryptolebias marmoratus*;

*Corresponding author: Ahmad Daud Om, Fisheries Research Institute (FRI), Tanjong Demong, 22200 Besut, Terengganu, Malaysia
E-mail: ahmaddaudom@yahoo.com

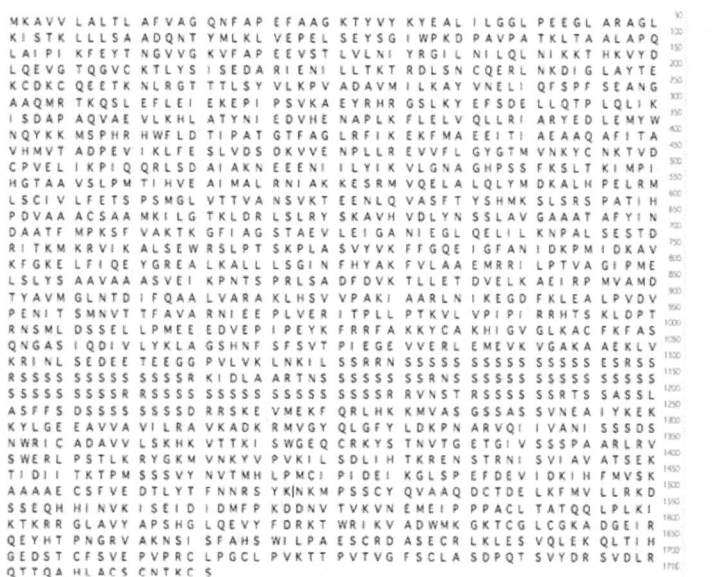

Figure 1: Amino acid sequence (1,704) in single letter code of Giant grouper vitellogenin peptide showing sequence coverage after trypsin digest.

GenBank; AAQ16635.1), striped bass (*Morone sexatillis*; GenBank; ADZ57172.1), rainbow trout (*Oncorhynchus mykiss*; GenBank; CAA63421.1), zebra fish (*Danio rerio*; GenBank; NP 001157843.1), and Japanese eel (*Anguilla japonica*; GenBank; AAV48826.1).

The deduced amino acid sequences were aligned using the ClustalW [4] program hosted by the DNA Data Bank of Japan (http://clustalw. ddbj.nig.ac.jp/top-j.html) and subjected to ClustalW analysis to construct a phylogenetic tree using the bootstrapped neighbor-joining method [5]. The sequence obtained was exported to FASTA format in notepad and then, was edited using Bioedit software to remove the unwanted and vector sequences to identify the location of the insert sequence. Multiple alignments from 14 fish peptide sequences of Vtg were conducted using eBiox 5.2.2 program and it were used in the phylogenetic analysis. The phylogenetic analysis was carried out using Molecular Evolutionary Genetic Analysis MEGA version 5.2.2 [6] with Maximum Likelihood and Neighbor Joining algorithms in order to estimate the phylogeny.

Domain

In order to verify the Vtg gene sequences obtained and elucidate structure-function relationship in Vtg, the conserved and essential domains and residues in Vtg and other members of the gene family such as Lipovitellin I (Lv-I) and II (Lv-II), phosvitin (Pv), polyserine track (PT), von Willebrand-factor type-D domain (VWD) were determined by using DELTA BLAST (http://blast.ncbi.nlm.nih.gov/ Blast.cgi.). The molecular characterization of primary structure Giant grouper Vtg gene such as protein domain, families and functional sites were determined by comparing the sequence to other fish.

Three dimensional (3D) structure prediction

Furthermore, in this study the prediction of 3-D structure of Giant grouper with others fishes from different orders were also viewed using protein homology/analogy recognition engine v 2.0 Phyre2 server http://www.sbg.bio.ic.ac.uk/ [7].

Results

Molecular phylogenic analysis

Result of phylogenetic analysis using maximum likelihood (ML) and neighbor joining (NJ) methods showed tree analysis generated two separated tree topology (Figure 2). In general, showed that *Epinephelus lanceolatus* Vtg is evolutionary more related to *Poecilia latipinna* and *Kryptolebias marmoratus*. It is noted that the distribution of Vtg phylogenetic was significantly different, between freshwater species (*Carassius auratus, Catla catla, Danio rerio* and *Clarias macrochepalus*) as one group of Vtg compare to seawater and euryhaline species (*Epinephelus lanceolatus, Poecilia latipinna, Kryptobias marmorata, Morone sexatilis, Thunnus thynnus, Fundulus heterocittus, Dicentrachus labrax* and *Anguilla japonica*) for another group. The constructed a phylogenetic tree that places closely related sequences under the same interior node and whose branch lengths closely reproduce the observed distances between sequences.

The results from evolutionary distance estimations are displayed in the distance matrix explorer (Table 1). Results describe the accuracy of pair wise alignment by Clustal under the specific simulation conditions and alignment parameters. Estimation of evolutionary distances between Vtg sequences is important for constructing phylogenetic trees (Figure 1), dating species divergences and understanding the mechanism of evolution of protein. Vtg sequence of giant grouper (*Epinephelus lanceolatus*) was closed (0.015) with *Poecilia latipinna* (Genbank: ACV65040.1) and very far from *Icthyomyzon unicuspis* (Genbank: AAA49327.1) (1.041). Estimating the number of nucleotide or amino acid substitutions needed to compute evolutionary distances is one of the most important subjects in molecular evolutionary genetics and comparative genomics. Estimation of evolutionary distance of Giant grouper Vtg with alignment of 13 other fish homologous sequence, revealed that Giant grouper Vtg was belongs to the marine fishes species rather than freshwater species group.

Domain architecture of Vtg

Study by Babin et al. [8], has proposed the domain architecture

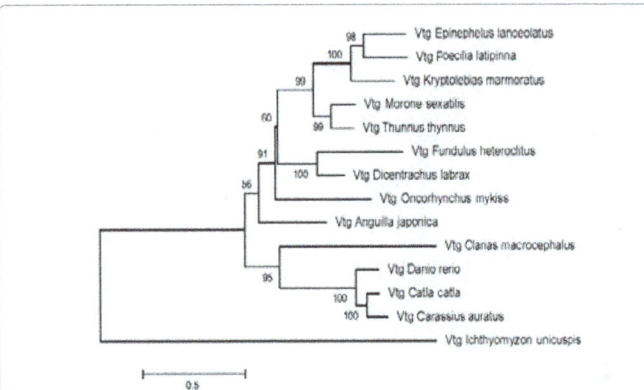

Figure 2: Polygenetic distributions of 14 species of Vtg sequences. Numbers besides nodes indicate the percent of bootstrap values for each branch of the tree in the 1,000 bootstrap trials.

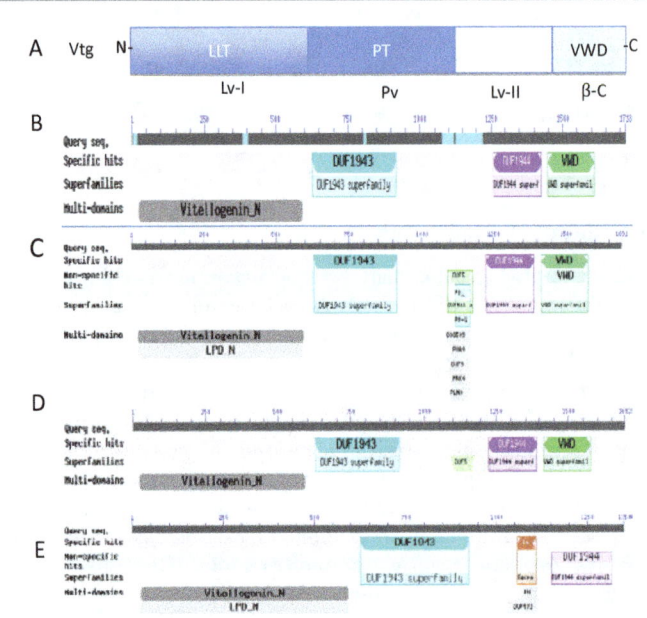

Figure 3: Schematic representation of the domain in Vtg peptide sequence, (A) Domain architecture of teleost fish Vtg (Babin, et al., 2007). Large lipid transfer (LLT) module also referred to as Vtg_N and LPD-N domain, polyserine track (PT) domain, and von Willebrand-factor type-D domain (VWD). The horizontal line indicates the receptor-binding region to VtgR. Lipovitellin I (Lv-I) and II (Lv-II), phosvitin (Pv), and β-component (β-C) are the yolk protein generated by the enzymatic cleavage of Vtg inside the oocyte. Conserved domains of (B) *Epinephelus lanceolatus*, (C) *Icthyomyzon unicuspis*, (D) *Dicentrachus labrax*, and (E) *Clarias macrochepalus* Vtg by using NCBI software.

and conserved sequence of teleost fish Vtg (Figure 3A). Based on the analysis, *Epinephelus lanceolatus* shows four main domains (Vitellogenin-N, DUF1943, DUF1944 and VMD) (Figure 3B), similarly found in *Dicentrachus labrax* but different compare to *Clarias macrochepalus*, *Catla catla*, and *Danio rerio* (figure not show) where VMD domains, was absent. This indicate characteristic of Vtg domain for freshwater species is control by present of VMD in Vtg. Analysis on domain of Vtg with PROSITE (http//www.expasy.org/prosite) clarified the amino acid sequence for Giant grouper was from sequence number 22 till number 660 (Figure 4).

The predicted secondary structures of Giant grouper Vtg are show in Figure 3. It was clearly seen that α-helix was predominantly present in the Giant grouper Vtg sequence and helix can be grouped into four major groups, which are located in domain region. Analysis indicated that the α-helix, β-sheet and the coil structure configurations have 39.96%, 25.54% and 34.48% respectively. As it can see in Figure 4, the 4-helics can recognize in the different color of domain region.

Three dimensional (3-D) structure prediction

Analysis of the 3-D structure found that *E. lanceolatus* Vtg gene shows this protein has the typical 4α-helices bundle protein that runs in anti-parallel (Figure 5A). Based on the color, its can categorized in 4 helix structure, which is Blue, Red, Light green and Green respectively. In the present study, the main structure of Vtg gene in Giant grouper from different species was similar at the 4-helic region (Figures 5B-5F). However, the difference can be seen in Helix-1 (blue) and Helix-4 (red) where the structure was totally different in Lamprey (*Icthyomyzon unicuspis*) but similar in Catfish (*Clarias macrocephalus*), Japanese Eel (*Anguilla japonica*) and European Seabass (*Dicentrachus labrax*). However, Zebra fish (Danio rerio) 3-D vtg structure was different with giant grouper in the position of reddish color Helix-4.

Discussion

Vitellogenin (Vtg) is an egg yolk precursor expressed in the females of nearly all oviparous species including fish, amphibians, reptiles, birds, most invertebrates, and the platypus. Vtg is the precursor of the lipoprotein and phosphoproteins that make up most of the protein content of yolk. There is potential of Vtg as a biomarker for measuring exposure of oviparous animals to estrogen or estrogen mimics, by using several fish species for which both *in vivo* and *in vitro* assays have been developed [9-11], Vtg functions as a nutritional source for the developing embryo, rather than as an important functional protein.

Three types of vitellogenin (Vtg) namely vitellogenin A (Vt gA), vitellogenin B (vtg B) and vitellogenin C (vtg C) have been identified in fishes. Paracanthopterygii and Achantopterygii generally express three types of Vtg at transcription level [12]. Advanced teleost fishes (Acanthomorpha) produced two complete types of Vtg (VtgAa and VtgAb) with five linear yolk protein domains organized from the amino-terminus as follows lipovitellin heavy chain (LvH), phosphitan (Pv), lipovitellin light chain (LvL), β-component (β'-c), and C-terminal peptide (C-t)) Reading [13].

Generally, assessment with Blast analysis showed that the Vtg amino acid protein sequences were similar with Vtg gene in the GeneBank database, it is very likely that 14 sequences obtained were Vtg gene fragments. Molecular characteristic of Vtg Giant grouper showed phylogenetic analysis by using maximum likelihood (ML) and neighbor joining (NJ) was generated two separated tree topology. The phylogenetic grouping showed the Giant grouper were closed to *Poecilia latipinna* and *Kryptolebias marmoratus* than other group fish at 98-100% similarity in terms of nucleotide and amino acids sequences respectively (Figure 1). It was shown that Giant grouper Vtg gene had the highest homology with *Poecilia latipinna*. These similarities (0.015 distance matric viewer) were closely related in terms of their Vtg gene sequence although from different environmental and ecological conditions.

Biological similarities are seen between lipoprotein and Vtg from the point of view of binding of hydrophobic molecules, cell specific uptake, and the possibility that these proteins may have a common ancestor [14-15]. In this investigation, the region of Giant grouper Vtg, which

(	1	2	3	4	5	6	7	8	9	10	11	12	13	14	)
(1)		(0.021)	(0.026)	(0.026)	(0.21)	(0.025)	(0.026)	(0.024)	(0.024)	(0.021)	(0.038)	(0.025)	(0.026)	(0.021)
(2)	0.435		(0.026)	(0.021)	(0.022)	(0.026)	(0.027)	(0.025)	(0.026)	(0.023)	(0.038)	(0.021)	(0.022)	(0.012)
(3)	0.614	0.604		(0.03)	(0.02)	(0.029)	(0.03)	(0.028)	(0.028)	(0.027)	(0.040)	(0.030)	(0.030)	(0.026)
(4)	0.602	0.443	0.734		(0.026)	(0.029)	(0.03)	(0.029)	(0.029)	(0.028)	(0.039)	(0.017)	(0.015)	(0.021)
(5)	0.426	0.474	0.417	0.614		(0.026)	(0.027)	(0.025)	(0.026)	(0.023)	(0.038)	(0.026)	(0.026)	(0.022)
(6)	0.565	0.608	0.701	0.706	0.619		(0.023)	(0.014)	(0.014)	(0.025)	(0.040)	(0.029)	(0.029)	(0.026)
(7)	0.614	0.652	0.76	0.755	0.647	0.514		(0.023)	(0.023)	(0.027)	(0.040)	(0.030)	(0.030)	(0.027)
(8)	0.544	0.583	0.695	0.709	0.586	0.207	0.5		(0.011)	(0.024)	(0.040)	(0.029)	(0.029)	(0.026)
(9)	0.556	0.608	0.692	0.723	0.616	0.225	0.516	0.146		(0.025)	(0.040)	(0.029)	(0.029)	(0.026)
(10)	0.433	0.503	0.632	0.672	0.496	0.57	0.629	0.541	0.591		(0.038)	(0.027)	(0.028)	(0.023)
(11)	1.043	1.041	1.083	1.064	1.036	1.085	1.099	1.104	1.102	1.018		(0.039)	(0.040)	(0.038)
(12)	0.589	0.428	0.743	0.34	0.595	0.713	0.736	0.698	0.709	0.635	1.059		(0.019)	(0.020)
(13)	0.607	0.464	0.733	0.237	0.625	0.711	0.753	0.723	0.729	0.676	1.083	0.382		(0.021)
(14)	0.443	0.162	0.625	0.422	0.464	0.611	0.64	0.62	0.62	0.52	1.032	0.391	0.441	

Table 1: Distance matrix viewer showing distance and their standard errors for sequence pairs. (Each number in bracket refer to each species below).(1)Vtg Oncorhynchus mykiss; (2)Vtg Morone sexatilis; (3)Vtg Fundulus heteroclitus; (4) Vtg Epinephelus lanceolatus; (5) Vtg Dicentachus labrax; (6)Vtg Danio rerio; (7) Vtg Clarias macrocephalus; (8) Vtg Catla catla ; (9) Vtg Carassius auratus; (10) Vtg Anguilla japonica; (11) Vtg Icthyomyzon unicuspis; (12) Vtg Krytolebias marmoratus; (13) Vtg Poecilia latipina; (14) Vtg Thannus thynnus.

Figure 4: Secondary structure of *Epinephelus lanceolatus* Vtg as predicted by the Phyre2 software. The green, blue color and the faint lines symbols represent α-helix, β-sheet and coil respectively.

showed homology, was limited to the N-terminal half of the molecule corresponding to the 660 domain profile. However, there is different in domain profile between seawater species and freshwater species. The catfish, (*Clarias macrochepalus)* was absent in von Willebrand-factor type-D domain (VWD), similarly finding [16] in Zebrafish. It should be noted that an additional DUF 1061 domain of unknown function was identified in the last region of Vtg peptide sequence.

The information from identification of Vtg (such as molecular mass and sequencing) could be useful during preparation of Vtg antibody production. Antibodies production is generated by *in vivo* or *in vitro* approaches, their identification relies mainly on screening of hybridoma supernatants or bacterially expressed antibody fragments. The molecular approach can be done on Giant grouper to understand the molecular respond towards fish growth and determine the individual of Giant grouper that has potential to increasing the Vtg production for increase eggs quality.

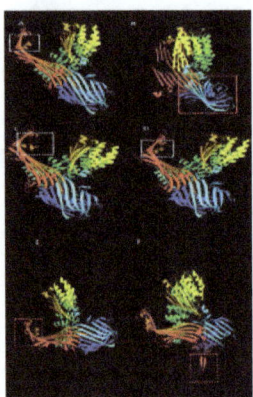

Figure 5: The 3-D structure of Vtg in fishes from different species. Figure A, B, C, D, E, and F are the 3-D structure of *Epinephelus lanceolatus*, *Ichthyomyzon unicuspis*, *Dicentrachus labrax*, *Clarias macrocephalus*, *Anguilla japonica* and *Danio rerio* respectively. The location of helix; blue: helix1; green: helix 2; light green: helix 3; red: helix 4.

Application of Vtg gene in aquaculture is promising in many aspects especially in molecular approach. This includes in the production of GMOs to improve the fish performance and gene regulation study to produce the high-quality eggs and determination of SNPs that can be used as genetic marker. Nutritional genomics is an area of science to studying how genes influence response of genetically to feed. Knowledge of these interactions and variations can be applied in the field of nutrigenetics to improved maturation diet for broodstock.

The results of this investigation will enable in further studies on the elucidation of the hormonal regulation of vitellogenesis including the physiological functioning with vitellogenin-stimulating ovarian hormone. This information can be used for improving the production of giant grouper for broodstock management.

Acknowledgment

This study is funded by the Ministry of Science, Technology and Innovation, Malaysia, under Intensified Research in Priority Areas (E-Science Fund, 004-07-05-06)

References

1. Pousis C, Santamaria N, Zupa R, Giorgi CD, Mylonas CC, et al. (2012) Expression of vitellogenin receptor gene in the ovary of wild and captive Atlantic bluefin tuna (*Thunnus thynnus*). Anim Reprod Sci 132: 101-110.

2. Hiramatsu, N, Chapman RW, Lindzey JK, Haynes MR, Sullivan CV (2004) Molecular characterization and expression of vitellogenin receptor from white perch (*Morone Americana*). Biol Reprod 70: 1720-1730.

3. Om AD, Safiah J, Nosrihah I, Yeong YS, Abol-Munafi AB (2013) Application MALDI-TOF on protein identification of vitellogenin in Giant grouper (*Ephinephelus lanceolatus*). Fish Physiol Biochem 39: 1277-1286.

4. Thompson JD, Higgins DG, Gibson TJ (1994) CLUSTALW: improving the sensitivity of progressive multiple sequence alignment through sequence weighting, position-specific gap penalties and weight matrix choice. Nucleic Acids Res 22: 4673-4680.

5. Saitou N, Nei M (1987) The neighbor-joining method: a new method for reconstructing phylogenetic trees. Mol Biol Evol 4: 406-425.

6. Tamura KD, Peterson N, Peterson G, Masatoshi SNM, Kumar S (2011) MEGA5: Molecular evolutionary genetics analysis using maximum likelihood, evolutionary distance, and maximum Parsimony methods. Mol Biol Evol 28(10): 2731-2739.

7. Kelley LA, Sternberg MJE (2009) Protein structure prediction on the web: a case study using the Phyre server. Nature Protocols 4: 363-371.

8. Babin PJ, Carnevali O, Lubzens E, Schneider WJ (2007) The fish oocyte: from basic studies to biotechnology applications.

9. Folmar LC, Denslow ND, Roa V, Chow M, Crain DA, et al. (1996) Vitellogenin induction and reduced plasma testosterone concentration in feral male carp (*Cyprinus carpio*) captured near a major metropolitan sewage treatment plant. Environ Health Perspect 104: 1096-1101.

10. Heppell SA, Denslow ND, Folmar LC, Sullivan CV (1995) "Universal" assay of vitelllogenin as a biomarker for environmental estrogen. Environ Health Perspect 103: 9-15.

11. Sumpter JP, Jobling S (1995) Vitellogenin as a biomarker for estrogenic contamination of the aquatic environment. Environ Health Perspect 103:173-178.

12. Hiramatsu N, Hara A, Hiramatsu K, Fukuda H, Gregory MW et al. (2002) Vitellogenin derived yolk proteins of white perch, *Morone Americana*: Purification, characterization, and Vitellogenin-receptor binding. Biol Reprod 67: 665-667.

13. Reading BJ, Sullivan CV (2011) Vitellogenesis in fishes.

14. Baker ME (1998) Is vitellogenin an ancestor of apoliporotein B-100 of human low-density lipoprotein and human lipoprotein lipase?. Biochem J 255: 1057-1060.

15. Babin PJ, Bogerd JFP, Kooiman WJ, Van Marrewijk DJ (1999) Apolipophorin II/I, apolipoprotein B, vitellogenin, and microsomal triglyceride transfer protein genes are derived from a common ancestor. J Mol Evol 49: 150-160

16. Wang HT, Yan JT, Tan, Gong Z (2000) A Zebrafish vitellogenin gene (Vg3) encodes a novel vitellogenin without a phosvitin domain and may represent.

28

Evaluation of Glutaraldehyde, Chloramine-T, Bronopol, Incimaxx Aquatic® and Hydrogen Peroxide as Biocides against *Flavobacterium psychrophilum* for Sanitization of Rainbow Trout Eyed Eggs

Alexandra Grasteau[1], Thomas Guiraud[1], Patrick Daniel[2], Ségolène Calvez[3], Valérie Chesneau[4] and Michel Le Hénaff[1]*

[1]*Bordeaux University, CNRS UMR EPOC, Talence, France*
[2]*Laboratoire des Pyrénées et des Landes, Mont de Marsan, France*
[3]*LUNAM University, Oniris, UMR INRA BioEpAR, Nantes, France*
[4]*Groupement de Défense Sanitaire Aquacole d'Aquitaine, Mont de Marsan, France*

Abstract

The effective conditions of glutaraldehyde, chloramine-T, bronopol, Incimaxx Aquatic® and hydrogen peroxide as some biocides commonly used by the aquaculture industry were investigated against *F. psychrophilum* in sanitization of rainbow trout eyed eggs. Bacteriostatic tests as well as bactericidal tests using ethidium monoazide bromide PCR assays were conducted *in vitro* on *Flavobacterium psychrophilum* while impacts of chemical treatments were studied *in vivo* on 240 [°C × days] rainbow trout eyed eggs. A 20-min contact time with bronopol (up to 2,000 ppm), chloramine-T (up to 1,200 ppm), glutaraldehyde (up to 1,500 ppm), hydrogen peroxide (up to 1,500 ppm) or with Incimaxx Aquatic® (up to 185 ppm, eq. peracetic acid) was effective against *F. psychrophilum* and did not affect the eyed eggs/fry viability. Collectively, the data obtained here clearly demonstrate that concentrations and duration of treatments commonly used to sanitize eyed eggs are widely overestimated in their effectiveness against *F. psychrophilum*. The new treatment conditions with the five studied biocides are bactericidal for *F. psychrophilum* and safe for rainbow trout eyed eggs. In this work, we developed an experimental approach to test some chemicals against fish pathogens to assist fish farmers in the effective and safe disinfection of eyed eggs.

Keywords: Aquaculture; Ethidium monoazide bromide; Disinfection susceptibility; *Flavobacterium psychrophilum*; Rainbow trout eggs; Viable qpcr

Introduction

Flavobacterium psychrophilum is the aetiological agent of 'rainbow trout fry syndrome' (RTFS) and 'bacterial coldwater disease' (BCWD), the two most significant systemic infections of primarily freshwater-reared salmonid fish [1] such as coho salmon (*Oncorhynchus kisutch*), rainbow trout (*Oncorhynchus mykiss*) and occasionally other fish species such as ayu (*Plecoglossus altivelis*) [2]. Several clinical manifestations have been described among which the most significant are mortality in juvenile fish (RTFS) and in adult, septicemia preceded by extensive necrotic lesions (BCWD) [3]. Consequently, considerable economic losses to fish aquaculture producers can occur (up to 90% in rainbow trout farmed in Norway [4]) and the erosion of tissue leading to a commercial downgrade of adult fish (for a review of *F. psychrophilum* biology, clinical signs and BCWD prevention and treatment, [5]). The control of *F. psychrophilum* infections is difficult and no effective vaccine is available yet despite numerous studies focused on the capability of some *F. psychrophilum* proteins to induce protection in fish. Potential targets identified for vaccine development include the OmpH-like surface antigen or the outer membrane glycoprotein OmpA [6,7] other immunogenic proteins such as trigger factor, ClpB, elongation factor G, gliding motility protein GldN and a conserved hypothetical protein [8]. Vaccination with FLAVO IPN and FLAVO AVM6, two mineral oil adjuvanted cocktails, induces responses that seemed capable of protecting rainbow trout against infections with *F. psychrophilum* [9]. However, recent study conducted with *F. psychrophilum* gliding motility N (GldN) protein underlines the importance of conducting multiple *in vivo* evaluations on potential vaccine(s) before any conclusions are drawn [10]. To date, the control of infections is yet achieved by antibiotic treatments using medicated feed (mainly florfenicol in 10 mg per kg of fish for 10 days) [11]. Some hatchery managers have expressed concerns about user safety

and the impact on the environment of such molecules. Indeed, these pharmaceuticals or their metabolic residues (i) may be found inside the fish flesh, (ii) may lead to the emergence of resistant strain pathogens and/or (iii) may have side effects on aquatic organisms accidentally exposed to them. Due to environmental constraints, the aquaculture industry seeks to limit the use of antibiotics and emphasizes a preventive approach based on the implementation of effective hygiene measures.

Infections in fish (as for most other livestock) with bacterial pathogens involve either horizontal transmission by direct spread from contaminated animals or from their environment polluted by secretions/excretions of other infected animals, or vertical transmission from the spawners to their offspring through eggs. Such contamination may occur in two ways: the first is true vertical transfer where pathogens from parent broodstock invade the gonads and possibly infect gametes and the future embryos. The second is pseudo-vertical transfer where the surface of the eggs after spawning constitutes a matrix for environmental pathogens and the larvae are contaminated during the hatching of the contaminated eggs. Many molecules have been tested for the surface disinfection of fertilized fish eggs to prevent the pseudo-

*Corresponding author: Michel Le Hénaff, Bordeaux University, CNRS UMR 5805 'Environments and Paleoenvironments Oceanic and Continental', Avenue of the Faculties, F-33405 Talence Cedex, France
E-mail: michel.lehenaff@agro-bordeaux.fr

vertical transfer of pathogens [12]. They reduce the spread of pathogens from parent broodstock farms to hatchery farms and improve the survival to hatch. The list of disinfectants includes glutaraldehyde [13,14], hydrogen peroxide, iodine and tannic acid [15,16], ozone [17] and numerous others. For most of them, CT values have been defined as the concentrations of disinfectants (C; mg/L) multiplied by the exposition time (T; min) for which antibacterial effects have been observed with no significant side effect in hatching ability of the eggs. Thus, it has been shown that concentrations of copper sulfate needed to eliminate *F. psychrophilum* (above 300 mg/L) were toxic for rainbow trout eggs and thus are not recommended for control of RTFS or BCWD [18]. Therefore, the objective of this work was to specifically reassess both bacteriostatic and bactericidal effects of five biocides commonly used by the aquaculture industry against *F. psychrophilum* in sanitization of rainbow trout eyed eggs.

Materials and Methods

Bacterial strains, media and growth conditions

F. psychrophilum strains used in this study were: the reference strain JIP02/86 (INRA) and some freshly isolated strains from rainbow trout showing clinical signs of the disease (Table 1). They were sampled in 2012-2013 from four French rainbow trout farms, where outbreaks of RTFS had been reported. Isolates were collected from organs of trout presenting clinical signs (brain, SESB02, MLEB15 and MTOB07; or kidney, PISK08, and ASOK05) and typed using qPCR [19,20] as well as pulsed-field gel electrophoresis [21]. The bacterial cells were cultivated in a modified FLP liquid medium [0.5% (w/v) tryptone, 0.05% (w/v) yeast extract, 0.02% (w/v) beef extract, 0.02% (w/v) sodium acetate (pH 7.2)] or in FLP solid medium (+ 15 g/L agar). Bacteria were incubated at 14°C under aerobic conditions (orbital stirring, 150 rpm). Purity of the bacterial suspensions was checked (i) by examination of Gram-stained smears and (ii) by qPCR using the universal primer or the *F. psychrophilum* specific set of 16S rDNA primers to calculate a specificity factor as indicated by Orieux et al. [20].

Trout hatchery

Egg samples were taken from a fish farm where recurring outbreaks of RTFS had occurred. The water sources were bore-hole water as well as surface water in a flow through system. Eggs were collected within an egg incubation tray stack 240 [°C × days] once every two days disinfected only with bronopol 50 ppm over a 1-hour period. The eggs were moved in a second farm for disinfection trials.

F. psychrophilum strains		JIP02/86	SESB02	PISK08	MLEB15	MTOB07	ASOK05
Origin		Rainbow trout, Aquitaine (France)	Freshly isolated from rainbow trout, Aquitaine (France)				
MIC (ppm)	Bronopol	3.1	3.1	1.6	3.1	6.3	1.6
	Hydrogen peroxide	7.8	6.2	3.1	31.3	62.5	3.1
	Glutaraldehyde	300.0	160.0	800.0	300.0	300.0	160.0
	Incimaxx Aquatic® (eq. peracetic acid)	125.0	62.5	125.0	62.5	62.5	31.2
	Chloramine-T	313.0	313.0	313.0	313.0	313.0	156.0

Table 1: *F. psychrophilum* strains and related type strains used in this study and the corresponding MIC (minimal inhibitory concentration) observed in the presence of disinfectant.

Preparation of biocide solutions

F. psychrophilum isolates were tested for sensitivity to five biocides commonly used in the fish industry: (i) glutaric dialdehyde or glutaraldehyde (Across Organics, Illkirch, France); (ii) tosylchloramide or *N*-chloro tosylamide, named chloramine-T (Merk Chimie, Fontenay-sous-Bois, France); (iii) 2-bromo-2-nitropropane-1,3-diol, named bronopol (Sigma, Saint Quentin Fallavier, France); (iv) Incimaxx Aquatic® (i.e., a mix of peroctanoïc acid, peracetic acid and hydrogen peroxide, 7 g/L, 83 g/L and 55 g/L, respectively; ECOLAB Food and Beverage Division, Issy-les-Moulineaux, France); and (v) hydrogen peroxide (Merck). They were freshly prepared as stock solutions by dilution in water: (i) glutaraldehyde (50,000 ppm or 0.5 mole/L); (ii) chloramine-T (50,000 ppm or 0.22 mole/L); (iii) bronopol (50,000 ppm or 0.25 mole/L); (iv) Incimaxx Aquatic® (830 ppm or 11 mmole/L in equivalent peracetic acid); and (v) hydrogen peroxide (50,000 ppm or 1.47 mole/L). All tested concentrations of biocides are expressed in ppm.

Antimicrobial assays

Minimal inhibitory concentrations MICs were determined by the broth micro-dilution method in 96-well microtiter plates (Corning's Life Science, Costar N° 3370, Grosseron, France) with *F. psychrophilum* grown to early-exponential phase (optical density at 600 nm [OD_{600}] = 0.020). Aliquots of the cell suspension (10 µL; about 50×10^6 bacteria / mL) were cultured *in triplicate* in 200 µL of two-fold serial dilutions of disinfectant in FLP medium placed in wells of 96-well microtitration plates. Growth and sterility controls were included for each isolate. Microtiter plates were incubated at 14°C and the growth was spectrophotometrically monitored (OD_{600}) for four days in a Dynex MRX-II Microplate Reader (Dynex Technologies, France). The MIC was defined as the lowest concentration of disinfectant in which no absorbance change was recorded over a 3-days period. Alternatively, the minimum bactericidal concentration (MBC) was determined to evaluate the cell viability after disinfectant treatments. Culture aliquots (0.5 ml; $OD_{600} = 0.05$) of *F. psychrophilum* were exposed for 0 to 40 min at room temperature to one of the five disinfectants assayed. Chemical agents were removed by centrifugation ($5,000 \times g$, 10 min, 4°C) and the cells were washed twice in PBS (50 mmol/L sodium phosphate buffer, 150 mmol/L NaCl, pH 7.4) and dispersed in PBS. Negative control (i.e. 100% of bacterial viability) was 0.5 mL of the working *F. psychrophilum* suspension untreated with any biocide while positive control (i.e. 100% of bacterial mortality) was 0.5 mL of the working *F. psychrophilum* suspension heat-treated at 95°C for 5 min using a standard laboratory heat block. No growth was observed after 5 days at 14°C when 50 µL of this suspension was spread on FLP solid medium. Then, the bacterial suspensions were subjected to EMA (ethidium monoazide bromide or phenanthridium, 3-amino-8-azido-5-ethyl-6-phenyl bromide, Sigma, Saint Quentin Fallavier, France) treatment to evaluate viable/dead cells according to Nocker and Camper [22]. Briefly, EMA dissolved in water (5 mg /mL) was added to *F. psychrophilum* suspensions to a final concentration of 2 µg/mL. A first 10-min incubation step in dark allowed to EMA to interact with DNA from permeabilized cells, only. The photoinduced cross linking EMA-DNA step was obtained by light exposition (2 cycles of 60 sec; 650 Watts halogen lamp) of samples on ice to avoid excessive heating. After EMA-treatment, cells were washed twice and dispersed in water for DNA analysis. The PCR experiment was performed with the *F. psychrophilum*-specific set of primers (Fp_16S1_fw and Fp_16Sint1_rev; [20]) and the amplified DNA was further analyzed by electrophoresis in a 2% agarose gel; the expected size of the amplicons was confirmed by comparison with DNA molecular weight markers (50 bp DNA step ladder, Promega, Charbonnières,

France). The EMA-qPCR reactions were performed *in triplicate* with a MX3000p Stratagene thermocycler (Agilent Technologies, Massy, France) as previously described [20]. Data were expressed as quantities of viable bacterial cells (± SD).

Determination of D- and Z-values

For all chemical agents tested in this study, D-value (the decimal reduction time, min) was defined as the exposure time required causing 90% (= one decimal logarithm, i. e., one $\log_{10}$) reduction of the initial population of *F. psychrophilum* cells, under specified concentrations. Consequently, the initial population of bacterial (about 1-10×10^7 cells) exposed to the chemical compound at time zero was quantified by EMA-qPCR as well as the survivors at time 10, 20, 30 and 40 min. Residual living cells were expressed as percentage of the initial population. D-values were determined from the negative reciprocal of the slopes of the regression lines using $\log_{10}$-transformed percentage of survivors *vs* time of exposure to the biocide solution [i. e., $\log_{10} N_S/N_O \times 100 = f(time)$, where N_S is surviving population and N_O is initial population]. Z value was defined as the increase in the concentration of a given biocide necessary to reduce the time of exposure to this biocide by a factor 10 (= one $\log_{10}$ reduction of the time). Practically, Z-value was determined from the negative reciprocal of the slope of the regression line using $\log_{10}$-transformed D-values *vs* the biocide concentrations [i.e., $\log_{10}$D-value = f([Disinfectant])].

Disinfection assays on eyed eggs

Triploid rainbow trout eggs were treated five days before hatching with one of the five disinfectants. Treatments were: (i) bronopol (50, 500 or 2,000 ppm), (ii) hydrogen peroxide (40, 1,000 and 2,500 ppm), (iii) glutaraldehyde (300, 1,500 and 2,000 ppm), (iv) Incimaxx Aquatic® (10, 150 and 185 ppm), and (v) choramine-T (50, 600 and 800 ppm). Untreated eggs were used as controls. Three replicate groups of 200 eggs were disinfected for each treatment in 500 mL beaker (400 mL of disinfectant solution). After a 20-min chemical treatment, eggs were rinsed in fresh hatchery water and placed on shelf (170 × 90 × 40 mm) as shown in Figure 1A and 1B, allowing the hatch and 5-7 days later, the fry development over a 5-weeks period (Figure 1C). Eggs and subsequent fry were daily observed after the disinfection step. Water parameters were the followings: (i) temperature: 12°C; (ii) pH 6.5; (iii) hardness below 3°fH. The cumulative percent mortality (CPM) was determined after 20 days, and the relative percent survival (RPS) was calculated using the following equation:

$$RPS = [1 - (CPM \text{ of disinfected eggs/fry)/(CPM of control eggs/fry)}] \times 100$$

Values from experiments (*n*=3 per treatment) were expressed as mean ± SE.

Statistical analysis

GraphPad PRISM® (GraphPad Software, USA) was used to analyze data from disinfection assays on eyed eggs. The significance of cumulative percentage mortality of eyed eggs/fry/juvenile fish was analyzed using a 1-way ANOVA and comparisons of all chemical treatments *vs* control were performed by Bonferroni's Multiple Comparison Test. Differences were considered significant at P < 0.05.

Results

Effectiveness of disinfectants to prevent the *in vitro* growth of *F. psychrophilum*

Glutaraldehyde, chloramine-T, bronopol, Incimaxx Aquatic® and

hydrogen peroxide were individually assayed to assess the capability of such products to inhibit the growth of five *F. psychrophilum* strains (Table 1). All of them were effective to control *F. psychrophilum* growth; among them, bronopol was the most effective with MICs less than 6.3 ppm. MICs recorded for the four other biocides ranged from 3.1-62.5 ppm for hydrogen peroxide, 160-800 ppm for glutaraldehyde, 31.5-1,000 ppm (eq. peracetic acid) for Incimaxx Aquatic® and 156-313 ppm for chloramine-T, respectively. Very slight differences in sensitivities to disinfectants were recorded with individual *F. psychrophilum* cultures suggesting possible different physiological states of the bacterial starting cells.

Optimization of EMA-qPCR to determine anti-*F. psychrophilum* susceptibility

The viable qPCR was used to quantify the susceptibility of *F. psychrophilum* to disinfectants. The effectiveness of 20 min-hydrogen peroxide exposition time was first tested by classical PCR with the reference strain JIP02/86 and five freshly isolates (Figure 2). The PCR amplification of DNAs using Fp_16S1_fw and Fp_16Sint1_rev primers generated an expected 146 bp-product from all untreated *F. psychrophilum* strains in the absence of EMA. Unlike this, no PCR product was observed from 95°C-heated cells in the presence of EMA indicating that most if not all of the bacterial cells were permeabilized in the course of the heat-treatment. Hydrogen peroxide displayed contrasted efficiencies when tested against the six strains

Figure 1: Pictures of the experimental device used to test the impact of different disinfectants on different batches of rainbow trout eggs.

(A) White arrows underline the water circulation inside all batches.
(B) A set of 200 eggs deposed onto shelf just after one biocide treatment.
(C) A set of survival fry 5-weeks post-treatment.

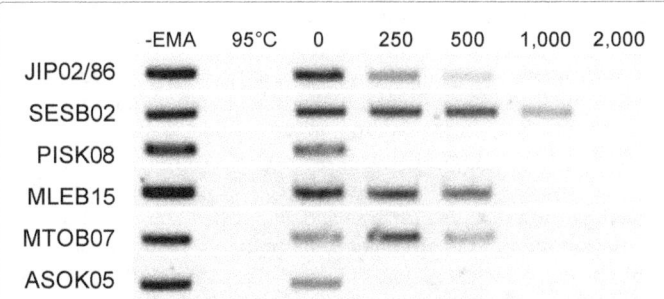

Figure 2: PCR products obtained by PCR using *F. psychrophilum*-specific primer set from EMA treated *F.psychrophilum* cells.

The cells had been previously exposed 20 min to hydrogen peroxide 250, 500, 1,000 or 2,000 ppm (250-2,000, respectively) or not-exposed (0). Positive and negative controls were EMA-untreated cells (-EMA) and 95°C-heated cells (95°C), respectively. *F. psychrophilum* strains were the type-strain JIP02/86 and five other strains freshly isolated from five French farms.

of *F. psychrophilum*. The strains PISK08 and ASOK05 were shown to be highly sensitive to the action of hydrogen peroxide because no amplification occurred for the weakest hydrogen peroxide concentration assayed here (250 ppm). On the other hand, the strain SESB02 exhibited a high resistance to hydrogen peroxide (1,000 ppm) while the three other strains, including the reference strain JIP02/86, were sensitive to concentrations above 500 ppm. Taking into account these results, *F. psychrophilum* JIP02/86 as moderately susceptible strain and reference strain was used to assess *F. psychrophilum* viability by EMA-qPCR after exposure to these biocides.

The next step was to evaluate the potential use of viable qPCR to investigate the hydrogen peroxide capability to kill *F. psychrophilum*. Stress gradients were tested in a preliminary screening over an assay period of 40 min with hydrogen peroxide concentrations ranging from 500 to 2,500 ppm (Figure 3). Increasing stress resulted in an increasing loss in *F. psychrophilum* viability during the first 20-minutes with a maximal three $\log_{10}$ unit reduction in the presence of 2,500 ppm hydrogen peroxide. This observation indicates clearly that not only the membrane integrity of *F. psychrophilum* was compromise by hydrogen peroxide treatment but so the effects observed were stress-dependant (i.e., the tested concentrations and the exposure times). Cell viability recorded after an incubation time above 20 minutes were not included in the exponential portions of the survivor curves for each of the hydrogen peroxide concentrations assayed and consequently not consistent with those observed at 10 and 20 minutes. A possible EMA oxidation with hydrogen peroxide could not be ruled-out; such a chemical alteration may have consequences in the EMA-capability to correctly-interact with DNA and therefore to inhibit further DNA amplifications. Based on this observation, an exposure time of 20 minutes only was selected in subsequent assays with other studied biocides considering that periods of 30 or 40 minutes were too long.

Bactericidal susceptibility of disinfectants against *F. psychrophilum*

The effectiveness of five antibacterial compounds was evaluated by of EMA-qPCR (Table 2). For each of the biocide concentrations tested, D-values were determined in order to calculate the Z-values. Data obtained with bronopol are shown for illustration in Figure 4. A reduction concentration-dependent in the *F. psychrophilum* population was recorded with bronopol treatments; about 0.8 to 3.0 $\log_{10}$ reduction units were observed in the course of 20-min treatments with bronopol 500 ppm to 2,500 ppm, respectively (Figure 4A). D-values derived from slopes of regression lines were found to be included between 25.7 min (bronopol 500 ppm) to 7.2 min only (bronopol 2,500 ppm) (Table 2). Z-value corresponding to bronopol was determined as 3,533 ppm (Figure 4B) indicating that theoretically, one treatment of *F. psychrophilum* with bronopol about 4,000 ppm (i. e., 500 + 3,533 ppm) is required to reduce the exposure time from 25.7 min to 2.6 min with same efficiencies against this pathogen. Alternatively, treatment durations corresponding to 5-$\log_{10}$ reduction in viability were included in a range of about two hours for bronopol 500 ppm to 36 min for bronopol 2,500 ppm. Similarly, D-values and Z-values were calculated for the four other chemical compounds as described above for bronopol (Table 2 and Figure 5). A 14-min exposition to hydrogen peroxide 500 ppm was needed to kill 90% of one suspension of *F. psychrophilum* while it was reduced to about 6 min with hydrogen peroxide 2,500 ppm. The theoretical Z-value was calculated to less than 5,900 ppm and 5-$\log_{10}$ reductions in bacterial viability were observed for treatment expositions between 1 h 30 min to 33 min for hydrogen peroxide 500 ppm and 2,500 ppm, respectively.

The glutaraldehyde treatments assayed (500 to 3,000 ppm) required exposition durations comprised between 1 h to 7 min to kill 90% of a *F. psychrophilum* population while they were between 5 h to less than 40 min to observe a 5-$\log_{10}$ reduction in viability. Z-value was estimated to about 3,000 ppm for glutaraldehyde. Similar efficient exposition times were observed for treatments with the peroctanoïc acid, peracetic acid, hydrogen peroxide based product (i.e., Incimaxx Aquatic˚) as well as with chloramine-T. However, the Incimaxx Aquatic˚ concentrations assayed here were 100 to 250 ppm eq. peracetic acid, only. The theoretical Z-value was calculated to about 200 ppm eq. peracetic acid. The efficient concentrations of chloramine-T were between 400 to 1,200 ppm and the Z-value was evaluated to 775 ppm.

Viability of rainbow trout eggs/fry after immersion disinfections

Triplicate groups of 200 rainbow trout eyed eggs were used in the study to evaluate the egg survival to hatch as well as the fry development after one 20-min period of egg chemical treatments (five disinfectants; three concentrations each). Data of viability studies collected at 5 weeks post-disinfection are summarized in Table 3. The viability of eggs/fry was not significantly impacted by disinfection treatments with four over the five chemical products used in this study. Indeed, no excess mortality was recorded when the eggs were treated with bronopol, hydrogen peroxide, Incimaxx Aquatic˚ as well as chloramine-T at the concentrations tested. Positive RPS values observed in the presence of bronopol and hydrogen peroxide suggest a protective effect of the two biocides relative to the negative control (e.g., untreated eggs) avoiding any multiplication of pathogens on the egg surface. On the other hand, negative RPS values were observed for eggs treated with Incimaxx Aquatic˚ at all concentrations tested. It could be that Incimaxx Aquatic˚ has a side effect on the eggs, very low when the disinfectant concentration is low and slightly stronger in the presence of higher concentrations. Similarly, a negative value RPS was obtained with chloramine-T 50 ppm while positive values were recorded for concentrations widely above 50 ppm suggesting the benefit of the disinfection. I could be that a treatment with chloramine-T 50 ppm was not included within the range needed for an efficient immersion disinfection of eggs. Unlike the previous observations, the viability of eggs/fry was clearly impacted by disinfection treatments with glutaraldehyde. Indeed, egg mortality was about 10% in glutaraldehyde 300 ppm, i.e. not different

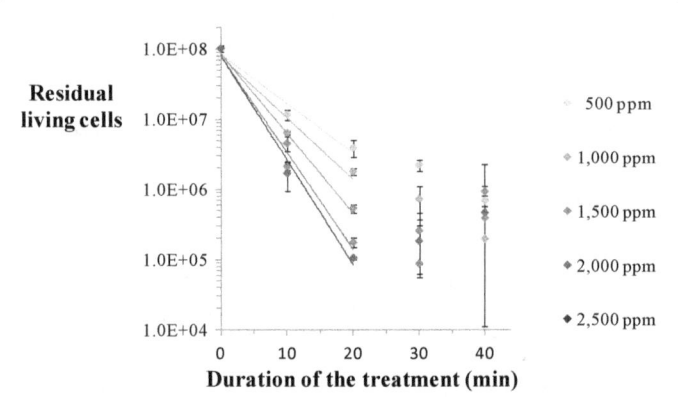

Figure 3: Impact of the hydrogen peroxide exposition on the *F. psychrophilum* JIP02/86 viability.

The bacterial cells were exposed to hydrogen peroxide 500, 1,000, 1,500, 2,000 or 2,500 ppm, treated with EMA and the residual living cells quantified by EMA-qPCR. Error bars represent standard deviation.

Chemical product	Concentration (ppm)	Duration of treatment[1]			D-value (min)	$t = n \times D$, $n = 5\text{-}\log_{10}$ (h:mm)	Z-value (ppm)
		0 min	10 min	20 min			
Bronopol	500	$2.92 \pm 0.79 \times 10^7$	$1.99 \pm 0.36 \times 10^7$	$4.86 \pm 1.45 \times 10^6$	25.7	2:08	
	1,000		$4.39 \pm 2.45 \times 10^6$	$1.88 \pm 1.12 \times 10^6$	16.8	1:24	
	1,500		$2.23 \pm 1.33 \times 10^6$	$4.68 \pm 0.43 \times 10^5$	11.1	0:56	3,571
	2,000		$1.28 \pm 0.57 \times 10^6$	$1.06 \pm 0.81 \times 10^5$	8.2	0:41	
	2,500		$4.32 \pm 3.00 \times 10^5$	$4.72 \pm 0.84 \times 10^4$	7.2	0:36	
Hydrogen peroxide	500	$10.2 \pm 0.90 \times 10^7$	$1.18 \pm 0.20 \times 10^7$	$3.97 \pm 1.10 \times 10^6$	14.2	1:11	
	1,000		$6.45 \pm 0.42 \times 10^6$	$1.78 \pm 0.18 \times 10^6$	11.4	0:57	
	1,500		$4.57 \pm 1.12 \times 10^6$	$5.36 \pm 0.82 \times 10^5$	8.8	0:44	5,882
	2,000		$2.11 \pm 0.26 \times 10^6$	$1.77 \pm 0.27 \times 10^5$	7.2	0:36	
	2,500		$1.69 \pm 0.76 \times 10^6$	$1.06 \pm 0.04 \times 10^5$	6.7	0:33	
Glutaraldehyde	500	$3.36 \pm 0.89 \times 10^7$	$2.23 \pm 1.98 \times 10^7$	$1.57 \pm 0.68 \times 10^7$	60.6	5:03	
	1,000		$7.36 \pm 1.62 \times 10^6$	$3.32 \pm 1.10 \times 10^6$	19.9	1:40	
	2,000		$8.27 \pm 3.03 \times 10^5$	$1.71 \pm 0.68 \times 10^5$	8.7	0:44	2,941
	3,000		$9.79 \pm 4.46 \times 10^4$	$8.25 \pm 4.74 \times 10^4$	7.7	0:38	
Incimaxx Aquatic® (eq. peracetic acid)	100	$1.19 \pm 0.34 \times 10^7$	$8.29 \pm 3.52 \times 10^6$	$4.93 \pm 0.43 \times 10^6$	52.4	4:22	
	150		$7.62 \pm 0.78 \times 10^6$	$3.29 \pm 1.46 \times 10^5$	12.8	1:04	194
	200		$1.66 \pm 0.70 \times 10^5$	$5.91 \pm 3.94 \times 10^4$	8.7	0:43	
	250		$1.39 \pm 1.07 \times 10^5$	$4.48 \pm 2.94 \times 10^4$	8.2	0:41	
Chloramine-T	400	$7.91 \pm 2.85 \times 10^7$	$4.90 \pm 1.22 \times 10^7$	$3.62 \pm 0.96 \times 10^7$	58.8	4:54	
	600		$2.08 \pm 0.51 \times 10^7$	$1.08 \pm 0.04 \times 10^7$	23.1	1:56	
	800	$8.27 \pm 4.81 \times 10^7$	$3.41 \pm 0.19 \times 10^6$	$4.39 \pm 1.07 \times 10^6$	15.7	1:18	775
	1,000		$3.42 \pm 0.52 \times 10^5$	$1.83 \pm 1.50 \times 10^5$	7.5	0:38	
	1,200		$1.09 \pm 1.32 \times 10^5$	$1.38 \pm 0.76 \times 10^4$	5.3	0:26	

[1]*Flavobacterium psychrophilum* suspensions were treated with the indicated chemical compound and the viability of the bacterial cells was evaluated by EMA-qPCR as described in the Material and Method section.

Table 2: Impact of chemical treatments on *F. psychrophilum* JIP02/86 viability.

Chemical products	Concentration (ppm)	Viability[1]	
		CPM[2] % (mean ± SE)	RPS[3] (%)
Control (PBS)		$11.36^a \pm 2.64$	0.00
Bronopol	50	$11.10^a \pm 3.60$	2.3
	500	$10.50^a \pm 2.82$	7.5
	2,000	$8.50^a \pm 3.56$	25.2
Hydrogen peroxide	40	$9.60^a \pm 2.77$	15.5
	1,000	$8.30^a \pm 2.14$	26.9
	2,500	$10.10^a \pm 3.12$	11.1
Glutaraldehyde	300	$10.30^a \pm 4.46$	9.3
	1,500	$27.30^a \pm 13.83$	-140.4
	2,000	$44.50^b \pm 15.83$	-291.8
Incimaxx Aquatic®	10	$11.60^a \pm 5.15$	-2.1
	150	$12.60^a \pm 4.98$	-10.9
	185	$14.30^a \pm 4.04$	-25.9
Chloramine-T	50	$14.10^a \pm 5.85$	-24.2
	600	$9.40^a \pm 3.75$	17.2
	800	$8.50^a \pm 2.95$	25.2

[1]Rainbow trout eggs (5 days before hatching) were treated 20 minutes with the corresponding chemical product and viability of eggs/fry was recorded over a 5-weeks period.
[2]CPM, the cumulative percentage mortality (n=5 per treatment).
[3]RPS, the relative percentage survival; it was determined relative to PBS treatment.
Mean CPM values with different superscripts indicate significant difference at $p < 0.05$.

Table 3: Impact of chemical treatments on rainbow trout eggs/fry viability.

to that observed for untreated eggs, while it was about 30% and 45% in glutaraldehyde 1,500 ppm and 2,000 ppm, respectively. However, the only significant difference was found between the treatment with glutaraldehyde 2,000 ppm and negative control and all almost other chemical treatments assayed in this work.

Discussion

Bronopol has been shown to be effective in protection against parasites infecting rainbow trout such as *Saprolegnia parasitica*, when administered as a daily bath/flush treatment at concentrations of 15 ppm and greater [23] or by *Ichthyophthirius multifiliis*, when exposed as long, low doses (24 h; 1 ppm) as well as short, high doses (30 min;

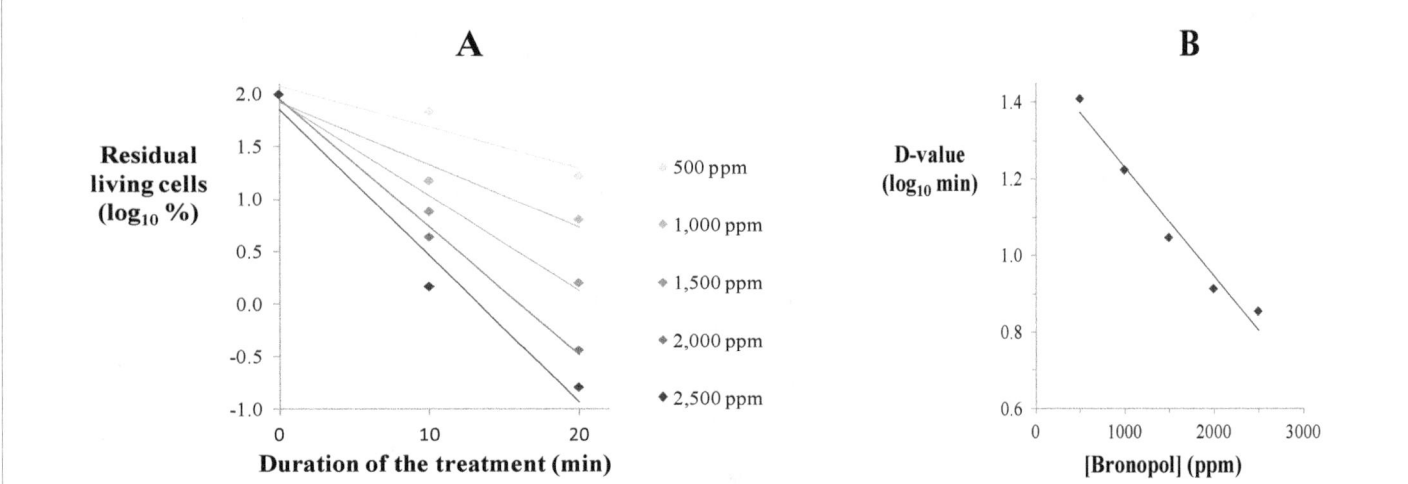

Figure 4: (A) Impact of the bronopol exposition on the *F. psychrophilum* JIP02/86 viability. The bacterial cells were exposed to different concentrations of bronopol, treated with EMA and the residual living cells were evaluated by qPCR. **(B)** D-values (i.e., the exposition time needed to destroy 90% of the initial cells) were determined for each of the bronopol concentrations assayed and were $\log_{10}$–transformed. A correspondence curve was drawn where each data point represents $\log_{10}$D-value determined for the corresponding bronopol concentration.

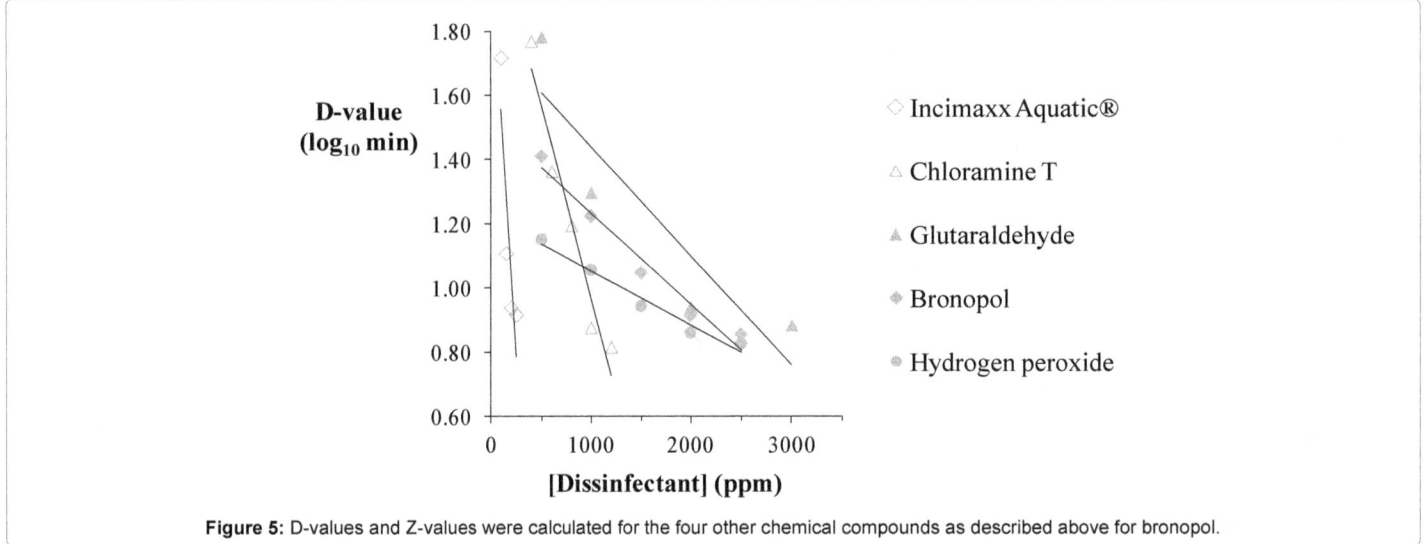

Figure 5: D-values and Z-values were calculated for the four other chemical compounds as described above for bronopol.

100 ppm). Accordingly with Birkbeck et al. [24] who determined MIC of bronopol for 13 bacterial pathogens isolated from marine fish, bronopol MIC ranging from 1.6 to 6.3 ppm were observed for the *F. psychrophilum*. However, none of them were bactericidal because *F. psychrophilum* growth was observed when samples of each mixture were taken two days post-treatment and spread on FLP-agar plate (Data not-shown). Consequently, bactericidal conditions [(i) concentration and (ii) treatment time] for bronopol were determined using the EMA-qPCR approach to identify the effective biocide conditions leading to the permeabilization of *F. psychrophilum* cells. Much higher bronopol concentrations than previously described were needed to observe an impact on the flavobacterial viability. Indeed, one decimal reduction of the viability was observed for treatments ranging from 500 to 2,500 ppm, for 26 to 7 min (D values), respectively. Such high concentrations or bronopol have been already recorded as effective for disinfection. Indeed, a bronopol concentration of 500 ppm was required for surface disinfection of Haddock eggs to achieve a significant reduction in bacterial numbers [25]. Such bronopol doses did not show any side effects on the rainbow trout egg survival: (i) the eggs viability was not

affected by 20 min-long baths in bronopol up to 2,000 ppm; and (ii) the hatching time was not modified significantly compared with controls. Similar observations have been recently made on the insensitivity of crustacean eggs at high bronopol concentrations (until 3,000 ppm administered for 15 min every second day on crayfish eggs) [26].

Chloramine-T is a biocide used worldwide as a disinfectant and antiseptic. Although, it is used in aquaculture against bacterial and protozoal infections eliciting little or no response of oxidative stress biomarkers from *Oncorhynchus mykiss* when exposed to chloramine-T 10 ppm for 20 min (3 days; 3 expositions per day; [27]), it is not yet approved by the U.S Food and Drug Administration (FDA) for use on fish [28]. Previously, chloramine-T 15 ppm was found to be effective within 10 min against some *Aeromonas* spp. bacteria (*A. hydrophila* and *A. salmonicida* subsp. *salmonicida*) while 60 min of contact time was required to be effective against *A. salmonicida* subsp. *achromogenes* [29]. In this work, *F. psychrophilum* is much more resistant to chloramine-T since 300 ppm were needed to inhibit the bacterial growth and only severe chloramine-T treatments ranging from 400 to

1,200 ppm (for about 1 hr to 5 min, respectively) were bactericidal. Experiments performed to evaluated chloramine-T as possible candidate for approval for use to control mortality in freshwater-reared salmonids caused by bacterial gill disease have been shown that the fry and fingerlings viability was unaffected by exposure to concentrations less than 100 and 60 ppm, respectively [30]. Clearly, rainbow trout eggs are much more resistant to chloramine-T since none of the 20-min treatments of chloramine-T up to 1,200 ppm did reduce significantly the egg viability.

Hydrogen peroxide as a strong oxidizing agent is widely used by aquaculture industry to treat fungal infections of fish with recommended concentration for bath treatments (500 ppm for 20 min; [31]) as well as to sanitize fish eggs for concentrations ranging from 500 until to 30,000 ppm for few minutes to 60 min [32,33]. While hydrogen peroxide is a useful and environmental friendly biocide, it can promote or boost, in some cases, fish infections (e.g., *Tenacibaculum maritimum* in turbot [34]). This has been smartly used to pre-stress rainbow trout fry with peroxide hydrogen (until 200 ppm; 60 min) to obtain a reproducible immersion model of *F. psychrophilum* infection [35]. Less than hydrogen peroxide 65 ppm was needed to specifically inhibit the growth of the six *F. psychrophilum* strains assayed here, while bactericidal effects on *F. psychrophilum* JIP02/86 were observed from hydrogen peroxide treatments ranging from 50 to 2,500 ppm for less than 15 min to 7 min, respectively. None of these hydrogen peroxide treatments (from 40 to 2,500 ppm; 20 min) on rainbow trout eyed eggs did modify significantly neither the hatching time nor the fry viability or deformity rates. Our data are in agreement with those reported by the Wagner's group. Indeed, short expositions of eyed eggs to hydrogen peroxide 30,000 ppm (1 min) or 6,000 rpm (5 min) reduced well the bacterial load on eggs but did not affect the subsequent development of eggs [15]. Because high hydrogen peroxide concentrations used, attention was focused on the need to maintain hydrogen peroxide solutions to pH values close to neutrality with additions of $NaHCO_3$. More recently, experiments with different hydrogen peroxide treated groups of trout eggs (from 10,000 ppm for 2 min to 500 ppm for 35 min) confirmed that mortalities did not significantly differ to that in untreated eggs; they pointed-out also that the bacterial abundance on control eggs was higher than treated eggs, with a prevalence of yellow colonies, possibly *F. psychrophilum* [33]. Taken into the whole, these results underline the effectiveness of peroxide against *F. psychrophilum* and its safety in disinfection process of rainbow trout eggs.

Incimaxx Aquatic˚ has been shown as promising formulation in aquacultural systems since at weaker doses than 8 ppm eq. peracetic acid, it is effective to control the free-living stages of the parasitic protozoae *Ichthyophthirius multifiliis* [36]. Due to the strong bactericidal, virucidal, fungicidal, and sporicidal activities, peracetic acid has been used as an effective therapeutic treatment against some fish pathogens including *Flavobacterium columnare* [37]. Reduction of the *in vitro* growth was observed for *F. columnare* with increasing peracetic acid concentration ranging from 1 to 10 ppm and concentrations higher than 15 ppm have been shown to be toxic for channel catfish eggs [38]. Unlike this, our findings underline the high resistance of *F. psychrophilum* to Incimaxx Aquatic˚ since no growth inhibition was observed from 31.2-125 ppm (eq. peracetic acid) and reduction of viability was observed for Incimaxx Aquatic˚ incubations 100-250 rpm for about one hour to 8 min, respectively. No toxicity for rainbow trout eggs was recorded by 20 min-long Incimaxx Aquatic˚ treatments up to 185 ppm. Due to the acidic character of Incimaxx Aquatic˚, it should be stressed that the water hardness has to be monitored and adjusted close to the neutrality to avoid deleterious effect of acidosis [39].

Glutaraldehyde is routinely considered as egg surface disinfectant for aquaculture. Doses of glutaraldehyde between 400–800 ppm for 5-10 min have been shown to improve the hatchability and larval survival in egg batches of Atlantic halibut when used for disinfection of egg surface [40]. However, these authors have cautioned that concentrations and contact times should be evaluated if disinfection with glutaraldehyde is to be applied to other fish species. Here, we found that such doses were bacteriostatic concentrations for *F. psychrophilum* and they were shown bactericidal *in vitro* (i.e., one $\log_{10}$ reduction of bioburden, only) for contact times ranging from 20 to 60 min. Less than 10 min were needed to observe similar effectiveness for glutaraldehyde 2,000-3,000 ppm. However, a 20-min treatment with glutaraldehyde 2,000 ppm did reduce significantly the survival of the rainbow trout eggs. Glutaraldehyde failed to reduce the survival of *F. psychrophilum* at concentrations that were safe for the rainbow trout eggs and some concerns about user safety (i.e. the possible long-term exposure for handlers causing irritations of the eyes, nose, throat, and skin and its potential adverse effects on the aquatic environment, [41]). Therefore there is a need for alternative methods using other chemicals for egg disinfection leading to the elimination of *F. psychrophilum*.

In conclusion, successful aquaculture requires knowledge of toxicity of applied chemicals, and acute toxicity tests are considered essential in selecting appropriate parameters of therapeutic baths. Here, we developed a new toolkit for assistance to fish farmers for the choice of effective chemicals against *F. psychrophilum* used to treat eyed eggs. However, treatments proposed in field condition have been employed in a particular situation of water quality (i.e., pH, temperature, hardness, organic matter.) and consequently they need to be adapted to different water qualities.

Acknowledgements

We would like to thank (i) Vincent Daubigné (Aqualande) and Sébastien Castillon (Aqualande/Viviers de France) for their contribution in management of eggs/fry in hatchery section, (ii) Simon Menanteau-Ledouble, Bastien Ipas and Annie Richard for their technical assistance and (iii) Alain Rives for his contribution for the design/manufacture of the investigational device to test egg/fry viability. This study was financially supported by grants from 'Région Aquitaine'/European Regional Development Fund (ERDF), 'Comité Interprofessionnel des Produits de l'Aquaculture' (CIPA) and by 'Bordeaux Sciences Agro'.

References

1. Borg AF (1960) Studies on myxobacteria associated with diseases in salmonid fishes. American Association for the advancement of science, Wildlife Disease, N° 8, Washington, DC.

2. Iida Y, Mizokami A (1996) Outbreaks of coldwater disease in wild ayu and pale chub. Fish Pathol 31: 157-164.

3. Bernardet JF, Bowman JP (2006) The Procaryotes, an evolving electronic resource for the microbiological community: the genus Flavobacterium. (3rdedn), Springer-Verlag, New-York.

4. Nilsen H, Olsen AB, Vaagnes O, Hellberg H, Bottolfsen K, et al. (2011) Systemic Flavobacterium psychrophilum infection in rainbow trout, Oncorhynchus mykiss (Walbaum), farmed in fresh and brackish water in Norway. J Fish Dis 34: 403-408.

5. Barnes ME, Brown ML (2011) A review of Flavobacterium psychrophilum biology, clinical signs, and bacterial cold water disease prevention and treatment. Open Fish Sci J 4: 40-48.

6. Dumetz F, Duchaud E, LaPatra SE, Le Marrec C, Claverol, S, et al. (2006) A protective immune response is generated in rainbow trout by an OmpH-like surface antigen (P18) of Flavobacterium psychrophilum. Appl Environ Microbiol 72: 4845-4852.

7. Dumetz F, LaPatra SE, Duchaud E, Claverol S, Le Hénaff M (2007) The Flavobacterium psychrophilum OmpA, an outer membrane glycoprotein, induces a humoral response in rainbow trout. J Appl Microbiol 103: 1461-1470.

8. LaFrentz BR, LaPatra SE, Call DR, Wiens GD, Cain KD (2011) Identification of

immunogenic proteins within distinct molecular mass fractions of *Flavobacterium psychrophilum*. J Fish Dis 34: 823-830.

9. Fredriksen BN, Olsen RH, Furevik A, Souhoka RA, Gauthier D, et al. (2013) Efficacy of a divalent and a multivalent water-in-oil formulated vaccine against a highly virulent strain of *Flavobacterium psychrophilum* after intramuscular challenge of rainbow trout (*Oncorhynchus mykiss*). Vaccine 31: 1994-1998.

10. Plant KP, LaPatra SE, Call DR, Cain KD (2014) Attempts at validating a recombinant *Flavobacterium psychrophilum* gliding motility protein N as a vaccine candidate in rainbow trout, *Oncorhynchus mykiss* (Walbaum) against bacterial cold-water disease. FEMS Microbiol Lett 358: 14-20.

11. Michel C, Kerouault B, Martin C (2003) Chloramphenicol and florfenicol susceptibility of fish-pathogenic bacteria isolated in France: comparison of minimum inhibitory concentration, using recommended provisory standards for fish bacteria. J Appl Microbiol 95: 1008-1015.

12. Kumagai A, Nawata A (2010) Prevention of *Flavobacterium psychrophilum* vertical transmission by iodophor treatment of unfertilized eggs in salmonids. Fish Pathol 45: 164-168.

13. Escaffre AM, Bazin D, Bergot P (2001) Disinfection of *Sparus aurata* eggs with glutaraldehyde. Aquacult Int 9: 451-458.

14. Katharios P, Agathaggelou A, Paraskevopoulos S, Mylonas CC (2007) Comparison of iodine and glutaraldehyde as surface disinfectants for red porgy (*Pagrus pagrus*) and White Sea bream (*Diplodus sargus sargus*) eggs. Aquac Res 38: 527-536.

15. Wagner EJ, Oplinger RW, Arndt RE, Forest AM, Bartley M (2010) The safety and effectiveness of various hydrogen peroxide and iodine treatment regimens for rainbow trout egg disinfection. N Am J Aquacult 72: 34-42.

16. Wagner EJ, Oplinger RW, Bartley M (2012) Evaluation of tannic acid for disinfection of rainbow trout eggs. N Am J Aquacult 74: 80-83.

17. Can E, Karacalar U, Saka S, Firat K (2012) Ozone disinfection of eggs from gilthead seabream *Sparus aurata*, sea bass *Dicentrarchus labrax*, red porgy, and common dentex *Dentex dentex*. J Aquat Anim Health 24: 129-133.

18. Wagner EJ, Randall W (2013) Toxicity of copper sulfate to *Flavobacterium psychrophilum* and rainbow trout eggs. J Aquat Anim Health 25: 125-130.

19. Del Cerro A, Mendoza MC, Guijarro JA (2002) Usefulness of a TaqMan-based polymerase chain reaction assay for the detection of the fish pathogen *Flavobacterium psychrophilum*. J Appl Microbiol 93: 149–156.

20. Orieux N, Bourdineaud JP, Douet DG, Daniel P, Le Hénaff M (2011) Quantification of *Flavobacterium psychrophilum* in rainbow trout *(Oncorhynchus mykiss)* tissues by qPCR. J Fish Dis 34: 811-821.

21. Siekoula Y (2012) Etude de la variabilité génétique de *Flavobacterium psychrophilum*, pathogène de salmonidés. Thèse de doctorat. Université de Nantes.

22. Nocker A, Camper AK (2006) Selective removal of DNA from dead cells of mixed bacterial communities by use of ethidium monoazide. Appl Environ Microbiol 72: 1997-2004.

23. Pottinger TG, Day JG (1999) A *Saprolegnia parasitica* challenge system, for rainbow trout: assessment of Pyceze as an anti-fungal agent for both fish and ova. Dis Aquat Organ 36: 129-141.

24. Birkbeck TH, Reid HI, Darde B, Grant AN (2005) Activity of bronopol (Pyceze®) against bacteria cultured from eggs of halibut, *Hippoglossus hippoglossus* and cod, *Gadus morhua*. Aquaculture 254: 125-128.

25. Treasurer JW, Cochrane E, Grant A (2005) Surface disinfection of cod *Gadus morhua* and haddock *Melanogrammus aeglefinus* eggs with bronopol. Aquaculture 250: 27-35.

26. Gonzalez A, Celada JD, Melendre PM, Carral JM, Saez-Royuela M, et al. (2013) Effects of different bronopol treatments on final survival rates in the artificial incubation of crayfish eggs (*Pacifastacus leniusculus*, Astacidae). Aquac Res 44: 354-358.

27. Tkachenko H, Kurhaluk N, Grudniewska J (2015) Biomarkers of oxidative stress and antioxidant defences as indicators of different disinfectants exposure in the heart of rainbow trout (*Oncorhynchus mykiss* Walbaum). Aquac Res 46: 679-689.

28. Bowker JD, Carty D, Trushenski JT, Bowman MP, Wandelear N, et al. (2013) Controlling mortality caused by external columnaris in largemouth bass and bluegill with Chloramine-T or Hydrogen Peroxide. N Am J Aquacult 75: 342-351.

29. Mainous ME, Kuhn DD, Smith SA (2011) Efficacy of common aquaculture compounds for disinfection of *Aeromonas hydrophila*, *A. salmonicida* subsp. *salmonicida*, and *A. salmonicida* subsp. *achromogenes* at various temperatures. N Am J Aquacult 73: 456-461.

30. Bowker JD, Carty D, Smith CE, Bergen SR (2011) Chloramine-T margin-of-safety estimates for fry, fingerling, and juvenile rainbow trout. N Am J Aquacult 73: 259-269.

31. Burridge L, Weis JS, Cabello F, Pizarro J, Bostick K (2010) Chemical use in salmon aquaculture: A review of current practices and possible environmental effects. Aquaculture 306: 7-23.

32. Wagner EJ, Oplinger RW, Bartley M (2012) Effect of single or double exposures to hydrogen peroxide or iodine on salmonid egg survival and bacterial growth. N Am J Aquacult 74: 84-91.

33. Wagner EJ, Oplinger RW, Bartley M (2012) Laboratory and production scale disinfection of salmonid eggs with hydrogen peroxide. N Am J Aquacult 74: 92-99.

34. Avendano-Herrera R, Magarinos B, Irgang R, Toranzo AE (2006) Use of hydrogen peroxide against the fish pathogen *Tenacibaculum maritimum* and its effect on infected turbot (*Scophthalmus maximus*). Aquaculture 257: 104-110.

35. Maria M, Henriksen M, Madsen L, Dalsgaard I (2013) Effect of hydrogen peroxide on immersion challenge of rainbow trout fry with *Flavobacterium psychrophilum*. PLoS ONE 8: e62590. doi:10.1371/journal.pone.0062590.

36. Picon-Camacho SM, Marcos-Lopez M, Beljean A, Debeaume S, Shinn AP (2012) In vitro assessment of the chemotherapeutic action of a specific hydrogen peroxide, peracetic, acetic, and peroctanoic acid-based formulation against the free-living stages of *Ichthyophthirius multifiliis* (Ciliophora). Parasitol Res 110: 1029-1032.

37. Marchand PA, Phan TM, Straus DL, Farmer BD, Stuber A, Meinelt T (2012) Reduction of in vitro growth in *Flavobacterium columnare* and *Saprolegnia parasitica* by products containing peracetic acid. Aquac Res 43: 1861-1866.

38. Straus DL, Meinelt T, Farmer BD, Mitchell AJ (2012) Peracetic acid is effective for controlling fungus on channel catfish eggs. J Fish Dis 35: 505-511.

39. Marchand PA, Straus DL, Wienke A, Pedersen LF, Meinelt T (2013) Effect of water hardness on peracetic acid toxicity to zebrafish, *Danio rerio*, embryos. Aquacult Int 21: 679-686.

40. Salvesen I, Oie G, Vadstein O (1997) Surface disinfection of Atlantic halibut and turbot eggs with glutaraldehyde: Evaluation of concentrations and contact times. Aquacult Int 5: 249-258.

41. Leung HW (2001) Ecotoxicology of glutaraldehyde: review of environmental fate and effects studies. Ecotoxicol Environ Saf 49: 26-39.

29

Effects of Supplementation Diet Containing *Microcystis* aeruginosa on Haematological and Biochemical Changes in *Labeo rohita* Infected with *Aeromonas hydrophila*

Jyotirmayee Pradhan[1] and Basanta Kumar Das[2]*

[1]*Government College (Autonomous), Angul-759143, Odisha, India*
[2]*Central Institute of Freshwater Aquaculture (CIFA), Kausalyaganga, Bhubaneswar-751 002, Odisha, India*

Abstract

An experiment of 100 days duration was conducted to test the effect of *Microcystis aeruginosa* in *Labeo rohita* against *Aeromonas hydrophila*. The rohu fingerlings (22 ± 2 g) were fed with the experimental diets incorporated with different concentrations of *M. aeruginosa* @ 0 g kg^{-1}, 0.5 g kg^{-1}, 1 g kg^{-1} and 5 g kg^{-1}. Replicate groups of fish were fed for three month daily @4% body weight. At an interval of 30 days blood and serum samples were assayed for different haematological [total erythrocyte count (TEC), total leucocyte count (TLC), haemoglobin] and biochemical [blood glucose, serum aspartate aminotransferase (AST), serum alanine aminotransferase (ALT) and alkaline phosphatase (ALP)] parameters. Significantly ($p \leq 0.05$) increased haemoglobin content and TLC was observed in *Microcystis* treated group. It was observed that the serum AST activity was significantly ($p \leq 0.05$) decreased to all the treated groups of fish as compared to control on entire assay period. Serum ALT activity was significantly ($p \leq 0.05$) different to all the treated groups of fish on day 60, on day 90 and on day 10 (except group B) bacterial post challenge as compared to control. After 90 days, fish were challenged with *A. hydrophila* and mortality (%) was recorded up to day 10 post challenge. Highest percentage of survival (72%) was noticed in the group fed 1.0 g *Microcystis* kg^{-1} dry diet. The present study suggests that the administration of bluegreen algae, *Microcystis aeruginosa* supplementation diets for 100 days protects the hematological and biochemical parameters in *L. rohita* from *A. hydrophila*.

Keywords: *Aeromonas hydrophila*; ALP; AST; Haemoglobin; *Microcystis aeruginosa*; *Labeo rohita*

Introduction

Nutrient supplementation in fish diets has been an economically promising method for improving the performance of different intensive fish production systems. To enhance a more economically sustainable aquaculture in the current millennium, many feed ingredient alternatives to fish meal at varying levels are now being sought. Thus functional feed additives strategy has recently gained considerable attention. From nutritional point of view, it does not only provide the essential nutrients required for normal physiological functioning, but also serve as a medium by which fish receive other components that may positively affect their health [1]. Several sectors of the aquaculture industry would benefit if cultured organisms were conferred with enhanced feed efficiency, growth performance, and disease resistance without environmental conflicts [2]. The other methods of preventing disease include immunostimulation through alteration of the diet and feeding practices [3]. In aquaculture, there are many studies reporting a variety of substances including bacterial [4], algal [5], animal and plant products [6-9] can be used as immunostimulants to enhance non-specific immune system of cultured fish species. With a detailed understanding of the efficacy and limitations of immunostimulants, they may become powerful tools to control fish diseases.

Microcystis aeruginosa (Kützing) is a unicellular, colonial blue-green alga (family Chroococcaceae [10]. Ingestion of waters containing high concentrations of *Microcystis* can cause abdominal stress in humans leading to precautionary beach closures and can kill dogs and farm animals if they drink significant quantities of the bloom waters. However, due to their adverse effects on higher organisms, several cyanobacterial metabolites are regarded as health-threatening toxins and have caused serious concern among water authorities worldwide. But *Microcystis* has been recognized in recent years as a producer

of a high number of secondary metabolites. *Microcystis aeruginosa* shows antibacterial and anti viral activity against the Gram-positive bacterium *Staphylococcus aureus*, some selected Gram-negative fish pathogens and influenza A virus [11-14]. It has been also reported that *Microcystis* stimulates the immunity and makes *L. rohita* more resistant to infection by *A. hydrophila* when fed in dried form in feed [15].

Fish should be fed with a balanced diet as nutritional deficiency can have an adverse impact on disease resistance. The analysis of blood indices has proven to be a valuable approach for analyzing the health status of farmed animals as these indices provide reliable information on metabolic disorders, deficiencies and chronic stress status before they are present in a clinical setting [16]. Haematological study is of immense importance when diagnostically evaluating fish health as in human health [17]. Enzymes are biochemical macromolecules that control metabolic processes of organisms, thus a slight variation in enzyme activities would affect the organism [18]. Thus, by estimating the enzyme activities in an organism, we can easily identify disturbances in its metabolism. In this study we monitored the disturbance of metabolism in fingerlings of *L. rohita* by feeding them

*****Corresponding author:** Das BK, Fish Health Management Division, Central Institute of Freshwater Aquaculture (CIFA),P. O. Kausalyaganga, Bhubaneswar-751 002, Odisha, India, E-mail: basantadas@yahoo.com

to *Microcystis* diet. This study was designed to evaluate the effect of feeding of *Microcystis* in the raw forms at different dietary levels on some haematological and biochemical profile of infected fish in *L. rohita* against *A. hydrophila*.

Material and Methods

Microcystis aeruginosa

Microcystis bloom was collected from Bindusagar, Bhubaneswar, India, with the help of plankton net made of bolting silk cloth (mesh size, 20 μ). Collected samples were washed three times in MilliQ (Millipore, USA) water to remove suspended particles adhered to it and finally filtered (Whatman Filter paper Size 40). Harvested *Microcystis* (514 g wet weight) was dried under room temperature for 2-3 days and the dry weight of the algae was 110 g. Then it was powdered. For each experiment, the required quantity of *Microcystis* powder was included in the feed.

Pathogen

Aeromonas hydrophila (ATCC 49140) was cultured in nutrient broth (Himedia) for 24 h at 37°C. The culture broth was centrifuged at 3000xg for 10 min. The supernatant was discarded and the pellet was resuspended in phosphate buffered saline (PBS, pH 7.4), and the OD of the solution was adjusted to 1.5 at 456 nm, which corresponded to 1×10^7 cells ml^{-1}. These bacterial suspensions were serially diluted using standard dilution technique with PBS and used for the challenge experiment.

Preparation of fish feed

The proximate composition of the basal diet was 39.4% crude protein, 7.4% lipids, 14.6% ash, 7.1% moisture and 3% fibre as per the method described by Misra et al. [19] (Table 1). Three experimental diets were formulated by including the required proportions of different feed ingredients such as ground nut oil cake, rice bran, fish meal, soyabean meal, vitamin & mineral mixture [7]. Powdered forms of *M. aeruginosa* were added to the above formulation at the rate of 0.1 g, 0.5 g and 1.0 g Kg^{-1} feed. Dry ingredients were mixed thoroughly and 1% binder was added. Water was added and mixed thoroughly in a mixer for 20 min. The resulting dough was pelleted, dried at room temperature for 48 h and then stored in airtight containers until fed.

Experimental design

For experiment, total 240 numbers of acclimatized rohu fingerlings were taken and divided into four groups (A, B, C and D) with feeding of *Microcystis* at the rate of 0, 0.5, 1.0 and 5.0 g kg^{-1} feed for 90 days. Duplicate tanks containing 30 numbers of fish were maintained for each group. After 90 days of feeding trail, 15 fish of duplicate tanks of each group were challenged intraperitonially with bacterial pathogen, *A. hydrophila* at the rate of 1×10^5 CFU per fish. Blood and serum samples were collected from each group and examined for the following parameter, such as total erythrocyte count, total leucocyte count, haemoglobin, glucose, serum aspartate amino transferase, serum alanine amino transferase and alkaline phosphatase.

Sampling

Feed was suspended from fish for 24h before blood samples were collected. From randomly picked fish (n=20 from each subgroup) at 30-day intervals, after anaesthetizing with 0.1 ppm MS-222. Blood was collected from the caudal vein with a 1-mL plastic syringe ringed with heparin and stored at 4°C and used the same day. Blood samples were also collected without heparin, allowed to clot, centrifuged at 7000 g and sera collected and refrigerated. From each group twelve and eight fishes were sampled for serum and blood, respectively, and kept in a separate tank. Sera and blood were pooled into six groups, depending upon volume, for estimation of immunological and biochemical parameters [15].

Determination of haematological parameters

The collected blood samples were immediately subjected to haematological analysis. The haemoglobin content of different blood samples collected was measured by cyanomethemoglobin method as per Van Kampen and Zijlstra [20]. Reagent solution (5 ml) containing potassium hexacyanoferrate (III) solution (potassium hexacyanoferrate III 0.6 mmol/l, potassium phosphate buffer 0.5 mmol/l; pH 7.2), potassium cyanide solution (potassium cyanide 0.75 mmol/lit, potassium phosphate buffer 2.5 mmol/l; pH 7.2 and detergent 0.1 g/l) was taken in a cleaned and dry test tube and then 0.002 ml of blood was added to it. Simultaneously a blank reading was taken. It was mixed thoroughly and incubated at 20-25°C. Absorbance was taken at an optical density 546 nm using the Bio-Rad Spectrophotometer. The haemoglobin was expressed as g%. The blood was diluted with appropriated diluting fluids for total erythrocyte count and total leucocyte counts were determined using improved Neubauer haemocytometer and calculated [21]. Replicated counts were made for each blood samples. The TEC and TLC count was expressed as cells per mm^3.

Determination of blood/serum biochemical Parameters

The different sera samples collected earlier were analysed for AST and ALT following the procedure of Wallnofer et al. [22] using diagnostic kits (Bayer Diagnostics, Baroda, India). Serum ALP was

Ingredients	Quantity included (g kg^{-1} diet)
Groundnut oil cake	400
Fish meal	250
Rice bran	200
Soya bean meal	120
Vitamins and minerals mixture	20
Starch	10
Composition	**Quantity (kg^{-1}) vitamin-mineral mixture**
Vitamin A	20 00 000 IU
Vitamin D3	4 00 000 IU
Vitamin B2	0.8 g
Vitamin E	300 IU
Vitamin K	0.4 g
Calcium pathtothernate	1 g
Nicotinamide	40 g
Vitamin B12	2.4 g
Choline chloride	60 g
Calcium	300 g
Mangenese	11 g
Iodine	0.4 g
Iron	3.0 g
Zinc	6.0 g
Copper	0.8 g
Cobalt	0.18 g

Calculated crude protein: 40%; estimated crude protein: 39.6%; calculated lipid content: 5%; estimated lipid content: 7.2%.

Supplies the above vitamin–mineral premix (Suplevite M) used in feed formulation per kg of feed were procured from Sarabhai Chemicals, Wadi, Baroda, India.

Table 1: Percentage inclusion of ingredients in basal diet with desired crude protein and lipid level [19].

determined by the procedure of Rosalki et al. [23]. Blood glucose content was estimated following the procedure of Schmidt [24] using standard kits (as per the manufacture's instructions; Roche diagnostic).

Challenge of fish

After 90 days of feeding, 15 fish from each subgroup was challenged intraperitoneally with a lethal dose of *A. hydrophila* (1×10^5 CFU per fish) and observed for a 10-day period for mortality. Cumulative mortality percentage was determined over 10 days. Haematological and biochemical parameters were assayed in post challenged groups of survived fish as per the methods described earlier.

Statistical analysis

All experiments were performed in duplicate. All statistical analyses were performed by the Statistical Analysis System (SAS) program package. Data were expressed as mean ± standard error (S.E.). The data were analyzed by the one-way analysis of variance (ANOVA) and mean differences among experimental groups were evaluated using Duncan's multiple range tests (DMRT) at the $p \leq 0.05$ significance level [25].

Results

Haematological indices

Significantly ($p \leq 0.05$) different haemoglobin content was observed in group of fish fed with 5 g kg⁻¹ *Microcystis* (D) on day 60, B on day 90 and C on day 10 bacterial post challenge period. However, the group of fish fed with *Microcystis* powder showed insignificant (p>0.05) difference haemoglobin level as compared to control on day 30 day of exposure period.

Insignificant difference (p>0.05) of TLC was found in *Microcystis* fed fish on day 30 of exposure period. The entire treatment group showed significantly ($p \leq 0.05$) different TLC as compared to control on day 90 and on day 100 (i.e. day 10 post challenge) exposure period. Similarly the significant ($p \leq 0.05$) difference was observed in B and D as compared to control on day 60 of observation period.

The group of fish fed with *Microcystis* showed significantly increased ($p \leq 0.05$) TEC in C on day 90 and on day 10 post challenge period as compared to their respective control. An insignificant (p>0.05) difference TEC was found in the entire treatment groups as compared to control on day 30 and on day 60 day of exposure period (Table 2).

Serum/blood biochemical indices

The group of fish fed with *Microcystis* was found a significantly ($p \leq 0.05$) decrease of blood glucose level in the treatment groups as compared to control at all the day of exposure period (except B on day 60 and day 90). There is no significant change of blood glucose level was observed in the entire experimental group in comparison to control on day 90 (Figure 1).

The level of AST was significantly ($p \leq 0.05$) different to all the treated groups of fish on day 60 (except group D), day 90 and after bacterial post challenge as compared to control. But no significant difference was found within treatment groups (except B) on day 30 of observation period (Figure 2).

The group of fish fed with *Microcystis* showed a significantly ($p \leq 0.05$) different in ALT activity on day 60, day 90 and on day 10 post challenge (except group of fish fed 0.5 g kg⁻¹) as compared to their respective control. But there was no significant difference was noticed between treated groups and control on day 30 of exposure period

Parameters	Groups	Pre challenge			Post challenge
		30 days	60 days	90 days	100 days
HB (g%)	A	8.20 ± 0.63ᵃ	7.00 ± 0.69ᵇ	7.55 ± 0.95ᵇ	8.01 ± 0.22ᵇ
	B	7.65 ± 0.32ᵃ	8.24 ± 0.46ᵃᵇ	10.25 ± 0.3ᵃ	7.45 ± 0.60ᵇ
	C	9.22 ± 0.46ᵃ	8.35 ± 0.52ᵃᵇ	9.22 ± 0.41ᵃᵇ	9.99 ± 0.38ᵃ
	D	8.02 ± 0.62ᵃ	9.65 ± 0.27ᵃ	8.002 ± 0.3ᵇ	8.25 ± 0.42ᵇ
TLC	A	14.02 ± 0.60ᵃ	13.22 ± 0.57ᶜ	14.57 ± 0.56ᵇ	15.96 ± 0.55ᶜ
	B	13.98 ± 1.07ᵃ	19.25 ± 0.59ᵃ	20.00 ± 0.67ᵃ	19.22 ± 0.40ᵇ
	C	15.21 ± 0.47ᵃ	15.02 ± 0.41ᵇᶜ	19.11 ± 1.00ᵃ	18.19 ± 0.73ᵇ
	D	14.52 ± 0.72ᵃ	16.66 ± 0.72ᵇ	20.21 ± 0.39ᵃ	22.33 ± 0.69ᵃ
TEC	A	0.71 ± 0.12ᵃ	0.81 ± 0.05ᵃ	0.73 ± 0.08ᵇ	0.94 ± 0.09ᵇ
	B	0.82 ± 0.05ᵃ	0.92 ± 0.06ᵃ	0.82 ± 0.04ᵇ	0.90 ± 0.03ᵇ
	C	0.73 ± 0.04ᵃ	0.99 ± 0.08ᵃ	1.22 ± 0.15ᵃ	1.50 ± 0.05ᵃ
	D	0.78 ± 0.09ᵃ	1.23 ± 0.21ᵃ	0.87 ± 0.08ᵇ	0.92 ± 0.19ᵇ

Note: Data are expressed as mean ± S.E. Superscript column wise on right hand side for particular treatment group are significantly (P<0.05) different from the control group. MaC-Control group, Ma1-0.5 gkg⁻¹, Ma2-1.0 gkg⁻¹ and Ma3-5.0 gkg⁻¹ *Microcystis aeruginosa* fed group.

Table 2: Effect of oral feeding of *Microcystis* powder on haematological parameters of *L. rohita* followed by i. p. challenge of *A. hydrophila* after 90 days.

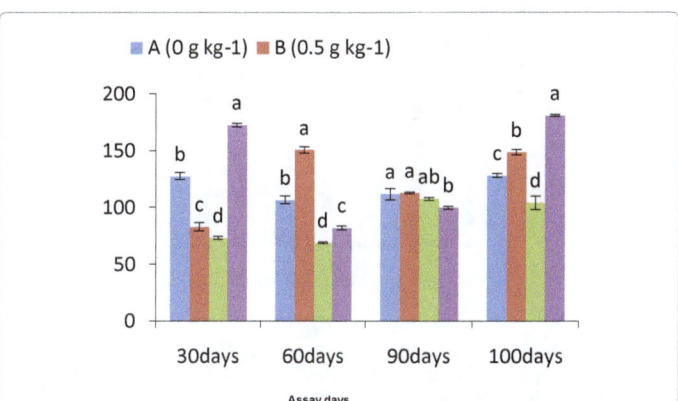

Figure 1: Effect of *Microcystis* on blood glucose level (g%) of *Labeo rohita* on different assay days during the experimental period. Bars bearing common superscripts are not significant at 5% level in comparison to each other (n=6).

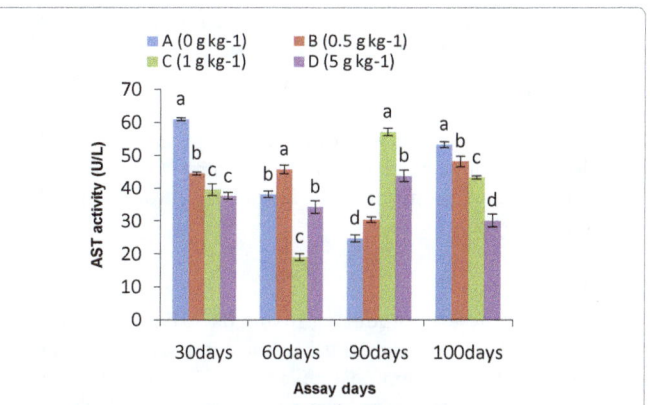

Figure 2: Effect of *Microcystis* on AST activity (U/L) of *Labeo rohita* on different assay days during the experimental period. Bars bearing common superscripts are not significant at 5% level in comparison to each other (n=6).

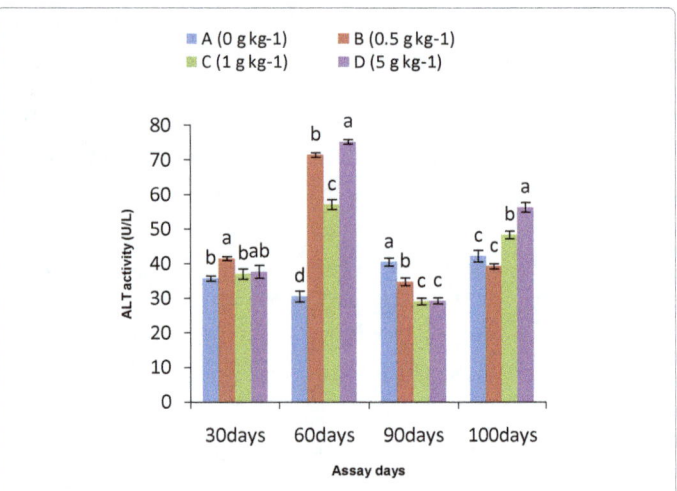

Figure 3: Effect of *Microcystis* on ALT activity (U/L) of *Labeo rohita* on different assay days during the experimental period. Bars bearing common superscripts are not significant at 5% level in comparison to each other (n=6).

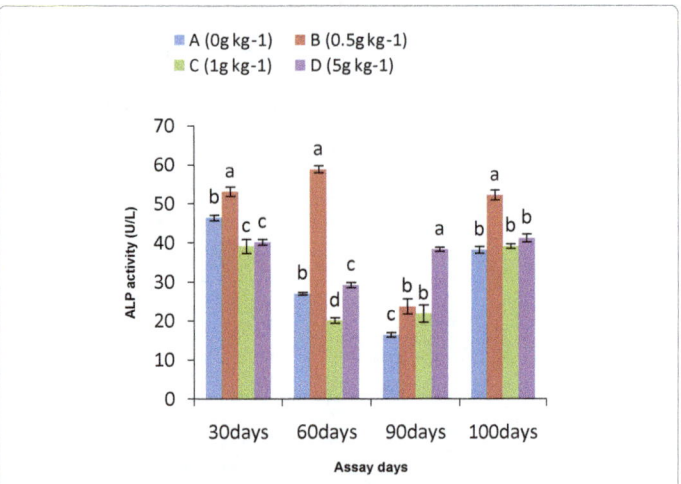

Figure 4: Effect of *Microcystis* on ALP activity (U/L) of *Labeo rohita* on different assay days during the experimental period. Bars bearing common superscripts are not significant at 5% level in comparison to each other (n=6).

(Figure 3). A significant (p ≤ 0.05) difference in ALP activity was found in all pre challenge treatment groups as compared to control in each assayed period as shown in Figure 4. But there was insignificant (p> 0.05) difference of ALP level was found on day 10 bacterial post challenge (except B) as compared to control.

Disease resistance

After challenging the fish with *A. hydrophila* the mortality was recorded up to 10th day. There was no mortality of fish up to 18 h post infection. The maximum survival was shown in group of fish fed 0.5g kg⁻¹ followed by 1.0g kg⁻¹.

Discussion

Microcystis besides producing toxins like Microcystin LR [26] it was found that it has potential use in the control of fish and shellfish diseases and also as immunostimulant [13,15]. Haematological and biochemical parameters have been acknowledged as valuable tools for monitoring fish health. So it is important to monitor the health status of fish after administration of *Microcystis* as feed additive in basic diet of fish.

Results of the haematological parameters of rohu fingerlings in this study showed that there were significant differences (p ≤ 0.05) among different dietary groups. It has been shown that variables such as age, sex, dietary state and stress alter blood values [27,28]. In present study the total leucocyte count significantly (p ≤ 0.05) increased in *Microcystis* fed group. This result is supported by another study [18], which found that there was an increase in the WBC count when *L. rohita* fingerlings were fed with *Mango kernel*. The increase in total leucocyte, neutrophils, lymphocytes and monocytes following feeding of algal and herbal diets supports the notion of antimicrobial properties of the algae [14,29] and traditional herbal medicine [17,30]. The white blood cells of rohu are known to increase following immunostimulation [31,32] and feeding with algal diets for 90 days [5]. The result of present investigation shows that the total erythrocyte counts increased in group of fish fed with 1g kg⁻¹ *Microcystis* on day 90 and on day 10 after postchallenge cab be correlated with the observation in Nile tilapia, *O. niloticus* that showed decreased RBC number after bacterial inoculation [33]. No significant difference of total erythrocyte count among the groups. The erythrocyte count increased with the administration of *Microcystis*, which might indicate an immunostimulant effect. The findings conform to those by Duncan and Klesius [34], who reported that the number of erythrocytes was significantly (p ≤ 0.05) greater in channel catfish fed with a diet containing β-glucan.

Glucose is one of the most important sources of energy for the animals. It has been reported as an indicator of stress caused by physical factors [35]. Blood glucose level in all the group of fish treated with *M. aeruginosa* supplemented diet showed no significant variation with the treatment groups on day 90 and after post challenge on day 10 (except group B and D). It shows that the fish were not under stress when fed with optimum level of *Microcystis* supplemented diets. Similar types of observation with different types of dietary supplementation were observed [36].

The present study was targeted to find out its impact on the fish by evaluating the major stress enzyme to liver taking a model fish rohu which contributes about more than 80% to the freshwater aquaculture production in India. The ALT and AST are indicates for the diagnosis of liver function [37] and damage [38]. Decreased AST and ALT activity in serum showed that oxaloacetate and glutamate are not available to Kreb cycle through this root of transmission [39]. Microcystins has been reported to impact the liver of fish [40,41]. ALT activity in the liver of fish fed the high algae meal diet was lower than that of the fish fed the control diet in the present study. Rabergh et al. [40] and Navratil et al. [42] reported that ALT and AST activities in blood plasma increased after common carp (*Cyprinus carpio L.*) received intraperitoneal injections of Microcystin-LR. However in our present study, dietary *Microcystis* did not increased the liver enzymes e.g. AST and ALT which might attributed either absence or less availability of Microcystin-LR. Increased activity of ALP was marked in group of fishes fed with *Microcystis* over the days and a significantly (p ≤ 0.05) higher ALP activity was observed in group of fish fed with 0.5 gkg⁻¹ dose for 60 days. Increased phosphatase activity indicates higher breakdown of energy reserved which are utilized for growth and survival of fishes. All these results indicate that *Microcystis* increases the resistance of *L. rohita* so that it can withstand the adverse conditions of a challenge. However, appropriate field trials remain necessary before using *Microcystis* as a feed additive in aquaculture farm.

Acknowledgement

The authors wish to thank the Director of the Central Institute of Freshwater

Aquaculture for encouragement and for providing facilities to carry out the work. The work is carried out partly funding source of ICAR in the form of AP Cess grant and partly from the Department of Science and Technology in the form of Women Scientist-A Scheme.

References

1. Ibrahem MD, Fathi M, Mesalhy S, Abd El-Aty AM (2010) Effect of dietary supplementation of inulin and vitamin C on the growth, hematology, innate immunity, and resistance of Nile tilapia (Oreochromis niloticus). Fish Shellfish Immunol 29: 241-246.

2. Gatlin DM, Li P, Wang X, Burr GS, Castille F, et al. (2006) Potential application of prebiotics in aquaculture, 8th International symposium on aquaculture nutrition: 371-376.

3. Baulny MOD, Quentel C, Fournier V, Lamour F, Gourvello RL (1996) Effect of long term oral administration of β-glucan as an immunostimulant or an adjuvant on some nonspecific parameters of the immune response of turbot, Scophthalmus maximus. Dis Aquat Org 26: 139-147.

4. Goetz FW, Iliev DB, McCauley LAR, Liarte CQ, Tort LB, et al. (2004) Analysis of genes isolated from lipopolysaccharide-stimulated rainbow trout (Oncorhynchus mykiss) macrophages. Mol Immunol 41: 1199-1210.

5. Das BK, Pradhan J, Sahu S (2009) The effect of Euglena viridis on immune response of rohu, Labeo rohita (Ham.). Fish Shellfish Immunol 26: 871-876.

6. Rao YV, Das BK, Jyotirmayee P, Chakrabarti R (2006) Effect of Achyranthes aspera on the immunity and survival of Labeo rohita infected with Aeromonas hydrophila. Fish Shellfish Immunol 20: 263-273.

7. Sahu S, Das BK, Mishra BK, Pradhan J, Sarangi N (2007a) Effect of Allium sativum on the immunity and survival of Labeo rohita infected with Aeromonas hydrophila. J Appl Ichthyol 23: 80-86.

8. Sahu S, Das BK, Pradhan J, Mohapatra BC, Mishra BK, et al. (2007b) Effect of Mangifera indica as feed additive on immunity and resistance to Aeromonas hydrophila in Labeo rohita fingerlings. Fish Shellfish Immunol 23: 109-118.

9. Ardo L, Yin G, Xu P, Va´radi L, Szigeti G, et al. (2008) Chinese herbs (Astragalus membranaceus and Lonicera japonica) and boron enhance the non-specific immune response of Nile tilapia (Oreochromis niloticus) and resistance against Aeromonas hydrophila. Aquaculture 275: 26-33.

10. Shameel M (2008) Change of divisional nomenclature in the Shameelian Classicification of algae. Int J Phycol Phycochem 4: 225.

11. Ishida K, Matsuda H, Murakami M, Yamaguchi K (1997) Kawaguchipeptin B, an antibacterial cyclic undecapeptide from the cyanobacterium Microcystis aeruginosa. J Nat Prod 60: 724.

12. Zainuddin EN, Mundt S, Wegner U, Mentel R (2002) Cyanobacteria a potential source of antiviral substances against influenza virus. Med Microbiol Immunol 191: 181-182.

13. Das BK, Pradhan J (2010) Antibacterial properties of selected freshwater microalgae against pathogenic bacteria. Indian J Fish 57: 61-66.

14. Pradhan J, Sahu S, Nilima P M, Mishra BK, Das BK (2011) Antibacterial properties of freshwater Microcystis aeruginosa (Kütz) to bacterial pathogen–a comparative study of bacterial bioassays. Ind J Animal Sci 81: 79-100.

15. Das BK, Pradhan J, Sahu S, Marhual NP, Mishra BK, et al. (2013) Microcystis aeruginosa (Kütz) incorporated diets increase immunity and survival of Indian major carp Labeo rohita (Ham.) against Aeromonas hydrophila Infection. Aquacult Res 44: 918-927.

16. Bahmani M, Kazemi R, Donskaya P (2001) A comparative study of some hematological features in young reared sturgeons (Acipense rpersicus and Huso huso). Fish Physiol Biochem 24: 135-140.

17. De Pedro N, Guijarro AE, Lopez-Patino MA, Marinez-Alvarez R, Delgado M (2005) Daily and seasonal variation in haematological and blood biochemical parameters in tench Tinca tinca. Aquacult Res 36: 1185-1196.

18. Roy SS (2002) Some toxicological aspects of chlorpyrifos to the intertidal fish Boleopthalmus dussumieri. University of Mumbai, India.

19. Misra CK, Das BK, Mukherjee SC, Pradhan J (2007) Effects of dietary vitamin C on immunity, growth and survival of Indian major carp Labeo rohita, fingerlings. Aquacult Nutr 13: 35-44.

20. Van Kampen EJ, Zijlstra WG (1961) Recommendations for haemoglobinometry in Human blood. Br J Haematol 13: 71.

21. Blaxhall PC, Daisley KW (1973) Routine haematological methods for use with fish blood. J Fish Biol 5: 771-781.

22. Wallnofer H, Schmidt E, Schmidt FW (eds) (1974) Synopsis der Leberkrankheiten. Georg Thieme Verlag. Stuttgart.

23. Rosalki RS (1993) Boerhringer Mannheim Gmblt analysis protocol. Clin Chem 39: 648.

24. Schmidt FH (1974) Methodender, Harn- and Blutzucker Bestimmung II. In: Boehringer Mannheim GmbH analysis protocol.

25. Duncan DB (1955) Multiple range and multiple F-tests. Biometrics 11: 1-42.

26. Oudra B, Loudiki M, Sbiyyaa B, Martins R, Vasconcelos V, et al. (2001) Isolation, characterization and quantification of microcystins (heptapeptides hepatotoxins) in Microcystis aeruginosa dominated bloom of Lalla Takerkoust lake-reservoir (Morocco). Toxicon 39: 1375-1381.

27. Barnhart RA (1969) Effects of certain variables on haematological characteristics of rainbow trout, Salmo gairdneri (Richardon). Trans Am Fish Soc 98: 411-418.

28. McCarthy DH, Stevensom JP, Roberts MS (1973) Some blood parameters of the rainbow trout (Salmo gairdneri Richardson). I. The kamloops variety. J Fish Biol 5: 1-8.

29. Das BK, Pradhan J, Pattnaik P, Samantaray BR, Samal SK (2005) Production of antibacterials from the freshwater alga Euglena viridis (Ehren). World J Microbiol Biotechnol 21: 45-50.

30. Parry RM, Chandan RC, Shahani KM (1965) A rapid and sensitive assay of muramidase. Proc Soc Exp Biol (NY) 119: 384-386.

31. Misra CK, Das BK, Mukherjee SC, Pattnaik P (2006a) Effect of long term administration of dietary β-glucan on immunity, growth and survival of Labeo rohita fingerlings. Aquaculture 255: 82-94

32. Misra CK, Das BK, Mukherjee SC, Pattnaik P (2006b) Effect of multiple injections of β-glucan on non-specific immune response and disease resistance in Labeo rohita fingerlings. Fish Shellfish Immunol 20: 305-319.

33. Ranzani-Paiva MJT, Ishikawa CM, Eiras AC, Silveira VR (2004) Effects of an experimental challenge with Mycobacterium marinum on the blood parameters of Nile Tilapia, Oreochromis niloticus (Linnaeus, 1757). Brazilian Arch Biol Technol 47: 945-953.

34. Duncan PL, Klesius PH (1996) Dietary immunostimulants enhance nonspecific immune responses in channel catfish but not resistance to Edwardsiella ictaluri. J Aquat Anim Health 8: 241-248.

35. Manush SM, Pal AK, Das T, Mukherjee SC (2005) Dietary high protein & vitamin C mitigate stress due to chelate claw ablation in Macrobrachium rosenbergii males. Comp Biochem Physiol- Part A 142: 10-18.

36. Das R, Raman RP, Saha1 H, Singh R (2013) Effect of Ocimum sanctum Linn. (Tulsi) extract on the immunity and survival of Labeo rohita (Hamilton) infected with Aeromonas hydrophila. Aquacult Res 1: 1-11

37. Ozaki H (1978) Diagnosis of fish health by blood analysis.

38. Oda T (1990) The biology of liver.

39. Shakoori AR, Mughal AL, Iqbal MJ (1999) Effects of sublethal dose of fenvalarate (a synthetic pyrethroid) administered for continuously for four weeks on the blood, liver and muscles of a fresh water ¢sh, Ctenopharyngodon idella. Bull Environ Contam Toxicol 57: 487-494.

40. Rabergh CMI, Bylund G, Eriksson JE (1991) Histopathological effects of microcystin-LR, a cyclic peptide toxin from the cyanobacterium (blue-green alga) Microcystis aeruginosa, on common carp (Cyprinus aeruginosa L.). Aquat Toxicol 20: 131-145.

41. Rodger HD, Turnbull D, Edwards C, Codd GA (1994) Cyanobacterial (blue-green algal) bloom associated pathology in brown trout (Salmo trutta L.) in Loch Leven, Scotland. J Fish Dis 17: 177-181.

42. Navratil S, Palikova M, Vajcova V (1998) The effect of pure microcystin LR and biomass of blue-green algae on blood indices of carp (Cyprinus carpio L.). Acta Vet (Brno) 67: 273-279.

30

Genetic Evidence for Sympatric Populations of Yellow Perch (*Perca flavescens*) in Lake Saint-Pierre (Canada): the Crucial First Step in Developing a Fishery Management Plan

Christelle Leung[1,2], Pierre Magnan[1,3] and Bernard Angers[1,2]*

[1]*Group for Interuniversity Research in Limnology and Aquatic Environment (GRIL)*
[2]*Department of biological sciences, Université de Montréal, C.P. 6128, Succursale Centre-Ville, Montreal, Quebec, Canada H3C 3J7*
[3]*Department of Chemistry-Biology, Université du Québec à Trois-Rivières, 3351 Boulevard des Forges, C.P. 500, Trois-Rivières, QC, Canada*

Abstract

To ensure the persistence of species facing anthropogenic pressures, an understanding of the dynamics and mechanisms that structure populations is of major importance. Determining the presence of distinct populations is a first step when sympatric populations are suspected. Yellow perch (*Perca flavescens*) is exploited by sport and commercial fishing in Lake Saint-Pierre (Quebec, Canada). Because habitat characteristics are spatially structured and this species is known to display natal site fidelity, this study's aim was to assess whether sympatric populations of yellow perch coexist in Lake Saint-Pierre. Low genetic differentiation is predicted due to the recent colonization of the system (<8,000 years). Simulations were first performed to confirm that population differentiation is better depicted using AFLP than microsatellite markers. A survey of the variation throughout the entire genome was then performed using the AFLP approach. To link individuals to their natal site, recently emerged larvae from different cohorts captured at different stations were analyzed. Results from three distinct AFLP surveys indicated a correlation between the genetic composition of individuals and geographic sites. These results confirmed the presence of multiple sympatric populations in Lake Saint-Pierre, resulting from natal site fidelity. While the genetic differentiation is very low, the management of this species should take into account the existence of distinct population structures.

Keywords: AFLP; Genetic differentiation; Fishery management plan; Natal site fidelity; Yellow perch; Lake Saint-Pierre

Introduction

The sustainable management of important sport and commercial fish species is a major concern. Human impacts on the environment can result in a loss or fragmentation of habitats, species introductions, pollution, climate change, and overexploitation of populations [1]. Population assessment provides managers with critical information that can be used to elaborate sustainable management programs. Each independent population of a given species has specific recruitment, growth, and mortality rates that determine their own population dynamics. An important prerequisite for elaborating management strategies is identifying the number of reproductively distinct populations [2-4].

Physical boundaries limiting adult dispersal in aquatic or marine environments may be cryptic or absent even though distinct reproductive units may exist [5-7]. Behaviour such as natal site fidelity, which is characterized by an individual's return to and reproduction at its birth site [8], may result in distinct reproductive units even in the absence of physical barriers. Such a behaviour is known to be common in fish [9-12].

While numerous methods can be used to discriminate sympatric populations [13-15], genetic assessments are generally required to confirm the presence of reproductively distinct populations [16-19]. Significant differences in allele frequencies will be detected when populations are isolated for a sufficient number of generations [20]. However, a failure to detect genetic differences between populations does not necessarily indicate that reproductively isolated populations do not currently exist: differentiation is a process that extends over many generations [21]. Because the different genetic markers may display different characteristics in terms of dominance, variability and abundance, the choice of the method to be used is critical [22-24].

The yellow perch, *Perca flavescens*, is a freshwater fish widely distributed in temperate and subarctic areas of North America [25]. This species has been subjected to sport and commercial fishing in many systems since the beginning of the 20th century [26-28]. There is evidence that yellow perch, as well as its sister species, the Eurasian perch (*Perca fluviatilis*), displays natal-site fidelity [29-33].

In Lake Saint-Pierre (LSP), yellow perch represents an important socio-economic resource. This fluvial lake of St. Lawrence River has a surface area of 350 km² and was designated as a wetland of international significance under the Ramsar Convention in 1998 and a biosphere reserve by UNESCO in 2000. Previous studies suggest the existence of several populations of yellow perch in LSP. Different stable carbon isotope ratios and parasite infections are observed at different localities of LSP [34,35]. In addition, different growth rates are observed between shores [36], suggesting differential habitat use and low contact among populations from different areas of the lake. However, the lake was open to colonization by freshwater fishes only recently. Following glacial retreat at the end of the Pleistocene, marine water invaded most of the St. Lawrence lowland, including LSP; the retreat of the Champlain Sea occurred only ca. 8,000 years ago. In addition, St. Lawrence River

*Corresponding author: Bernard Angers, Department of Biological Sciences, Université de Montréal. C.P. 6128, Succursale Centre-Ville, Montreal, Quebec, Canada H3C 3J7.
E-mail: bernard.angers@umontreal.ca

yellow perch is known to originate from a unique founder group since no marked genetic structure has been detected [37]. Low genetic differentiation among populations is therefore expected.

This study aims to determine whether several populations of yellow perch coexist in LSP. AFLP analysis reveals a high number of low variable loci, thus this method is known to be better at detecting population differentiation than microsatellite markers in some cases [22,38]. The first specific objective of this study was to use simulations to confirm the usefulness of AFLP markers compared to microsatellites. The second specific objective was to assess genetic differentiation among five putative populations within LSP. We characterized the genetic diversity of each spawning site by capturing individuals in the first two weeks following their emergence. Because larvae of this species have limited capacity to move actively [39], they better depicted reproductive processes than adults. To reduce the probability of capturing kin, different years and several points were sampled for each site of LSP. Genetic diversity was screened using the AFLP method, considering the simulation results.

Materials and Methods

Simulations

To investigate the differences between AFLP and microsatellite markers, five populations of 3,000 individuals were simulated using EASYPOP software, version 2.0.1 [40]. For each population, 4,000 generations were produced, the first 1,000 of which were used to homogenate allele frequencies by imposing a high migration rate (m = 0.5). The simulation assumed a unique large founder group. The next 3,000 generations were followed using various migration rates (varying from m = 3×10^{-4} to m = 10^{-1}) and a stepping stone model to enable population differentiation; this takes into account that yellow perch mature at 2–3 years of age [25] and colonized LSP 8,000 years ago. This also allowed the simulation of different N_Em because the effective size of natural populations is unknown. Ten replicates were computed for each migration rate.

Two genetic markers were simulated. On the one hand, 10 microsatellite loci were generated with a stepwise mutation model (SMM) with two mutation rates, $\mu = 5 \times 10^{-4}$ and $\mu = 10^{-3}$, and 10 possible allelic states; these parameters are common in microsatellite studies [41-43]. On the other hand, 100 AFLP loci were generated using a K-allele model (KAM) with two possible allelic states and a mutation rate of $\mu = 10^{-9}$. The dominance of AFLP markers was taken into account by changing the genotype of heterozygotes "presence/absence" in homozygotes "absence/absence". Samples of 20 individuals per population were used to compare both markers.

Sampling

Larvae were sampled following their emergence in May and June. Sampling was conducted by the Quebec Ministry of Natural Resources and Wildlife (MRNF; Quebec, Canada) using push nets [44] at numerous points around the lake. Five sites, separated by 5 to 12 km, were chosen (Figure 1) according to the abundance of larvae (unpublished data). For each area, two years (2003–2004 or 2008–2009) and two points, separated by ca. 0.5 to 3.40 km, were selected to avoid sib sampling (Figure 1, Table 1). In addition to LSP samples, two populations from distinct lakes [45] were included to assess differentiation in the absence of migration.

AFLP fingerprinting

Yellow perch larvae were identified using morphological characteristics by MRNF personnel and preserved in 95% ethanol. Total DNA was extracted by proteinase K digestion followed by phenol-chloroform purification and ethanol precipitation [46].

Code	Sampling date		Geographic coordinates		# of individuals AFLP1	AFLP2	AFLP3
Site D	2008	12-Jun	46.20794	-72.65937	-	5	5
	2008	03-Jun	46.18998	-72.69617	-	-	6
	2009	06-Jun	46.19915	-72.65859	-	2	7
	2009	06-Jun	46.19317	-72.69134	-	-	4
Site F	2003	27-May	46.14660	-72.81855	5	-	-
	2004	02-Jun	46.14323	-72.80480	5	-	-
	2008	30-May	46.14580	-72.80933	-	5	8
	2009	26-May	46.14101	-72.81177	-	4	10
Site G	2008	02-Jun	46.13729	-73.01412	-	5	6
	2008	02-Jun	46.14727	-73.00002	-	-	6
	2009	15-Jun	46.13901	-72.99918	-	3	8
Site M	2004	27-May	46.19117	-73.00322	6	-	-
	2008	29-May	46.19321	-72.97397	-	-	4
	2008	28-May	46.19321	-72.97397	-	3	7
	2009	04-Jun	46.20885	-72.95030	-	5	8
Site Y	2004	28-May	46.26333	-72.85480	5	-	-
	2008	04-Jun	46.26074	-72.83183	-	6	6
	2008	04-Jun	46.26423	-72.81647	-	-	6
	2009	10-Jun	46.26040	-72.83172	-	5	5
	2009	10-Jun	46.26678	-72.82529	-	-	6
				Total	21	43	102

Table 1: Sampling dates, geographic coordinates (latitude, longitude), and number of individuals analyzed with AFLP for each site of LSP.

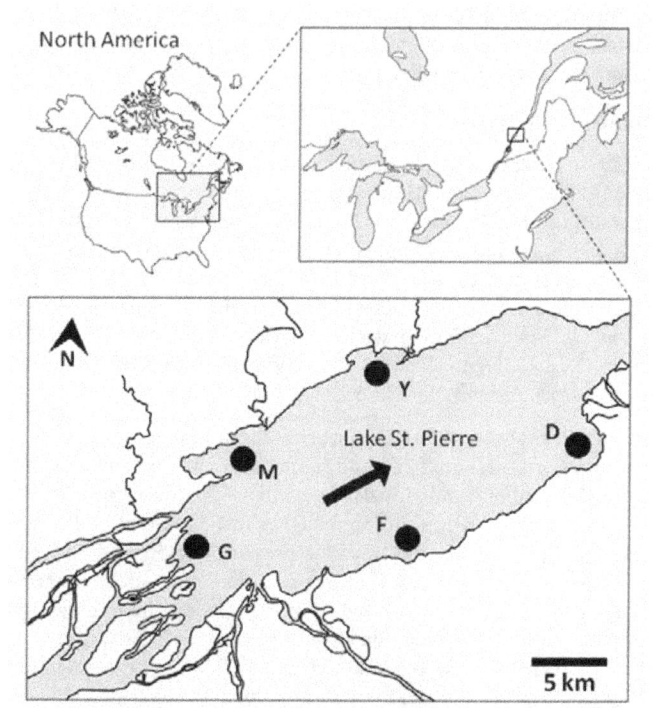

Figure 1: Location of Lake Saint-Pierre (Canada) and areas sampled. The arrow indicates the direction of water flow.

	AFLP1		AFLP2		AFLP3	
Restriction enzymes	EcoRI – MseI		EcoRI – MseI		EcoRI – TaqαI	
Selective PCR *	9		6		1	
Post-amplification restriction enzymes	-		-		Rsal - Hinfl - Mbol Msel - NlaIII - Alul	
Sampling years	2003-2004				2008-2009	
	n	Loci	n	Loci	n	Loci
Site D	-	-	7	322 (141)	22	152 (77)
Site F	10	750 (385)	9	330 (164)	18	154 (75)
Site G	-	-	8	317 (154)	20	159 (81)
Site M	6	759 (301)	8	315 (148)	19	156 (75)
Site Y	5	641 (195)	11	329 (159)	23	157 (75)
Total LSP	**21**	**842 (476)**	**43**	**349 (214)**	**102**	**160 (99)**
Lake M2	-	-	-	-	7	160 (42)
Lake O2	-	-	-	-	7	174 (86)
All lakes	**21**	**842 (476)**	**43**	**349 (214)**	**116**	**187 (136)**

*Selective combinations used for AFLP 1 (EcoANNxMseCNN) ACxAG, ACxTC, GCxAG, ACxGC, CGxGC, CGxAG, GGxTT, AGxCC, GCxCC; AFLP 2 (EcoANNxMseCNN) ACxTC, CGxGC, AGxCC, GCxTT, ACxTT, AGxAG; AFLP 3 EcoACGxTaqCTA

Table 2: Characteristics of the different AFLP surveys performed in this study. For each survey, the number of individuals and number of loci surveyed are indicated; polymorphic loci are in parentheses.

Genetic variability was surveyed using the AFLP procedure [47]. Three different DNA fingerprints were performed using two sets of restriction enzyme: AFLP 1 (MseI × EcoRI), AFLP 2 (MseI × EcoRI), and AFLP 3 (TaqαI × EcoRI), and different numbers of selective combinations were performed according to the restriction enzymes (Table 2). However, when using the TaqαI and EcoRI restriction enzymes, an additional digestion was performed by adding three restriction enzymes to the selective PCR product (RsaI - HinfI - MboI or MseI - NlaIII - AluI). Contrary to the TE-AFLP method [48], this procedure provides additional restriction fragments. Such distinct surveys of the genome help to minimize the chance that loci under selection were responsible for cluster similarities.

Electrophoresis was performed on a denaturing 6% polyacrylamide (19:1 acrylamide:bis-acrylamide) gel for AFLP products. Silver nitrate staining was used to visualize polymorphisms [49].

Statistical analyses

Characterization of population diversity of sampled individuals: Allele frequencies were estimated from the presence–absence of AFLP loci assuming Hardy-Weinberg Equilibrium and using the square root method [50]. An index of population differentiation was thus estimated by calculating the F_{ST} [51], and significance was assessed by 1,000 permutations. These parameters were estimated with AFLP-SURV, version 1.0 [52]. To ensure that all used markers were neutral, BayeScan version 2.01 [53] and Mcheza [54] software were used to identify candidate loci under natural selection.

Cluster analyses without a priori assumptions about population structure: For both simulated and sampled populations, phylogenetic relationships among individuals were inferred to determine whether genetic similarities within sites were higher than among sites. Euclidian distance [55] was estimated from the presence–absence of AFLP loci while allele sharing distance [56] was used for microsatellite markers. Genetic distances were then used to infer a phylogenic tree using the neighbour-joining (NJ) method [57]. A phylogenetic parsimony score (PPS) [58] for each NJ tree was calculated to assess the geographic homogeneity of the cluster. Because a different number of individuals were used for each batch of sampled data, PPS were standardized as [PPS - (number of populations - 1)] / (number of individuals - 1). A cluster exclusively composed of individuals from a single site has a standardized PPS of zero while a random organization will tend to be close to 1. Allele sharing distance and NJ trees were computed using POPULATIONS, version 1.2.30 [59], and Euclidian distance and PPS were calculated with R version 1.12.2 software.

Cluster analyses with an a priori assumption about population structure: The proportion of individuals correctly and unambiguously reassigned to their population or spawning site (p-value > 0.05 only for the source population) according to allele frequencies was calculated for microsatellite and AFLP loci using GENECLASS2 [60] and AFLPOP, version 1.1 [61], respectively.

For samples from LSP, the cluster analysis is based on the fact that individuals from a given spawning site were assumed to belong to the same population. Redundancy analyses (RDA) were computed to have constrained ordination between genetic polymorphism and spatial distribution of larvae for AFLP loci. The regression was tested using 1,000 permutations with R version 1.12.2 software.

Results

Simulations

For both AFLP and microsatellite markers, population differentiation decreased when N_Em increased. At a low rate of migration ($N_Em = 0.0003$), population differentiation was characterized by high F_{ST} values and low parsimony scores, indicating that individuals were correctly clustered according to their population. We observed a high percentage of reassignment of individuals to their population of origin (Table 3). When migration rate increased, a gradual homogenization of populations occurred: an increasing parsimony score (more clustering errors) was observed as a function of migration rate; F_{ST} values and the correct reassignment rate also decreased (Figure 2A and 2C).

Simulation results revealed that for any N_Em used, F_{ST} values were higher for AFLP than for microsatellites (Figure 2A). This difference between markers increased as N_Em decreased. When populations were homogenized (i.e., N_Em ranging from 30 to 300), no population differentiation was detected with AFLP or microsatellite markers (high standardized PPS; values ranging from 0.31 to 0.63). Moreover, both markers had low F_{ST} values and no significant differentiation was observed for either (Figure 2A). Individuals also failed to be correctly reassigned to their population of origin, although the rate of reassignment was slightly higher when using AFLP than microsatellites markers (mean of 15% vs. 3%, respectively; Figure 2C).

In contrast, when $N_E m$ was lower than 9, ALFP markers allowed better clustering of individuals according to their population (lowest parsimony scores) than did microsatellite markers. As a result, AFLP performed better for both clustering of individuals and reassignments (Figure 2B and 2C). The same observations were made for both mutation rates used for microsatellite markers (1×10^{-3} or 5×10^{-4} mutations per generation) (Figure 2).

AFLP data from LSP

AFLP 1 was performed as a preliminary analysis. Numerous loci (n=842) were scored for a limited number of individuals (n=21) sampled in 2003 and 2004, with 476 polymorphic loci (56.53%). A high F_{ST} value was obtained (0.2316, *p-value* < 0.001) and a perfect cluster of individuals according to their sampling site was observed, with a standardized PPS of zero (Figure 3A). This result is confirmed by RDA analysis ($R^2 = 0.205$, *p-value* = 0.001, Figure 3B).

Analyses of additional individuals from two other sampling years

	AFLP1	AFLP2	AFLP3
With all loci			
F_{ST}	0.2316*	0.1610*	0.1010*
Standardized PPS	0	0.1667	0.1287
Reassignment rate	90.47 %	74. 42 %	43.14 %
Without loci detected as under positive selection			
Detected outlier loci	12 (2.02 %)	18 (5.16 %)	4 (2.50 %)
F_{ST}	0.2141*	0.1501*	0.0884*
Standardized PPS	0	0.1905	0.1862
Reassignment rates	90.48 %	76.74 %	22.55 %

*p-value < 0.001

Table 3: Results of the different AFLP performed in this study. Partition of the genetic diversity (F_{ST}), standardized parsimony scores (PPS), and reassignment rate are given over all loci and without loci presumably under positive selection (outlier loci detected with the Mcheza software).

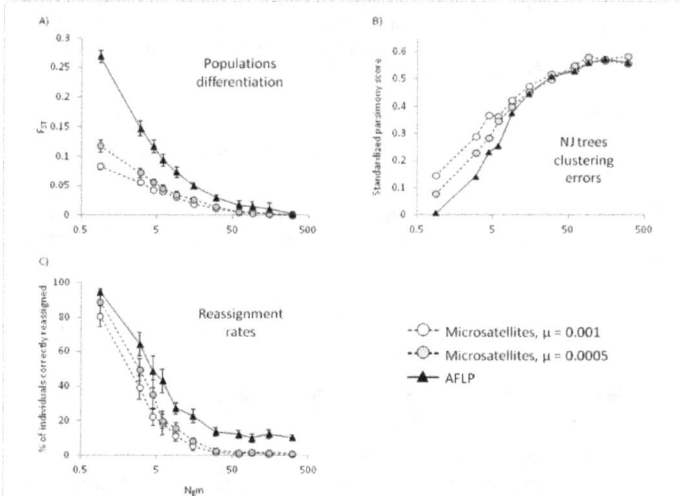

Figure 2: Results of AFLP and microsatellite marker simulations. (A) Population differentiation, (B) standardized parsimony scores, and (C) proportion of correct reassignment as a function of $N_E m$ for microsatellite (circle) and AFLP (triangle) markers.

(2008 and 2009) provide a total of 349 and 160 loci for AFLP2 and AFLP3, respectively. Even though different endonucleases were used, the proportion of polymorphic loci was similar between the two AFLP analyses, with 61.32% (214 loci) for AFLP2 and 61.88% (99 loci) for AFLP3. Significant FST values were estimated for AFLP2 (0.1610, *p-value* < 0.001) and AFLP3 (0.1010, *p-value* < 0.001). Reassignment rates were 74.42% and 43.14%, with standardized PPS values of 0.16 and 0.13 for AFLP2 and AFLP3, respectively (Figure 3C and 3E). This indicates that individuals clustered according to sampling site no matter what year they were sampled. RDA supported the spatial organization of individuals for AFLP2 ($R^2 = 0.213$, *p-value* = 0.001) and AFLP3 ($R^2 = 0.139$, *p-value* = 0.001) (Figure 3D and 3F).

A comparison between LSP sites and two geographically distant lakes confirmed the very low differentiation among LSP sites. Using both analyses with and without a priori assumption showed no ambiguity in individuals clustering according to their lake of origin: a null standardized PPS was observed, correct reassignment was as high as 81%, and differences between lakes were also confirmed with RDA analysis ($R^2 = 0.146$, *p-value* < 0.001).

Few loci were detected as possibly being under positive selection for AFLP1 (12 loci; 2.02%), AFLP2 (18 loci; 5.16%), and AFLP3 (4 loci; 2.50%) when using the Mcheza program. However, analyses with Bayescan software were found to be more conservative, with only one or two loci detected as outlier loci for each independent AFLP procedure. Furthermore, similar F_{ST} values and standardized PPS were obtained

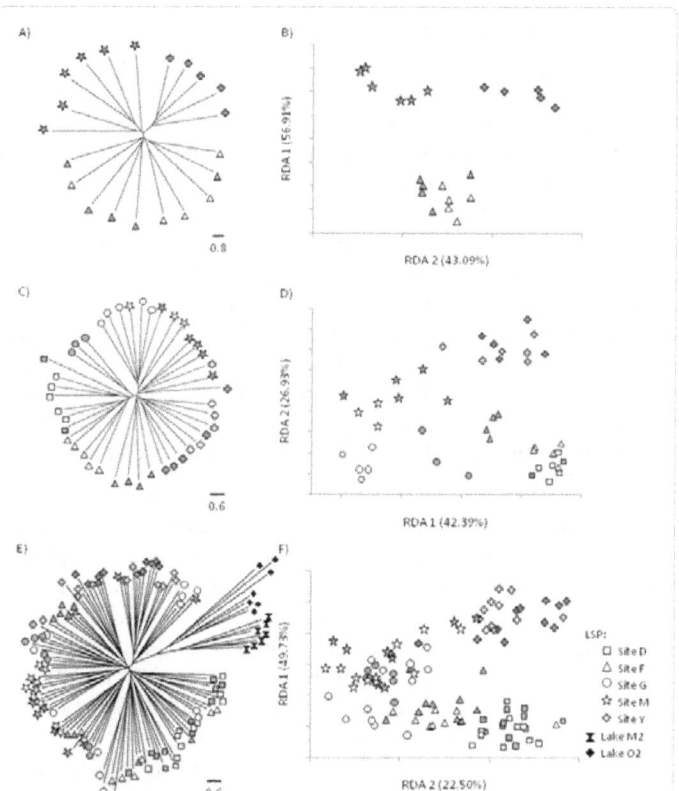

Figure 3: Individual relationships inferred from the different AFLP surveys. Unrooted NJ trees obtained for AFLP1 (A), AFLP2 (C), and AFLP3 (E); RDA plots obtained for AFLP1 (B), AFLP2 (D), and AFLP3 (F) as a function of sampling sites within LSP. The percentages of constrained inertia for each corresponding axis are in parentheses. White and grey symbols indicate individuals sampled respectively in 2003 and 2004 for AFLP1 and in 2008 and 2009 for AFLP 2 and AFLP3.

when excluding these loci (Table 2). RDA analysis also indicated the same organization for AFLP1 (R^2 = 0.186, *p-value* = 0.001), AFLP2 (R^2 = 0.200, *p-value* = 0.001), and AFLP3 (R^2 = 0.123, *p-value* = 0.001).

Discussion

Genetic differentiation of sympatric populations in LSP

The present study shows that yellow perch in LSP are more genetically similar within a small area than among geographically distant ones. Furthermore, no marked differentiation was observed among years. These results confirm the presence of multiple sympatric populations in the lake. Individuals representative of a given area were collected over two years and at two different sites, and were sampled on different dates to reduce the probability of sampling related individuals. Consequently, it appears extremely unlikely that sib sampling was responsible for the genetic similarity observed within a region.

Similarly, loci under putative positive selection were detected because of their higher-than-expected F_{ST} over all loci. However, even removing these loci from the analyses resulted in similar population differentiation. This indicates that the clustering of individuals by site did not rely on a few loci, but rather that several loci revealed a similar trend. Interestingly, the fact that some loci displaying a strong geographic signal should be under selection would represent further evidence of spatial organization through time. Considering the alternative hypothesis, that larvae are from a panmitic population, such a geographic signal over tens of loci would be the result of selection on individuals at each generation. This would be unlikely, since it would imply an extremely high mortality rate.

Natal site fidelity

The presence of multiple sympatric populations in the LSP is consistent with yellow perch natal site fidelity. The species is known to be generally sedentary, with displacement of adults generally not exceeding 4 km, although some vagrant individuals have been found tens of kilometres from their marking site [26]. Furthermore, local characteristics, such as otolith chemical concentration, migration time, or ultraviolet radiation tolerance, vary according to environmental condition [62-65]; this is also seen in the Eurasian perch [66]. Both species display spawning site fidelity, as shown by mark–recapture experiments [67-70]. Within LSP, yellow perch display different growth rates [36], and Bertrand et al. [35] have shown that feeding range does not exceed 2 km in LSP, suggesting a non-random distribution of individuals.

Natal site fidelity behaviour of yellow perch has been studied using egg mass manipulations on a specific spawning site of Lake Lochaber, Nova Scotia, Canada. Four years of systematic removal of egg masses resulted in a declining number of eggs laid on that site years after, even though the site remained a potential spawning ground [32,33]. This behaviour has also resulted in genetically distinct sympatric populations of Eurasian perch [29,71], and the same pattern of differentiation was observed over years [30]. Natal site fidelity could be based on olfactory signals: it was shown that Eurasian perch are able to recognize related individuals [72].

AFLP and larvae

AFLP markers are known to be more powerful than microsatellites in individual-based population assignment because the higher number of dominant loci compensates for the high variability of co-dominant markers for assessing population differentiation [22,23,73]. Simulated data revealed similar trends using either AFLP or microsatellite loci. When population differentiation is well defined (low migration rates), no significant differences between AFLP or microsatellite markers are observed. Indeed, the genetic organization of populations can be detected with both methods, as reported in previous studies [74,75]. However, both methods failed to detect genetic organization when high migration occurs between populations. A difference between the methods is observed at intermediary migration rates, where AFLP markers allow better detection of population differentiation than microsatellites. This may explain, at least partially, why a previous study using microsatellite markers on adult yellow perch failed to detect genetic differentiation among sample sites within LSP [37].

The use of the larval stage not only allowed the localization of birth site, but also provided certitude that individuals belonged to the same cohort. Indeed, using adults may result in a group of individuals from different cohorts, since yellow perch life span can reach 21 years [25]. Genetic differences occurring over time may result from variability in recruitment and reproductive success as well as effective population size. Consequently, genetic differentiation of a population cannot be detected when pooling individuals from different cohorts even if it is observed in separate cohorts [76-78].

In conclusion, this study revealed multiple sympatric populations of yellow perch coexisting in LSP, which is consistent with ecological observations in terms of local adaptations and the evidence of natal site fidelity observed in this species. Recent colonization of LSP and expected straying among sympatric populations leading to gene flow resulted in very low differentiation when compared to populations from different lakes. Nonetheless, each population represents a distinct demographic unit with specific recruitment, growth, and mortality rates. These results have implications for management and should be considered when multiple spatially organized populations occur in a same fishery [79-81].

Acknowledgements

We thank M Prud'homme and V Jaouen for their laboratory assistance; A Bertolo, P Legendre, and FJ Lapointe for helpful advice; and the Quebec Ministry of Natural Resources and Wildlife for providing samples. This research was supported by research grants from the Natural Sciences and Engineering Research Council of Canada (NSERC) to BA and PM and from the Group for Interuniversity Research in Limnology and Aquatic Environment (GRIL).

References

1. Soule ME (1991) Conservation: tactics for a constant crisis. Science 253: 744-750.

2. Ferguson A (1989) Genetic differences among brown trout, *Salmo trutta*, stocks and their importance for the conservation and management of the species. Freshwater Biol 21: 35-46.

3. Rossiter SJ, Jones G, Ransome RD, Barratt EM (2000) Genetic variation and population structure in the endangered greater horseshoe bat *Rhinolophus ferrumequinum*. Mol Ecol 9: 1131-1135.

4. Waples RS (1991) Pacific salmon, *Oncorhyncus* spp., and the definition of "species" under the Endangered Species Act. Mar Fish Rev 53: 11–22.

5. Moore IT, Bonier F, Wingfield JC (2005) Reproductive asynchrony and population divergence between two tropical bird populations. Behav Ecol 16: 755-762.

6. Han YS, Sun YL, Liao YF, Liao IC, Shen KN, et al. (2008) Temporal analysis of population genetic composition in the overexploited Japanese eel *Anguilla japonica*. Mar Biol 155: 613-621.

7. Dittman A, Quinn T (1996) Homing in Pacific salmon: mechanisms and ecological basis. J Exp Biol 199: 83-91.

8. Blair GR, Quinn TP (1991) Homing and spawning site selection by sockeye salmon (Oncorhynchus nerka) in Iliamma lake, Alaska. Can J Zool 69: 176-181.

9. Massicotte R, Magnan P, Angers B (2008) Intralacustrine site fidelity and nonrandom mating in the littoral-spawning northern redbelly dace (Phoxinus eos). Can J Fish Aquat Sci 65: 2016-2025.

10. CE, Hamilton DJ, McCarthy I, Wilson AJ, Grant A, et al. (2006) Does breeding site fidelity drive phenotypic and genetic sub-structuring of a population of Arctic charr? Evol Ecol 20: 11-26.

11. Waters JM, Epifanio JM, Gunter T, Brown BL (2000) Homing behaviour facilitates subtle genetic differentiation among river populations of Alosa sapidissima: microsatellites and mtDNA. J Fish Biol 56: 622-636.

12. Nielsen EE, Hansen MM, Loeschcke V (1999) Genetic variation in time and space: microsatellite analysis of extinct and extant populations of Atlantic salmon. Evolution 53: 261-268.

13. Pradel R (1996) Utilization of capture-mark-recapture for the study of recruitment and population growth rate. Biometrics 52: 703-709.

14. Burnham KP, Overton WS (1979) Robust estimation of population size when capture probabilities vary among animals. Ecology 60: 927-936.

15. Begg GA, Waldman JR (1999) An holistic approach to fish stock identification. Fish Res 43: 35-44.

16. Miller LM, Kallemeyn L, Senanan W (2001) Spawning-site and natal-site fidelity by northern pike in a large lake: mark–recapture and genetic evidence. Trans Am Fish Soc 130: 307-316.

17. Freedberg S, Ewert MA, Ridenhour BJ, Neiman M, Nelson CE (2005) Nesting fidelity and molecular evidence for natal homing in the freshwater turtle, Graptemys kohnii. Proc R Soc B 272: 1345-1350.

18. AB, Bowen BW, Avise JC (1990) A genetic test of the natal homing versus social facilitation models for green turtle migration. Science 248: 724-727.

19. Varnavskaya NV, Wood CC, Everett RJ, Wilmot RL, Varnavsky VS, et al. (1994) Genetic differentiation of subpopulations of sockeye salmon (Oncorhynchus nerka) within lakes of Alaska, British Columbia, and Kamchatka, Russia. Can J Fish Aquat Sci 51: 147-157.

20. Waples RS (1998) Separating the wheat from the chaff: patterns of genetic differentiation in high gene flow species. J Hered 89: 438-450.

21. Crandall KA, Bininda-Emonds ORP, Mace GM, Wayne RK (2000) Considering evolutionary processes in conservation biology. Trends Ecol Evol 15: 290-295.

22. Campbell D, Duchesne P, Bernatchez L (2003) AFLP utility for population assignment studies: analytical investigation and empirical comparison with microsatellites. Mol Ecol 12: 1979-1991.

23. Mariette S, Le Corre V, Austerlitz F, Kremer A (2002) Sampling within the genome for measuring within-population diversity: trade-offs between markers. Mol Ecol 11: 1145-1156.

24. Ferguson MM, Danzmann RG (1998) Role of genetic markers in fisheries and aquaculture: useful tools or stamp collecting? Can J Fish Aquat Sci 55: 1553-1563.

25. Craig J (1987) The biology of perch and related fish. Croom Helm, London.

26. Thorpe JE (1977) Morphology, physiology, behavior, and ecology of Perca fluviatilis L. and P. flavescens Mitchill. J Fish Res Board Can 34: 1504-1514.

27. Malison JA (2000) A white paper on the status and needs of yellow perch aquaculture in the North Central Region. US Department of Agriculture, North Central Regional Aquaculture Center, Lansing, Michigan.

28. Bronte CR, Selgeby JH, Swedberg DV (1993) Dynamics of a yellow perch population in western Lake Superior. N Am J Fish Manag 13: 511-523.

29. Gerlach G, Schardt U, Eckmann R, Meyer A (2001) Kin structured subpopulations in Eurasian perch (Perca fluviatilis L.). Heredity 86: 213-221.

30. Bergek S, Bjöerklund M (2009) Genetic and morphological divergence reveals local subdivision of perch (Perca fluviatilis L.). Biol J Linn Soc 96: 746-758.

31. Bergek S, Olsson J (2009) Spatiotemporal analysis shows stable genetic differentiation and barriers to dispersal in the Eurasian perch (Perca fluviatilis L.). Evol Ecol Res 11: 827-840.

32. Aalto SK, Newsome GE (1989) Evidence of demic structure for a population of yellow perch (Perca flavescens). Can J Fish Aquat Sci 46: 184-190.

33. Aalto SK, Newsome GE (1990) Additional evidence supporting demic behaviour of a yellow perch (Perca flavescens) population. Can J Fish Aquat Sci 47: 1959-1962.

34. Bertrand M, Marcogliese DJ, Magnan P (2010) Effect of wetland enhancement on parasites of juvenile yellow perch. Wetlands 30: 300-308.

35. M, Cabana G, Marcogliese DJ, Magnan P (2011) Estimating the feeding range of a mobile consumer in a river–flood plain system using 13C gradients and parasites. J Anim Ecol 80: 1313-1323.

36. Glemet H, Rodriguez MA (2007) Short-term growth (RNA/DNA ratio) of yellow perch (Perca flavescens) in relation to environmental influences and spatio-temporal variation in a shallow fluvial lake. Can J Fish Aquat Sci 64: 1646-1655.

37. Leclerc EM, Mailhot Y, Mingelbier M, Bernatchez L (2008) The landscape genetics of yellow perch (Perca flavescens) in a large fluvial ecosystem. Mol Ecol 17: 1702-1717.

38. Gaudeul M, Till-Bottraud I, Barjon F, Manel S (2004) Genetic diversity and differentiation in Eryngium alpinum L. (Apiaceae): comparison of AFLP and microsatellite markers. Heredity 92: 508-518.

39. Whiteside MC, Swindoll CM, Doolittle WL (1985) Factors affecting the early life history of yellow perch, Perca flavescens. Environ Biol Fish 12: 47-56.

40. Balloux F (2001) EASYPOP (version 1.7): a computer program for population genetics simulations. J Hered 92: 301-302.

41. Balloux F, Brunner H, Lugon Moulin N, Hausser J, Goudet J (2000) Microsatellites can be misleading: an empirical and simulation study. Evolution 54: 1414-1422.

42. Nauta MJ, Weissing FJ (1996) Constraints on allele size at microsatellite loci: implications for genetic differentiation. Genetics 143: 1021-1032.

43. Slatkin M (1995) A measure of population subdivision based on microsatellite allele frequencies. Genetics 139: 457-462.

44. Paradis Y, Mingelbier M, Brodeur P, Magnan P (2008) Comparisons of catch and precision of pop nets, push nets, and seines for sampling larval and juvenile yellow perch. N Am J Fish Manag 28: 1554-1562.

45. Gagnon MC, Angers B (2006) The determinant role of temporary proglacial drainages on the genetic structure of fishes. Mol Ecol 15: 1051-1065.

46. Sambrook J, Fritsch EF, Maniatis T (1989) Molecular cloning: a laboratory manual (2ndeds) Cold Spring Harbor Laboratory Press, New York.

47. Vos P, Hogers R, Bleeker M, Reijans M, Lee T, et al. (1995) AFLP: a new technique for DNA fingerprinting. Nucleic Acids Res 23: 4407-4414.

48. Van der Wurff AWG, Chan YL, Van Straalen NM, Schouten J (2000) TE-AFLP: combining rapidity and robustness in DNA fingerprinting. Nucleic Acids Res 28: e105.

49. Bassam BJ, Caetano-Anollés G, Gresshoff PM (1991) Fast and sensitive silver staining of DNA in polyacrylamide gels. Anal Biochem 196: 80-83.

50. Krauss SL (2000) Accurate gene diversity estimates from amplified fragment length polymorphism (AFLP) markers. Mol Ecol 9: 1241-1245.

51. Weir BS, Cockerham CC (1984) Estimating F-statistics for the analysis of population structure. Evolution 38: 1358-1370.

52. Vekemans X (2002) AFLP-surv version 1.0. Distributed by the author. Laboratoire de Génétique et Ecologie Végétale, Université Libre de Bruxelles, Belgium.

53. Foll M, Gaggiotti O (2008) A genome-scan method to detect selected loci appropriate for both dominant and codominant markers: a Bayesian perspective. Genetics 180: 977-993.

54. Antao T, Beaumont MA (2011) Mcheza: a workbench to detect selection using dominant markers. Bioinformatics 27: 1717-1718.

55. Danielsson PE (1980) Euclidean distance mapping. Comput Graph Image Process 14: 227-248.

56. Jin L, Chakraborty R (1994) Estimation of genetic distance and coefficient of gene diversity from single-probe multilocus DNA fingerprinting data. Mol Biol Evol 11: 120-127.

57. Saitou N, Nei M (1987) The neighbor-joining method: a new method for reconstructing phylogenetic trees. Mol Biol Evol 4: 406-425.

58. Fitch WM (1971) Toward defining the course of evolution: minimum change for a specific tree topology. Syst Zool 20: 406-416.

59. Langella O (2007) Populations 1.2. 30: Population genetic software.

60. Piry S, Alapetite A, Cornuet JM, Paetkau D, Baudouin L, et al. (2004) GENECLASS2: a software for genetic assignment and first-generation migrant detection. J Hered 95: 536-539.

61. Duchesne P, Bernatchez L (2002) AFLPOP: a computer program for simulated and real population allocation, based on AFLP data. Mol Ecol Notes 2: 380-383.

62. Post JR, McQueen DJ (1988) Ontogenetic changes in the distribution of larval and juvenile yellow perch (Perca flavescens): A response to prey or predators? Can J Fish Aquat Sci 45: 1820-1826.

63. Williamson CE, Metzgar SL, Lovera PA, Moeller RE (1997) Solar ultraviolet radiation and the spawning habitat of yellow perch, Perca flavescens. Ecol Appl 7: 1017-1023.

64. Brazner JC, Campana SE, Tanner DK (2004) Habitat fingerprints for Lake Superior coastal wetlands derived from elemental analysis of yellow perch otoliths. Trans Am Fish Soc 133: 692-704.

65. JC, Campana SE, Tanner DK, Schram ST (2004) Reconstructing habitat use and wetland nursery origin of yellow perch from Lake Superior using otolith elemental analysis. J Great Lakes Res 30: 492-507.

66. Balling TE, Pfeiffer W (1997) Location dependent infection of fish parasites in Lake Constance. J Fish Biol 51: 1025-1032.

67. Muncy RJ (1962) Life history of the yellow perch, Perca flavescens, in estuarine waters of Severn River, a tributary of Chesapeake Bay, Maryland. Chesap Sci 3: 143-159.

68. Glover DC, Dettmers JM, Wahl DH, Clapp DF (2008) Yellow perch (Perca flavescens) stock structure in Lake Michigan: an analysis using mark-recapture data. Can J Fish Aquat Sci 65: 1919-1930.

69. Willemsen J (1977) Population dynamics of percids in Lake IJssel and some smaller lakes in the Netherlands. J Fish Res Board Can 34: 1710-1719.

70. Kipling C, Le Cren ED (1984) Mark recapture experiments on fish in Windermere, 1943–1982. J Fish Biol 24: 395-414.

71. Bergek S, Björklund M (2007) Cryptic barriers to dispersal within a lake allow genetic differentiation of Eurasian perch. Evolution 61: 2035-2041.

72. Behrmann-Godel J, Gerlach G, Eckmann R (2006) Kin and population recognition in sympatric Lake Constance perch (Perca fluviatilis L.): can assortative shoaling drive population divergence? Behav Ecol Sociobiol 59: 461-468.

73. Bensch S, Åkesson M (2005) Ten years of AFLP in ecology and evolution: why so few animals? Mol Ecol 14: 2899-2914.

74. Mariette S, Chagné D, Lézier C, Pastuszka P, Raffin A, et al. (2001) Genetic diversity within and among Pinus pinaster populations: comparison between AFLP and microsatellite markers. Heredity 86: 469-479.

75. Ravel S, Monteny N, Olmos DV, Verdugo JE, Cuny G (2001) A preliminary study of the population genetics of Aedes aegypti (Diptera: Culicidae) from Mexico using microsatellite and AFLP markers. Acta Trop 78: 241-250.

76. Papetti C, Susana E, La Mesa M, Kock KH, Patarnello T, et al. (2007) Microsatellite analysis reveals genetic differentiation between year-classes in the icefish Chaenocephalus aceratus at South Shetlands and Elephant Island. Polar Biol 30: 1605-1613.

77. Papetti C, Susana E, Patarnello T, Zane L (2009) Spatial and temporal boundaries to gene flow between Chaenocephalus aceratus populations at South Orkney and South Shetlands. Mar Ecol Progr Ser 376: 269-281.

78. Maes GE, Pujolar JM, Hellemans B, Volckaert FAM (2006) Evidence for isolation by time in the European eel (Anguilla anguilla L.). Mol Ecol 15: 2095-2107.

79. Lande R (1988) Genetics and demography in biological conservation. Science 241: 1455-1460.

80. MacLean JA, Evans DO (1981) The stock concept, discreteness of fish stocks, and fisheries management. Can J Fish Aquat Sci 38: 1889-1898.

81. Dionne M, Caron F, Dodson JJ, Bernatchez L (2009) Comparative survey of within-river genetic structure in Atlantic salmon; relevance for management and conservation. Conserv Genet 10: 869-879.

31

Effects of Zinc Amino Acid in Walking Catfish (*Clarias macrocephalus*) Female Broodstock First Sexual Maturation

Siti-Ariza Aripin[1]*, Orapint Jintasataporn[2] and Ruangvit Yoonpundh[2]

[1]*School of Fisheries and Aquaculture Sciences, Universiti Malaysia Terengganu, 21030, Terengganu, Malaysia*
[2]*Faculty of Fisheries, Kasetsart University, 10900, Bangkok, Thailand*

Abstract

This study examines the effects of zinc amino acid (ZnAA) to the first sexual maturity stage in female broodstock of the Walking catfish, *Clarias macrocephalus*. The different ZnAA levels of Control (0 ppm ZnAA), ZnAA1 (100 ppm ZnAA) and ZnAA2 (200 ppm ZnAA) in the diet was applied to the first sexual maturation female catfish (Availa®Zn, Zinpro Corporation, Eden Prairie, MN USA). ZnAA accumulation, broodstock maturation analysis and breeding performance were evaluated. The ZnAA treatment has significant different in serum, meat and ovary ZnAA accumulation. The ZnAA treatment increased the fecundity, gonadosomatic index, egg diameter and development of oocytes at tertiary yolk stage. In comparison, the ZnAA treatment was insignificant in estradiol level. During artificial fertilization, the ZnAA treatment enhanced the fertilization rate and the larval survival rate. During recovery breeding, ZnAA treatment significantly increased the egg production and larval hatching rate. The optimum level to enhance the *Clarias macrocephalus* female broodstock first sexual maturation is ZnAA1.

Keywords: Zinc amino acid; Maturation; Reproduction; Female; Catfish

Introduction

Walking catfish (*Clarias macrocephalus*) is one of the favourite aquaculture species in Southeast Asia especially Thailand. It has a high market value due to its tender flesh and delicious flavour [1,2]. Despite being an important aquaculture species, *C. macrocephalus* population continues to decline due to several issues. *C. macrocephalus* population has become endangered due to the introduction of *Clarias* hybrid species and limited supply of wild broodstock caused by over exploitation in aquaculture farming that claimed the natural habitat of this species [1,3]. As a result, it has taken a toll in the supply of mature broodstock catfish for artificial propagation. Zinc is an essential trace element that plays an important role as a co-factor of enzymes and is a component of many important metallo enzymes. Hence, zinc is fundamentally important for the functioning of reproductive system [4]. Zinc deficiency will cause retarded growth, delayed sexual development, impaired reproduction in males and females, congenital abnormalities, and low hatching rate [4,5]. According to Salgueiro et al. [6], zinc supplementation is able to improve the infertility in female. Supplemented zinc can be absorbed by intestine and delivered to the liver and becomes zinc protein or zinc metallo thionein [7]. The zinc protein in the liver or vitellogenin is transported via the blood to the ovary in order to enhance the oocytes growth for the developing embryo and larvae after fertilization [8]. Zinc is also involved in the production and secretion of luteinizing hormone (GTH-II), follicle-stimulating hormone (GTH-I) and prolactin [6,9]. Zinc amino acid (ZnAA) is zinc that bind to amino acids ligand. According to Formigoni et al. [10], organic minerals including zinc, is capable to bind with ligand such as amino acids, peptides and proteins. In addition, the organic zinc has higher retention, bioavailability and absorption rate compared to inorganic zinc such as $ZnSO_4$ or ZnO [11]. Information on the occurrence and metabolic roles of ZnAA in broodstock maturation and reproductive performance is important in order to initiate and enhance the first sexual maturity of female *C. macrocephalus* ZnAA fundamental mechanism remains unclear even though there are strong evidences of its roles in enhancing the reproductive performance. In order to enhance the maturation and early embryonic development, it is important to investigate the effects of ZnAA administration to the first sexual maturity stage in female broodstock of the *C. macrocephalus*.

Materials and Methods

Fish and culture condition

The maiden *Clarias macrocephalus* female broodstock were obtained from the Fisheries Station of Kham Pheng Phet, Department of Fisheries, Ministry of Agriculture and Cooperative, Thailand. The experiment trial was carried out at the Laboratory of Nutrition and Aquafeed, Department of Aquaculture, Faculty of Fisheries, Kasetsart University, Bangkok, Thailand. Approximately eighteen weeks old catfish were acclimatized and maintained in 500L tanks at the density of 15 ind/m²/fish/tank and were fed with control feed for two weeks prior to the experiment. A total of 45 females (initial weight 63.85 ± 4.97 g) were subjected to normal photoperiod (12 hours day light) prior to treatment and fed at a level equivalent to 3% of their body weight. The diet was divided into two equal feedings per day. The fishes were randomly distributed in three treatments (Control, ZnAA1 and ZnAA2) and with three replicates. The experiment tanks were continuously aerated to maintain the oxygen supply and the duration of experiment was for eight week.

Experimental diets

The basal diet was formulated from practical ingredients that contained approximately 22% fishmeal, 35% soybean, 1% spirulina, 12% wheat flour, 11.8% tapioca, 5% ricebran, 2% fish oil, 3% soy oil, 1.2% mineral premix, 2% soy lecithin, 1.5% calcium phosphate, 1% attractant, 2% binder and 0.5% vitamin premix (Table 1). The diet

*Corresponding author: Siti-Ariza Aripin School of Fisheries and Aquaculture Sciences, Universiti Malaysia Terengganu, Malaysia, E-mail: siti.ariza@umt.edu.my

Ingredient	Dry weight (%)
Fishmeal	22.0
Soybean	35.0
Spirulina	1.0
Wheat flour	12.0
Tapioca	11.8
De oil ricebran	5.0
Tuna fish oil	2.0
Soya oil	3.0
Mineral premix	0.5
Soy lecithin	2.0
Calcium phosphate	1.5
MgSO4.7H2O	0.1
KCl	0.6
Attractant	1.0
Binder	2.0
Vitamin premix	0.5

Table 1: Ingredients of the basal diet.

Proximate analysis (%)	Control	ZnAA1	ZnAA2
Moisture	3.6	3.0	3.0
Crude protein	37.7	38.0	37.8
Crude Fiber	9.3	9.3	9.3
Ash	1.5	1.7	1.6
Calcium	11.6	11.7	11.8
Phosphorus	1.8	1.9	1.9
Ether extract	1.2	1.3	1.4

Table 2: Proximate analysis for different ZnAA level in the diet.

also consisted of 37% crude protein and 9.3% crude lipid (Table 2). Diets containing ZnAA (one zinc ion bound to one amino acid ion) were prepared by adding different levels of ZnAA (Availa®Zn, Zinpro Corporation, Eden Prairie, MN USA) to the basal diet. These ZnAA concentrations were 0 ppm (control), 100 ppm (ZnAA1) and 200 ppm (ZnAA2) per kilogram in the diet according to NRC [12].

Zinc analysis

The ZnAA analysis was performed by using Inductive Couple Plasma–Optical Emission Spectrophotometer (ICP–OES) at Central Laboratory (Thailand) Company Limited. 0.25-2 g of samples were prepared for the analysis. All determinations were made in three replicates. Weighted triplicate of the samples were mixed with 7 ml nitric acid and 1 ml hydrogen peroxide in each flask. Microwave digestion was applied at 220°C for 45 minutes. After cooling, the resulting solutions were diluted up to 25 ml in volumetric flasks with deionised water. Blanks were prepared in the same way as the sample but excluding the samples. The prepared samples were injected and analysed in ICP–OES machine. The results were compared with standard curve to determine the ZnAA concentration [13].

Growth performance

All female *C. macrocephalus* were anaesthetised with clove oil and were individually weighted prior and after the experiment to measure the fish growth by using the following formula:

Weight gain (%) = (Final body weight – Initial body weight) ÷ Initial body weight × 100

Histology

The histology method was in accordance to Drury and Wallington

[14]. The ovary samples were fixed in 10% buffered formalin and dehydrated in graded alcohol series. The ovaries were embedded in paraffin wax, cut to four micrometer, stained with hematoxylin-eosin and observed under Motic microscope (Motic BA210 Digital Laboratory Microscope with Moticam 1000 camera). Staging of female oocytes was performed in accordance to Lubzens et al. [15].

Gonadosomatic index

The Gonadosomatic Index was determined: GSI (%) = 100 × ovary weight ÷ body weight

Estradiol analysis

Chemiluminometric enzyme immunoassay of estradiol was determined by accordance to Ibrahim and Harabawy [16]. Estradiol analysis was conducted using commercially available immunoassay kit (IMMULITE® Estradiol, Siemens Medical Solution Diagnostic and United Kingdom).

Artificial breeding

Artificial breeding was conducted with the remaining female *C. macrocephalus* from each treatment to evaluate the reproductive performance of ZnAA experiment. The female broodstock were lightly anaesthetized with clove oil and were individually inspected to observe matured fish indication including papilla colouring, abdominal swelling and swollen papilla. A mixture of 30 µg buserelin acetate (LHRH analogue) and 10 mg domperidone (dopamine analogue) were induced into one kilogram female broodstock (0.1ml mixture for 100 g females) [17]. Male African catfish milt was obtained from dissected testis as the control male to standardize the male semen quality. The eggs and milt were mixed for fertilization and the fertilized eggs were then transferred to the hatching tanks for incubation.

Fish reproductive characteristics

Egg production, fertilization rate, hatching rate, gonadosomatic index (GSI) and survival rate (seven days old) of the larvae were investigated after the artificial fertilization. Egg production was estimated by direct counting of sub-sample of fertilized eggs in the female ovaries [18]. Oocytes diameters were measured from fresh ovarian tissue with Motic microscope (Motic BA210 Digital Laboratory Microscope with Moticam 1000 camera). Fertilization rate, larval hatching rate and larval survival rate were determined in accordance to Unuma et al. [19]. A recovery breeding was conducted a month later to assess the recovery breeding performance with similar parameters used in the initial breeding session.

Statistical analysis

Statistical analysis was performed using SPSS software. Data were analysed by one-way ANOVA (analysis of variance) and by the Duncan test which analyses the significant differences among means. The means comparisons significance was tested at P<0.05 [20].

Results

The *C. macrocephalus* female broodstock were subjected to ZnAA accumulation analysis at the end of the experiment trial. ZnAA concentrations in serum, meat and ovary were significantly different between the treatments with P value of 0.006, 0.013 and 0.03 respectively (Figure 1 and Table 3). However, there were no statistical differences for liver, bone, egg and total tissue with the P value at 0.07, 0.9, 0.1, and 0.8, respectively (Table 3). Dietary ZnAA treatment did not significantly affect the weight gain with the mean weight gain

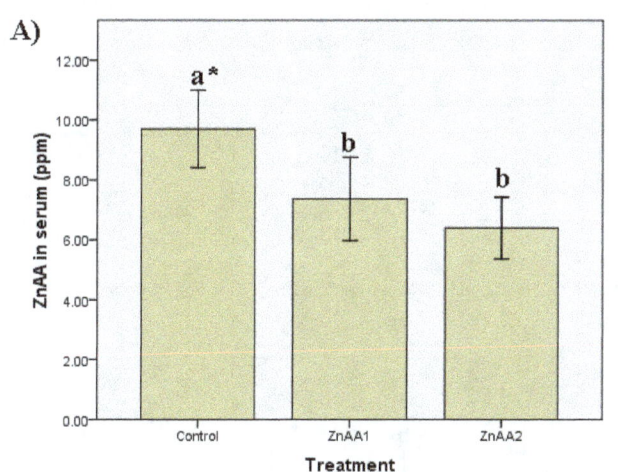

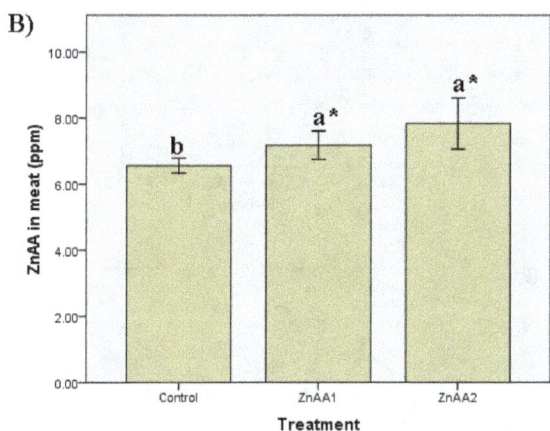

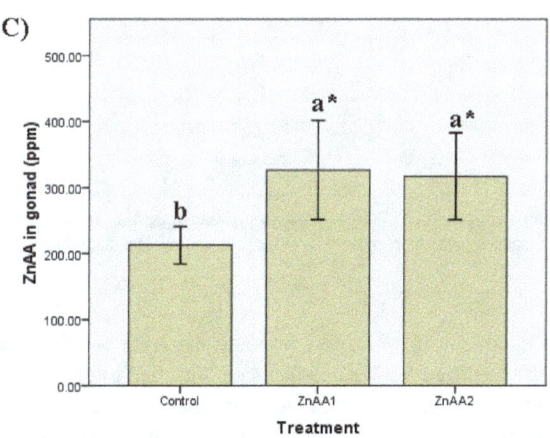

Figure 1: ZnAA concentration in serum (A), meat (B) and ovary (C) of female *C. macrocephalus* after eight weeks of ZnAA treatment. Values are expressed as mean ± SEM p<0.05.

ranged from 10.3–14.2% with the value at p=0.6. Similar result was found in estradiol profile (p=0.4) where the estradiol profile was not significantly different between treatments (Table 4). In histological analysis, most ovaries in the control group contained the highest percentage of peri-nucleolus stage (PNS) where the percentage was 51.5% (control), 28.7% (ZnAA1) and 20.8% (ZnAA2) (Figure 2 and

Table 4). Differences in ovarian histology between the ZnAA treatments after eight weeks were highest in the ZnAA2 group for tertiary yolk stage (TYS). The well-developed oocytes at tertiary yolk stage (TYS) were 15.5% (control), 50.5% (ZnAA1) and 54.6% (ZnAA2) (Figure 2 and Table 4). After the experiment trial, there were significant increase in the gonadosomatic index (GSI), fecundity, and egg diameter in the presence of ZnAA treatment with the P value at 0.002, 0.025, and 0.001, respectively (Figure 3 and Table 4). The significant difference in GSI, fecundity, egg diameter and also the prominent evidence in histology

Control	ZnAA1	ZnAA2	P value
9.70[a] ± 1.5	7.37[b] ± 1.7	6.40[b] ± 1.2	0.006
6.56[b] ± 0.3	7.18[a] ± 0.5	7.84[a] ± 0.9	0.013
25.34 ± 0.9	29.60 ± 6.5	23.51 ± 3.2	0.07
30.43 ± 4.4	31.07 ± 3.2	30.83 ± 17.8	0.9
55.09 ± 7.0	52.65 ± 2.8	49.44 ± 2.4	0.1
212.81[b] ± 34.5	326.88[a] ± 92.2	317.20[a] ± 80.6	0.03
140.35 ± 29.6	142.42 ± 36.2	133.35 ± 35.7	0.8

[a,b] Values with different superscripts in a row differ significantly (*P*<0.05).

Table 3: ZnAA concentration in serum, meat, liver, bone, egg and ovary with different ZnAA levels (mean ± SD).

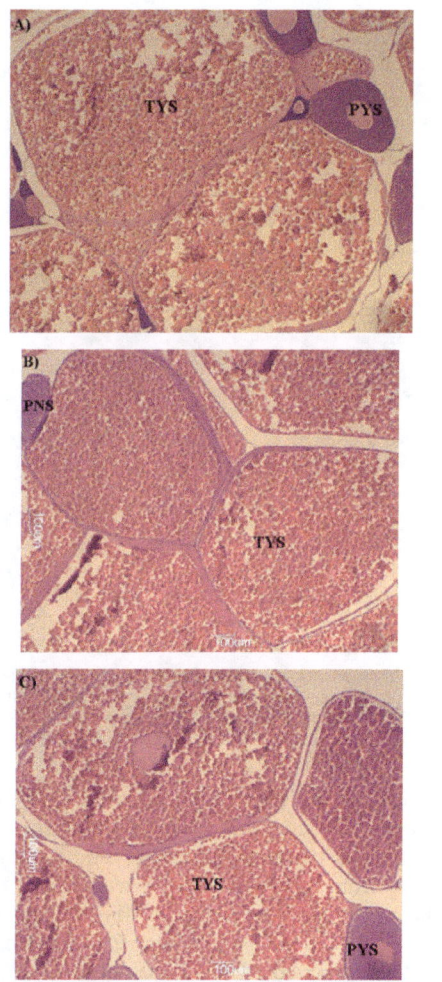

Figure 2: Effect of ZnAA treatment on ovarian histology of the *C. macrocephalus*. Cross section of an ovary of fish treated with control (A), ZnAA1 (B) and ZnAA2 (C). PNS; peri-nucleolus stage, PYS; primary yolk stage, TYS; tertiary yolk stage. Scale bar=100 µm.

Treatment	Control	ZnAA1	ZnAA2	P value
Weight gain (%)	10.3 ± 13	12.8 ± 10	14.2 ± 8	0.6
Estradiol (pg/ml)	486.7 ± 151	634.7 ± 191.5	1812 ± 2215	0.4
Histology tertiary yolk stage (TYS)	15.5%	50.5%	60.3%	-
Histology peri-nucleolus stage (PNS)	51.5%	28.7%	20.8%	-
Gonadosomatic index (%)	5.61[b] ± 0.44	9.03[a] ± 2.14	10.22[a] ± 2.46	0.002
Fecundity (egg/g)	3034.1[b] ± 329	4955.3[a] ± 2028	5756.5[a] ± 1766	0.025
Egg diameter (µm)	1448.4[b] ± 79	1520.3[a] ± 132.8	1530.5[a] ± 81.0	0.001

[a,b]Values with different superscripts in a row differ significantly ($P<0.05$)

Table 4: Maturation analysis of female *C. macrocephalus* with different ZnAA levels (mean ± SE).

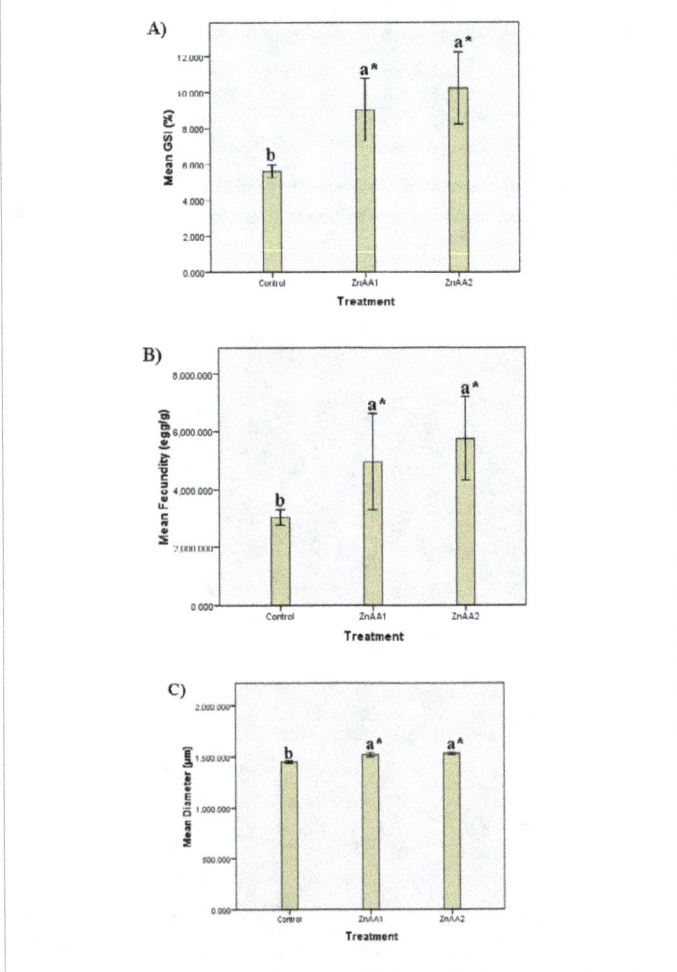

Figure 3: Mean gonadosomatic index (A), fecundity (B) and egg diameter (C) of female *C. macrocephalus* after eight weeks of ZnAA treatment. Values are expressed as mean ± SEM (GSI N: 6 animals/replicate, fecundity N: 6 animals/ replicate, egg diameter N: 180 eggs/replicate). p<0.05.

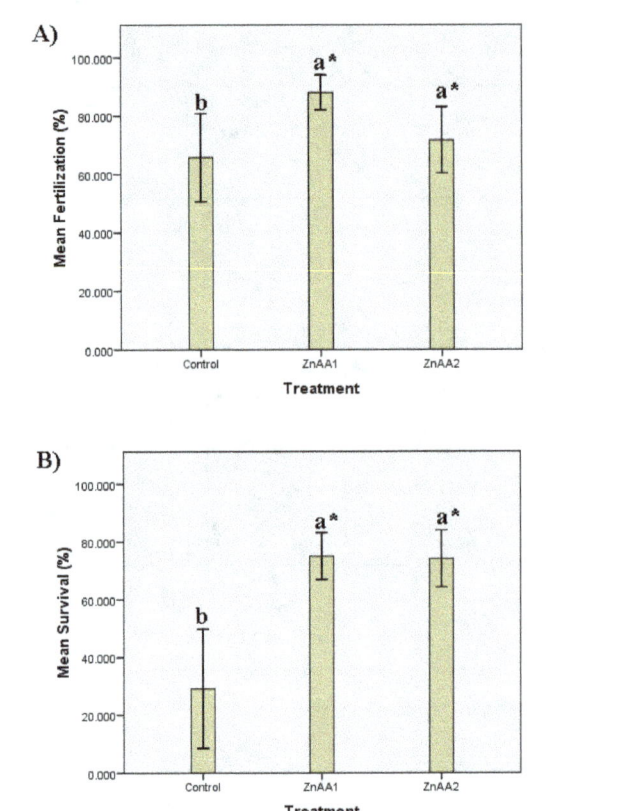

Figure 4: Mean spawning fertilization rate (A) and survival rate (B) after eight weeks of ZnAA treatment. Values are expressed as mean ± SEM p<0.05.

analysis indicated that a higher number of vitellogenic and matured follicles in ovaries were found in ZnAA treated treatment (ZnAA1 and ZnAA2). After eight weeks of ZnAA experiment trial, the female broodstock were artificially fertilized with semen from control males to evaluate the reproductive performance. The fertilization rate of the ZnAA treated fertilized female ZnAA1 and ZnAA2 were significantly higher (p=0.045) compared to the control groups (Figure 4A and Table 5). There was higher number of larval survival rate observed in females broodstock exposed in ZnAA treatment for *C. macrocephalus* with a P value at 0.001 (Figure 4B and Table 5). However, the ZnAA treatment has no significant different in the fecundity and hatching rate

with the P value at 0.2, and 0.7, respectively. After the first breeding session, the fertilized female broodstock continued to be fed with trial feed for another four weeks and were artificially fertilized with semen from control males to evaluate the recovery reproductive performance. There was higher percentage of female that ready to breed observed in ZnAA exposed treatment with the percentage at 50% (Control), 71.4% (ZnAA1) and 87.5% (ZnAA2) (Table 6). The fecundity and hatching rate of the all ZnAA treatments were significantly higher with the P value at 0.001, and 0.025, compared to the control groups (Figure 5 and Table 6). However, the ZnAA treatment has no significant effects in the larval survival rate (p=0.3) (Table 6).

Discussion

In this study, ZnAA accumulations were found to be significantly

Treatment	Control	ZnAA1	ZnAA2	P value
Fertilization Rate (%)	65.75[b] ± 21	88[a] ± 7.9	72[a] ± 16	0.045
Survival rate (%)	29.1[b] ± 17.9	75[a] ± 11	74[a] ± 13	0.001
Hatching rate (%)	10.5 ± 8	14.3 ± 14	16 ± 13	0.7
Egg production (egg/kg)	2105 ± 1045	2715 ± 973	3371 ± 1997	0.2

[a,b]Values with different superscripts in a row differ significantly (P<0.05)

Table 5: Breeding performance of female *C. macrocephalus* with different ZnAA levels (Mean ± SE).

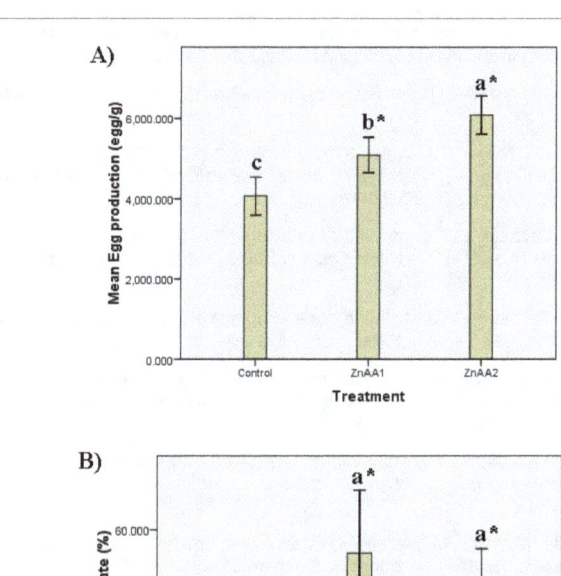

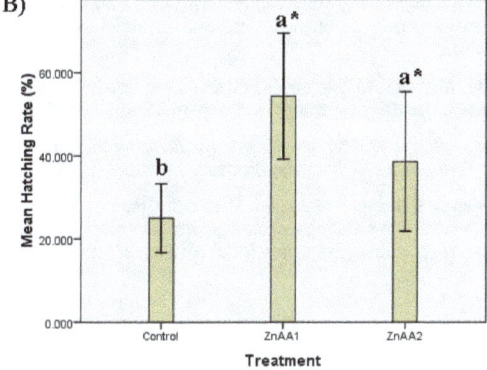

Figure 5: Mean egg production (A) and hatching rate (B) from the recovery breeding session. Values are expressed as mean ± SEM p<0.05.

Treatment	Control	ZnAA1	ZnAA2	P value
Breeding broodstock (%)	50%	71.4%	87.5%	-
Egg production(egg/kg)	4069.5[c] ± 472	5644.8[b] ± 493	6499[a] ± 630	0.001
Hatching rate (%)	25[b] ± 36	54[a] ± 29	39[a] ± 30	0.025
Survival Rate (%)	40.7 ± 9	54.3 ± 27	46.5 ± 25	0.6

[a,b,c]Values with different superscripts in a row differ significantly (P<0.05)

Table 6: Recovery breeding performance of female *C. macrocephalus* with different ZnAA levels (Mean ± SE).

different in serum, meat and ovary. According to Thompson et al. [21], exogenous ZnAA was then absorbed by intestine to the liver and was passed on to the ovary during reproductive development. This mechanism is regulated with metal-binding protein metallo thionein in the liver where the primary function of metallo thionein is to control a pool of loosely ligated ZnAA within the cell. In other study, Thompson et al. [21] stated that ZnAA is accumulated in the liver and transported by the bloodstream to other organs. Thus, it explains the significant increase of ZnAA in ovary in the current study. The ZnAA accumulation in other parts of the treated female such as bone

suggested that zinc stimulates bone formation [22]. Zinc also helps bone mineralization by acting as a cofactor for alkaline phosphatase [23]. In this study, the estradiol profile in the *C. macrocephalus* female broodstock was not significant with ZnAA treatment compared to the control group. Usually, estradiol is secreted into the bloodstream to the liver and stimulates the synthesis of vitellogenin production by hepatocytes [24]. In teleost, estradiol concentration gradually elevates during vitellogenesis and declines in response of luteinizing hormone (GTH-II) as oocytes begin their maturation [9,25]. Thus, it explains the insignificant difference of estradiol profile in this study. The level of GSI, fecundity, egg diameter and mature cells in histology analysis from the ZnAA treated group demonstrated a significant difference compared to the control group. Previous studies demonstrated that the estradiol induced vitellogenesis process in the liver [24]. Vitellogenesis is a process where vitellogenin is transported from the liver via the bloodstream to the ovary. In ovary, vitellogenin is incorporated by a receptor-mediated process into subsequently nourishing the developing embryo [21,26,27]. Vitellogenin is also known as zinc bounded protein [27]. The role of ZnAA protein is essential as cofactor for enzymes which are involved in DNA, RNA and protein synthesis as well as a requirement for membrane and polyribosome stability [8]. During vitellogenesis, oocytes increase in size, thus it explains the significant increase in GSI, fecundity, egg diameter and prominent mature cells in histology analysis. ZnAA metallo thionein influences the reproductive cycle by being involve in the hepatic synthesis of a yolk precursor (vitellogenin) and induces oocytes growth prior to fertilization. According to Banks et al. [8], vitellogenin are transported via the blood to the growing oocytes and processed, for the developing embryo and larvae after fertilization. In the present study, the fertilization rate and survival rate of the ZnAA treated fertilized female were significantly higher compare to the control group. According to Riggio et al. [28], the role for zinc during embryonic development had to act as a regulator of cell division and morphogenesis. While zinc application enhances the cell proliferation, zinc deprivation reduces cell division and stimulates congenital abnormalities of foetal organs derived from ectodermal, mesodermal and endodermal germ lines [5,28]. In recovery reproductive performance, the ZnAA treatment groups demonstrated a higher percentage of female that prompt to spawn. The other parameter such as fecundity and hatching rate showed a significant difference from the control group. The female broodstock in recovery experiment were fed with ZnAA treatment have longer duration. This condition favours more ZnAA accumulation in the liver thus, increased the levels of ZnAA metallothionein for vitellogenesis. The vitellogenesis produces vitellogenin in the liver and transports it to the ovary and nourishing the developing embryo [21].

Conclusion

The recent study indicated that the optimum zinc amino acid treatment in enhancing the *Clarias macrocephalus* female broodstock first sexual maturation and improving the reproductive performance is ZnAA1 (100 ppm ZnAA).

Acknowledgement

The authors would like to thank the members of Laboratory of Nutrition and Aquafeed, Department of Aquaculture, Faculty of Fisheries, Kasetsart University, Bangkok for the assistance during this study. This study was funded by Ministry of Education, Malaysia.

References

1. Petkam R, Moodie GEE (2001) Food particle size, feeding frequency, and the use of prepared food to culture larval walking catfish (*Clarias macrocephalus*). Aqua 194: 349-362.

2. Areerat S (1987) *Clarias* culture in Thailand. Aqua 63: 355-362.

3. Na-Nakorn U, Kamonrat W, Ngamsiri T (2004) Genetic diversity of walking catfish, *Clarias macrocephalus*, in Thailand and evidence of genetic introgression from introduced farmed *C. gariepinus*. Aqua 240: 145-163.

4. Baker DH, Ammerman CB (1995) Zinc bioavailability. In Bioavailability of nutrients for animals: amino acids, minerals and vitamins.

5. Black RE (2001) Micronutrients in pregnancy. Br J of Nutr 85: 193-197.

6. Salgueiro MJ, Zubillaga M, Lysionek A, Sarabia MI, Care R, et al (2000) Zinc as an Essential Micro Nutrient: A Review. Nutr Res 20: 737-755.

7. Thompson ED, Mayer GD, Balesaria S, Glover CN, Walsh PJ, et al. (2003) Physiology and endocrinology of zinc accumulation during the female squirrelfish reproductive cycle. Comp Biochem Physiol part A 134: 819-828.

8. Banks SD, Thomas P, Baer KN (1999) Seasonal variations in hepatic and ovarian zinc concentrations during the annual reproductive cycle in female channel catfish (*Ictalurus punctatus*). Comp Biochem Physiol Part C 124: 65-72.

9. Aizen J, Kowalsman N, Niv MY, Levavi-Sivan B (2014) Characterization of tilapia (*Oreochromis niloticus*) gonadotropins by modelling and immunoneutralization. General and Comparative Endocrinology 207: 28-33.

10. Formigoni A, Fustini M, Archetti L, Emanuele S, Sniffen C, et al. (2011) Effects of an organic source of copper, manganese and zinc on dairy cattle productive performance, health status and fertility. Animal Feed Science and Technology 164: 191-198.

11. Garg AK, Mudgal V, Dass RS (2008) Effect of organic zinc supplementation on growth, nutrient utilization and mineral profile in lambs. Animal Feed Science and Technology 144: 82-96.

12. NRC (2011) Nutrient Requirements of Fish. National Academies Press, Washington.

13. Xie N, Huang J, Li B, Cheng J, Wang Z, et al. (2015) Affinity purification and characterisation of zinc chelating peptides from rapeseed protein hydrolysates: Possible contribution of characteristic amino acid residues. Food Chemistry 173: 210-217.

14. Drury RA, Wallington EA (1967) Carleton's Histological Technique. Oxford University Press, Oxford.

15. Lubzens E, Young G, Bobe J, Cerda J (2010) Oogenesis in teleost: how eggs are formed. General and Comparative Endocrinology 165: 367-389.

16. Ibrahim ATA Harabawy ASA (2014) Sublethal toxicity of carbofuran on the African catfish *Clarias gariepinus*: Hormonal, enzymatic and antioxidant responses. Ecotoxicology and Environmental Safety 106: 33-39.

17. Fermin AC (1991) LHRH-a and domperidone-induced oocytes maturation and ovulation in bighead carp, *Aristichthys nobilis* (Richardson). Aqua 93: 87-94.

18. Bagenal TB (1978) Aspects of Fish Fecundity. In Ecology of Freshwater Fish Production, Blackwell Scientific Publications, Oxford.

19. Unuma T, Kondo S, Tanaka H, Kagawa H, Nomura K, et al. (2004) Determination of the rates of fertilization, hatching and survival in the Japanese eel, *Anguilla japonica*, using tissue culture microplates. Aqua 241: 345-356.

20. Gomez KA, Gomez AA (1984) Statistical procedures in agricultural research, Wiley, New York.

21. Thompson ED, Mayer GD, Walsh PJ, Hogstrand C (2002) Sexual maturation and reproductive zinc physiology in the female squirrelfish. The Journal of Experimental Biology 205: 3367-3376.

22. Sa MVC, Pezzato LE, Lima MMBF, Padilha PM (2004). Optimum zinc supplementation level in Nile tilapia *Oreochromis niloticus* juveniles diets. Aqua 238: 385-401.

23. Yamaguchi M (1998) Role of zinc in bone formation and bone resorption. The Journal of Trace Elements in Experimental Medicine 11: 119-135.

24. Sun B, Pankhurst NW (2006) In vitro effect of vitellogenin on steroid production by ovarian follicles of greenback flounder *Rhombosolea tapirina*. Comp Biochem Physiol Part A 144: 78-85.

25. Devlin RH, Nagahama Y (2002) Sex determination and sex differentiation in fish: an overview of genetic, physiological, and environmental influences. Aqua 208: 191-364.

26. Falchuk KH, Montorzi M, Vallee BL (1995) Zinc uptake and distribution in *Xenopus laevis* oocytes and embryos. Biochem 34: 16524-16531.

27. Montorzi M, Falchuk KH, Vallee BL (1995) Vitellogenin and lipovitellin: zinc proteins of *Xenopus laevis* oocytes. Biochem 34: 10851-10858.

28. Riggio M, Filosa S, Parisi E, Scudiero R (2003) Changes in zinc, copper and metallothionein contents during oocytes growth and early development of the teleost *Danio rerio* (zebrafish). Comp Biochem Physiol Part C 135: 191-196.

32

Monitoring a Massive Escape of European Sea Bass (*Dicentrarchus Labrax*) at an Oceanic Island: Potential Species Establishment

Besay Ramírez[1]*, Leonor Ortega[1,2], Daniel Montero[3], Fernando Tuya[1] and Ricardo Haroun[1]

[1]*Research Group on Biodiversity and Conservation, Center for Biodiversity and Environmental Management, University of Las Palmas de Gran Canaria, Las Palmas 35017, Spain*
[2]*Philippe Cousteau "Union of the Ocean" Foundation, C/ General Oraá 26, 28006 Madrid, Spain*
[3]*Aquaculture Research Group, University of Las Palmas de Gran Canaria, P.O. Box 56, 35200 Telde, Spain*

Abstract

The post-escape behavior of aquaculture escapees is a growing topic of research. We monitored a massive escape event of the European sea bass, *Dicentrarchus labrax*, which occurred at a sea-cage fish farm off the oceanic island of La Palma, Canary Islands, eastern Atlantic. Stomach contents and gonadal development of escapees were analyzed from two islands (Gran Canaria and La Palma) in order to assess the degree of post-escape establishment,. We also tested (at both islands), the suitability of fatty acid profiles as biomarkers of aquaculture escapes, processing recaptured escaped fish at a range of distances away from aquaculture facilities. Escaped European sea bass concentrated within breakwaters and decreased in abundance through time after the massive escape at La Palma. Decapod crustaceans (particularly *Percnon gibbesi* and *Rhynchocinetes sp*) were the main diet constituents of escapees, followed by fishes (mainly the parrotfish, *Sparisoma cretense*). Only one spawner male was found. Crude lipid, oleic acid, linoleic acid, linolenic acid, eicosapentaenoic acid, $\sum$n-9 fatty acids and $\sum$monounsatured fatty acids showed higher values in cultured or escaped individuals near cages relative to fish far away from farms. Arachidonic acid, docosahexaenoic acid, $\sum$n-3, satured fatty acids, $\sum$n-3/$\sum$n-6 ratio and Palmitic acid showed the opposite pattern. Our data showed that escaped European sea bass is able to exploit natural recourses, altering their fatty acid profiles relative to farmed conspecifics. The usefulness of fatty acids as biomarkers is, however, limited to a short period of time after escape events.

Keywords: Escapees; Stomach contents; Fatty acids; Gonadal development; Canary islands

Introduction

Among environmental issues facing aquaculture, escapes are widely regarded as a major problem in the marine environment, including genetic interactions through inbreeding, transfer of pathogens, prey predation, introduction of alien species, habitat alteration, etc. [1-8]. The European sea bass, *Dicentrarchus labrax*, is a voracious predator, feeding mainly on crustaceans, molluscs and fishes [9-13], and its diet changes with fish size [13-15] for escaped individuals, the diet also depends on the time at liberty [13].

Escapes at some locations may be particularly problematic, principally where local populations are reduced, or in areas outside the species' natural distribution range [7]. Current populations of the European sea bass in studied islands (Gran Canaria and La Palma) (Figure 1) are related to aquaculture escapes because no native populations had existed [12,16]. In these islands, for example, escaped European sea bass diet overlaps with other top predators and may become a new competitor for local species [12,13,17]. Yet, there is no reported evidence of reproduction of escaped European sea bass from studied islands [12,17], although developed gonads have been found [12].

Several morphological and physiological indicators have been proposed as useful tools for the identification of escapes, in which the FA profile has been claimed to be a good bio-indicator [18-20]. The FA profile has received increasing attention due to changes in aqua feed ingredients over recent years. Traditionally, fish were fed with diets based on fishmeal and fish oil as the main ingredients to ensure a competitive price and adequate content in some essential FAs for fish, such as docosahexaenoic acid (DHA) (22:6n−3), eicosapentaenoic acid (EPA) (20:5n−3) and arachidonic acid (ARA) (20:4n−6). These FAs,

which have a plethora of very important functions, are considered essential for marine fish, since these species do not have the ability to bio-convert shorter FAs into these FAs [21]. Given that the world production of fish oil is stagnated with the consequent increase in cost, there is a strong trend for the use of vegetable oils as fish oils substitutes [22]. These FAs precursors are incorporated into the tissues of farmed fish, and can also be transferred to wild fish [23]. Wild fishes around offshore aquaculture cages may feed on pellets released from farms, resulting in changes in body condition and FA profiles, bringing their body composition and that of other organisms in different trophic levels closer to that of cultivated fish [23-25]. For this reason, the presence of certain FAs, such as LA, ARA and oleic acid (OA) (18: 1n−9), in wild organisms has been proposed as an indicator of the influence of aquaculture on marine ecosystems [23,25-27].

The goal of this paper was, firstly, to analyze the spatio-temporal variability in the population structure of escaped European sea bass after a massive escape (approximately 1,500,000 fish, 400,000 kg, 30-45 cm of total length at the moment of escape) that occurred on February 2010 at La Palma Island (Canary Islands, eastern Atlantic)

*Corresponding author: Besay Ramírez, Research Group on Biodiversity and Conservation, Center for Biodiversity and Environmental Management, University of Las Palmas de Gran Canaria, Las Palmas 35017, Spain
E-mail: besay.ramirez@gmail.com

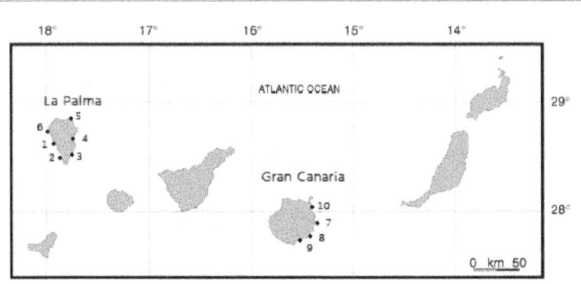

Figure 1: Map of the study area detailing sampling locations at La Palma Island: 1 Tazacorte (Acuipalma S.L. farm); 2 El Remo (marine reserve); 3 La Salemera; 4 Santa Cruz de la Palma; 5 Puerto Talavera; 6 Puntagorda; and Gran Canaria Island: 7 Melenara (ADSA S.L. and Canexmar S.L. farms); 8Arinaga; 9 Castillo del Romeral (ADSA S.L. and Playa Vargas S.L. farms); 10 (San Cristobal).

(Figure 1). We secondly determined whether escaped European sea bass diet, reproductive potential and FA profile changed with distance from aquaculture cages. We hypothesize that: (1) escaped fish can redistribute around the entire island perimeter after a massive escape, (2) being able to adapt to the wild by consuming local prey, (3) lacking reproduction and (4) altering their FA profiles depending on distance from the source of escapees.

Materials and Methods

Sampling design and study locations

To monitor the massive escape of European sea bass that occurred between the 20 and 26th February 2010 at La Palma Island, we selected six locations throughout the entire island perimeter (Figure 1). One location was immediately adjacent to the escape point (a sea-cage fish farm off Tazacorte, Figure 1) and the rest were selected at different distances away from this point, northward and southward around the entire island perimeter. One location (El Remo) was set within a marine reserve (*Reserva Marina Isla de La Palma*). No location was selected in the north face of the island due to prevailing swells from the NW that impedes regular sampling. This protocol was repeated six times (i.e. sampling campaigns): August 2010, September 2010, October 2010, November 2010, August 2011 and October 2011. Locations encompassed a range of habitats at shallow water (*ca.* 5-10 m depth).

To study stomach contents, gonad development and FAs profile, escaped European sea bass were captured by spear-fishing throughout the entire study, i.e. from August 2010 to October 2011, at both La Palma and Gran Canaria. Although some of them presented unusable tissues due to shaft impacts, this was the most appropriate and effective capture method because local fishermen do not fish close to the surf, where European sea bass are often grouped [12]. Catches were kept in ice until processed in the laboratory, where individuals were weighed (accuracy to 0.1 g) and measured (accuracy to 0.1 cm). The gonads, stomachs, muscle tissue and livers were removed from each individual; the gonads were preserved in formaldehyde and the stomachs preserved in 70% ethanol; muscle tissue and livers were frozen for subsequent lipid analysis. Samples for lipid analyses were categorized according to 3 distances from aquaculture facilities: 'Cages': 0 km (inside cages), 'Near cages': <10 km and 'Far away from cages': >10 km. In addition to samples collected at La Palma Island (i.e. where the escape event took place), samples were simultaneously taken at Gran Canaria Island (Figure 1) to assess whether patterns with distance from aquaculture cages for a range of descriptors (see below) were consistent between the

two islands. By increasing the spatial replication, the robustness of the study is enhanced if results were consistent between both islands. In La Palma Island, there is only one off-shore farm, while at Gran Canaria there are four off–shore fish farms located at two locations (Figure 1).

Fish surveys

At each location and time, fish were counted by means of visual census techniques through four replicated 50 m long transects, which were haphazardly laid out during daylight hours. The abundance and size of fish was recorded on waterproof paper by a SCUBA diver within 2 m of either side of transects, according to standard procedures implemented in the study region [28]. We also recorded the type of habitat under each transect (%), which was categorized according to 'big boulders' (>2 m of diameter), 'small boulders' (<1 m of diameter), 'sand', 'breakwaters' (i.e. artificial man-made constructions) and 'bare rock' [28].

Stomach contents

To analyze the diet composition of escapees, stomachs were weighed and the content removed. Total food items from each stomach were placed on filter paper to eliminate excess moisture and subsequently weighed. Prey items were identified to the lowest possible taxonomic level; the total number of prey items were counted and weighed for each stomach. The percentage composition by number and weight was then calculated for each prey category to calculate the indices of importance by number (IN), wet mass (IW) and global importance (IG).

$IN=[(\% \ composition \ by \ number) \ X \ (\% \ occurrence)]^{1/2}$

$IW=[(\% \ wet \ mass) \ X \ (\% \ occurrence)]^{1/2}$

$IG=(IN\% +IW\%)/2$

Gonadal development

For each fish, gonads were macroscopically examined for sex differentiation to establish the stage of gonadal development, by using a visual scale of five stages of maturity based on color and the relative size of gonads: I undeveloped; II developing; III mature; IV spawn; V post-spawn [29].

Lipid and FA profiles

A total of 60 livers and 60 muscles samples were analyzed. 10 for each distance ('Cages', 'Near Cages', and 'Far away from Cages') from both La Palma and Gran Canaria islands (*n*=10). Biochemical assays followed standard procedures [30]. Moisture content was determined by drying the sample at 105°C, until achieving a constant weight. Crude lipid content was extracted following the method of Folch et al. [31]. FAs from total lipids (stored under nitrogen atmosphere at −80°C) were prepared by transmethylation, as described by Christie [32] and FA methyl esters separated by gas chromatography following Izquierdo et al. [33]. All analyses were conducted in triplicate.

Data analysis

Fish survey data and FA profile composition were analyzed by means of ANOVA [34]. We tested for differences in escapees abundance between locations at varying distances from the escape point around La Palma Island and the sampling times through a 2-way ANOVA that incorporated the factors: (1) "Locality" (fixed factor with six levels, corresponding to the 6 locations) and (2) "Time" (fixed factor with six levels and orthogonal to the previous factor); "Time" was considered fixed, as sampling dates were equidistantly separated.

A χ^2 tested for differences in diet composition (by considering the percentages of the different prey items) between islands, and a Wald-Wolfowitz test contrasted differences in fish length between islands. To test for differences in FA profile composition (total lipid, OA, LA, ALA, palmitic acid (PA) (16:0), ARA, EPA, DHA, Σsaturated, Σmonounsatured, Σn-9, Σn-6, Σn-3 F and Σn-3/Σn-6 ratio) with varying distance from cages, 2-way ANOVAs incorporated the factors: (1) "Island" (fixed factor with two levels: La Palma and Gran Canaria); (2) "Distance" (fixed factor with three levels corresponding to the three distances from the aquaculture sea farms: 'Cages', 'Near cages' and 'Far away from cages'. Before the analyses, the Cochran's test was used to check for homogeneity of variances. If the test detected heterogeneous variances (Cochran's test, P<0.01), data transformation was performed. In some cases, variances remained heterogeneous despite transformations; the significance level was then set at the more conservative 0.01 value instead of the conventional 0.05 level to decrease a type I error [34]. If ANOVA detected significant differences, further analyses were performed by using the SNK *a posteriori* multiple comparison test [34]. When data did not achieve homogeneous variances, the Games-Howell post hoc test was used. The SIMPER analysis identified the main contributors to differences in FA profiles between islands at varying distance from cages. A Chi square (X^2) tested for departures from a 1:1 sex ratio. Non-metric multidimensional scaling (MDS) ordination plots were implemented to visualize differences in the FA profiles (OA, LA, ALA, PA, ARA, EPA, DHA, Σsaturated, Σmonounsatured, Σn-9, Σn-6, Σn-3 FAs and Σn-3/ n-6 ratio) from muscle and liver tissue between islands and the three distances from the aquaculture sea farms: 'Cages', 'Near cages' and 'Far away from cages'. The SPSS (v. 15.0), PRIMER (v. 5.2.4) and PERMANOVA (v.1.6) software were used in these statistical analyses.

Results

Spatial-temporal distribution of escaped European sea bass after the escape event

The mean abundance of escaped European sea bass at La Palma Island was 4.9 ± 14.3 ind 100 m^{-2} (mean ± SE, *n*=144 transects) during the study period, which varied between 0 fish in several transects to a maximum of 100 ind 100 m^{-2}. All locations had escaped European sea bass, at least one individual during the study period (Appendix 1). The ANOVA demonstrated that escaped European sea bass densities decreased significantly (p<0.01) through the study period, particularly in October 2010 (Figure 2 and Table 1) and that Tazacorte (i.e. the point of massive release) had significantly (p<0.01) higher densities (17.0 ± 29.1 ind 100 m^{-2}) than the other locations (Figure 2). The largest densities of escaped European sea bass from La Palma were observed in breakwaters (Game-Howell test, p<0.05, Figure 3). The length range of observed European sea bass varied between 15 and 60 cm (Figure 4); the majority of fish, however, were between 35 and 50 cm (Appendix 1).

Stomach contents of escaped European sea bass

A total of 101 individuals were collected (Appendix 2); a total of 77 stomachs were analysed (24 stomachs were broken by the spear impact during field collections); twenty stomachs (25.97%) were empty and 19 stomachs (24.67%) contained unidentifiable items. Decapod crustaceans were the main prey, followed by osteichthyes and inorganic matter (Figure 5). The most commonly preyed crustaceans were *Percnon gibbesi* and *Rhynchocinetes sp*, while the main fish species was the parrotfish, *Sparisoma cretense* (Figure 5). Stomachs from La Palma (n=54) mainly contained crustaceans, while stomachs from Gran Canaria (n=20) were dominated by unknown osteichthyes. This

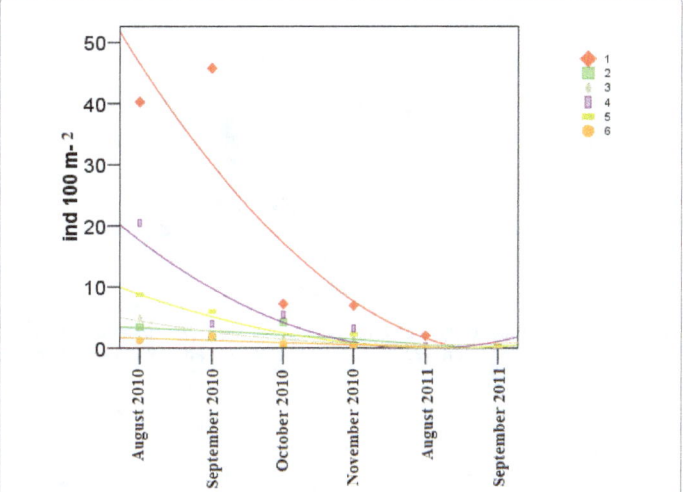

Figure 2: Mean abundances (ind 100 m^{-2}) of escaped sea bass, *Dicentrarchus labrax*, at each time and location at La Palma Island. **1** Tazacorte (Acuipalma S.L. farm); **2** El Remo (marine reserve); **3** La Salemera; **4** Santa Cruz de la Palma; **5** Puerto Talavera; **6** Puntagorda.

Source of variation	df	MS	F	P
Time	5	204.909	79.096	0.0002
Locality	5	150.132	57.951	0.0004
Time x Locality	25	29.859	11.526	0.2918
Residual	108	25.907		

Table 1: ANOVA results of the effect of 'Time' and 'Locality' on the mean abundance of sea bass (*Dicentrarchus labrax*) at La Palma Island.

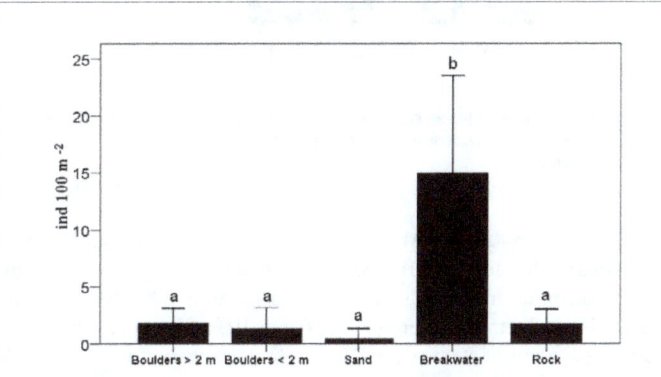

Figure 3: Mean abundances (ind 100 m^{-2}) of escaped sea bass, *Dicentrarchus labrax*, at different substratum types at La Palma Island (data pooled from the different locations). Alphabetic superscripts indicate significant differences.

resulted in significant differences in diet between La Palma and Gran Canaria (χ^2=100.2421, d.f.=16, p<0.0001), despite a lack of significant differences in fish length between islands ($Z_{adjusted}$=-2.5266, p=0.8407). However, similar percentages of *Sparisoma cretense* and inorganic matter (sand and plastics) were presented in fish stomachs from both islands.

Gonadal development of escaped European sea bass

A total of 87 escaped European sea bass were examined; gonads from 14 individuals were not analyzed due to spear impact during collection. The sex ratio did not deviate from a theoretical 1:1 ratio (χ^2

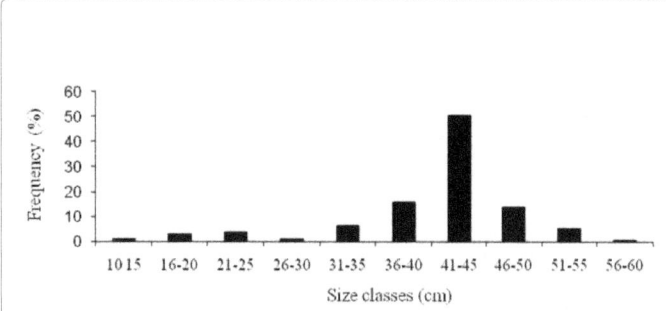

Figure 4: Size-frequency distributions of sea bass, *Dicentrarchus labrax*, escaped in La Palma on February 2010 (data pooled from the different locations).

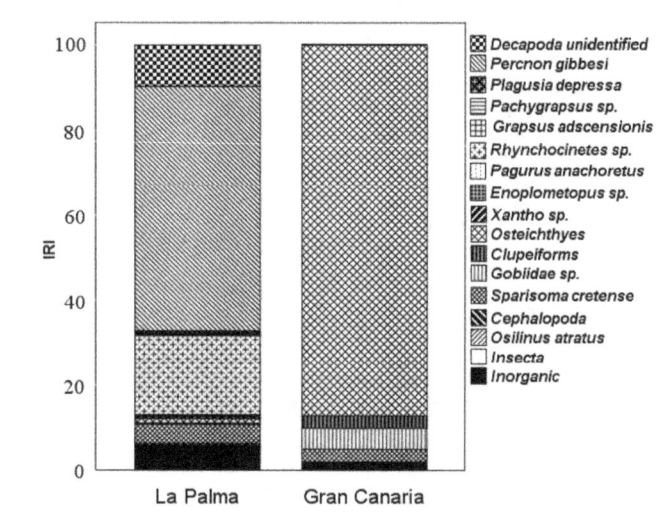

Figure 5: Percentage Index of Relative Importance (% IRI) of different prey in the diet of sea bass, *Dicentrarchus labrax*, escaped from La Palma and Gran Canaria islands.

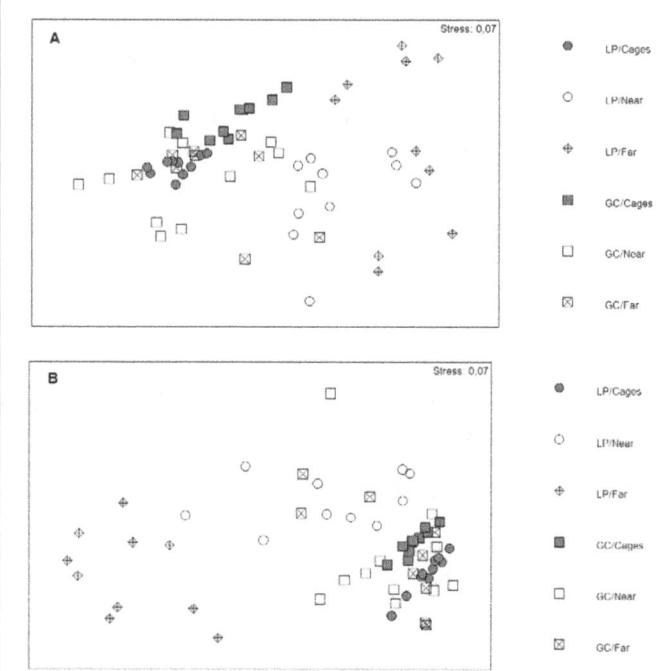

Figure 6: Non-metric multi-dimensional scaling ordination of (A) muscle and (B) liver fatty acids profiles of sea bass, *Dicentrarchuslabrax*,from 'Cages', 'Near' (<10 km away from the cages) and 'Far' (>10 km away from the cages) at La Palma (LP) and Gran Canaria (GC) islands.

Source of variation	df	MS	F	P	Pairwise tests
Island (Is)	1	6.875.190	220.042	0.0002	
Distance (D)	2	5.596.024	179.102	0.0002	
Is x D	2	5.857.735	187.478	0.0002	LP: 'Cages' ≠ 'Near' ≠ 'far away' GC: 'Cages' = 'Near' = 'far away'
Residual	54	312.450			

Table 2: Result of multivariate analysis of variance (MANOVA) testing for differences in the fatty acids profiles (including liver and muscle tissue) between island and varying distance from farms. LP: La Palma, GC: Gran Canaria.

=2.400, d.f.=1, p=0.121). About 20.7% of escaped European sea bass were sexually undefined. The 87.5% of males and 91.7% of females presented immature gonads (stages I and II); only 8.3% (considering both males and females) were mature (stage III) and exclusively 1 male were in a spawning state (state IV). No females were found in the IV stage. No fish (both male and female) were found in stage V.

FA profiles of escaped European sea bass

For Gran Canaria Island, the nMDS plot did not show a separation of individuals according to distance from cages for both muscle and liver tissues (Figure 6a and 6b). In contrast, at La Palma Island, individuals separated according to their distance from cages (Figure 6a and 6b). This resulted in an inconsistent pattern in FA profiles among groups at varying distance from cages between islands (MANOVA, significant 'Island x Distance' interaction, Table 2). The SIMPER analysis showed that, for muscle tissue, the main differentiating variables between localities at varying distance from the cages were: total lipid, DHA, LA, Σsatured, Σmonounsatured and Σn-3 FAs. For liver tissue, the main contributors to dissimilarities were total lipid, DHA, OA, Σsatured, Σmonounsatured and Σn-3 FAs.

The majority of FAs showed a similar pattern with varying distance

from cages at both islands (Appendix 3). At La Palma Island, we typically detected significant differences in the percentage of FA with varying distance from the farm; this pattern, however, was minored at Gran Canaria Island, where we exclusively detected significant differences with varying distance from cages in the percentage of Σn-3/Σn-6 ratio (Figure 7.13 and 7.14), muscle total lipid (Figure 7.1), OA from muscle (Figure 7.3) and EPA from muscle (Figure 7.9). A posteriori SNK tests demonstrated that crude lipid from individuals captured at La Palma Island, including both muscle (Figure 7.1) and liver (Figure 7.2) tissues, had significantly larger values in samples coming from 'cages' than 'far away' from cages; 'Near' samples did not differ from 'far away' samples for muscle tissue (Figure 7.1). At La Palma Island, the concentration of OA was significantly higher in samples from 'cages' than 'near' cages or 'far away', for both muscle (Figure 7.3) and liver (Figure 7.4) samples. Cultured ('cages') European sea bass from La Palma had significantly higher LA (Figure 7.5 and 7.6) and ALA (Figure 7.7 and 7.8) than 'near' and 'far away' samples, for both muscle and liver; fish from 'near' and 'far away' distances did not show significant differences. The percentage of EPA in muscle was larger at 'cages' relative to 'near' and 'far away' samples at Gran Canaria

(Figure 7.9); at La Palma, however, there was no difference between 'cages' and 'far away' samples (Figure 7.9). The percentage of EPA in liver tissue (Figure 7.10) differed between 'near cages' and 'far away' samples collected from La Palma. Σn-9 and ΣMonounsatured FAs from La Palma followed the same pattern: significantly larger values for cultured fish than 'far away', for both muscle (Figure 7.15 and 7.17, respectively) and liver (Figure 7.16 and 7.18, respectively) tissue samples.

In contrast, several FAs showed an overall increase in their concentration with distance away from cages. For example, ARA showed higher values in 'far away' than 'cages' and 'near' samples, for both muscle (Figure 7.19) and liver (Figure 7.20) for La Palma, and exclusively for muscle tissue at Gran Canaria (Figure 7.19). At La Palma, the percentage of DHA was significantly higher in 'far away' and 'near' samples than cultured fish ('cages') for muscle samples (Figure 7.11). For liver samples (Figure 7.12), however, there was no difference between 'near' and 'cages'. The concentration of Σn-3 FAs from La Palma, for both muscle (Figure 7.21) and liver (Figure 7.22), were larger in 'far away' than 'cages' samples. At Gran Canaria, we detected a larger concentration of Σn-3 FAS for muscle samples at 'cages' (Figure 7.21). No difference in the level of PA was registered between 'cages' and 'far away' samples from both islands, for both muscle (Figure 7.23) and liver (Figure 7.24). The Σn-3/Σn-6 ratio and satured FAs from La Palma, for both muscle (Figure 7.13 and 7.25, respectively) and liver (Figure 7.14 and 7.26, respectively), were significant higher from 'near' and 'far away' than 'cages' samples.

Finally, it is worth noting that the percentage of all FAs of cultured fish (i.e. 'cages') differed between islands (Figure 7, Appendix 3), except the fat percentage, DHA, satured, ARA and PA in muscle tissue, and monounsatured, Σn-9, satured, OA and PA in liver tissue.

Discussion

Populations of European sea bass in the studied islands are related to aquaculture escapes [12,16]. Our results agree with this idea, because European sea bass density over time was much higher adjacent to the sea-cage fish farm at Tazacorte, as also demonstrated by Toledo-Guedes et al. [35], than at the other locations, where most fish disappear through time after the massive release. Moreover, the size class distribution of European sea bass from La Palma clearly indicated that the majority of individuals are aquaculture escapees, as also indicated by Toledo-Guedes et al. [13,35] for the same escape event. The European sea bass may reach any location around the islands perimeter after a massive escape, showing therefore a great capacity of dispersion as proposed by previous studies [13,16,35-37]. Small, chronic, escapes, however, have not caused a high dispersion of escapees; escaped fish are typically located around cages or in the near coast [12,16,38].

Over time, escaped fish tend to approach shallow waters [13,16,18,35,37]. Despite Toledo-Guedes et al. [12] showed a preference by the European sea bass for bottoms covered by boulders on shallow waters, our study demonstrated that escaped European sea bass prefers breakwaters; in particular, higher European sea bass densities were observed in those locations (Tazacorte and Santa Cruz) where breakwater do exist. In this sense, Santa Cruz was the second location with a higher density of fish, in spite of being far away from the escape point (Figure 1). This result contrasts with those observations reported by Toledo-Guedes et al. [35], which demonstrated lower densities in the east face of the island (contrary to the escape location) and so a clear correlation between the distance from the fish farm and European sea bass density. It is noteworthy that this result was despite

our sampling and that performed by Toledo-Guedes et al. [13,35] were carried out within a similar temporal window.

Previous studies have concluded that escaped European sea bass is able to exploit natural resources in the Canary Islands [12,16,36]. The principal preys identified in all of these studies are common on subtidal bottoms of the Canary Islands. We corroborated this idea, showing an overlap with the diet indicated by Toledo-Guedes et al. for the same escape event, who highlighted *Percnon gibessi* as the principal prey. However, these authors did not found *Rhynchocinetes sp* as a relevant diet constituent of escaped fish, although much of escaped fish were caught practically in the same locations and time. These differences could be due to variations in sample size [11,13,41] and, of course, the somehow opportunistic feeding behavior of the European sea bass [10]. Nonetheless, our overall diet results agree with previous studies [15,39] that showed crustacean decapods and osteichthyes as the principal prey of wild European sea bass.

Toledo-Guedes et al. [13] showed that the diet of fishes from massive escape events typically differs as a result of the 'time at liberty'. Previous studies [13,40] highlighted that these differences in escaped fish diet could reflect a 'hunting learning' period, i.e. recent escapees predate mainly over crustaceans that are less mobile and 'more time escaped fish' predate over fish. Lorenzen et al. [7] concluded that there are critical uncertainties on the effects of different domestication strategies on the fitness of cultured fish in the wild. We observed that diets of escaped European sea bass differed between La Palma and Gran Canaria; *Percnon gibessi* was the main prey for escapees from La Palma and osteichthyes for Gran Canaria. However, in the light of FA results, 'more time escaped fish' (those from La Palma) mainly fed over crustaceans, while recently escaped fish (those from Gran Canaria) principally fed over fish. These preys are common in both islands, so prey availability does not seem to be the cause of these differences. In any case, we may overestimate this issue, since we have not analyzed a high number of stomachs from Gran Canaria, due to difficulties to find escaped fish.

Escaped European sea bass diet includes a high percentage of inorganic items [12,16]. Carrillo and Castillo and Toledo-Guedes et al. [12] showed a mean 82.63% and 50% (respectively) of stomach vacuity, proving evidence of a low adaptation to wild conditions of escaped individuals. Yet, we recorded a mean stomach vacuity of 25.97%, suggesting a high variability of this parameter. In any case, inorganic matter (sand and plastics) was the third main item in the stomachs of escaped European sea bass in our study, highlighting the difficulties of escaped fish to wild feeding conditions. Toledo-Guedes et al. [13] reported a similar value for long-term escaped fish (33.9% of vacuity), while the value was lower for recently escaped fish (12.5% vacuity); as a result, this contradicts the previous idea and suggest that the European sea bass is able to actively exploit available resources in natural habitats.

European sea bass requires low salinities (<35‰) in its natural habitat to trigger gametogenesis; in turn, the last phase of gonadal development require a *ca.* 35‰ of salinity and, concurrently, larval survival increases at low salinities. Spawning optimal thermal range is 13-15°C [41]. Recruitment is typical in estuarine habitats across its distribution range [10,15,42]. During eggs incubation, the optimal temperature range is between 13°-17°C [43-45] and larval development lasts 46 days at 16.5°C [46]. European sea bass first maturation is reached at a range of length between 23-46 cm [47] and the reproduction often occurs between December-March [15,48-50].www.fishbase.org In our study, only a very low number of individuals, including both males and females, showed mature gonads; this contrasts with a larger number

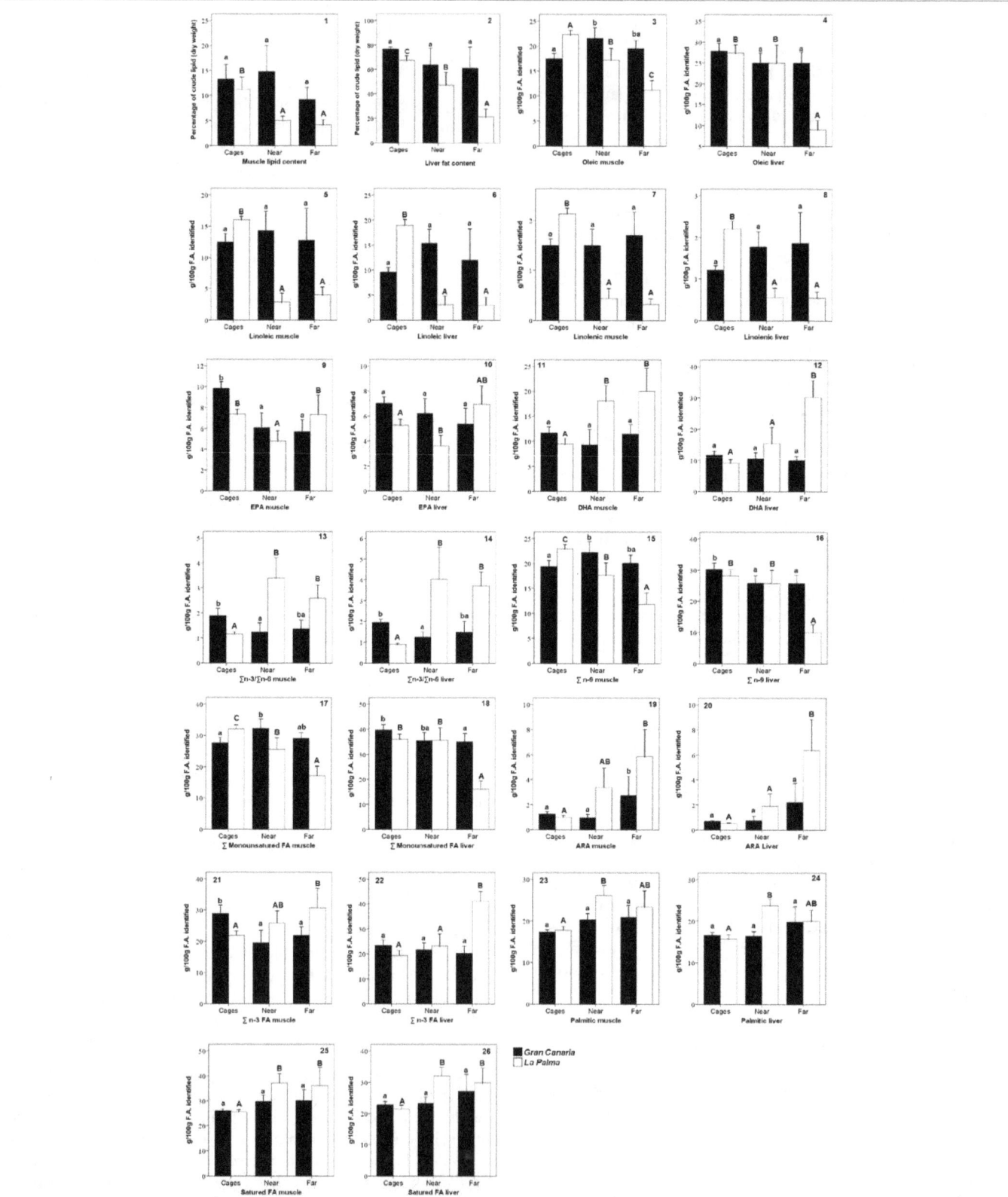

Figure 7: Percentage of lipid content and fatty acids in muscle and liver of escaped sea bass, *Dicentrarchuslabrax*,from 'cages', 'near' (<10 km away from the cages) and 'far' (>10 km away from the cages). Alphabetic superscripts denote significant differences among distances away from farms. **1)**muscle lipid content; **2)** liver fat content; **3)** muscle oleic acid (18:2n−9); **4)** liver oleic acid (18:2n−9); **5)** muscle linoleic acid (18:2n−6); **6)** liver linoleic acid (18:2n−6); **7)** muscle linolenic acid (18:3n−3); **8)** liver linolenic acid (18:3n−3); **9)** muscle eicosapentaenoic acid (20:5n−3); **10)** liver eicosapentaenoic acid (20:5n−3); **11)** muscle docosahexaenoic acid (22:6n−3); **12)** liver docosahexaenoic acid (22:6n−3); **13)** muscle ratio n−3/n−6; **14)** liver ratio n−3/n−6; **15)** muscle $\sum$ n-9 FA; **16)** liver $\sum$ n-9 FA; **17)** muscle $\sum$ monounsatured FA; **18)** liver $\sum$ monounsatured FA; **19)** muscle arachidonic acid (20:4n−6); **20)** liver arachidonic acid (20:4n−6); **21)** muscle $\sum$ n-3 FA; **22)** liver $\sum$ n-3 FA; **23)** muscle palmitic acid (16:0); **24)** liver palmitic acid (16:0); **25)** muscle $\sum$ satured FA; **26)** liver $\sum$ satured FA.

of mature gonads found by Toledo-Guedes et al. [12]. However, both studies have not found evidence of reproduction. This is noteworthy, because individuals were in their sexual maturity period. In addition, only one male at a spawning state was observed. This could be because the range of temperature and salinity necessary for gonadal development [39] is not present in the studied islands (including higher temperatures). Even if the European sea bass would spawn, neither incubation of eggs conditions, nor term-haline conditions for larval development [44-46], are present in the Canary Islands. The sex ratio (1:1) was, moreover, not optimal (4:1 according to Arias [11]) and no juvenile European sea bass had been reported [12,13,16]. Therefore, there is no evidence of European sea bass reproduction at La Palma and Gran Canaria islands, as similarly concluded by Toledo-Guedes et al. [12] and Carrillo and Castillo [16]. In any case, further and more specific studies would be necessary in order to know escaped European sea bass capacity of reproduction in Canary Islands.

After the mass escape event, the number of European sea bass decreased through time, almost disappearing, providing evidence that no more escape events occurred at La Palma Island. A similar conclusion was highlighted by Toledo-Guedes et al. [12,35]. This decrease is principally due to a high fishing pressure, failure to adapt (starvation, deformities) and predation by large-sized fish (e.g. *Seriola sp, Sphyraena viridensis, Pomatomus saltatrix*) [4,8,12,35,37,38,51-53]. Nevertheless, Arechavala-López et al. [8] showed low recapture rates by local fishermen on European sea bass inform the Mediterranean Sea (including both recreational and professional captures). As previous studies, we also recognize fishing pressure as a significant driver of the progressive decay in fish abundances through time. About 63 tons of escaped European sea bass were fished by this professional (artisanal) fleet during the study period in La Palma ('Consejería de Agricultura y Pesca del Gobierno de Canarias', pers. comm.) and recreational fishermen captures have been reported as larger than 100 fish, or 100 kg, per angler in just one day in the study region [37]. If we assume that recreational captures were larger than those performed by professional fishermen, the overall capture was far from reaching *ca.* 400 tons of escaped European sea bass. We then hypothesize that remaining fish were predated or non-adapted. Moreover, the great decrease occurred during October, matching the arrival at coastal areas of migratory, large-sized, predators (tuna, yellowtails, *Sphyraena viridensis*, dolphins and sharks) [54]. Beside this, it is plausible that escaped European sea bass may compete with native species, such as *Dicentrarchus punctatus*, juveniles of *Sphyraena viridensis, Synodus sp., Serranus sp.*, which have a similar ecological niche and feeding habits, as concluded by González-Lorenzo et al. [36], Toledo-Guedes et al. [12] and Tavares and González .

In the light of overall FA results, it seems clear that escaped fish from Gran Canaria were recently escaped, because no difference existed with varying distance from aquaculture cages. In particular, the 18:C FAs would tend to be eliminated or washed-out progressively from the muscle, as soon as those animals finish to feed aquafeeds and begin to feed on wild prey [55,56]. This emphasizes the rapid dispersion that European sea bass presents after an escape event as we observed at La Palma by means of visual census. FA profiles of escapees changes over time after an escape by wash-out [27,53] so it seems clear that escaped European sea bass at La Palma have been more time escaped than at Gran Canaria. Our initial hypothesis aimed to relate distance with time (more distance, more time to disperse, more wash out), but we somehow failed to detect this, mainly because this species has a quick dispersion behavior. The FA composition of muscular tissue is often related to dietary FA composition [57], so some FAs can be

used as bioindicators of escaped fish [23,25,26,58-61]. Normally, reared European sea bass present higher proportions of OA and LA and lower proportions of PA, ARA, EPA, and DHA than wild European sea bass [23,27,62]. Additionally, percentages of total saturated, as well as the n-3/n-6 ratio, are higher in wild than cultured European sea bass [23,62]; these results match the outcomes of this study. FA composition, however, may not be reliable biomarkers of aquaculture activities since other human activities, including discharges of urban waters, may induce similar changes [53], directly by feeding or due to the relatively high conservation of FA composition throughout the food web [23]. This is particularly relevant for European sea bass, as a result of its quick dispersion behaviour. We also observed differences in the profile of FAs between cultured fish from Gran Canaria and La Palma; this is a consequence of continuous changes in aquafeeds formulae that depends on the ingredients availability and fluctuating market prices [63]. DHA, OA and LA seem to be selectively retained in European sea bass muscle [57]. When fish that has been fed with vegetable oil containing diets is subjected to a period of fish oil re-feeding, the amount of 18°C FAs (particularly LA in those fish fed previously with soybean oil containing diet) remains higher and EPA lowers [57].

To determine if a FA is a good aquaculture biomarker, it would be necessary to perform a study at the same region, where the FA composition should be in similar from both culture and escaped fish and distinct from wild fish. In the present study, however, there was no wild fish, so we were able to exclusively assess if FA composition changed with varying distance from farms. Contrary to Fernández-Jover [23], FA profile cannot be used as biomarkers, because there were significant differences between distance groups ('cages'-'near'-'far') in this study. Moreover there is an inconsistency for FAs highlighted among several studies [27,53,61,64,65], so the usefulness of FAs as biomarkers for escaped fish is doubtful. We found different FAs than those proposed by Arechavala-López et al. [66] (i.e. LA and ARA) as possible biomarkers. The current study could suggests PA (liver) and EPA (both at liver and muscle) as the candidates as possible biomarkers, because these FAs were not different with varying proximity from the farms; however further studies are necessary to confirm its reliability. This matches the result showed by Montero et al. [57], where EPA values did not become to control diet (fish oil) after a re-feeding period in the laboratory. In the Canary Islands, Ramírez et al. [53] concluded that ALA is a possible aquaculture biomarker for bogue, *Boops boops*, a zoo-planktivorous and opportunistic fish that aggregates around sea-cage fish farms in the Canary Island and the Mediterranean; this is because samples taken around sewage discharge points did not increase the ALA percentage. However, this study demonstrated that ALA cannot be used as a biomarker for escaped European sea bass. FAs could be used to identify escaped fish when matching similar FA profiles from cultured fish. However, if FA profiles do not match those of cultured fish, it is impossible to work out whether a fish has been born in the wild or, alternatively, is an old escapee that has progressively suffer a wash-out.

In summary, despite escaped fish being able to exploit natural resources, the density of escapees decreased through time. Those fish able to adapt and use natural resources altered their FA profiles in comparison to cultured fish, denoting the poor usefulness of FA profile as a good bio-indicator, limiting their potential to a very short period of time after escape events.

Acknowledgement

Authors would like to thank María López Ruano for her help to produce this document. Our thanks also go to Tamia Brito, manager of La Palma Marine

Protected Area. The staff at Acuipalma S.L., for their collaboration. David Jiménez Alvarado and Yamilet Cárdenes are acknowledged for their assistance during field sampling. This work was funded by Prevent Escape project (nº 226885).

References

1. CBD (Secretariat of the Convention on Biological Diversity) (2004) Solutions for sustainable mariculture avoiding the adverse effects of mariculture on biological diversity.

2. Molina L, Vergara J (2005) Impacto ambiental de jaulas flotantes: estado actual de conocimientos y conclusiones prácticas. Boletín del Instituto Español de Oceanografía 21: 75-81.

3. Naylor R, Hindar K, Fleming IA, Goldburg R, Williams S, et al. (2005) Fugitive salmon: assessing the risks of escaped fish romnet-pen aquaculture. Bioscience 55: 427-437.

4. Vergara JM, Haroun R, González MN, Molina L, Briz MO, et al. (2005) Environmental Impact Assessment of Cage Aquaculture in Canarias Pesquera.

5. Jensen O, Dempser T, Thorstad EB, Uglem I, Fredheim A (2010) Escapes of fish from Norwegian sea-cage aquaculture: causes, consequences and preventions. Aquaculture Environmental Interactions 1: 71-83.

6. Grigorakis K, Rigos G (2011) Aquaculture effects on environmental and public welfare; The case of Mediterranean mariculture. Chemosphere 85: 899-919.

7. Lorenzen K, Mangel M (2012) Cultured fish: integrative biology and management of domestication and interactions with wild fish. Biological Reviews of the Cambridge Philosophical Society 87: 639-660.

8. Arechavala-Lopez P, Izquierdo-Gómez D, Sánchez-Jerez P, Bayle-Sempere J (2014) Simulating escapes of farmed European sea bass from Mediterranean open sea-cages: low recaptures by local fishermen. Journal of Applied Ichthyology 30: 185-188.

9. Tortonese E (1986) Moronidae.

10. Laffaille P, Lefeuvre JC, Schricke M, Feunteun E (2001) Feeding ecology of 0-group sea bass, Dicentrarchus labrax, in salt marshes of Mont Saint Michel Bay (France). Estuaries and Coasts 24: 116-125.

11. Leitao F, Santos M, Erzini K, Monteiro C (2008) The effect of predation on artificial reef juvenile demersal fish species. Marine Biology 153: 1233-1244.

12. Toledo-Guedes K, Sánchez-Jerez P, González-Lorenzo G, Brito A (2009) Detecting the degree of establishment of a non-indigenous species in coastal ecosystems: sea bass Dicentrarchus labrax escapes from sea cages in Canary Islands (Northeastern Central Atlantic). Hydrobiologia 623: 203-212.

13. Toledo-Guedes-Guedes K, Sánchez-Jerez P, Benjumea M, Brito A (2014) Farming-up coastal fish assemblages through a massive aquaculture escape event. Marine Environmental Research 98: 86-95.

14. Kennedy M, Fitzmaurice P (1972) The biology of the bass, Dicentrarchus labrax (Linné 1758), in Irish waters. Journal of the Marine Biological Association of the United Kingdom 3: 39-68.

15. Arias A (1980) Growth, diet and reproduction of the golden (Sparusaurata L.) and sea bass (Dicentrarchus labrax L.) in the estuaries. Fisheries Research 44: 59-83.

16. Carrillo J, Castillo R (2001) Estudio del Impacto producido por la introducción de las lubinas (Dicentrarchus labrax Linnaeus, 1758) y doradas (Sparus aurata, Linnaeus, 1758), escapadas de los cultivos en jaulas flotantes, en el ecosistema marino litoral de Gran Canaria. Informe inédito. Gran Canaria: 2001.72 pp. Viceconsejería de Pesca del Gobierno de Canarias.

17. Carrillo M, Zanuy S, Blazquez M, Ramos J, Piferrer F, et al. (1995) Sex control and diploid manipulation in European sea bass.

18. Dunn RJK, Welsh DT, Teasdale PR, Lee SY, Lemckert CJ, et al. (2008) Investigating the distribution and sources of organic matter in surface sediment of Coombabah Lake (Australia) using elemental, isotopic and fatty acid biomarkers. Continental Shelf Research 28: 2535-2549.

19. Hu J, Zhang G, Li K, Peng P, Chivas AR González-Lorenzo (2008) Increased eutrophication offshore Hong Kong, China during the past 75 years: evidence from high-resolution sedimentary records. Marine Chemistry 110: 7-17.

20. Maazouzi C, Masson G, Izquierdo MS, Pihan JC (2008) Chronic copper exposure and fatty acid composition of the amphipod Dikerogammarus villosus: results from a field study. Environment Pollution 156: 221-226.

21. Izquierdo MS, Koven W (2011) Lipids.

22. Tacon AGJ, Metian M (2008) Global overview on the use of fish meal and fish oil in industrially compounded aquafeeds: trends and future prospects. Aquaculture 285: 146-158.

23. Albert GJT, Marc Metianb (2011) Global overview on the use of fish meal and fish oil in industrially compounded aquafeeds: Trends and future prospects. Aquaculture Environment Interactions 285: 146-158.

24. Skog T, Hylland K, Torstensen BE, Berntssen M (2003) Salmon farming affects the fatty acid composition and taste of wild saithe Pollachius virens L. Aquaculture Research 34: 999-1007.

25. Fernández-Jover D, Lopez-Jimenez J, Sanchez-Jerez P, Bayle-Sempere J, Gimenez-Casalduero F, et al. (2007) Changes in body condition and fatty acid composition of wild Mediterranean horse mackerel (Trachurus mediterraneus, Steindachner, 1868) associated to sea cage fish farms. Mar Environt Res 63: 1-18.

26. Rueda FM, Hernández MD, Egea MA, Aguado F, García B, et al. (2001) Differences in tissue fatty acid composition between reared and wild sharpsnout sea bream, Diplodus puntazzo (Cetti, 1777). British Journal of Nutrition 86: 617-622.

27. Arechavala-López P, Fernández-Jover D, Black K, Ladoukakis E, Bayle-Sempere J, et al. (2013) Differentiating the wild or farmed origin of Mediterranean fish: a review of tools for sea bream and European sea bass. Reviews in Aquaculture 5: 137-157.

28. Tuya F, Boyra A, Sánchez-Jerez P, Haroun RJ, Barberá C (2004) Relationships between rocky-reef fish assemblages, the sea urchin Diadema antillarum and macroalgae throughout the Canarian Archipelago. Marine Ecology Progress Series 278: 157-169.

29. Arias A (1979) Experimental Biology and cultivation of the golden Sparus aurata L., and sea bass Dicentrarchus labrax L.

30. AOAC. Association of Official Analytical Chemists (2000) Official Methods of Analysis.

31. Folch J, Lees M, Sloane-Stanley GH (1957) A simple method for the isolation and purification of total lipids from animal tissues. J Biol Chem 226: 497-509.

32. Christie W (1982) Lipid Analysis.

33. Izquierdo MS, Watanabe T, Takeuchi T, Arakawa T, Kitajima C (1990) Optimum EFAlevels in Artemia to meet the EFA requirements of red sea bream (Pagrus major).

34. Underwood AJ, (1997) Experiments in ecology. Their logical design and interpretation.Cambridge University Press, UK.

35. Toledo-Guedes K, Sánchez-Jerez P, Brito A (2014) Influence of a masive aquaculture escape event on artisanal fisheries. Fisheries Management and Ecology 21: 113-121.

36. González-Lorenzo G, Brito A, Barquín J (2005) Impacts of fish escapes cage mariculture in the Canaries. Vieraea 33: 449-454.

37. Tavares D, González N (2009) Characterization of events exhaust Sparus aurata and Dicentrarchus labrax in Gran Canaria by interviewing anglers.

38. Arechavala-Lopez P, Uglem I, Fernández-Jover D, Bayle-Sempere J, Sánchez-Jerez P (2011) Immediate post-escape behaviour of farmed European sea bass (Dicentrarchus labrax) in the Mediterranean Sea. Journal of Appl Ichthyol 27: 1375-1378.

39. FAO (Food and Agriculture Organisation) (2012) Fish finder.

40. Arechavala-Lopez P, Uglem I, Fernandez-Jover D, Bayle-Sempere JT, Sanchez-Jerez P (2012) Post-escape dispersion of farmed seabream (Sparus aurata L.) and recaptures by local fisheries in the Western Mediterranean Sea. Fish Res 121-122: 126-135.

41. Moretti A, Pedini G, Fernandez-Criado M, Cittolin G, Guidastri R (1999) Hatchery production procedures. In: Manual on hatchery production of seabass and gilthead seabream (vol 1). FAO, Rome, Italy.

42. Serrano L (1989) Age, Growth and sexuality of sea bass, Dicentrarchus labrax (Linnaeus758) (Perciformes, Moronidae) from Aveiro lagoon, Portugal. Scientia Marina 53: 121-126.

43. Dufour V, Cantou M, Lecomte F (2009) Identification of sea bass (Dicentrarchus labrax) nursery areas in the north-western Mediterranean Sea. Journal of the

Marine Biological Association of the United Kingdom 89: 1367-1374.

44. Devauchelle N, Coves D (1988) The characteristic of European sea bass (Dicentrarchuslabrax) eggs: descripction, biochemical composition and hatching performances. Aquatic Living Resources 1: 223-230.

45. Saka S, Firat K, Kamaci H (2001) The development of European sea bass (Dicentrarchuslabrax L., 1758) eggs in relation to temperature. Turkish Journal of Veterinary and Animal Sciences 25: 139-147.

46. Houde E, Zastrow C (1993) Ecosystem and taxon-specific dynamic and energetics properties of fish larvae assemblages. Bulletin of Marine Sciences53: 290-335.

47. Froese R, Pauly D (2006) Fish base.

48. Do Chi T, Hoai T (1971) Croissance différentielle of Dicentrarchus labrax (Linnaeus 1758).

49. Barnabé G (1976) Contribution to the connaissance of biologie du loup Dicentrarchus labrax (L.) (Poisson Serranidae).

50. González A (2003) Aromatase cytochrome P450 activity in the sea bass (Dicentrarchus labrax).

51. Sanchez-Jerez P, Fernandez-Jover D, Bayle-Sempere J, Valle C, Dempster T, et al. (2008) Interactions between blue fish (Pomatomus saltatrix.) and coastal sea-cage farms in the Mediterranean Sea. Aquaculture 282: 61-67.

52. Arechavala-Lopez P, Uglem I, Sanchez-Jerez P, Fernandez-Jover D, Bayle-Sempere JT, et al. (2010) Movements of grey mullets (Liza aurataand Chelonlabrosus) associated with coastal fish farms in the western Mediterranean Sea. Aquaculture Environment Interactions 1: 127-136.

53. Ramírez B, Montero D, Izquierdo M, Haroun R (2013) Aquafeed imprint on bogue (Boopsboops) populations and the value of fatty acids as indicators of aquaculture-ecosystem interaction: Are we using them properly? Aquaculture 414-415: 294-302.

54. Bas C, Castro J, Hernández-Gracía V, Lorenzo JM, Moreno T, et al. (1995) Fishing in the Canary Islands and areas of influence.

55. Torstensen BE, Froyland L, Ornsrud R, Lie O (2004) Tailoring of a cardio protective muscle fatty acid composition of Atlantic salmon (Salmosalar) fed vegetable oils. Food Chemistry 87: 567-580.

56. Izquierdo MS, Montero D, Robaina LE, Caballero MJ, Rosenlund G, et al. (2005) Alteration in fillet fatty acid profile and flesh quality in gilthead sea bream (Sparusaurata) fed vegetable oils for a long period. Recovery of fatty acid profiles by fish oil feeding. Aquaculture 250: 431-44.

57. Montero D, Robaina L, Caballero MJ, Gines R, Izquierdo MS (2005) Growth, feed utilization and flesh quality of European seabass (Dicentrarchuslabrax) fed diets containing vegetable oils: a time-course study on the effect of a re-feeding period with a 100% fish oil diet. Aquaculture 248: 121-134.

58. Bell JG, Henderson RJ, Tocher DR, McGhee F, Dick JR, et al. (2002) Substituting fish oil with crude palm oil in the diet of Atlantic salmon (Salmosalar) affects muscle fatty acid composition and hepatic fatty acid metabolism. Journal of Nutrition 132: 222-230.

59. Blanchet C, Lucas M, Julien P, Morin R, Gingras S, et al. (2005) Fatty acid composition of wild and farmed Atlantic salmon (Salmosalar) and rainbow trout (Oncorhynchusmykiss). Lipids 40: 529-531.

60. Megdal PA, Craft NA, Handelman GJ (2009) A simplified method to distinguish farmed (Salmosalar) from wild salmon: Fatty acid ratios versus astaxanthin chiral isomers. Lipids 44: 569-576.

61. Mnari A, Bouhlel I, Chraief I, Hammami M, Romdhane MS, et al. (2007) Fatty acids in muscles and liver of Tunisian wild and farmed gilthead sea bream, Sparusaurata. Food Chemistry 100: 1393-1397.

62. Alasalvar C, Taylor K, Zubcov E, Shahidi F, Alexis M (2002) Differentiation of cultured and wild sea bass (Dicentrarchuslabrax): total lipid content, fatty acid and trace mineral composition. Food Chem. 79: 145-150.

63. Gunstone FD (2010) The world's oil and fats.

64. Grigorakis K, Alexis MN, Taylor KD, Hole M (2002) Comparison of wild and cultured gilthead sea bream (Sparusaurata); composition, appearance and seasonal variations. International Journal of Food Science and Technology 37: 477-484.

65. Arechavala-Lopez P, Sanchez-Jerez P, Bayle-Semper J, Fernandez-Jover D, Martinez-Rubio L, et al. (2011) Direct interaction between wild fish aggregations at fish farms and fisheries activity at fishing grounds: a case study with Boops boops. Aquaculture Research 42: 996-1010.

66. Arechavala-Lopez P, Sánchez-Jerez P, Bayle-Sempere J, Uglem I, Mladineo I (2013) Reared fish, farmed escapees and wild fish stocks - a triangle of pathogen transmission of concern to Mediterranean aquaculture management. Aquaculture Environment Interactions 3: 153-161.

33

The Effect of Lipid Composition in Diets on Ovicell Generating of the Russian Sturgeon Females

Fedorovykh JV*, Ponomarev SV, Bakaneva JM, Bakanev NM, Sergeeva JV, Bakhareva AA, Grozesku JN and Egorova VI

Astrakhan State Technical University (ASTU), Department of Aquaculture and Water Bioresources, 414056, Rus-sian Federation, Astrakhan, Tatishev st, 16, Russia

Abstract

This research is aimed to study the effect of lipid composition in diets on ovicell generating of the Russian sturgeon females. Addition of fish oil in number of 9% (with 12-13% total fat content in mixed fodder) proved to be an opti-mal rate of mixed fodder supplement with fish oil for both commodity sturgeon farming and pre-spawning mainte-nance of breeders with the view of improving the quality of the live fertilized roes and caviar alike. According to the data of biochemical analyses of the sturgeons body composition, hematological indices, growth rates and general condition of the cultivated fish the effect was rather positive. At a minimum expenditure of fodder the body weight gain of the fish body in these series of experiments amounted 3.2% with maximum possible level of survivability (100%). Besides being highly nourishing fish oil also attracting.

Keywords: Dry feed; Fish oil; Fatty acids; Females; Gonads; Lipids; Sturgeon

Abbreviations: CP: Crude Protein; i-AA: Indispensable Amino Acids; ECR: Erythrocyte Sedimentation Rate; PUFA: Polyunsaturated Fatty Acids

Introduction

Artificial sturgeon farming has received a new incentive to development caused by invention and wide use of modern qualitative diets so it has become possible to get about 200 kilograms per square meter of fine fish commodity and solve the problem of import replacement. The quality of complete feed in use is one of the determinative factors of successful commercial fish-farming [1-3]. It ensures intensive growth and development of the fish without natural forage reserve, as well as availability of optimal basic nutrients balance particularly protein and fat, provision with such indispensable nutrients as amino and fatty acids, carotinoids, macro- and microelements, vitamins [4,5]. For juveniles and tradable sturgeon (both type and hybrid species) farming it is necessary to use dry diets which is known to be the basis for normal breeder's gonad products development as well as for optimal growth and evolution of various age fish. However a great deal of untrimmed fodder contains about 20% of fat and fat content in fodders of various trademarks can be different [6,7] and inconsistent with physiological needs of fish for unsaturated fatty acids. It can cause deranged development of ovicells, tissues and internal of fish [8-10]. This research is aimed to study the effect of lipid composition in diets on ovicell generating of the Russian sturgeon females (Figures 1 and 2).

Material and Methods

During the experiment with efficiency of the sturgeons' growth on combifodder -aimed for commodity farming- with different fat content and fatty acids composition juveniles of the Russian sturgeon were used as a subject of the research. The collection of samples for conducting biochemical and histological gametal cells analysis of Russian sturgeon females was taken at sturgeon fish hatcheries of Astrakhan oblast during spawning campaign. We carried out experimental work at the innovative centre of Astrakhan State Technical University (ASTU) "Biowater park – aquaculture scientific and technical centre". The fish was bred in direct-flowed water basins of 2×2×0,7 meters size. Water temperature in basins during the experiment was 16,5-21,5°C,

oxygen concentration – 7,8-8,2 milligrams per liter, pH - 7,3-7,5. Since a female reaches the age of puberty the observance of diet feeding is of a great importance as it stimulates roe development and has an effect on fertility. This problem is rather complex - that's why diets is necessary to contain a great amount of assailable, food valued proteins, balanced amino acids required for the normal development of roe [11,12]. The optimal level of fat and linolenic (ω3) fatty acids allows to get high-quality piscicultural and food roe without accumulating excess mass of fat [13-15]. The experimental variants of mixed fodder recipes are presented in Table 1. Animal origin ingredients, foremost fish flour, are the richest in protein and indispensable amino acids. The abovementioned compounding contained fish flour produced of fresh anchovy sprat. It is distinguished with high level of lysine, methionine, arginine and other scarce amino acids determining normal growth and development of fish [16]. The protein and indispensable amino

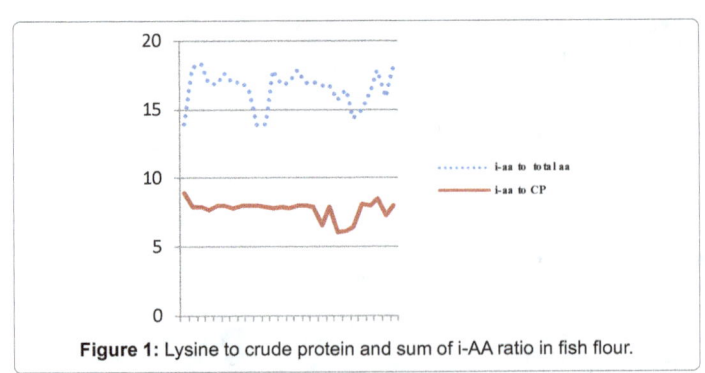

Figure 1: Lysine to crude protein and sum of i-AA ratio in fish flour.

*Corresponding author: Fedorovykh JV, Astrakhan State Technical University (ASTU), Department of Aquaculture and Water Bioresources, 414056, Russian Federation, Astrakhan, Tatishev st, 16, Russia E-mail: jaqua@yandex.ru

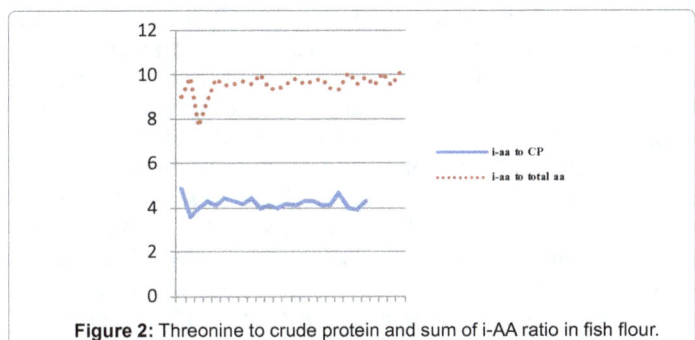

Figure 2: Threonine to crude protein and sum of i-AA ratio in fish flour.

acids content in fish flour produced of fresh anchovy sprat is presented in Table 2. At present crude protein percentage in fish flour doesn't necessarily mean high quality of this element [17,18]. In order to really estimate the quality of crude protein and subsequently make out an appropriate recipe it is necessary to know exact percentage of not "crude" protein (CP) but of "amino-acid protein" and indispensable amino acids (i-AA) especially in such macro ingredients with high protein content as fish flour. These graphs indicate the ratio of four basic indispensable amino acids – lysine, methionine, threonine and cystine to crude protein and sum of indispensable for fish flour amino acids. The ratio of the each four indispensable amino acids both to crude protein and to sum of indispensable amino acids is apparently predictable value positioned on rather narrow limits. The equilibration of amino acid content in a diet in compliance with organism needs is an essential element of the sturgeons full nutrition during pre-spawning period [19,20]. When determining these needs the types and quantity of indispensable amino acids are established to appear in the forage for growth and vital activity of fish. Shortage of any essential amino acid inevitably confines use of other amino acids for protein synthesis thus reducing its efficiency [21,22]. Amino acid content in developed recipes is presented in Table 3. The basic indices of fish productivity were estimated from their body enlargement speed and muscle bulk extension. Measuring of fish was conducted by the following formulae.

Pure gain was determined by the formula [2].

$$Pg = M_f - M_i,$$

where M_f is final mass of the juvenile, g; M_i is initial mass of the juvenile, g.

Mean daily gain was determined by the formula [2,14]

$$P = (M_f - M_i)/t,$$

where P is mean daily gain, g; M_f and M_i – are final and initial mass of the fish respectively, g; t – term of breeding, days.

Mean daily growth rate was determined by the compound interest formula [12]:

$$A = [(M_\kappa/M_o)1/t - 1]\cdot100\ (\%),$$

where A is mean daily growth rate, %; M_κ и M_o – respective mass of fish in the beginning and in the end of experiment; t – duration of the experiment, days.

Comparative assessment of fattiness was determined by Fultone coefficient expressed in formula [8]:

$$Q_F = (P*100)/L^3,$$

where Q_F is fattiness ratio;

P – mass of fish (mg.);

L – length of fish (mm.) from the snout to the overall height of the caudal fin.

Feed ratio was determined as proportion of expended fodder to mass extension of fish during the experiment.

Feed expense were determined by the formula [8]:

$$K = C_f/(m_f - m_i),$$

where C_f is amount of fodder expensed during the breeding term,

m_f is final mass of the fish,

m_i - initial mass of the fish.

Physiological state of juvenile Russian sturgeon was determined by indications of total protein concentration and hemoglobin concentration. Erythrocyte sedimentation rate (ECR) showed whether there was any pathology. We took blood specimen by means of medical syringe from tail vein, farther settled it in the vial till total segregation of plasma and corpuscles, then poured the serum by means of medical syringe into another clean vial. Cells, tissue and organ morphology of the juvenile Russian sturgeon s was explored on

Ingredients	Content, %	
Fish flour	47	38
Fish oil	9	18
Flour from crustaceans	5	5
Blood flour	5	5
Wheat gluten	10	10
Corn gluten	8	8
Wheat germ flakes	10	10
Fodder yeast	5	5
Premix (VMP PO-5) (vitamin-mineral supplement)	1	1
Overall content, %		
Crude protein	52	45
Crude fat	12	22
Crude carbohydrates	16,7	16,4
Crude cellulose	0,9	0,9
Crude ash	10,0	7,4
Total energy, mJ/kg	17,4	19,9

Table 1: A scheme of prosecution of experiment in evaluation of the efficiency of mixed fodder usage for commodity sturgeon farming, with various contents of fish oil.

Elements	%
Crude protein	68.0
Lysine	5.44
Methionine	2.04
Methionine ± Cystine	2.72
Tryptophane	0.75
Arginine	4.01
Histidine	1.56
Phenylalanine	2.79
Threonine	2.79
Valine	3.33
Leucine	5.10
Isoleucine	2.99

Table 2: The protein and indispensable amino acids content in fish flour produced of fresh anchovy sprat

Amino acid	Protein to fat ratio		Amino acid need of fish %
	52/12	45/22	(Tscherbina and Gamygin 2006)
Indispensable			
Lysine	4.49	4.2	4.1
Methionine	1.45	1.49	1.0
Tryptophane	0.75	0.68	-
Arginine	4.71	4.38	3.1
Histidine	1.89	1.76	0.7
Leucine	4.83	4.4	4.8
Isoleucine	3.9	2.6	3.5
Phenylalanine	3.2	2.9	2.6
Threonine	2.92	2.5	2.9
Valine	3.74	3.4	3.3
Replaceable			
Cystine *	1.03	0.75	0.9
Glycocoll	2.9	2.4	2.1
Tyrosine *	1.4	1.2	1.1

* - conditionally replaceable

Table 3: Amino acid content of experimental variants of diets for pre-spawning maintenance of sturgeon breeders for the purpose of improving the quality of fertilized hard and food roe

series of histologic specimen prepared by standard methods [23]. After fixing the samples underwent dehydration in alcohol with increasing concentration (60°, 70°, 80°, 90°, 96°, absolute alcohol). The exposition term in each alcohol was 24 hours. Then the specimens were for 2-3 days steeped into photoxylin-castor oil (1:1). After that the material was successively placed in chloroform for 3 hours; chloroform-paraffin mixture at 37°C for night; in pure paraffin at 56°C for 30-40 minutes. After that the material was poured into paraffin and quenched in cold water. We examined gametal cells structure by slicing paraffin blocks into sagittal pieces on standard microtome. We straightened series of slices 4-5 micro tons depth in a drop of warm water and stuck onto the object-plates polished with protein-glycerin mixture. Before imbuing the plates with slices had been dewaxed in xylene, xylene was removed by alcohols with falling concentration [7]. In total we made 62 microscopic sections. For imbuing the specimens we used Heidenhein's iron hematoxylin and Mallory azan stain. Then we put imbued slices into Canada balsam under the cover glass. The slices were assayed under a microscope Biolam Lomo. Specimens were examined under a microscope OLYMPUS BX40. The pictures were taken with digital ocular camera for microscope DSM. The results were subjected to statistical review. All the digital experiment data were processed on IBM PS/AT assisted with integrated package Statistica v 6.0, Microsoft Office Excel 2007 was also in use. The veracity of differences in compared indices was determined by Student criterion [24] (Figures 3 and 4).

Research Results

With all-round reduction in population level of the Sturgeon in natural bodies of water opportunity of productive herds rejuvenating at the expense of naturally generated breeders is almost completely limited. Therefore broodstock management is based on "from egg to egg" raising of mature individuals or on domesticating of males and females sourced from wild population if it is possible. Possibility of acquiring mature individuals, frequently with excess fat gonad uptake, grown at commodity farms is not excluded. It causes breeding potential decrease of such fish and complexity in process of roe produced for food deriving [23,24]. At sturgeon hatcheries of Astrakhan oblast

winter females' hard roe, stocking of which decreases every year, was compared with domesticated breeders roe and some sufficient differences in fat buildup level both in roe and in egg were registered. It testifies of gradual obesity and affected fertilization and survivability of eggs (Table 4). High fat in roes of domesticated females results from dietary features and application of fat-laden feed compound (18% and more) particularly. Such roes were larger in size and noted for high fat and lower protein level. During spawning campaign those roes were poorly fertilized, seldom got off the ovaries, badly ovulated under gonadotropic hormone. Oocytes histological studies on the second stage (or second fat stage) are represented on (Figures 5 and 6). Considerable part of examined fish turned out to have oocyte structure abnormalities appeared in tunic destruction. Outer divergences were caused by changes in protein and lipid composition of oocytes resulted from disbolism and misbalance in the organism because of unbalanced diet. Most females had ovary hyperemia. Their ovaries were blood-filled what was presumably caused by blood stream deceleration after "obesity of the roes" (Figure 7). Moreover nucleolus hypertrophy, occurrence of additional centers in cytoplasm and appearance of indistinctive stippling accounted for necrosis were noticed (Figure 8). Significantly deformated cells also occurred. Deformation of oocytes results from turgor impairment in membranes and decrease of their strength. Such oocytes had indefinite shape (Figure 9). Different kinds of oocytes abnormalities appeared to be caused by direct influence of misbalanced feeding of females during pre-spawning maintenance. Revealed abnormalities in some indices of lipid and protein composition in the roes of Russian sturgeon breeders were accounted for non-observance of farming conditions during pre-spawning period and applying of misbalanced diets. The most frequent problems of commodity sturgeon farming are the following: breakdown in biotechnology of aquacultural objects cultivation, absence of feeding schedule tight control, applying of feed compound with misbalanced nutrients composition, excess fish-holding density per unit of farming area. Figure 10 represents adipose degeneration of gonads of Russian sturgeon female on the third maturity. Such roes are unapt for food. Aforementioned factors cause occurrence of fat accumulation in gonads that result in troubles with intravital roe processing, short extraction

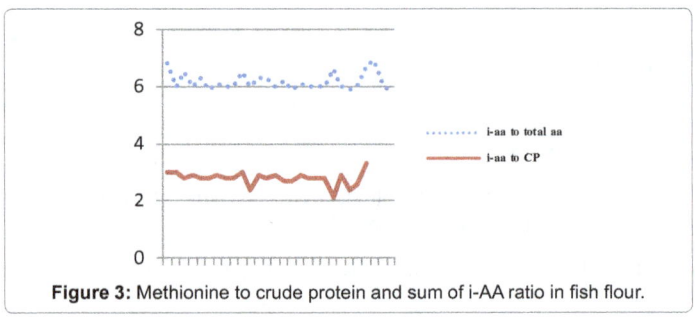

Figure 3: Methionine to crude protein and sum of i-AA ratio in fish flour.

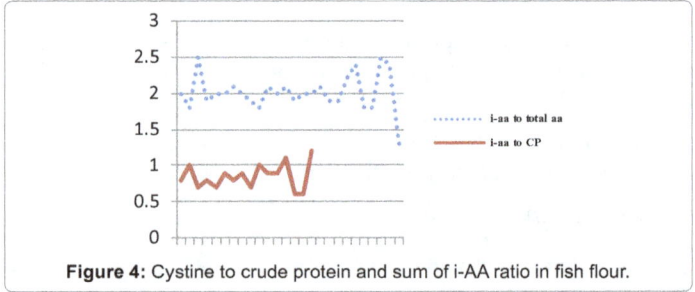

Figure 4: Cystine to crude protein and sum of i-AA ratio in fish flour.

Indices	Domesticated females		Winter females
	Matured in artificial conditions for the first time	Matured in artificial conditions for the 3-4-th time	
Dry matter	39.4 ± 1.1	38.1 ± 2.5	37.4 ± 4.4
Protein	25.8 ± 0.5	24.2 ± 2.1	27.3 ± 3.5
Fat	13.3 ± 1.5	12.9 ± 1.0	8.7 ± 1.2
Mineral matters	1.2 ± 0.4	1.6 ± 0.2	1.6 ± 0.3

Table 4: Biochemical structure of Russian sturgeon roe (% of raw matter)

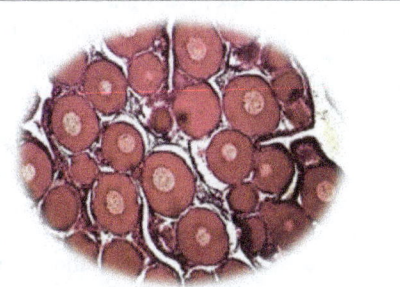

Figure 5: Russian sturgeon ovary in the second maturity (normal feature) (magnification 22x10).

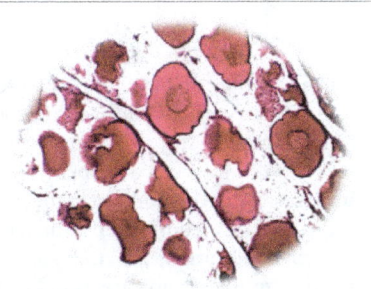

Figure 6: Oocytes degeneration in the Russian sturgeon ovary in the second maturity (magnification 22 x10).

of gonad products from body weight, difficulties in processing such gonads for food. Fat in sturgeons' nutrition has been long estimated only in respect to energy value. Later on biological value of fat proved to depend on presence of indispensable (essential) polyunsaturated fatty acids. Among PUFAs are linolenic and linolic acids as well as their derivatives – eicosapentaenoic, docosahexaenoic and arachidonic acids generated over series of intermediate stages by prolongation of hydrocarbon chain and desaturation at the expense of olefinic linkages increase:

1. Linolenic (18:3 ω-3)	Eicosapentaenoic (20:5 ω -3)	Docosahexaenoic (22:6 ω -3)
2. Linolic (18:2 ω -6)	Arachidonic (20:4 ω -6)	

Distillation characteristic of fundamental polyunsaturated fatty acids for the sturgeons was ascertained to content within the 9% mark of fish oil application into feed composition (Figure 11). Such fat is processed from maritime commercial fishes dwelling in northern seas and whose total lipids comprise a lot of PUFA ω 3: pentaenic and hexaenic AA. PUFA is an important factor of reproductive process of the sturgeons. The deficit of ω3 acids in fish forage causes fertility reduction, roe quality degradation, reduction of uncombined embryo exclusion, increasing number of deformed larvae and their death. Working on estimating of fish oil level influence on efficiency

of commercial cultivation we divided all the fish into control and experimental groups. During the experiment we apply dry feed with fat content of 9 and 18%. On the first work stage it was necessary to estimate influence of 18% (total content – 21%) fish fat composition in the forage on pisciculture-biological indices of sturgeon farming in reservoirs without application of live food. When carrying out the experiment we use herring sardine fish oil which contained 20% of unsaturated fatty acids, 45% of oleinic acid (ω9), 25% of fatty acids of ω3 tier and 10% of fatty acids of ω6 tier. The quality of oil was high (peroxide value – 0.12 unit, acid number - 15 units). The research results generally proved that excess increase in fish oil quantity in forage results in decrease of pisciculture biological indices. The best growth indices were achieved in control variant with fish oil content of 12% (totally with fish flour fat) in the food mix. The growth of body weight of fishes was 27 grams more than experimental variant with double the amount of fish oil (18%) in forage, totally with fish flour fat 21%. Survivability of fish was 100% in both variants. However the individuals in experimental group behaved less active, they were marked with low mobility and slow growth rate under increased feed expenses (Table 5). Coefficient of weight accumulation of the control variant fish with 9% of fish oil contents in the forage was high and amounted 0,035 units while the index of experimental variant with 18% content of fish oil was about 0,008 units, it corroborates natural growth rate reduction. However the question of the function lipid structures of fodder fat including unsaturated fatty acids perform remains obscure. Thus as the result of the experiment supplement of production forage with 18% of fish oil (totally 21%) was determined to cause some reduction of growth rate. It relates to the fact that excess fat in fish mixed fodders induces hepatic accumulation of decomposition products that may cause fatty degeneration of liver. Growth rate of fish on mixed fodders supplemented with 9% of fat (totally 12%) was consistently high during all the time of experiment. It is a matter of fact that in order to improve

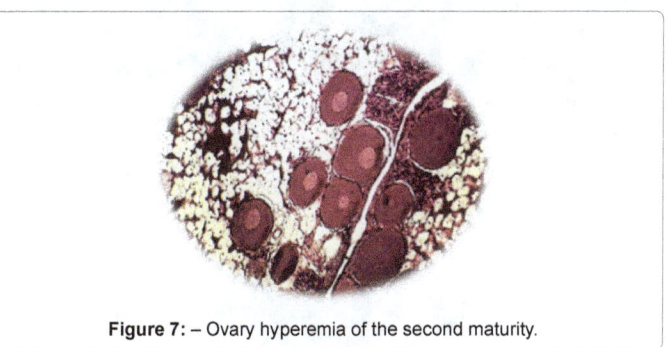

Figure 7: – Ovary hyperemia of the second maturity.

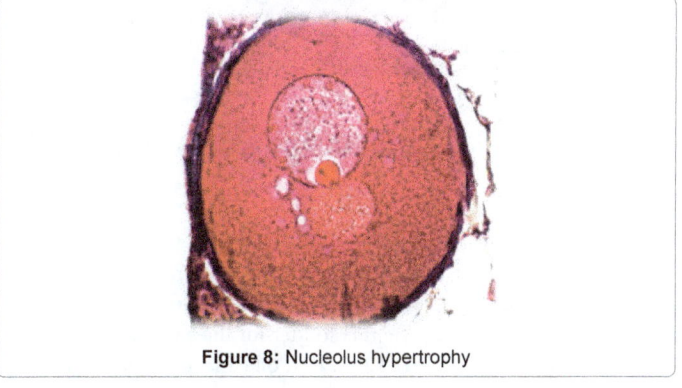

Figure 8: Nucleolus hypertrophy

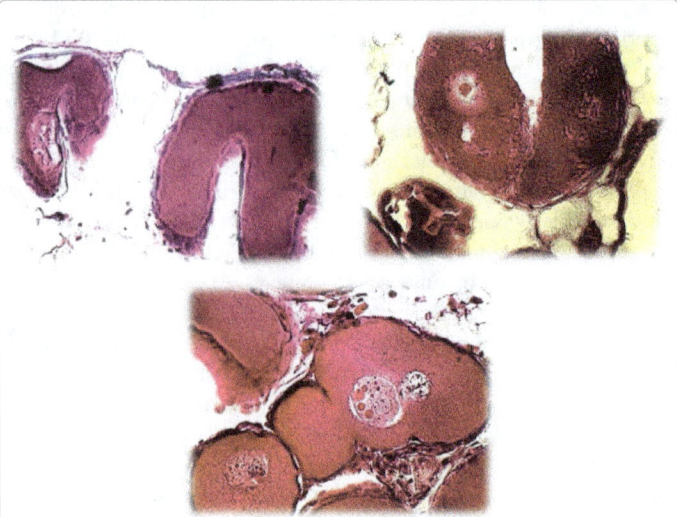

Figure 9: Deformated gametal cell of Russian sturgeon female in pre-spawning period.

Figure 10: Gonads of Russian sturgeon entirely covered with fat following misbalanced diet.

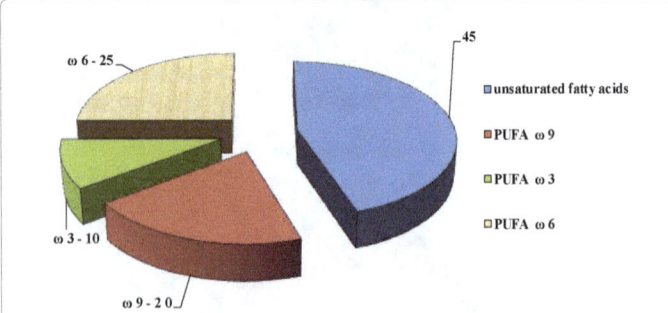

Figure 11: Contents of unsaturated fatty acids in fish oil in feed % of total amount.

21%) results in reducing of all abovementioned pisciculture-biological indices of farmed fish growth and maturation. At the same time determined hematologic indices of the farmed sturgeons which had been fed with production fodder with various content of supplementary fish oil allowed to calculate the best physiological status of fish (Table 6). Higher hemoglobin and erythrocytes content in the blood of fish with 9% of fat variant is evidence of positive effect of its lower contents in fodder. On females age of two and three histological research of their gonad products was carried out (Figures 12 and 13). The gametal gonads of all the examined three years old species were of the second maturity stage. The oviparous laminae were well-formed. Histoplogical specimens indicated numerous oocytes at the various stages of protoplasmatic growth which formed elder gametal cells generation. Their development proceeded without apparend deviations (with the exception of some few deformated ones). Expanded sanguineus blood vessels and blood cells clumps beyond the vascular bed were found in some sites. Thus conducted complex analysis of pisciculture-biological and physiological findings as the experimental results allowed us to set this standard. At the same time recalculation of fatty acids of ω3 and ω6 tiers allowed to determine their contents in quantity of 3 and 1,2 (in total lipids), percentage of fodder mass. The values of PUFA of linolenic (ω3) and linolic (ω6) tiers – 1-1,5% are similar to the ones of salmon family (1,5-2,0).

Discussion

Addition of fish oil in number of 9% (with 12-13% total fat content in mixed fodder) proved to be an optimal rate of mixed fodder supplement with fish oil for both commodity sturgeon farming and pre-spawning maintenance of breeders with the view of improving the quality of the live fertilized roes and caviar alike. According to the data of biochemical analyses of the sturgeons body composition, hematological indices, growth rates and general condition of the cultivated fish the effect was rather positive. At a minimum expenditure of fodder the body weight gain of the fish body in these series of experiments amounted 3.2% with maximum possible level of survivability (100%). Besides being highly

Indices	Variants	
	Control (9%)	Experiment (18%)
Initial mass, grams	243.59 ± 4.4	242.44 ± 4.35
Final mass, grams	288.04 ± 2.87***	260.35 ± 4.16***
Pure gain, grams	45	18
Average daily gain, grams	1.5	0.6
Average daily gain, %	5.44	4.94
Survival rate, %	100	100
Feed expense	1.1	1.5
Breeding time, days.	30	30

Note: differences are clean under *** - $P \leq 0.001$

Table 5: Pisciculture-biological indices of hybrid sterlet x beluga breeding on the forage with various fish oil contents

Indices	Experimental variants	
	Control (9%)	Experiment (18%)
Hemoglobin, g/dL	78.34 ± 0.72***	73.64 ± 1.2***
Hematocrit, %	28.24 ± 0.44***	24.52 ± 0.89***
Erythrocytes , million per mm³	0.894 ± 0.012	0.840 ± 0.03
Hemoglobin content of an erythrocytes, µg /erythrocyte.	6.8 ± 0.11*	6.39 ± 0.12*
Serum protein, g%	3.8 ± 0.098	2.37 ± 0.1

Note: differences are clean under * - $P \leq 0.05$; *** - $P \leq 0,001$

Table 6: Hematologic indices of Russian sturgeon bred on production fodder with various fish oil contents.

pisciculture-biological indices of sturgeon mixed foddered breeding meant for commercial farming fat-containing ingredients should be expediently added subject to standards [8,21]. Optimal rate of herring sardine fish oil addition to the mix fodder for the sturgeons equals 9% (with total quantity in fodder 12%). Fish oil concentration to 18% (i.e.

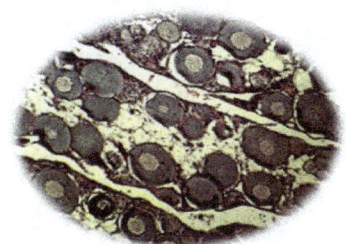

Figure 12: Ovary of Russian sturgeon female. The second stage of gonads maturity. (2 years old). Hematoxylin and eosin stain. Magnification 22x10. 1.

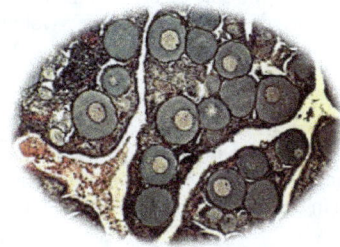

Figure 13: Ovary of Russian sturgeon female. The second stage of gonads maturity. (3 years old). Hematoxylin and eosin stain. Magnification 22x10. 1.

nourishing fish oil also attracting.

References

1. Zagrebina ON, Khasanipur AA, Kokoza AA (2014) Morphophysiological indices of juvenile Russian sturgeon and its hybrid with Siberian sturgeon. Harmonious exploration and preservation of aquatic bioresources. Rostov-on-Don publishers, Russia.

2. Ostroumova IN (1987) The improvement of recipe of diets and methods of fish feeding in industrial farming.

3. Cowey CB (1980) Protein metabolism in fish.

4. Ostroumova IN (2012) Biological basics of feeding.

5. Glencross BD, Booth M, Allan GL (2007) A feed is only as good as its ingredients - a review of ingredient escalation strategies for aquaculture feeds. Aquaculture Nutrition 13: 17-34.

6. Ponomarev SV, Grozesku JN, Bakhareva AA (2013) Fodders and feeding of fish in aquaculture: Manual. Moscow.

7. Gamygin EA, Bagrov AM (2013) Some aspects of feeding and fodder production for fish at the present days.

8. Bakaneva JM, Bakanev NM, Fedorovykh JV (2014) Effect of disbalanced diet on roe quality of the sturgeon females.

9. Halver JE (1978) Lipids and fatty acids.

10. Ponomarev SV (2010) The estimation of the diets with various fat contentions for sturgeon.

11. Watanabe T (1979) Sparing action of lipids on dietary. Protein in protein fish-low protein diet with high calorie content.

12. Volkova OV, Eletsky YK (1982) Foundations of histology with histological technique.

13. Romeis B (1954) Microscopic technique.

14. Halver JE (1972) Fish nutrition.

15. Watanabe T (1982) Lipid Nutrition in Fish. Comp Biochem Physiol 73: 3-15.

16. Laqin GF (1990) Biometrics. Vysshaya shkola, Moscow.

17. Ponomareva EN, Bakhareva AA (1999) New polyvitamin premix for sturgeons. The materials of the Second International symposium "Resource saving technology in aquaculture".

18. Sherbina MA, Gamygin IA, Salkova IA (1996) The influence of extrusion on the nutrition of feed raw materials for fish.

19. Castell JD, Tiews K (1979) Report of the EIFAC, IUNS and ICES Working Group on the standardization of the methodology in fish nutrition research.

20. Corraze G (1994) Nutrition lipidique des poisons: imprortance et consequenced. La piscicult franc 117: 25-36.

21. Hasan MR (2001) Nutrition and feeding for sustainable aquaculture development in the third millennium.

22. Jauncey K (1995) Advances freshwater fish nutrition.

23. Woodgate SL (2004) Creating alternative protein sources for aquafeeds using applied enzyme technologies. Nutritional Biotechnology in the Feed and Food Industries, Proceedings of Alltech's Twentieth Annual Symposium, Kentucky, USA.

24. Xu R, Hung SSO, German JB (1996) Effects of dietary lipids on the fatty acid composition of triglycerides and phospholipids in tissues of white Sturgeon. Aquas Nutr 2: 101-109.

34

On the Biology of *Siganus rivulatus* Inhabits Bitter Lakes in Egypt

El-Drawany*

Department of Zoology, Faculty of Science, Zagazig University, Egypt

Abstract

The present work was carried out to study the age, growth, length-weight relationships, spawning season, length at first sexual maturity and mortality of the commercial Rabbit fish, Siganus rivulatus, inhabits Bitter Lakes in Egypt. Fish samples which were collected during the period from January to December 2011 were selected to represent all fish size categories in the catch. Each fish specimen was measured, weighted and then dissected to detect the maturity stage of the gonads. The annual rates of total, natural and fishing mortality were calculated as 0.8840, 0.2214 and 0.6626^{yr-1} respectively. Current exploitation rate 'E' was estimated at 0.75. The length-weight relationships for males and females were estimated, respectively as:

$$W=0.01042 \times L^{3.0101} \text{ and } W=0.00952 \times L^{3.042}.$$

The age data derived from the otolith readings were used to estimate the growth parameters of the von Bertalanffy growth equation. The estimated parameters were: $L_\infty=35.5$ cm, K=0.0849 and $t_0=-0.843$ for females, otherwise they were: $L_\infty=36.5$ cm, K=0.0786 and to =-1.00382 for males. It was found that, both males and females matured at a total length of about 15.4 cm. The natural spawning period of this fish species is in the summer, from May to July.

Keywords: Environment; Fishes; Phytoplankton; Fishery production

Introduction

The Siganidae form a small family of herbivorous, widely distributed fishes in the Indo-West Pacific Ocean [1]. They are of economic importance for the fishery production in several countries in the Indo Pacific and Middle East regions. *S. rivulatus* has established large populations in its new environment and can be considered as one of the most successful Lessepsian fish [2-4]. It was found in several different overgrown habitats (rocks with algae, sand with algae, and grass with algae) it is considered to be strictly herbivorous [5,6]. *Siganus rivulatus* are mainly found in shallow water, usually less than 15 meters. This shallow depth should be predictable for species that is almost entirely herbivorous where algae are the more widespread at these depths. The fishes begin their life as a phytoplankton feeder on many small diatoms, then ingest zooplankton consisting of copepods species [7]. The type and amount of food influence significant the body composition and its nutritional value [8]. Family Siganidae have a single row of flattened close-set teeth for rasping at meatier seaweeds. The mouth is specially designed to aid in the removal of algae from in between rock crevices, or coral branches. The upper jaw is fixed, with only the tip of the mouth being able to move up or down creates a nibbling action. Five species of Siganids are known in the northwestern area of Red Sea [9]. Two of which have immigrated through the Suez Canal and now are established itself successfully in the eastern Mediterranean up to the Aegean sea and along the coast of Egypt and Libya up to Tunisia in the southern Mediterranean [10]. The first species is dominant in the Egyptian waters of Mediterranean, in the coastal area off Alexandria [11] especially in Abu-Qir Bay [12]. Information about *Siganus rivulatus* age and pertinent growth rate were studied in both the Red Sea [9-15] and in the Mediterranean [16-19]. An existing data mainly concern the reproductive biology and rearing experiments [20-24] have been reported. The Rabbit fish Siganus rivulatus Forsskal [25] hold particular promise for marine aquaculture development by virtue of their herbivorous/omnivorous feeding habits and consequent ability to feed low on the aquatic food chain [26], their high tolerance to environmental factors, rough handling and crowding [27] fast growth and possibility to obtain their seeds from the wild or by artificial propagation [28]. Therefore, the aim of the present study is concerned with the updating biological aspects of Siganus rivulatus which is one of the dominant fish species occurring in the artisanal fishery of the studying area. Such information could be employed for best planning of fishing and culturing this fish in the Egyptian waters moreover managing its stock in the area of study.

Materials and Methods

As regards to 420 specimens of Rabbit fish, *Siganus rivulatus*, were collected monthly around 2011, from the landing site at Bitter lakes. Fishes were put immediately in crushed ice and transported to the laboratory, where they were subjected investigation. Date of capture, total fish length (mm) and total fish weight (0.1 g) were recorded for each fresh specimen. Fishes were dissected to define their sex and gonad maturity stages. Gutted fish were weighed to the nearest 0.1 g and gonads were weighed to the nearest 0.00l g. Both otoliths (Sagittae) of each fish specimens were removed, cleaned in ethanol and stored for age determination after drying. The otoliths then after were treated with xylene and moistened with chamomile oil [19] to be clearly visible under reflected light microscope in order to read the annual rings. Counts of rings and measurements were always performed in the same direction, from the nucleus to the edge of the otolith. Back-calculation of total length, were obtained by Lee's equation. The growth parameters L_∞, k and t_0 of *Siganus rivulatus* were estimated by means of von Bertalanffy plot (Sparre and Venema [29] and application of von Bertalanffy equation von Bertalanffy [30].

$$L_t = L_\infty (1 - e^{-k(t-to)})$$

Where L_t is the total length at time t; k is a growth constant; L_∞

***Corresponding author:** El-Drawany, Department of Zoology, Faculty of Science, Zagazig University, Egypt, E-mail: samy_drawany@yahoo.com

is the asymptotic length; and tθ describes the theoretical age where L_t is zero [31]. The length weight relationships were estimated from the allometric equation, $W = a L^b$ Ricker [32] where W is total body weight (g), L the total length (cm), a and b are the coefficients of the functional regression between (W) and (L) and they were obtained using the Newton algorithm from the Microsoft® Excel Solver routine.

The spawning season was determined by the curvilinear average values of monthly Gonadosomatic index (GSI) for both males and females where:

GSI= 100 [gonads weight (g)/gutted weight (g)]

Total mortality rate (Z) was estimated based on the length at first capture methods evaluated by Beverton and Holt 1957 [33].

Z=K ($L_∞$ - Lm/ Lm - Lc)

Where:

Lm=the average total length of the entire catch.

Lc=the length at which 50% of the fish entering the gear

Natural mortality rate (M) was estimated by using the equation derived by Ursin [34] based on the mean total length where:

$M = W^{-(1/b)}$

W=mean total length.

b=constant of length weight relationship.

Fishing mortality rates (F) were calculated as the difference between Z and M where Z=F + M.

The annual exploitation rate (E) was obtained according Sparre et al. where: E=F/Z

The total mortality coefficient "Z" was estimated using the method of Pauly [35].

Results

Total length frequency distribution

The total length of all individuals (n=420) collected in the present study ranged from 9 to 28.7 cm, and the most frequent length was 18 cm as illustrated in Figure 1 and Table 1.

The age composition and growth

The age distribution of samples ranged from I to VI years for *S. rivulatus*, based on the results of otolith reading (Figure 2). The age group III was dominant (27.6%) for male followed by age groups II

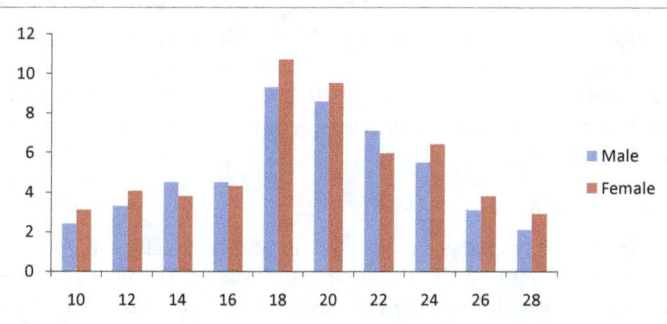

Figure 1: Length frequency distribution of male and female of *Siganus rivulatus*.

Age group	Mean number	Observed Length (mm)	Back calculated lengths in mm					
			1	2	3	4	5	6
I	34	11.0	10.46					
II	41	15.5	10.69	14.15				
III	61	18.9	11.50	15.20	17.68			
I V	36	21.7	11.35	15.37	16.98	20.19		
V	33	23.9	10.81	15.85	18.81	20.90	23.44	
VI	12	26.4	10.75	15.34	19.52	22.02	23.69	25.36
Average total	217		10.92	15.18	18.95	21.44	23.57	25.36
Increment of length			10.92	4.26	3.77	2.49	2.13	1.79
% of annual Increment			43.1	16.88	14.87	9.82	8.40	7.06

Table 1: Average back-calculated lengths (cm) of male *Siganus rivulatus* from the Bitter lakes, Egypt.

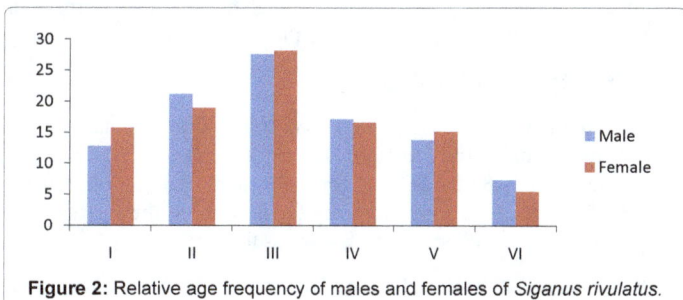

Figure 2: Relative age frequency of males and females of *Siganus rivulatus*.

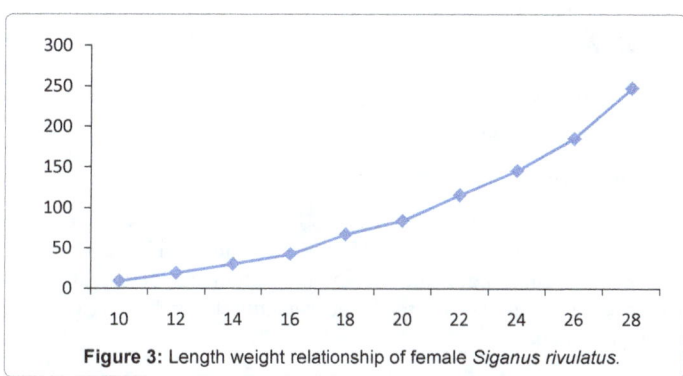

Figure 3: Length weight relationship of female *Siganus rivulatus*.

(21.2%), IV (17.2%), V (13.8), I (12.8) and VI (7.4%), While the age group III was dominant (28.1%) for female followed by age groups II (18.9%), IV (16.6%), I (15.7%), V (15.2%) and VI (5.5%). The growth of the studied fish was described by the von Bertalanffy model based on the back-calculated length at age data (Figure 3). The estimated growth functions were:

$L_t = 36.5$ cm $(1 - e^{-0.0786(t+1.0038)})$ for males and

$L_t = 35.5$ cm $(1 - e^{-0.0849(t+0.8430)})$ for females

Growth in both sexes is almost similar (K=0.0786 for males and K=0.0849 for females). However, the maximum theoretical length were 36.5 cm and 35.5 cm for males and females, respectively.

Growth in length

The otolith radius-total length relationship of males and females are described by the following equations:

L=5.3227 + 4.0465 S for males (r^2=0.998) and

L=4.8465 + 4.01173 S for females (r^2 =0.995)

Where: L is the total length in centimeters, S is the otolith radius in millimeters and r is the correlation coefficient.

Back-calculations

The following formula was derived to obtain the back-calculated total length at the end of each year of life for males and females respectively.

$$Ln=(L - 5.3227) \; Sn \, / \, S + 5.3227$$

$$Ln=(L - 4.8465) \; Sn \, / \, S + 4.8465$$

Where: Ln is the length at the end of nth year, Sn is the radius of the otolith to n^{th} annulus, S is the total radius of the otolith and L is the total length at capture. From the data given in Tables 1 and 2, it is obvious that, S. rivulatus attains its highest growth rate in terms of length during the first year of life, after which a gradual decrease in growth increments was noticed with further increase in age.

Length-weight relationship

The total length of S. rivulatus varied from 9.0 to 28.5 cm and 9.0 to 28.9 cm for males and females respectively, while the total weights ranged between 10.83 to 246.7 g. for males and 10.68 to 248.3 g for females. The equations were extracted for describing the relationship between weight and length for both sexes as follows:

$W=0.00952 * L^{3.042}$ (r^2=0.98)   or Log W=- 2.02136 + 3.0424 Log L for female and $W=0.01042 * L^{3.010}$ (r^2=0.98) or Log W=- 1.9821 + 3.010 Log L for female

Where:

W is the total weight (gm)

L is the total length (cm)

R is the correlation coefficient.

The high values of r^2 indicate a good measure for the strength of these equations and closeness of observed and calculated values of fish weight. The length and weight measurements of the analyzed specimens used to describe the length-weight relationship are given in Figures 3 and 4.

Condition factor

The present results of condition factor revealed that the (Kn) values fluctuated between 0.98-1.04 with an average of 0.999 for males and

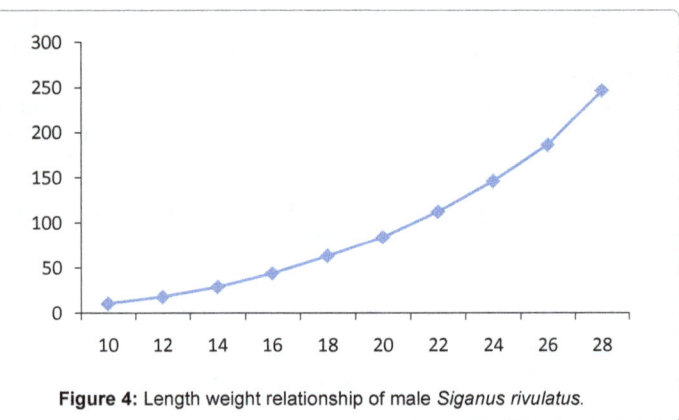

Figure 4: Length weight relationship of male *Siganus rivulatus*.

Age group	Mean number	Back calculated weights in (g)					
		1	2	3	4	5	6
I	34	11.55					
II	41	13.76	39.82				
III	61	14.98	45.38	91.16			
I V	36	13.49	38.13	80.06	113.82		
V	33	13.72	42.21	76.62	104.14	134.41	
VI	12	14.34	44.11	80.44	115.73	144.56	168.44
Average total	217	13.64	41.93	82.07	111.23	139.49	168.44
Increment of weight		13.64	28.29	40.14	29.16	28.26	28.95

Table 3: Calculated weights (g) of male *Siganus rivulatus* from the Bitter lakes, Egypt.

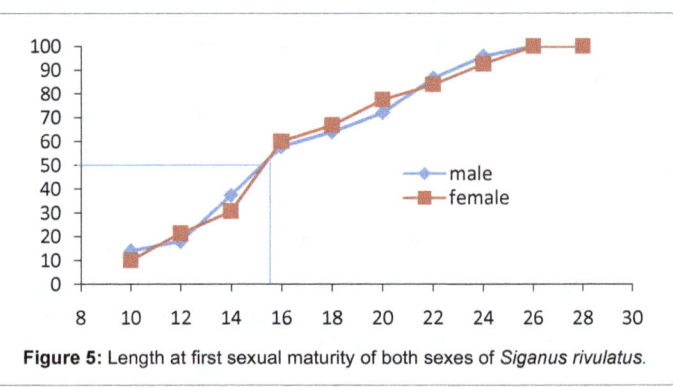

Figure 5: Length at first sexual maturity of both sexes of *Siganus rivulatus*.

between 0.92-1.07 with an average of 1.00 for females (Table 3).

Length at first maturity

To determine the length at first maturity, males and females were grouped into 10 mm size groups and the percentage occurrence of fish at the different maturity stages in each size group was calculated. Examination of the male and female maturity stages indicated that males and females of *S. rivulatus* matured at about 15.5 cm total length (2 year old) (Figure 5).

Gonadosomatic Index (GSI)

The monthly changes in GSI values of individuals of both sexes are given in Figure 6. It was observed that the GSI values of males and females were low during February, March, and April 2011. Index values began to increase after April to reached maximum values in July, and then began to decrease indicating that its breeding season is in July.

Age group	Mean number	Observed length (mm)	Back calculated lengths in mm					
			1	2	3	4	5	6
I	26	10.8	10.32					
II	43	15.3	10..93	15.50				
III	56	18.5	11.24	16.18	20.35			
IV	35	21.4	10..86	15.28	19.50	21.89		
V	28	24.7	10..92	15.80	19.22	21.26	23.12	
VI	15	26.2	11.08	16.03	19.53	22.01	23.68	24.97
Average total	203		10..89	15.76	19.65	21 .72	23.40	24.97
Increment of length			10..89	4..87	3.89	3.07	1.68	1.57
% of annual Increment				43.6	19.50	15.58	8.29	6.73	6.29

Table 2: Average back-calculated lengths (cm) of female *Siganus rivulatus* from the Bitter lakes, Egypt.

Growth in weight

The calculated weights at the end of each year of life of *S. rivulatus* were estimated by applying the corresponding length-weight equation to the back calculated lengths. The resulting values are given in Tables 3 and 4. The obtained results for males and females indicated that the growth rate in weight was slow during the first year of life. Then the annual growth increment in weight increased with further increase in age until it reached its maximum value at the end of the third year of life, after which a decrease in the growth increment was observed.

Age group	Mean number	Back calculated weights in (g)					
		1	2	3	4	5	6
I	34	12.21					
II	41	13.04	30.32				
III	61	16.24	37.61	81.83			
IV	36	15.61	38.89	72.91	105.16		
V	33	13.48	42.66	83.47	112.89	138.51	
VI	12	13.26	38.66	77.89	114.76	143.00	169.37
Average total	217	13.97	37.63	79.03	110.94	140.76	169.37
Increment of weight		13.97	23.66	41.40	31.91	29.82	28.61

Table 4: Calculated weights of female *Siganus rivulatus* from the Bitter lakes, Egypt.

Sex ratio

As presented in Table 6, the percentage of females to males (sex ratio) of *Siganus rivulatus*, in Bitter lakes is fluctuated around the year. It was obvious that sex-ratio deviate significantly from 1:1 among the size classes. However, the overall sex ratio (Males: Females) of this species during the year was 1: 1.079 and did not differ significantly from 1:1. It ranged between 48.4 in June as a minimum value to 55.8 in December, maximum one. However, the number of both sexes is equal with percentage 50.0% in May. But in July (spawning period) the males somewhat predominated females (percentage of sex ratio is 48.5%).

Mortality rates

By using the cumulative curve of S. rivulatus (Figure 7) illustrating length at first capture at 50 % and applying the method of Sparre et al. the total mortality coefficient (Z) was estimated. This coefficient was found to be 0.8840 $^{year-1}$. But the Natural mortality coefficient "M" which obtained from the mean total length was 0.2214 $^{year-1}$. Using the estimated (M) and (Z) the fishing mortality (F) was obtained (0.6626 $^{year-1}$), where Z=M + F.

Exploitation rate (E)

According to Gulland [36] who suggested that the optimum exploitation rate in an exploited stock should equal approximately 0.50.

Author	Location	Year	sex	L∞(cm)	K (year^{-1})	to (year)
El-Gammal [13]	Red Sea (Egypt	1988	M	31.5	0.501	-0.09373
			F	34.2	0.434	-0.1002
EL-Okda [23]	Red Sea (Egypt	1991	M	30.6	0.513	-0.07638
			F	31.2	0.499	-0.04007
			C	30.9	0.505	-0.05562
EL-Okda [23]	Mediterranean(Egypt)	1998	C	32.0	0.299	-0.09786
Bilecenoglu and Kaya [18]	Med. (Turkey)	2002	M	21.1	0.345	-0.537
			F	22.6	0.267	-0.473
			C	22.3	0.279	-0.503
El-Ganainy and Ahmed [15]	Red (Egypt)	2002	C	29.4	0.735	-0.220
Mehanna and Abdallah [39]	Red (Egypt)	2002	C	37.1	0.397	-0.186
Bariche [43]	Mediterr. (Lebanon)	2005	C	31.9	0.225	-1.307
Present study	**Bitter Lakes Egypt**	**2011**	M	36.5	0.0786	-1.0038
			F	35.5	0.0849	-0.843

Table 5: Von Bertalanffy parameters from the literature and the present study for male (M), female (F) and combined sex (C) of *Siganus rivulatus*.

Authors	Location	year	sex	a	b	r
Hashem [9]	Red Sea (Saudi Arab.)	1983	combined	0.021	3.071	-
El-Gammal [13]	Red Sea (Egypt)	1988	Male	0.012	2.838	0.99
			Female	0.011	2.841	0.99
Mohamed [17]	Mediterranean Sea (Egypt)	1991	Male	0.018	2.830	0.99
			Female	0.017	2.866	0.99
			Combined	0.012	2.934	0.99
EL-Okda [23]	Mediterranean Sea (Egypt)	1998	Combined	0.016	2.872	-
Taskavak and Bilecenoglu [18]	Mediterranean Sea (Turky)	2001	Combined	0.047	3.203	0.98
Mehanna and Abdallah [39]	Mediterranean Sea (Egypt)	2002	Combined	0.022	2.820	0.93
Bilecenoglu and Kaya [18]	Mediterranean Sea (Turky)	2002	Male	0.075	3.135	0.95
			Female	0.064	3.221	0.95
			Combined	0.071	3.179	0.95
Mehanna and Abdallah [39]	Red Sea (Egypt)	2002	Combined	0.012	3.020	-
Tharwat and Al-Owafeir	Red Sea (Saudi Arb.)	2003	Combined	0.013	2.990	0.99
Bariche et al. [43]	Mediterranean Sea (Lebanon)	2005	Male	0.020	3.323	0.95
			Female	0.010	3.011	0.96
			Combined	0.010	3.037	0.99
Present study	**Bitter Lakes (Egypt)**	**2011**	**Male**	**0.0104**	**3.010**	**0.98**
			Female	**0.0095**	**3.042**	**0.98**

Table 6: Estimated parameters of *S. rivulatus* recorded by some other authors in different regions.

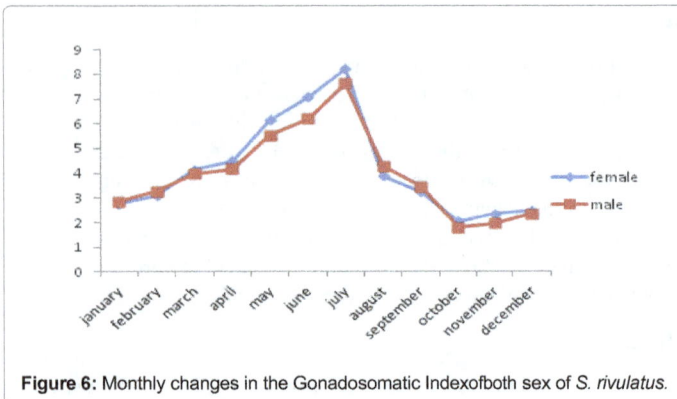

Figure 6: Monthly changes in the Gonadosomatic Indexofboth sex of *S. rivulatus*.

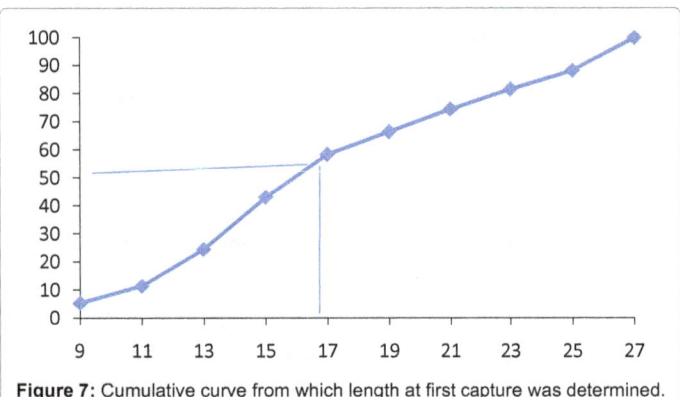

Figure 7: Cumulative curve from which length at first capture was determined.

The current exploitation rate "E" was estimated at 0.75. Accordingly, the high value of the current exploitation rate indicates that the stock of S. rivulatus in the Red Sea is subjected to overfishing.

Length at first capture (Lc)

The length at first capture L50% (the length at which 50% of the fish are first exposed to capture) was estimated as a component of the length converted catch curve analysis (FiSAT), was found to be 16.4 cm which corresponds to an age of 2.37 years.

Discussion

The most abundant size class (18 cm) for both sexes of S. rivulatus, is composed of individuals that have already reproduced (15.5 cm). This observation is closed to that reported for the same species at Libyan coast [25]. The same authors stated that, the age is necessary to assess population dynamics and the state of exploited resources. The opaque zone on the Otoliths of *S. rivulatus* corresponds to a rapid growth but the translucent zone to a slow growth of the fish. This phenomenon is observed in a large number of fish species living in subtropical and temperate regions in tropical regions Brothers, [37] and in other Lessepsian species [38]. The maximum ages for males and females of *Siganus rivulatus* were observed at six years in the present study. However, Mohamed stated that, the maximal age of both males and females of *Siganus rivulatus* were four years in Alexandria region, while Bariche estimated it to be six years in Batrun, Lebanon. On the other hand, in Red Sea, El-Gammal estimated the maximal age to be four and five years for males and females respectively whereas Mehanna and Abdallah [39] estimated five years for combined sex for the same fish species using the same method. Otherwise, Hashem estimated six years in Mediterranean and Red Sea respectively for *Siganus rivulatus* using the Peterson (length frequency) method. Hussein recorded six years

while Bilecenoglu and Kaya determined eight years of Mediterranean Sea for the same fish using posterior body scale reading. Also, Shiekh-eldin recorded six years in the Red Sea, using the same method and El-Okda determined four years as maximum age using the vertebrae. These differences between the age groups may be returned to the way of samples collection and its length range or the method used in age determination [19]. The maximum theoretical length is found to be slightly higher but in accordance with those available in the literature for *S. rivulatus*, except for those calculated by Mehanna and Abdallah (Table 5). The differences between parameters of the von Bertalanffy growth equation in different locations (Mediterranean and Red Sea) found in Table 5 may be resulted from different factors such as sampling, ageing method used or even geographical differences of habitats and fish condition. Von Bertalanffy parameters from the literature and the present study for male (M), female (F) and combined sex (C) of *Siganus rivulatus* (Table 6). The sex ratio of females to males for all individuals of *S. rivulatus* was about 1.079: 1. However this ratio varied in different length groups. These changes in percentage of females to males were most probably due to spawning and feeding migration [40]. The increase in ratio of females in February to April and the decrease of such ratio in September and October may be due to feeding migrations. In the other hand, the relatively equal percentage in May, June and July, which is accepted as spawning period, all population members come together and form sex ratio of about 1 : 1 (spawning migrations). The spawning season of S. rivulatus in Bitter Lakes extended from May to July (i.e. three months). These observation is closed to that reported for the same species at Turkey [41], at Israel [42] and at Lebanon [43] and slightly differed from that mentioned in the Mediterranean coast of Turkey by Torcu [44] and Yeldan and Avsar [40] from April to August and of Alexandria by Hussein from July to August as well Mohamed [17] and El-Okda [23] from May to August. In the other hand, the spawning season of Siganus rivulatus in Red Sea begins earlier than in Mediterranean as reported by Popper et al. [45] at the Gulf of Aqaba, also Hashem [9] from March to April and by Amin [20] from February to April at Jeddah region of the Red Sea. The value of total mortality obtained from the present study (Z=0.884 year[-1]) is agree with that reported by El-Gammal [13] in the Red Sea at Ghardaqa region where Z=0.82 year[-1]. Nonetheless, it is lower than that estimated by Mehanna and Abdallah [39] in the Egyptian sector of the Red Sea (Z=1.270 [year-1]), and El-Ganainy and Ahmed in the eastern side of Suez Gulf (Z=3.15 [year-1]). On the other hand, the natural and fishing mortalities (M=0.2214 and F=0.6626 [year-1]) in the present study are lower than that obtained by El-Ganainy and Ahmed [15] in the eastern side of Suez Gulf (M=1.43 and F=1.72) and nearly equal to that reported by Mehanna and Abdallah [39] (M=0.26 and F=1.01 [year-1]). Concerning mortality estimates, comparison is so difficult due to shortage of data and the total mortality coefficient is not a species-specific parameter, but an area specific parameter.

The exploitation rate of *Siganus rivulatus* in the present study (E=0.75) is higher than 0.50. of Gulland suggested that as a rule of thumb a fish stock is optimally exploited at a level of fishing mortality that generates E=0.50 where optimum fishing mortality equal the natural mortality (F=M). However, Pauly [35] proposed a lower optimum fishing mortality (F=0.40). Therefore, the stock of rabbit fish (*Siganus rivulatus*) from the Bitter lakes is being seriously over-exploited by the nylon trammel nets used, such observation noted also in Egyptian sector of the Red Sea by Mehanna and Abdallah where E=0.80, and the eastern side of Suez Gulf by El-Ganainy and Ahmed [15] where E=0.55. It can be concluded that the *S. rivulatus* stock in the Egyptian Bitter Lakes is in a circumstances of overexploitation and to maintain this valuable fisheries resource some management measures, including reduction of the present level of fishing mortality and increase in the length at first capture should be applied.

References

1. Woodland DJ (1983) Zoogeography of the Siganidae (Pisces): an interpretation of distribution and richness patterns. Bulletin of marine science 33: 713-717.

2. Ben-Tuvia A (1985) The impact of the Lessepsian (Suez Canal) fish migration on the eastern Mediterranean ecosystem. In Mediterranean Marine Ecosystem (Moraitou-Apostolopoulou, M. and Kiortsis, V., eds): 367-375.

3. Papaconstantinou C (1990) The spreading of Lessepsian fish migrants into the Aegean Sea (Greece). Scientia marina 54: 313-316.

4. Bariche M, Letourneur Y, Harmelin-Vivien M (2004) Temporal fluctuations and settlement patterns of native and Lessepsian herbivorous fishes on the Lebanese coast (eastern Mediterranean). Environmental Biology of Fishes 70: 81-90.

5. Azzurro E, Andaloro F (2004) A new settled population of the Lessepsian migrant Siganus luridus (Pisces: Siganidae) in Linosa Island, Sicily Strait. Journal of the Marine Biological Association United Kingdom 84: 819-821.

6. Shakman E, Boedeker C, Bariche M, Kinzelbach R (2009) Food and feeding habits of the Lessepsian migrants Siganus luridus Rüppell, 1828 and Siganus rivulatus Forsska'l, 1775 (Teleostei: Siganidae) in the southern Mediterranean (Libyan coast). Journal of Biological Research-Thessaloniki 12: 115-124.

7. Crompton DWT (1985) Reproduction. In Biology of the Acanthocephala: 213-271.

8. Papoutsoglou SE, Papaparaskeva Papoutsoglou EG (1978) Comparative studies on body composition of rainbow trout (Salmo gairdneri R) in relation to type of diet and growth rate. Aquaculture 13: 235-243.

9. Hashem MT (1983) Biological studies on Siganus rivulatus (Forsk.) in the Red Sea. J. Fac. Mar. Sci., Jeddah 3: 119-127.

10. Ben-Tuvia A (1978) Immigration of fishes through the Suez Canal. Fish Bull 76: 249-255.

11. Abdel-Maguid SA (1997) Biological and hematological studies on marine coastal fishes, family. Gobiidae in Alexandria (Egypt). Ph.D. Thesis, Fac. Sci. Alex. Univ.

12. Philips AE, Akel EH (2003) Investigation of beach seine catches of Abu-Qir Bay (Egypt). Bull. Nat. Inst. Oceanogr. And Fish A R E 1: 79-91.

13. El-Gammal FI (1988) Age, growth and mortality of the rabbitfish Siganus rivulatus (Forssk.1775) from the Red Sea. Bull. Inst. Oceanogr. Fish. Egypt, 74: 13-21.

14. Shiekh-eldin MY (1988) Biological studies on certain marine teleosts. M. Sc. Faculty of Science, Ain Shams Univ.

15. El-Ganainy AA and Ahmed AI (2002) Growth, mortality and yield-per-recruit of the rabbit fish, siganus rjvula tus, from the eastern side of the Gulf of Suez, sinai coast, red se. Egypt J Aquai Biol Fisk 6: 67-81.

16. Hussein KA (1986) Timing of Spawning and Fecundity of Mediterranean Siganus rivulatus Forsskal. Bull. Inst. Oceanogr. Fish. Cairo vol 12: 175-186.

17. Mohamed NI (1991) Biological and Biochemical Studies of Some Siganid Fishes from the Mediterranean Waters off Alexandria, M.Sc. Thesis, Fac. Sci., Alexandria University.

18. Taskavak E, Bilecenoglu M (2001) Length-weight relationship for 18 Lessepsian (Red Sea) immigrant fish species from the eastern Mediterranean coast of Turkey. J Mar Biol Assoc U K 18: 895-896.

19. Bariche M (2005) Age and growth of Lessepsian rabbitfish from the eastern Mediterranean. Journal of Applied Ichthyology 21: 141-151.

20. Amin EM (1985) Reproductive cycle of male Siganus rivulatus Forsskal with indication to Gonosomatic and Hepatosomatic Indices. Bull. Inst. Oceanogr. Fish. Cairo 11: 149-164.

21. Lundberg B, Lipkin Y (1992) Seasonal, grazing site and fish size effects on patterns of algal consumption by the herbivorous fish, Siganus rivulatus, At Mikhmoret (Mediterranean, Israel). Environm. Quality, Ecosy. Stability, Vol. V/B.

22. Lundberg B, Golani D (1995) Diet adaptations of Lessepsian migrant rabbitfish, Siganus luridus and S. rivulatus, to the algal resources of the Mediterranean Coast of Israel. Mar Ecol – P.S.Z.N.I 16: 73-89.

23. El-Okda NI (1998) Comparative studies on certain biological aspects of Siganus in marine waters of Egypt. Ph. D. Thesis, Fac. Sci., (Benha) Zagazig. Univ.

24. Yeldan H, Avşar D (2000) A preliminary study on the reproduction of rabbit fish, Siganus rivulatus (Forsskål, 1775), in the northeastern Mediterranean. Turk J Zool 24: 173-182.

25. Shakman E, Winkler H, Oeberst R, Kinzelbach R (2008) Morphometry, age and growth of Siganus luridus Rüppell, 1828 and Siganus rivulatus Forsskål, 1775 (Siganidae) in the central Mediterranean (Libyan coast). Revista de Biología Marina y Oceanografía 43: 521-529.

26. Tacon AG, Rausin N, Kandari M, Cornelis P (1990) The food and feeding of marine fin fish in floating net cages at the National Sea Farming Development Centre, Lampung, Indonesia: Rabbitfish Siganus canaliculatus (Park), Aquaculture and Fish. Management 21: 375-339.

27. Carumbana EE, Luchavez JA (1979) A comparative study of the growth rates of Siganus canaliculatus, S. spinus and S. guttatus reared under laboratory and semi-natural conditions in Southern Negros Oriental, Philippines. Silliman Journal 26: 187-209.

28. Beckman DW, Wilson CA (1995) Seasonal timing of opaque zone formation in fish otoliths. In: Recent developments in fish otolith research. D. H. Secor; J. M. Dean and S. E. Campana (Eds). University of South Carolina Press, Columbia.

29. Sparre P, Venema SC (1992) Introducion to tropical fish stock assessment. Part I: Manual. FAO Fisheries Technical Paper No. 306.

30. Von Bertalanffy L (1938) A quantitative theory of organic growth. Human Biology 10: 181-213.

31. Ricker WE (1973) Linear regressions in fishery research. Journal of the Fisheries Research Board of Canada 30: 409-434.

32. Ricker WE (1975) Computation and interpretation of biological statistics of fish populations. Bulletin of the Fisheries Research Board of Canada 191: 1-382.

33. Beverton RJ, Holt SJ (1957) On the dynamics of exploited fish populations. U. K. Min. Agric Fish Fish Invest 19: 230-233.

34. Sparre P, Ursin E, Venema SC (1989) Introduction to tropical fish stock assessment. Part 1. Manual. FAO Fisheries Technical Paper No: 306.

35. Pauly D (1983) Length-converted catch curves. A powerful tool for fisheries research in -tropics. Parti. ICLARM Fishbyte 2: 17-19.

36. Gulland JA (1971) The fish resources of the Ocean. West Byfleet, Surrey, Fishing News (Books), Ltd., for FAO: 255 p.

37. Brothers EB (1979) Age and growth studies on tropical fishes. In Stock Assessment for Tropical Small-Scale Fisheries (B. S. Saila and P.M. Roedel, eds), pp. 119-1 36. Rhode Island: International Center for Marine Resource Development, University of Rhode Island.

38. Golani D, Ben-Tuvia A (1985) The biology of the Indo-Pacific squirrelfish, Sargocentron rubrum (Forsska° I), a Suez Canal migrant to the eastern Mediterranean. J Fish Biol 27: 249-258.

39. Mehanna SF, Abdallah M (2002) Population dynamic of the rabbitfish, Siganus rivulatus, from the Egyptian sector of the Red Sea. J K A U Mar Sci 13: 1-170.

40. Yeldan H, Avşar D (2000) A preliminary study on the reproduction of rabbit fish, Siganus rivulatus (Forsskål, 1775), in the northeastern Mediterranean. Turk J Zool 24: 173-182.

41. Akşiray F (1987) Türkiye Deniz Baliklari ve Tayin Anahtari (The identification sheets for the Turkey's marine fishes). Istanbul Üniv.Rektorüğü lyayinlari. No: 3490. II: Baski.811p.

42. Golani D (1990) Environmentally-induced meristic change in Lessepsian fish migrants, a comparison of source and colonizing populations. Bulletin de l´Institut Océanographique de Monaco 7: 143-152.

43. Bariche ML, Harmelin MV, Quignard JP (2003) Reproductive cycle and spawning periods of two Lessepsian siganid fishes on the Lebanese coast. J Fish Biol 62: 129-142.

44. Torcu H (1994) Indo Pacific-fish species distributed along Southern Aegean Sea and Mediterranean, and the studies on the biology and ecology of gatfish (Upeneus moluccensis, Bleeker, 1885) and Lizardfish (Saurida undosquamis, Richardson, 1848). Selçuk Üniversitesi Fen Bil. Enst. Doktora Tezi. Koya.124p.

45. Popper DM, Gordin H, Kissil GW (1973) Fertilization and hatching of rabbitfish Siganus rivulatus.Aquaculture 2: 37-44.

35

Study on the Monthly Variation in Hydro biological Condition and Its Relation to Fish Production of a Sewage Fed Bheri System at Suburban Kolkata

Chatterjee NR*, Sahoo D and Chetri C

Department of Aquaculture, West Bengal University of Animal and Fishery Sciences, Belgachia, Kolkata-700037, India

Abstract

The concept of resource recovery systems and waste recycling involving the wetlands of Kolkata are gradually loosing interest due to unplanned urban expansion without understanding the ecological, environmental and economic benefit of the century old sewage fed systems. The water quality parameters, noted during the study, exhibited marked seasonal variation and some were indicative of productivity. The BOD was optimum but DOM (Dissolved Organic Matter) was significantly lower. Among the nutrients, the values of phosphorous were higher and the values of all other nutrients found to increase with temperature. Overall production decreased in the successive years. Total plankton production, nitrate-nitrogen, and DOM and BOD showed negative co-relation with total fish production.

Keywords: Bheri; Variation; Hydrobiological condition; Production

Introduction

The world is facing a diverse variety of interrelated problems such as food scarcity, nonrenewable energy source, unemployment and urbanization. The current dilemma occurs because of our strategies for development contravenes the basic philosophy of ecology. The world ecology is increasingly becoming more unstable through an ever dependence on finite resources of fossil fuel, which has been brought sharply into focus by recent hike in the price of crude oil. Food production system in the developing countries, both agriculture and aquaculture, are energy intensive and thus are not considered suitable for developing countries. An alternative strategy, for aquaculture and agriculture development, is dependence on locally available resources of renewable energy and resources. A significant impact could be made through recycling of organic waste and it is suggested that organic waste should be recycled into fish whenever feasible, since fish raised in such a way may be cheapest source of animal product [1].

Sewage can be defined as a cloudy fluid arising out of domestic wastes containing mineral and organic matter either in solution or in suspension or in colloidal or in pseudocolloidal form in a dispersed state. The daily rate of sewage production in India is estimated to be about 100 lt/capita. Thus about 3-4 million m³ of sewage is produced daily in the cities as against a total flow of wastewater of 10 million m³/day. It is increasingly being recognized that domestic sewage which is rich in nutrient resources (like N, P, K and organic matter) can be suitably utilized for productive purpose [2]. The farmers around kolkata city are using domestic sewage for fish culture almost a century ago and which is widely used to meet the growing demand of fish in this thickly populated Indian city. The system appears to have started long back although large scale use of sewage for fish culture began in 1930s.The area under this unique system of culture peaked at 12000 ha, but in recent years there happens to be steep decline in the area due to increasing urbanization. Currently the area under sewage fed culture system has been reduced to less than 4,000 ha and the people dependent on these wetlands for their livelihood have been seriously affected. These sewage fed fish ponds, which are locally known as "Bheries" are usually large and as big as 40 hector in size, these sewage fed fish ponds are generally shallow and vary from 50 cm to 150 cm. In general farming here constitutes five distinct phases covering pond preparation, primary fertilization, fish stocking, secondary fertilization and fish harvesting. Currently 148 units of such fisheries are reportedly operating in Eastern Kolkata with the current production of 4-5 ton/ha./yr. These sewage fed Bheries around suburban kolkata are well managed by different co-operatives. The sewage fed Bheri under present study is managed by and occupy an area about 75 ha the average depth is around 1 mt. Management by co-operative started in mid 1980s with the support of Govt. The annual transaction of this society are large is magnitude with financial transaction amounting to move than Rs. 13 million. There exist also an integration of pig and duck. This sewage fed system is subjected to stress and strain in the form of less availability of nutrient as the good quality of raw sewage are not coming to the systems. The scientific investigation of this vast water body was undertaken with the following objectives.

➤ To determine seasonal variation in the hydrobiological condition of this 'Bheri' system.

➤ To evaluate the production performance of fish per hector water body.

➤ To correlate the fish production/month with consequent variation in hydro biological parameters.

➤ To investigate the importance and effectiveness of this water body in the treatment of Kolkata city sewage.

Materials and Methods

The present study was carried out to investigate the seasonal

*Corresponding author: Chatterjee NR, Department of Aquaculture, West Bengal University of Animal and Fishery Sciences, Belgachia, Kolkata-700037, India
E-mail: nrchatterjee40@gmail.com

variability of water and soil quality status and fisheries potentialities of Bheri No. 4 receiving sewage effluent at weekly interval. This sewage fed Bheri situated at suburban Kolkata, which is under East Kolkata wetland area. The detailed survey on Physico-chemical status of water, sediment quality and biological status was carried out for a period of 6 months. Standard method of APHA [3] was followed for this purpose. In order to study the water body, three sampling sites were selected covering entire stretch of the Bheri. Fortnightly collection of water samples were done for six month period from January 2006 to June 2006. Planktons and soil samples were collected once in a month. All the experiments were conducted between 10.00 am to 1.00 pm. Annual fish landing data were taken from the official records of the Co-Operative.

Field studies

Collection of samples for soil/water quality and plankton analysis: For water quality estimation, water samples were collected from three sampling sites without disturbing the bottom sediments. Soon after the collection, temperature was recorded using a mercury thermometer to the nearest 0.10°C. Samples were fixed following standard method as in APHA. During collection of water samples, cautions were taken so as to prevent air bubbling, which might influence water parameters such as dissolve oxygen, free carbon-di-oxide etc. Soil samples and plankton are also collected following standards methods.

Pond preparation and Management pattern: In general pond management includes five phases such as pond preparation, primary fertilization, stocking, secondary fertilization and fish harvesting.

Pond preparation: Generally after every 5-6 years ponds are drained, dried and silt are removed from the topsoil of pond bottom, the bottom is tilled, lime applied on the top surface and dikes are renovated. The pond preparation generally undertaken during winter months.

Primary fertilization: Sewage is drawn from central canal into the pond and allowed to stabilize for 15-20 days. The photosynthetic activities in the pond is the basis of biological purification of the sewage. Once the water turns complete green, fish are stocked.

Fish stocking: Large size fingerlings are stocked after every 3 months in a year as harvesting continues throughout the year.

Periodic fertilization: Following fertilization, sewage is fed to ponds @ 5% of the total volume of water at interval throughout the culture period. Continuous inflow and outflow is maintained (Figure 1).

Dike protection: Dyke is protected by water hyacinth which provide shelter to the fish and remove heavy metals.

Fisheries: Fish catches in every effort were observed during the entire investigation period. Percentage contribution of different group of species were computed and prevalence of diseases was also carefully noted.

Laboratory studies

Physico-chemical analysis of water: Different physico-chemical parameters of water were estimated by following standard methods as in APHA 1998 (Figure 2).

Estimation of Soil parameters: pH and organic carbon was measured by following standard methods (APHA, 1998).

Statistical analysis

The data obtained in respect of different parameters were statistically

Figure 1: Plankton collection at one of the sampling site.

Figure 2: One ft. deep pond mud being cut and stacked on the bank.

Figure 3: Central sewage canal fed to different *Bheris*.

analyzed following Analysis of variance to derive the significant difference between the sampling sites. Correlation coefficient (r) values between physico chemical parameters of water and phytoplankton were also worked out.

Results

Physico-chemical properties of the Bheri system

Hydrobiology: The seasonal variation in physico-chemical parameters of the Bheri system affects the productivity chain and as a result of this the entire aquatic productivity equilibrium is disturbed (Figure 3). Depending on the incoming waste materials into the system, the hydrobiological properties varied. Besides, the high percentage of chemical waste results in substantial variation in physico chemical properties (Figure 4). With this understanding, the hydro biological characteristics of the Bheri were fortnightly estimated.

Temperature: Temperature varied from a minimum of 21.3°C

Figure 4: Periphery of this *Bheri* covered with *Eichhornia* plant to protect from windaction.

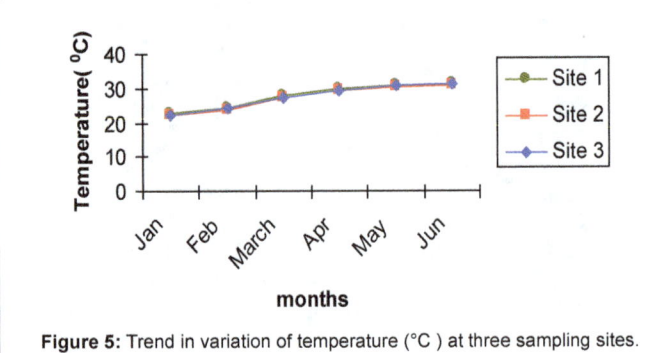

Figure 5: Trend in variation of temperature (°C) at three sampling sites.

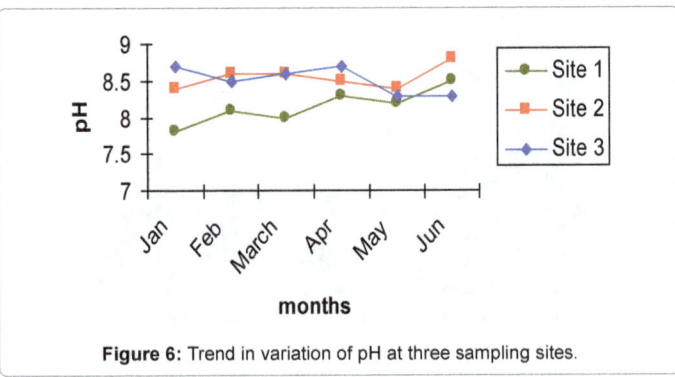

Figure 6: Trend in variation of pH at three sampling sites.

(January, Site 2) to a maximum of 32.3°C (June, Site 3) (Figure 5), means an increasing trend from January to June.

pH: Normally the pH ranged between the minimum values of 7.5 (January, Site 1) to the maximum value of 8.9 (June, Site 3) (Figure 6). All three sites showed highly alkaline pH but Site 2 showed slightly higher pH value.

Dissolved Oxygen: All three sampling site showed marked variation in dissolved oxygen content (Figure 7) and ranged from a minimum of 1.50 mg/lt (May, Site 2) to a maximum of 4.04 mg/lt (February, Site 3).

Free Carbon dioxide: Figure 8 indicate that, the CO_2 content was nil except in June with a minimum of 1.64 mg/lt (Site 3, June) and maximum of 2.20 mg/lt respectively.

Biological Oxygen Demand (BOD): As indicated in Figure 9 the

BOD value ranged between 20.76 mg/lt (January, site 1) to 10.70 mg/lt (site 3) showed higher value.

Total alkalinity: Total Alkalinity fluctuated widely at three sampling sites (Figure 10) and ranged from 208.7 mg/lt (January, Site 3) to 315.3 mg/lt. Site 1 showed higher value due as the raw sewage enter through it.

Ammonia-Nitrogen: Ammonia–Nitrogen ranged from 0.740 ppm (January, Site 3) to the maximum of 1.402 ppm (June, site 1) (Figure 11). Higher value at site 1 due to entry of raw sewage.

Nitrate-Nitrogen: Nitrate -nitrogen content varied at different site (Figure 12). Site 3 showed higher values than the other sites and ranged between 0.120 ppm (June, site 1) to 0.282 ppm (January, site 3).

Phosphate–Phosphorus: The fortnightly variations was noticed at three sites (Figure 13) with a maximum of 1.826 mg/lt (March, Site 1)

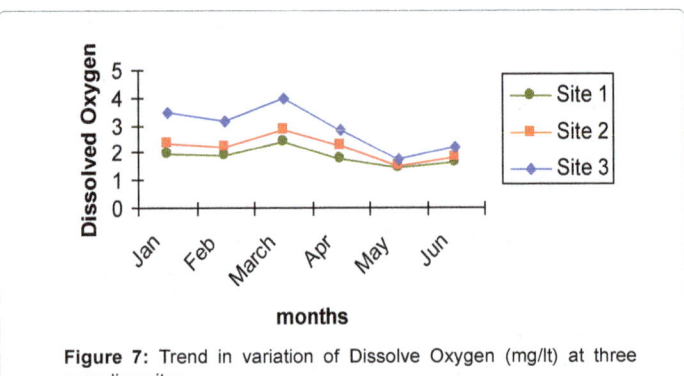

Figure 7: Trend in variation of Dissolve Oxygen (mg/lt) at three sampling sites.

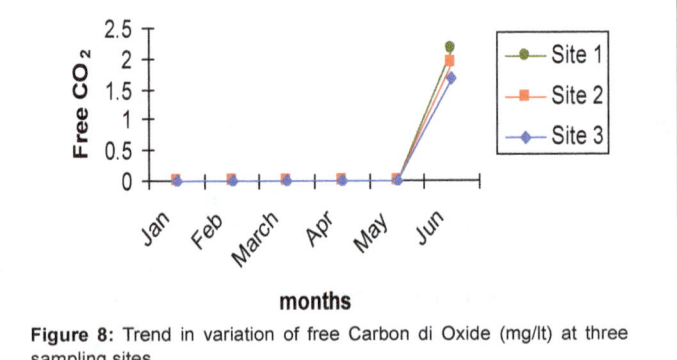

Figure 8: Trend in variation of free Carbon di Oxide (mg/lt) at three sampling sites.

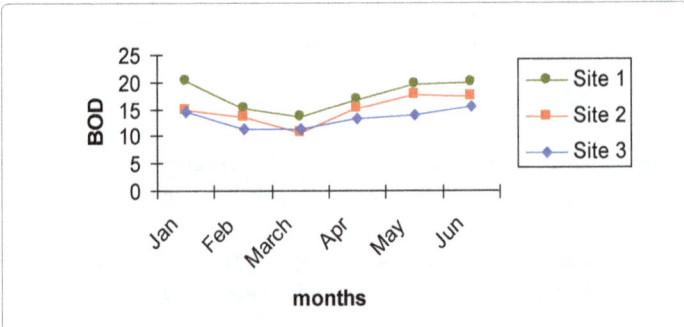

Figure 9: Trend in variation of Biological Oxygen Demand (mg/lt) at threesampling sites.

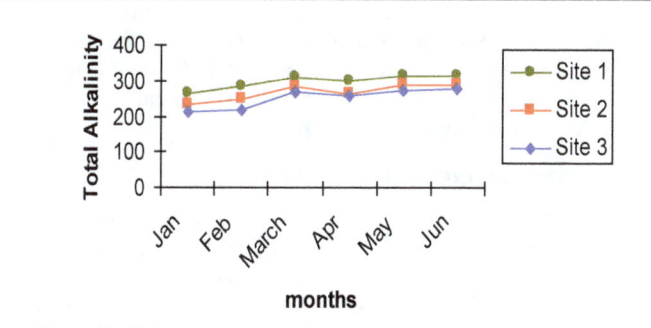

Figure 10: Trend in variation of Total Alkalinity (mg/lt) at three sampling sites.

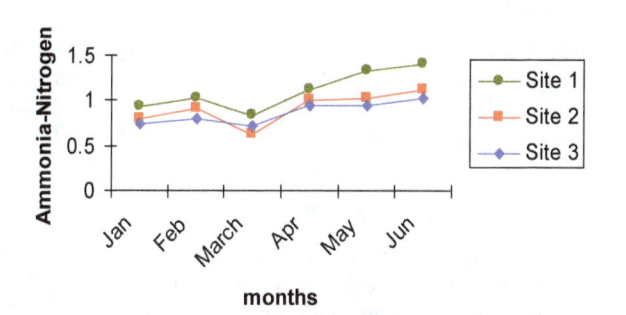

Figure 11: Trend in variation of Ammonia-Nitrogen (mg/lt) at three sampling sites.

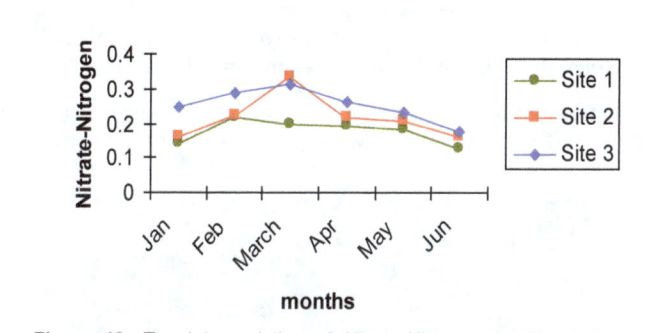

Figure 12: Trend in variation of Nitrate-Nitrogen (mg/lt) at three sampling sites.

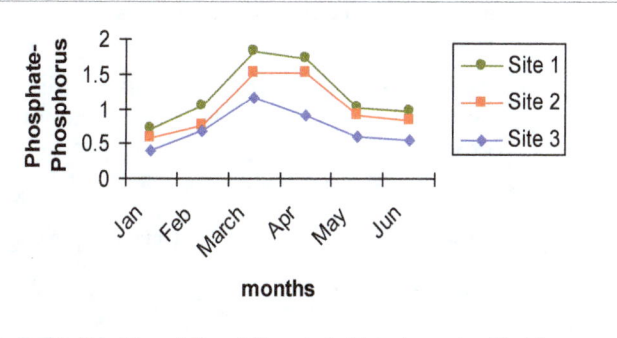

Figure 13: Trend in variation of Phosphate-Phosphorus (mg/lt) at three sampling sites.

to a minimum of 0.401 mg/lt (January, Site 3). The overall trend was moreover same.

Dissolved organic matter: Dissolved organic matter showed a lower range and fluctuates between 3.542 mg/lt (March, site 3) to 6.416 mg/lt (may, site 1). Organic matter content was rich in site 1 than the other two sites (Figure 14) as usual.

Soil

pH: Varied between 6.9 (March, site 2) to 7.3 (May, site 2) values at 3 sites were comparable (Figure 15).

Organic carbon: The range at three sites were between 1.32% (January, site 3) to 2.41% (May, site 1). Steady increase in organic carbon content (%) was noticed at all the sites from January to June (Figure 16).

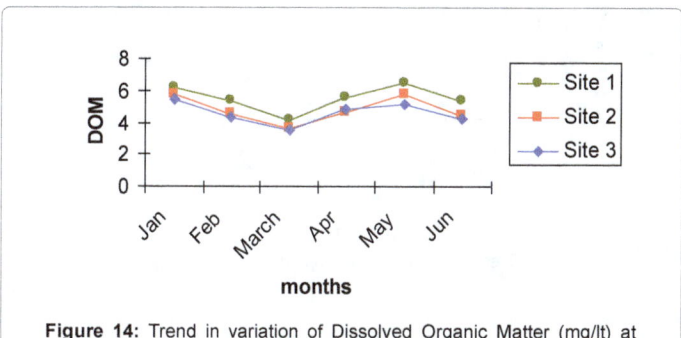

Figure 14: Trend in variation of Dissolved Organic Matter (mg/lt) at three sampling sites.

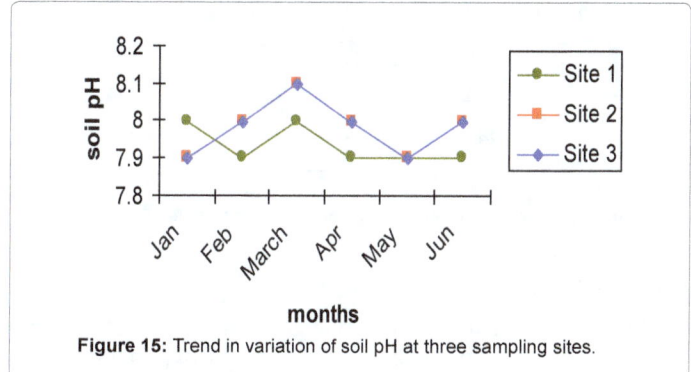

Figure 15: Trend in variation of soil pH at three sampling sites.

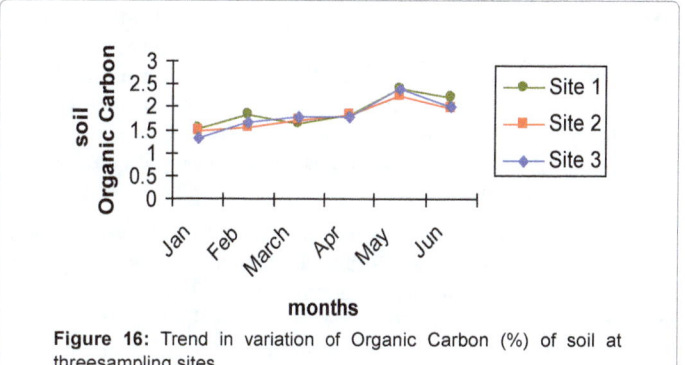

Figure 16: Trend in variation of Organic Carbon (%) of soil at threesampling sites.

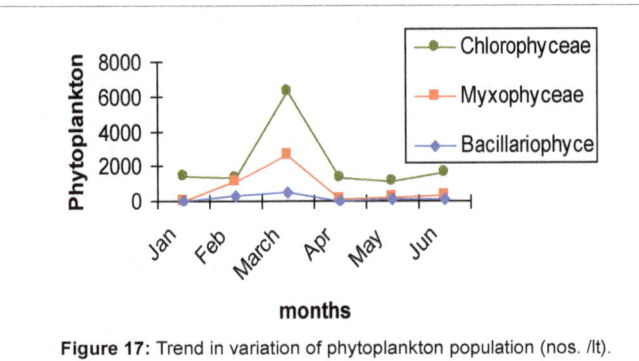

Figure 17: Trend in variation of phytoplankton population (nos. /lt).

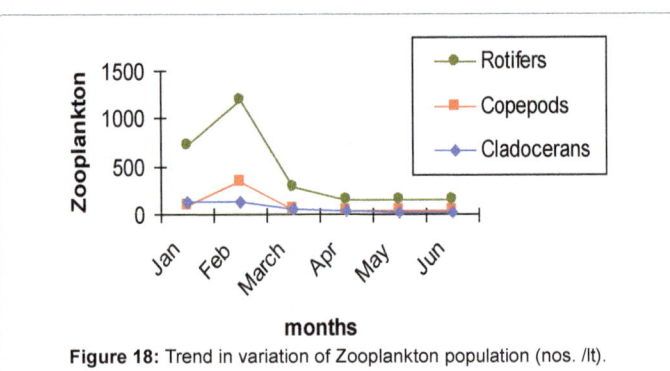

Figure 18: Trend in variation of Zooplankton population (nos. /lt).

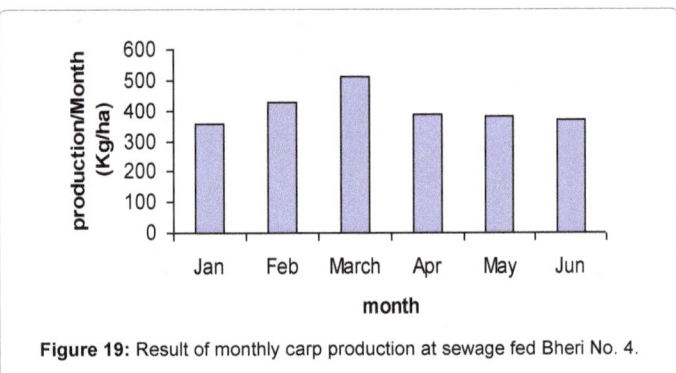

Figure 19: Result of monthly carp production at sewage fed Bheri No. 4.

Planktons

As the sewage fed system carry high nutrient load, so it serves as a ready source for plankton growth. Dense phytoplankton bloom noticed in some places.

Phytoplankton: The total phytoplankton count varied from a minimum of 800 no /lt (May) to a maximum of 9434 no /lt (March) (Figure 17). The major group was chlorophyceae, Myxophyceae and Bacillariophyceae showing variable intensity with time. Chlorophyceae dominates the other two groups are represented by *Coelastrum, Pediastrum, Spirogyra, Oscillatoria* and *Spirulina*. The next dominating group was Myxophyceae while Bacillariophyceae represented by few Navicula and Synedra.

Zooplanktons: The major group represented by Rotifers, Copepods and Cladocerans (Figure 18). Rotifers dominated the other two groups and represented by *Brachionus, Keratella* and *Cyclops* while copepod represented by *Diaptomum*. Cladocerans mainly dominated by Moina and *Daphnia* spp. Rotifer population was maximum in February (1203 nos/lt) and minimum in June (157 nos/lt). Rotifer population

dominated in Feb (1203 nos/lt) and minimum during June (157 nos/lt). Minimum count of Copepods was obtained in the month of May (30 nos/lt) and maximum in the month of February (334 nos/lt). Cladocerans showed its peak in the month of February (134 nos./lt) and minimum in the month of June (22 nos/lt).

Fish stocking and production of the Bheri

Both Indian and exotic carps are stocked in Bheris, with a preference over Indian carps due to their herbivorous feeding behavior and high growth rate and includes Catla (*Catla catla*), Rohu (*Labeo rohita*), Mrigal (*Cirrhinus mrigala*) and Bata (*Labeo bata*) but mrigal outnumber other carps. Generally fingerlings and stocked to reduce mortality rate [4-7] (Figure 19). The culture of Tilapia (*Oreochromis nilotica* and *O. mossambicus*) is getting popularity and constitute about 5-10% of the stock. Fingerlings are stocked at three months interval.

Disease occurrence

Only ulcers (Epizootic Ulcerative Syndrome, EUS) were found to a lesser extent on the body of mrigal and on silver carps in winter months. Carp lice (*Argulus*) and parasitic infection Lernea (anchor worm) were recorded on mrigal and catla in the month of February (Figure 20). Occurrence of other diseases were not noticed.

Fish harvesting and marketing

Harvesting is done every day in early morning in between 4 am to 6 am at different Live fish marketing is becoming more popular, and is considered a kind of value addition to fetch more price (Tables 1 and 2). The captured fish are stocked in a depuration pond for varying length of time. Depurated fish are marketed alive location by drag net.

Statistical Analysis

The data obtained were statistically analyzed by employing analysis of variance (ANOVA one-way) and Pearson's correlation method. The summary of these ANOVA is presented in appendix 1.

Relationship of water and sediment quality parameters with fish production (for site 1, 2 and 3) were done using method of Backward

Figure 20: Carps harvested from the sewage fed *Bheri*.

Indian major Carp; Minor carps; Exotic carps	12000 kg fingerlings at a time each three months interval
Other fishes	Natural entry

Table 1: Rate of Stocking.

Year	Production (metric ton)
2001-2002	547.50
2002-2003	412.26
2003-2004	403.80
2004-2005	358.20

Table 2: Harvesting of fishes in the last four years.

elimination technique of Multiple regression because of less number of sample size with probability of F to enter as 0.05 and F to removal as 0.10.

(Following equations has been extracted taking least predictors to explain the growth):

1) Production Site-1 =318.62-10.23* CO_2+40.68** PO4-P+0.25** Bacilariophyceae R2=0.99, adj. R2=0.99, S.E. of Estimated=3.46

2) Production Site-2 =349.80 -1.32 water temp. -8.48* CO_2 +51.42* PO4-P+ 0.26** Bacilariophyceae R2=0.99, adj. R2=0.99, S.E. of Estimated = 0.52.

3) Production Site-3=315.48 -11.54 CO_2 +79.15 * PO4-P+ 0.22** Bacilariophyceae R2=0.99, adj. R2=0.99, S.E. of Estimated=4.27

In the above equations it has been assumed that production is time independent but actually it should be time related or autoregressive over the period of observations. Thus in the following equations autoregressive model along with predictor variables as selected by previous equations following Melard's algorithm for estimation for all sites and the ultimate equations were made.

1) Production Site-1 (n)=318.95 -0.85 production site 1 (n-1) -11.21 CO_2 +39.91* PO4-P+0.25* Bacilariophyceae Adj. S.S=13.16, Marquardt const.=1.00-E8=1.00x10-8, No. of iterations=6

#Note: PO4-P (Phosphate phosphorus) followed by bacillariophyceae have enhance the production but CO_2 have suppressed the production at site 1.

2) Production Site-2 (n)=323.53 -0.85 production site 2 (n-1) -14.93 CO_2 +40.86 PO4-P+0.27* Bacilariophyceae Adj. S.S.=27.07, Marquardt const. =0.10, No. of iterations=7

#Note: Bacilariophyceae have enhanced the production in site 2.

3) Production Site-3 (n)=316.24 -0.84 production site 3 (n-1) -13.53 CO_2+76.75 PO4-P+0.23* Bacilariophyceae Adj. S.S.=22.14, Marquardt const.=1.00- E7=1.00x10-7, No. of iterations=5

#Note: Bacilariophyceae have enhanced the production in site 3.

(*)= Result is Significant at 0.05 levels

(**)=Result is Significant at 0.01 levels

Discussion

As we know production in bheris is intimately related to physico-chemical properties at soil-water interface and sewage quality. During this study emphasis was given for characterization of water and sediment quality and assessment of suitable ecosystem, for optimum fish growth. The study was conducted covering leading Bheris of North Kolkata.

Hydrography

Temperature: The degree and annual variation in temperature of water body have a great bearing upon its productivity in general. Any fluctuation in water temperature affects other environmental parameters. Our data confirm that temperature positively co-related with the total alkalinity (0.940) but no significant co-relation existed between temperature and monthly fish production.

pH: The alkaline pH range in spite of considerable amount of organic load may be attributed to high rate of photosynthetic activity in the pond water and also to the prevalence of large amount of water-soluble bases. pH is considered favorable for good yield of fish. Higher pH value during summer month (March to May) may be due to the concentration of sewage resulted from high degree of evaporation. During rest of the season pH value was average. Presence of substantial amount of organic matter, increase CO_2 production, which in turn lowers the pH value. Oswald [8] and Seenayya [9] also reported higher pH value in sewage water. No significant co-relation was observed between pH and monthly fish yield (Appendix-I).

Dissolved Oxygen: Dissolved oxygen content is important for direct need affects the solubility and availability of many nutrients and therefore the productivity of aquatic ecosystem [10]. In the present investigation there was an erratic fluctuation of dissolved oxygen content in all the ponds throughout the study period. DO concentration around 5 ppm was found to be favorable for fish growth [11-13]. In our study dissolved oxygen exhibited a direct relationship (0.820) with total phytoplankton production and strong negative co-relation (-0.925) with ammonia nitrogen (NH_3-N).

Carbon di Oxide: The value of CO_2 fluctuate widely due to its capacity to combine with different cataions (Ca^{++}, Na^+, K^+) and other elements. During the study dissolved free CO_2 concentration was found nil except in the month of June (1.92 ppm to 2.21 ppm). The absence of free CO_2 might be due to the presence of high phytoplankton bloom. No significant co-relation between the presence of carbon dioxide and fish production was found. As indicated Goel and Trivedi the increase in organic matter results in high biological and chemical demand, with consequent decrease in the dissolved oxygen levels and increase in free carbon-di-oxide level.

Biochemical Oxygen Demand: Generally sewage fed Bheries exhibits higher values of Biochemical oxygen demand but we did not notice any definite change in BOD throughout the investigation although sitewise there was the marked variations in BOD level. In present investigation the BOD ranged from a minimum value of 11.52 ppm (February) to maximum of 20.3 ppm (June). In the present investigation the BOD values were found within the desired range Biochemical oxygen demand showed highly significant negative co-relation with Nitrate Nitrogen (0.944) and total phytoplankton production (-0.844) respectively.

Total alkalinity: Total alkalinity is related to the productivity of waterbody. In the present investigation fluctuation in total alkalinity at different centre was not significant. The average value ranged from 210-315.3 ppm. In all the sites peak was observed in June and minimum in January. Fish ponds having alkalinity of 50 mg/lt are considered productive [11,12]. High total alkalinity value may lead to the liberation of CO_2 following the process of decomposition of bottom deposit. The importance of alkalinity in view of its relationship with available CO_2 for photosynthesis was emphasized by wallen [14]. A highly productive water should have total alkalinity more than 100 ppm [15]. Since all the sites showed a total alkalinity value above 100 mg/lt, it is not likely to be a limiting factor for the productivity of a pond, but on the basis of total alkalinity value the existing farm may be classified as productive. No significant co-relation was noted between total alkalinity and fish production. But total alkalinity showed highly positive correlation with temperature.

Ammonia-Nitrogen (NH_4-N): Raw sewage is the source of higher concentration of ammonia -nitrogen. Although the ammonia is a major excretory product of aquatic animals but this nitrogen source is quantitatively minor in comparison to that generated by microbial decomposition. It is present in the aquatic system mainly as dissociated ionic NH_4^+. In all the sites a narrow range of ammonia nitrogen from 0.723 to 1.402 was noticed. Site 1 showed higher range of ammonia nitrogen as it serves as entry of raw sewage. Relative amount of ammonia nitrogen is higher than the nitrate nitrogen due to low oxygen concentration. At higher pH and temperature ammonia is toxic to fish [12]. In the afternoon period nitrification of ammonia to nitrate enhances to a considerable level [16]. Generally sewage fed ponds exhibit high level of ammonia. No adverse effect however, on the fish due to high concentration of ammonia was observed [6,17]. Our study confirms the earlier one. No co-relation between ammonia nitrogen (NH4-N) with monthly fish production was found to exist but NH4-N showed significant negative co relation (-0.925) with dissolved oxygen.

Nitrate-Nitrogen (NO$_3$-N): Nitrate- Nitrogen is an important water parameters and influence productivity of aquatic system. Welch opined that nitrate in natural water exists in a continuous changing state due to the relation of nitrate with nitrifying bacteria and demand by nitrate consuming organisms such as phytoplankton and higher aquatic plants. Nitrate-Nitrogen value varied considerably during the study period. Generally higher values observed during summer month and ranged from 0.125-0.336 ppm with a peak value in April, may be due to mineralization process at higher temperature. Saha et al. [18] reported a nitrate value from 0.08-1.80 ppm at Kulia beel during 1981-82. A similar trend was observed by Bhoumik [19] in beels and bours of West Bengal. A positive and highly significant co-relation (0.899) was found between nitrate-nitrogen content and monthly fish production. Nitrate nitrogen also showed significant direct relation (-0.944) with biochemical oxygen demand.

Phosphate-Phosphorus (PO$_4$-P): Phosphorus mainly occurs in the form of orthophosphates and is the most critical factor in maintaining pond productivity [20]. Phosphate value did not show any seasonal variation. A more favorable growth response of plankton population was observed at the time of higher concentration of phosphates in pond water. Higher value of phosphorus observed during March-April month (1.82-1.73 ppm) and minimum value was observed during January (0.423 ppm). A productive pond passes a phosphorus concentration of 0.02-0.05 ppm [20]. Significant co relation also could not found between phosphorus and other parameters. Direct influence of available phosphorus with productivity was established and confirms the findings of Ghosh [2,20].

Dissolved Organic Matter (DOM): The dissolved organic matter of the Bheri varied between 3.531-6.512 ppm with a higher value recorded during summer months. Dissolved Organic Matter content showed a significant indirect relation (-0.822) with fish production. No such relations were found with other parameters.

Soil

pH: Soil pH at neutral range (pH 6.5 to 7.5) is suitable for fish culture. PH value fluctuates depending on soil. In the present study the soil pH ranged between 6.9-7.3.

Organic carbon: Organic carbon was found very high in the fish pond and ranged between 1.32% to 2.41%. Ghosh et al. [20,21] found a similar trend of organic carbon range. Present study conforms the above findings of earlier study. Muds are water logged and often deficient in dissolve oxygen. Low concentration of dissolve oxygen prevents efficient utilization of organic matter by bacteria. In the present case accumulation of organic carbon may be attributed to heavy load of oxygen deficient mud.

Plankton

In sewage fed fishpond, plankton is the principle natural food component for the fishes and is directly linked with fish production. The data derived from the present investigation represent a distinct variations in composition of zooplankton and phytoplankton in different months. The maximum recorded in February- march (9834 nos. /lt) and minimum in May (1011 nos/lt) indicating a bimodal distribution of planktons in the sewage fed pond. Similar pattern of plankton distribution was reported by [22].

Phytoplankton: Phytoplanktons are primary producer and form the basis of an autotrophic food chain. In all the sewage fed fishponds phytoplanktons showed high concentration and are considered an important biotic component of an aquatic system. During present investigation phytoplankton population varied between 800 nos./lt (May) to 9432 nos./lt (March) in all the sites. The rise in phosphate concentration may be attributed to themineralization of nutrients which causes enhancement of phytoplankton abundance. Seven species of phytoplanktons identified belongs to three groups of algae such as Chlorophyceae, Myxophyceae and Bacillariophyceae. Among the Chlorophyceae the species were *Coelastrum, Pediastrum* and *Spirogyra*. Myxophyceae represented by *Oscillatoria, Spirulina* and Bacillariophyceae group was represented by *Navicula* and *Synedra*. Based on numbers the groups were ordered as of *Chlorophyceae >Myxophyceae >Bacillariophyceae*. In sewage irrigated pond, the nutrients rich effluent enhance the production of phytoplankton as also secondary and tertiary food organisms [20,23,24]. Ghosh et al. [20] while working on hydrobiological condition of sewage fed pond observed similar pattern of phytoplankton concentration. The dominating role of Chlorophyceae in sewage treatment ponds are also reported by [22,25]. Colder months are congenial for occurrence of phytoplanktons according to findings. This view confirms with findings of Saha [26].

Zooplankton: The importance of zooplankton in the eutrophication of freshwater ecosystem have been emphasized by many investigators [27]. The seasonal variation of zooplankton in the Bheri was similar to that of phytoplanktons and exhibited its peak in February month (1671 nos./lt) and minimum in the month May (211 nos/lt). The zooplanktons were represented by Rotifers, Cladocerans and Copepods.

Rotifers were dominating among all the Zooplanktons. It was represented by *Brachionus* and *Keratella*. Cladocerans were represented by *Cyclops* and *Diaptomus*. Compared to phytoplanktons the zooplankton density was lower. The seasonal occurrence and species composition in plankton production is directly or indirectly influenced by physico-chemical factors of water. In the present study plankton density showed highly significant relationship (1% level) with fish production, dissolved oxygen (5% level) and indirect relation with biochemical oxygen demand (5% level). Vittal Rao reported that BOD of sewage water has significant influence on algal succession, which also favors this investigation.

Fish production

Waste or sewage fertilized ponds produce high yield because of high abundance of natural food produced in wastes [24]. The overall production of this Bheris was 4800 kg/ha/yr. Monthly fish production was documented from cooperative record book. From the year 2001-02 there is constant decrease in fish production due to unavailability of good quality of sewage. High rate of siltation also found to reduce fish production.

Sreenivasan recorded fish production ranging from 1000-5486 kg/ha from Chennai city sewage fed ponds. Ghosh [22] observed the production of 5402-8619 kg/ha when opted for composite fish farming of Indian and exotic carps in sewage irrigated fishpond. Vass, mentioned the use of effluent from domestic septic tanks from Java, and reported that the yield of fish was 4000 kg/ha/yr without any supplementary feed. Saha [26] estimated annual fish production of 2.2 ton/ha from sewage fed fisheries of West Bengal. Schaperclaus [27] reported that the fish crop attributed to sewage ranged from 165-1037 kg/ha/yr from Berlin, Germany. A comparable production were observed in these Bheris [28].

References

1. Wohlfarth G (1978) Utilization of manure for fish farming. Proceeding of the Conference of on Fishfarming and Wastes, London.

2. Manna NK, Banerjee S, Bhoumik ML (2001) Evaluation of production performance of carp in a lentic freshwater water sewage-fed polyculture pond. Indian J Fish 48: 375-381.

3. American Public Health Association (1998) American Water Works Association and Water Pollution control Federation. Standard method for the examination of water and waste water. (20th edtn), Am Publ Hlth Assoc, Washington, USA.

4. Ghosh D, Sen S (1987) Ecological History of Calcutta's Wetland Conservation. Environ Conserv. 14: 219-226.

5. Krishnamurthi KP (1988) Present status of sewage fed fisheries in Maharastra in Wastewater-fed Aquaculture.

6. Prabhavalthy G, Arun GJ (1988) Fish culture in sewage fed ponds of Tamilnadu.

7. Jhingran AG, Ghosh A (1988) Aquaculture as a potential system of sewage disposal a case study. J Inland Fish Soc. India 20: 1-8.

8. Oswald WJ (1960) Light conversion efficiency of algal growth in sewage. J San Eng Div. Am Soc Civil Engr. 86: 71.

9. Seenayya G (1971) Ecological studies in the plankton for certain freshwater ponds of Hydrabad, India I Physico-chemical complexes. Hydrobiologia 37: 7-31.

10. Wetzel RG (1983) Limnology. Saunders College Publishing, USA.

11. Banerjee SM (1967) Water quality and soil condition of fish ponds in some states of India in relation to fish production. Indian J Fish 14: 115-144.

12. Boyd CE (1982) Water quality management for pond fish culture. Elsevier scientific publishing company, Amsterdam, Kingdom of the Netherlands.

13. Ellis MM (1937) Detection and measurement of stream pollution. US. Bull 58: 365-437.

14. Wallen IE (1955) Some limnological considerations in the productivity of Oklahoma farm ponds. J Wildlife Mgmt 19: 450-462.

15. Alikunhi KH (1957) Fish culture in India. Fm. Bull. Indian Coun Agri Res 20: 144.

16. Chow T (1958) A study of water quality in the fish ponds of Hong Kong University. Fish J 2: 7-2.

17. Sreenivasan A, Muthuswamy S (1979) Fish culture possibilities in sewage treated ponds of Madrash. J Fish 8: 140-142.

18. Saha SB, Bhagat MJ, Pathak V (1990) Ecological changes and its impact on fish yield of Kulia beel in Ganga basin. J Inland Fish Soc India 22: 7-11.

19. Manna NK, Banerjee SM, Bhoumik ML (1998) Live fish production through waste water utilization and recycling. J Inland Fish Soc India 30: 9-18.

20. Ghosh A, Rao LH, Banerjee SC (1974) Studies of hydrobiological conditions of a sewage fed pond with a note on their role in fish culture. J Inland Fish Soc India 6: 51-61.

21. Rai SP (1994) Impact of sewage effluent on soil variable of some non drainable fish pond. J Inland Fish Soc India 26: 83-88.

22. Chakraborty NM, Asthana A (1989) Plankton succession and ecology of a sewage fed fish pond in West Bengal. Envt Ecol 7: 549-554.

23. Hepher B, Schroeder GL (1977) Wastewater utilization in Israel aquaculture.

24. Allen GH, Hepher B (1979) Recycling of Waste through aquaculture and constraints to wider application.

25. Bhowmik ML, Sarkar UK, Pandey BK (1993) Plankton abundance and composition in sewage-fed fish ponds. J Inland Fish Soc India 25: 23-29.

26. Saha KC (1970) Sewage-fed Fisheries. In Fisheries of West Bengal. West Bengal Government Press, West Bengal.

27. Bilgrami KS, Duttamunshi JS, Bhowmik BN (1985) Biomonitoring of river Ganges at polluted sites in Bihar.

28. Schaeperclause W (1959) Die Karpfenteich wirtschft in der Denfschen Demokratischen republic, Stizungsher Dtsch Akal.

36

Variation in the Chemical Composition of *Saccharina Japonica* with Harvest Area and Culture Period

Jae-Ho Hwang[1], Nam-Gil Kim[2], Hee-Chul Woo[3], Sung-Ju Rha[1], Seon-Jae Kim[4] and Tai-Sun Shin[5]*

[1]College of Fisheries and Ocean Science, Chonnam National University, Yosu 550-749, Korea
[2]Department Marine Biology and Aquaculture, Gyeongsang National University, 445 Inpyeong-dong, Tongyeong-si, Gyeongsangnam, 650-160, Korea
[3]Department of Chemical Engineering, Pukyong National University, 365 Sinseon-ro, Yongdang-dong, Nam-gu, Busan, 608-739, Korea
[4]Department of Marine Bio Food, Chonnam National University, Yeosu 550-749, Korea
[5]Division of Food Nutrition Science, Chonnam National University, Gwangju 500-757, Korea

Abstract

Saccharina japonica is commercially important marine brown algae which grow as a single blade (reaching 10 meters in length) with a short stipe. In this study, the edible brown weed *Sacchaina japonica* was assessed for nutritional composition. Samples were collected monthly from seaweed farms at Kijang and Wando on the south coast of the Republic of Korea, during the 2011 culture season. *S. japonica* in Kijang and Wando showed the highest crude protein content in February and the highest carbohydrate content in July. Monthly changes in sugar, fatty acid, mineral, and total amino acid contents observed from February to July 2011. Fucose was the most abundant and galactose the second most abundant in the monosaccharide composition profiles, while mannose, glucose, xylose, ribose, and rhamnose were present in low quantities and lactose, mannitol, and arabinose were not detected. Significant increases of the major fatty acids in Kijang (C18:2 n-6 and C20:4 n-6) and Wando (C18:3 n-6) were observed as the culture period progressed. The highest mineral content of both Kijang and Wando samples is potassium and followed by sodium, calcium, magnesium, and so on. In the total amino acid contents, Kijang samples increased from February to April but decreased from May to July, while Wando samples increased on March but decreased from April to July.

Keywords: *Saccharina japonica*; Brown algae; Harvest area; Culture period; Chemical composition

Introduction

China, Japan, and the Republic of Korea are the largest consumers of edible seaweeds [1]. Seaweeds include high alginic acid, fucoidan, and laminara contents, so it is effective for hematocele and lipid metabolism improvement such as lowering blood pressure and cholesterol in the blood, and anti-cancer [2]. According to a survey conducted on worldwide production of aquatic plants, there are approximately 16 million tons of annual aquatic plants, of which 14.9 million tons produced by aquaculture [3]. Algal production in Korea is mainly limited to *Porphyra tenera*, *Saccharina japonica*, and *Undaria pinnatifida*, which comprise 94% of the total harvested seaweed [4]. *S. japonica* is very popular as a healthy food because of low calorie and abundant vitamin, mineral, dietary fiber, calcium, potassium, magnesium, phosphoric acid, and microelements and high iodine content as compared with other seaweeds [5].

In recent years, many studies on macro-algae have carried out and their proximate composition differs according to species, geographic origin, and seasonal conditions [6,7]. Growth change of laminaria closely related with culture period, most researchers studied to determine correlation between growth and nitrogen concentration [8]. Moreover, growth and chemical composition are various in different environments such as current, nutrients supply, fresh water inflow, and water temperature. Perennial *Saccharina japonica* generates alternately, and grows at subantarctic zone as well as temperate climate regions [9]. Cosson [10] reported that survival rate of *Laminaria digitata* spores is substantially lowered at over radiation intensity (about 170 $\mu E \cdot m^{-2} \cdot s^{-1}$). Kang and Koh [11] found that optimal growth temperature and light intensity of *Laminaria japonica* sporophyte were at 10°C and 70 $\mu E \cdot m^{-2} \cdot s^{-1}$.

To our knowledge, detailed studies have not conducted to evaluate

the effects of the culture period and harvest area on the chemical composition of *S. japonica*. This fundamental study performed to assess changes of proximate composition, sugar, fatty acid, mineral, and amino acid of *S. japonica* obtained from two sampling regions in Korea, Kijang and Wando, which had definitely different environment, and during the culture period from February to July.

Materials and Methods

Sampling

In order to observe variations in chemical composition during the harvest time, *S. japonica* was collected from an environmentally quite different seaweed farm at Kijang and Wando located on the southern coast of the Republic of Korea once a month from February to July 2011 (Figure 1). Both sporophytes of *S. japonica* transferred to the ocean at 0.5 m water depth in the same time (December 2010), and 5-20 individuals of whole *S. japonica* (blade, stem, and root) collected during the 2011 culture season. Freshly collected plants wrapped in paper towels with seawater, sealed in plastic bags, kept in an icebox, and transport to the laboratory where they washed with distilled water twice and freeze-dried. Each powered *S. japonica* (about 500 g) used for triplicate analysis.

*Corresponding author: Tai-Sun Shin, Division of Food Nutrition Science, Chonnam National University, Gwangju 500-757, Korea
E-mail: shints@chonnam.ac.kr

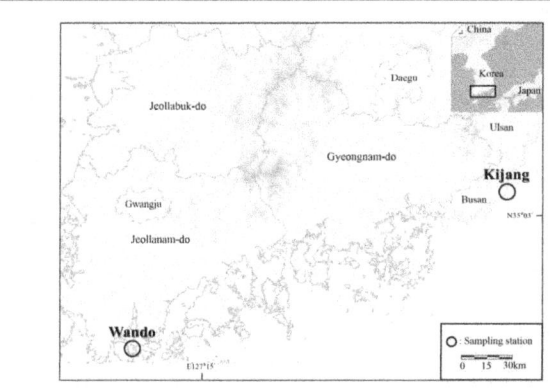

Figure 1: A map showing the site where *Saccharide japonica* were harvested during the 2011 culture period.

Condition	
Column	Shim-pack ISA-07 (4.0 mm×250 mm)
Mobile phase	A: potassium borate (pH 8.0) B: potassium borate (pH 9.0)
Flow rate	0.6 mL/min, gradient
Reagent	1% arginine in 3% boric acid (0.5 mL)
Reaction temperature	150°C
Detector	Fluorescence detector (Ex=320, Em=430)
Oven temperature	65°C

Table 1: HPLC operating conditions for component sugars.

General component analysis

Moisture, crude protein, crude lipid, and ash content were determined using the standard methods described by the Association of Official Analytical Chemists [12]. Protein content analyzed using the semi-Kjeldahl method. Lipids extracted with anhydrous diethyl ether using a Soxhlet apparatus. Moisture quantified by oven drying the samples at 105°C for 12 h. Ash was determined after incineration in a furnace at 550°C. Total carbohydrate content calculated by subtracting the sum of moisture, crude protein, crude lipid, and ash mass from that of the total sample [13].

Component sugar analysis

In order to extract component sugar, a test sample (100 mg) mixed in the 15 mL test tube with 5 mL of 2M HCl. The oxygen in the test tube replaced by nitrogen gas, sealed, and placed in a heating mantle at 100°C for 5 h [14]. Hydrolyzed sample cooled, neutralized by adding 5 mL of 2M NaOH, and centrifuged at 650 g for 30 min. 3 mL supernatant filtered through a Millipore membrane (0.45 μm pore size), and analyzed by operating conditions (Table 1) using HPLC (Prominence HPLC, Shimadzu Co, Ltd. Kyoto, Japan).

Fatty acid composition analysis

Bligh and Dyer extraction was performed using the following method [15]: Briefly, lipids were extracted from 5-g samples by homogenization with 100 mL of chloroform and 200 mL methanol. The samples were then filtered and evaporated to remove solvent. Fatty acid methyl esters (FAME) were prepared using boron trifluoride (BF3) according to a method described by the AOAC [12]. Quantitative analysis of FAME was carried out on a GC-2010 gas chromatograph(Shimadzu Co., Japan) equipped with a split/splitless capillary inlet system and a flame ionization detector (FID) using SP-

2560 capillary columns (0.20-μm stationary phase thickness, 100 mm (length)×0.25 mm (i.d.); Supelco, Inc., USA). The sample (0.5 μl) was injected in the split mode using an automatic injection system (AOC-20i, Shimadzu Co., Japan). The oven temperature was programmed to increase from 160 to 220°C at 1°C min^{-1} with an initial hold of 5 min and final hold of 40 min. The other operation parameters were as follows: injector temperature, 250°C; detector temperature, 250°C; helium carrier gas flow, 20 cm s^{-1}; split ratio, 1:50. The peak areas for the calibration curves and for calculation of fatty acid composition of oil samples were measured using a GC Solution system (Shimadzu Co., Japan).

Mineral contents

For the determination of mineral elements (calcium, copper, iron, potassium, magnesium, manganese, sodium, and zinc), samples were digested by dry ashing and dissolved in 1 M HCl [12]. The final diluted solution for calcium contained 1% lanthanum to overcome interferences. The concentration of the elements in *S. japonica* were determined with atomic absorption spectrophotometry (Perkin-Elmer, model 3110). Triplicate determinations for each element were carried out. The concentration of the elements were determined from calibration curves of the standard elements.

Amino-acid analysis

Samples (0.5 g) were acid-hydrolyzed with 3 mL of 6 N HCl in vacuum-sealed hydrolysis vials at 121°C for 24 h. Tubes were cooled after hydrolysis, opened, and placed in a rotary evaporator at 50°C to remove HCl from the sample. The residue was then adjusted to pH 2.2 with 0.2 M sodium citrate loading buffer (pH 2.2), diluted to a final volume of 10 mL with water, filtered through a Millipore membrane (0.2 μm pore size), and analyzed for amino acids using an amino-acid analyzer (Pharmacia Biochrom 20, Biochrom Ltd., UK).

Statistical analysis

All mean values were analyzed by one-way analysis of variance (ANOVA, SPSS 1999). Values are expressed as mean ± standard deviation (SD; n=3 replicates). Group means were considered to be significantly different at p<0.05.

Results

Changes in proximate composition with harvest area and culture period

The proximate compositions of Kijang and Wando samples are shown in Tables 2 and 3. There was a high variation in moisture, crude protein, ash and crude lipid content with culture period and harvest area among the Kijang and Wando samples collected at different months from February to July. *S. japonica* in Kijang and Wando showed the highest crude protein content in February and the highest carbohydrate content in July. In the crude lipid content, February samples in Kijang and Wando generally tended to decrease until July. There was a high variation in ash content with culture period and harvest area, ranging from 14.29 ± 1.47% to 19.39 ± 0.75% (Tables 2 and 3).

Changes in component sugar and fatty acid composition with harvest area and culture period

Component sugar compositions of Kijang and Wando samples are shown in Tables 4 and 5. Fucose was the most abundant and galactose the second most abundant in the monosaccharide composition profiles. Mannose, glucose, xylose, ribose, and rhamnose were present at low

Component	Culture period					
	Feb	Mar	Apr	May	Jun	Jul
Moisture	10.55 ± 0.51[a]	10.67 ± 0.45[a]	10.25 ± 0.49[a]	10.31 ± 0.98[a]	10.45 ± 1.41[a]	10.41 ± 0.22[a]
Crude protein	9.39 ± 0.45[a]	8.54 ± 0.36[b]	7.61 ± 0.34[c]	7.27 ± 0.70[cd]	6.62 ± 0.51[d]	5.72 ± 0.51[e]
Crude lipid	1.69 ± 0.08[a]	1.43 ± 0.06[b]	1.35 ± 0.06[bc]	1.09 ± 0.12[d]	1.23 ± 0.14[cd]	1.17 ± 0.02[d]
Ash	15.11 ± 0.73[b]	17.88 ± 0.72[ab]	18.39 ± 0.75[a]	17.86 ± 2.58[ab]	17.35 ± 2.04[ab]	16.51 ± 0.34[ab]
Carbohydrate[b]	63.26 ± 3.02[a]	61.48 ± 2.59[a]	62.40 ± 2.61[a]	63.47 ± 6.78[a]	64.35 ± 5.62[a]	66.19 ± 1.23[a]

[a]Values represent means ± standard error (n=3). Mean values in the same row followed by different letters differ significantly (p<0.05). [b]Carbohydrate content (%)=100-(% moisture + % protein + % lipid + % ash).

Table 2: Seasonal variation of proximate composition (%) in the dried sea tangle (S. japonica) cultured at Kijang area[a]

Component	Culture period					
	Feb	Mar	Apr	May	Jun	Jul
Moisture	10.38 ± 0.47[a]	10.51 ± 0.45[a]	10.45 ± 0.50[a]	10.12 ± 0.96[a]	10.34 ± 1.18[a]	10.46 ± 0.19[a]
Crude protein	8.20 ± 0.36[a]	8.20 ± 0.40[a]	7.51 ± 0.31[a]	6.54 ± 0.71[b]	5.58 ± 0.80[c]	5.15 ± 0.14[c]
Crude lipid	2.00 ± 0.09[b]	2.35 ± 0.09[a]	1.56 ± 0.07[c]	1.37 ± 0.21[cd]	1.26 ± 0.10[d]	1.23 ± 0.02[d]
Ash	16.68 ± 0.77[ab]	17.35 ± 0.73[a]	17.86 ± 0.80[a]	15.82 ± 2.39[ab]	14.29 ± 1.47[b]	14.69 ± 0.29[b]
Carbohydrate[b]	62.74 ± 2.89[a]	61.59 ± 2.61[a]	62.62 ± 2.91[a]	66.15 ± 7.29[a]	68.53 ± 7.22[a]	68.47 ± 0.49[a]

[a]Values represent means ± standard error (n=3). Mean values in the same row followed by different letters differ significantly (p<0.05). [b]Carbohydrate content (%)=100-(% moisture + % protein + % lipid + % ash).

Table 3: Seasonal variation of proximate composition (%) in the dried sea tangle (S. japonica) cultured at Wando area[a]

Sugar	Culture period					
	Feb	Mar	Apr	May	Jun	Jul
Rhamnose	0.11 ± 0.00[a]	0.08 ± 0.00[b]	0.08 ± 0.00[b]	0.04 ± 0.00[c]	0.04 ± 0.00[c]	0.03 ± 0.01[d]
Ribose	0.19 ± 0.00[a]	0.10 ± 0.00[b]	0.10 ± 0.00[b]	0.01 ± 0.00[d]	0.02 ± 0.00[c]	0.01 ± 0.01[d]
Mannose	0.66 ± 0.01[b]	0.55 ± 0.01[d]	0.65 ± 0.01[b]	0.72 ± 0.02[a]	0.60 ± 0.01[c]	0.56 ± 0.02[d]
Fucose	3.32 ± 0.08[c]	4.17 ± 0.13[b]	4.84 ± 0.12[a]	2.68 ± 0.06[d]	2.27 ± 0.06[f]	2.52 ± 0.16[e]
Galactose	2.31 ± 0.06[a]	1.92 ± 0.04[b]	1.97 ± 0.06[b]	0.80 ± 0.02[d]	0.88 ± 0.02[c]	0.71 ± 0.11[e]
Xylose	0.35 ± 0.01[a]	0.24 ± 0.01[b]	0.16 ± 0.00[c]	0.06 ± 0.00[d]	0.05 ± 0.00[e]	0.05 ± 0.00[e]
Glucose	0.55 ± 0.01[d]	0.64 ± 0.02[c]	0.68 ± 0.02[b]	0.73 ± 0.01[a]	0.72 ± 0.02[a]	0.53 ± 0.12[d]
Total	7.48 ± 0.17[b]	7.70 ± 0.17[b]	8.47 ± 0.21[a]	5.03 ± 0.13[c]	4.56 ± 0.11[d]	4.40 ± 0.14[d]

[a]Values represent means ± standard error (n=3). Mean values in the same row followed by different letters differ significantly (p<0.05).

Table 4: Seasonal variation of component sugar in the dried sea tangle (S. japonica) cultured at Kijang area[a]

Sugar	Culture period					
	Feb	Mar	Apr	May	Jun	Jul
Rhamnose	0.12 ± 0.00[b]	0.13 ± 0.00[a]	0.07 ± 0.00[c]	0.03 ± 0.00[d]	0.02 ± 0.00[e]	0.02 ± 0.00[e]
Ribose	0.19 ± 0.01[a]	0.10 ± 0.00[b]	0.09 ± 0.00[c]	0.01 ± 0.00[e]	0.02 ± 0.00[d]	0.01 ± 0.01[e]
Mannose	0.75 ± 0.02[a]	0.67 ± 0.02[c]	0.74 ± 0.02[a]	0.70 ± 0.02[b]	0.59 ± 0.01[d]	0.47 ± 0.08[e]
Fucose	3.49 ± 0.09[a]	2.98 ± 0.06[b]	2.92 ± 0.08[bc]	3.01 ± 0.07[b]	2.83 ± 0.05[c]	1.85 ± 0.54[d]
Galactose	2.21 ± 0.05[a]	1.65 ± 0.04[c]	1.81 ± 0.04[b]	0.80 ± 0.02[d]	0.80 ± 0.01[d]	0.63 ± 0.09[e]
Xylose	0.43 ± 0.01[a]	0.28 ± 0.01[c]	0.30 ± 0.01[b]	0.06 ± 0.00[d]	0.07 ± 0.00[d]	0.05 ± 0.01[e]
Glucose	0.67 ± 0.02[b]	0.67 ± 0.02[b]	0.73 ± 0.02[a]	0.72 ± 0.01[a]	0.42 ± 0.01[c]	0.42 ± 0.01[c]
Total	7.87 ± 0.21[a]	6.48 ± 0.14[b]	6.65 ± 0.21[b]	5.34 ± 0.16[c]	4.76 ± 0.11[d]	3.45 ± 0.73[e]

[a]Values represent means ± standard error (n=3). Mean values in the same row followed by different letters differ significantly (p<0.05).

Table 5: Seasonal variation of component sugar in the dried sea tangle (S. japonica) cultured at Wando area[a].

quantities, and lactose, mannitol, and arabinose were not detected.

The fatty acid compositions of Kijang and Wando samples are shown in Tables 6 and 7. Lignoceric acid (24:0) was the most abundant fatty acid, followed by arachidonic acid (20:4 n-6), oleic acid (18:1 n-9), and palmitic acid (16:0). Polyunsaturated fatty acid (PUFA) and monounsaturated fatty acid (MUFA) constituted about 54.9%, 52.3% of total fatty acids, and saturated fatty acids (SFA) represented 45.1%, 47.7% of the total fatty acids in the Kijang and Wando samples, respectively. The Kijang-Jul samples showed the highest PUFA composition (37.5%) among the samples, while Wando-Mar showed the lowest PUFA composition (30.1%), indicating that there was a high variation in fatty acid contents with the harvest area and culture period.

Changes in mineral content and total amino acid composition with harvest area and culture period

The mineral contents of Kijang and Wando samples are shown in Tables 8 and 9. The results show that S. japonica is rich in K and Na with moderate amounts of Ca and Mg whereas Cu, Fe, Mn, and Zn are present in small quantities. The total amino acid (TAA) compositions of Kijang and Wando samples are shown in Tables 10 and 11. Glutamic acid, aspartic acid, alanine, and leucine were the most common amino acids in all samples, while the percentage of cysteine was the lowest in the TAA profile. TAA of Kijang samples decreased during the harvest time from April to July while TAA of Wando samples decreased from March to July.

Fatty acid (%)	Culture period					
	Feb	Mar	Apr	May	Jun	Jul
12:0	0.18 ± 0.01[c]	0.27 ± 0.01[a]	0.21 ± 0.00[b]	0.03 ± 0.00[f]	0.05 ± 0.00[e]	0.07 ± 0.00[d]
14:0	9.50 ± 0.21[a]	9.66 ± 0.19[a]	7.11 ± 0.15[c]	4.73 ± 0.15[e]	7.68 ± 0.15[b]	6.02 ± 0.12[d]
16:0	13.53 ± 0.31[d]	16.25 ± 0.35[b]	17.22 ± 0.22[a]	17.34 ± 0.40[a]	14.51 ± 0.33[c]	11.07 ± 0.27[e]
16:1 n-7	3.08 ± 0.08[c]	3.78 ± 0.12[a]	3.44 ± 0.08[b]	3.06 ± 0.07[c]	3.49 ± 0.09[b]	3.20 ± 0.09[c]
18:0	0.86 ± 0.02[c]	1.00 ± 0.02[b]	1.16 ± 0.03[a]	0.69 ± 0.02[d]	0.87 ± 0.02[c]	0.89 ± 0.02[c]
18:1 n-9	19.20 ± 0.40[a]	16.83 ± 0.42[c]	15.67 ± 0.42[d]	13.30 ± 0.40[e]	16.27 ± 0.39[cd]	17.75 ± 0.42[b]
18:2 n-6	5.58 ± 0.15[b]	6.79 ± 0.19[a]	6.78 ± 0.15[a]	6.93 ± 0.14[a]	6.95 ± 0.16[a]	7.03 ± 0.16[a]
18:3 n-6	1.78 ± 0.04[e]	2.17 ± 0.05[d]	2.90 ± 0.07[c]	2.94 ± 0.07[c]	3.97 ± 0.09[a]	3.76 ± 0.09[b]
18:3 n-3	7.38 ± 0.15[a]	6.54 ± 0.16[c]	6.79 ± 0.21[b]	3.84 ± 0.09[d]	2.96 ± 0.07[e]	2.97 ± 0.06[e]
20:0	0.40 ± 0.01[c]	0.50 ± 0.01[b]	0.50 ± 0.01[b]	0.26 ± 0.01[d]	0.51 ± 0.01[ab]	0.53 ± 0.01[a]
20:2 n-6	1.30 ± 0.03[c]	1.76 ± 0.04[a]	1.59 ± 0.05[b]	1.66 ± 0.04[b]	1.80 ± 0.04[a]	1.78 ± 0.04[a]
20:3 n-6	1.42 ± 0.03[e]	1.95 ± 0.04[a]	1.84 ± 0.05[b]	1.67 ± 0.04[c]	1.53 ± 0.03[d]	1.62 ± 0.05[c]
20:4 n-6	13.35 ± 0.30[d]	14.92 ± 0.40[c]	15.37 ± 0.32[c]	19.19 ± 0.45[b]	19.40 ± 0.21[b]	20.34 ± 0.45[a]
C24:0	22.44 ± 0.44[b]	17.58 ± 0.40[d]	19.42 ± 0.54[c]	24.36 ± 0.64[a]	20.01 ± 0.21[c]	22.97 ± 0.54[b]
Saturates	46.91 ± 1.07[ab]	45.26 ± 1.29[bc]	45.62 ± 1.11[abc]	47.42 ± 0.62[a]	43.64 ± 0.93[c]	41.55 ± 1.26[d]
Monoenes	22.27 ± 0.59[a]	20.62 ± 0.46[bc]	19.11 ± 0.61[d]	16.35 ± 0.49[e]	19.77 ± 0.45[cd]	20.95 ± 0.48[b]
Polyenes	30.81 ± 0.67[d]	34.13 ± 0.80[c]	35.27 ± 0.76[bc]	36.23 ± 0.81[ab]	36.60 ± 1.02[ab]	37.50 ± 0.45[a]
P/S	0.66 ± 0.02[d]	0.75 ± 0.02[c]	0.77 ± 0.02[c]	0.76 ± 0.02[c]	0.84 ± 0.02[b]	0.90 ± 0.02[a]

[a]Values represent means ± standard error (n=3). Mean values in the same row followed by different letters differ significantly (p<0.05).

Table 6: Seasonal variation of fatty acid composition (percentage of weight) in the dried sea tangle (*S. japonica*) cultured at Kijang area[a].

Fatty acid (%)	Culture period					
	Feb	Mar	Apr	May	Jun	Jul
12:0	0.28 ± 0.01[b]	0.58 ± 0.01[a]	0.09 ± 0.00[d]	0.04 ± 0.00[e]	0.16 ± 0.00[c]	0.08 ± 0.00[d]
14:0	10.47 ± 0.25[b]	10.92 ± 0.13[a]	8.70 ± 0.17[c]	6.72 ± 0.18[e]	7.09 ± 0.15[d]	5.74 ± 0.18[f]
16:0	16.08 ± 0.40[c]	16.94 ± 0.33[b]	17.13 ± 0.45[b]	18.07 ± 0.37[a]	11.25 ± 0.14[d]	10.58 ± 0.24[e]
16:1 n-7	2.92 ± 0.07[d]	3.10 ± 0.08[c]	3.45 ± 0.07[b]	3.15 ± 0.05[c]	3.79 ± 0.09[a]	3.20 ± 0.07[c]
18:0	1.25 ± 0.02[a]	1.14 ± 0.02[b]	0.82 ± 0.01[c]	0.81 ± 0.02[c]	0.82 ± 0.02[c]	0.81 ± 0.02[c]
18:1 n-9	17.26 ± 0.19[a]	17.07 ± 0.34[a]	14.95 ± 0.30[c]	15.41 ± 0.43[bc]	15.71 ± 0.42[b]	15.95 ± 0.48[b]
18:2 n-6	5.24 ± 0.05[d]	5.46 ± 0.17[d]	6.03 ± 0.17[c]	6.38 ± 0.13[b]	6.86 ± 0.15[a]	6.51 ± 0.13[b]
18:3 n-6	1.75 ± 0.04[f]	2.22 ± 0.05[e]	2.34 ± 0.06[d]	2.65 ± 0.07[c]	3.05 ± 0.08[b]	3.67 ± 0.09[a]
18:3 n-3	6.85 ± 0.16[a]	5.70 ± 0.13[c]	6.26 ± 0.14[b]	4.70 ± 0.13[d]	4.15 ± 0.13[e]	4.00 ± 0.09[e]
20:0	0.40 ± 0.01[d]	0.47 ± 0.01[b]	0.43 ± 0.01[c]	0.35 ± 0.00[e]	0.42 ± 0.01[c]	0.49 ± 0.01[a]
20:2 n-6	1.49 ± 0.04[b]	1.71 ± 0.03[a]	1.40 ± 0.04[c]	1.68 ± 0.05[a]	1.26 ± 0.04[d]	1.52 ± 0.04[b]
20:3 n-6	1.58 ± 0.03[b]	1.77 ± 0.04[a]	1.63 ± 0.04[b]	1.51 ± 0.03[c]	1.59 ± 0.04[b]	1.49 ± 0.03[c]
20:4 n-6	13.55 ± 0.32[e]	13.19 ± 0.41[e]	16.82 ± 0.39[c]	15.60 ± 0.37[d]	17.55 ± 0.36[b]	18.96 ± 0.45[a]
C24:0	20.89 ± 0.60[c]	19.73 ± 0.45[d]	19.95 ± 0.50[cd]	22.93 ± 0.24[b]	26.30 ± 0.74[a]	27.00 ± 0.71[a]
Saturates	49.37 ± 1.56[a]	49.78 ± 0.98[a]	47.12 ± 1.16[b]	48.93 ± 1.04[a]	46.04 ± 1.12[bc]	44.71 ± 0.58[c]
Monoenes	20.17 ± 0.46[a]	20.17 ± 0.46[a]	18.40 ± 0.32[c]	18.56 ± 0.43[c]	19.50 ± 0.62[ab]	19.15 ± 0.57[bc]
Polyenes	30.46 ± 0.69[d]	30.05 ± 0.80[d]	34.48 ± 0.62[b]	32.51 ± 0.90[c]	34.46 ± 0.74[b]	36.14 ± 0.81[a]
P/S	0.62 ± 0.01[d]	0.60 ± 0.02[d]	0.73 ± 0.02[b]	0.66 ± 0.01[c]	0.75 ± 0.02[b]	0.81 ± 0.02[a]

[a]Values represent means ± standard error (n=3). Mean values in the same row followed by different letters differ significantly (p<0.05).

Table 7: Seasonal variation of fatty acid composition (percentage of weight) in the dried sea tangle (*S. japonica*) cultured at Wando area[a].

Discussion

There are big environmental differences between Wando and Kijang. Wando is semi-closed sea, and affected by big tide and fresh water inflow from many rivers around. Kijang has a small tide, but high temperature high salinity Tsushima current and low temperature low salinity North Korea current meets in this area. To our knowledge, this is the first study that evaluated differences in the nutritional composition of *S. japonica* with harvest area and culture period. We found that protein content of *S. japonica* was highest in February and the carbohydrate content was highest in July for the Kijang and Wando samples over the culture period from February to July 2011. A similar pattern was previously reported for the collection of *Laminaria japonica* [16]. Rosemberg and Ramus [17] found inverse relationships between carbohydrate and protein content in the red seaweed *Gracilaria cervicornis* during collection from July 2000 to June 2001. The seaweed protein content was lowest when photosynthetic activity and carbohydrate synthesis were highest. Shin et al. [18,19] found that carbohydrate content of *Porphyra yezoensis* increased with late culture period: Dec (39.4%), Feb (47.2%). However, the protein content decreased with late culture period: Dec (39.4%), Feb (34.6%). Lipid content was not affected by culture period. A positive correlation was also detected between carbohydrate and temperature, along with correlations with salinity and solar radiation, which indicated that carbohydrate synthesis and protein concentration are affected by several seasonal factors, including water temperature, nitrogen content, and light intensity [16,18]. The lipid content was low relative to the other chemical constituents. However, the lipid content observed

Mineral	Culture period					
	Feb	Mar	Apr	May	Jun	Jul
Ca	567.11 ± 16.71[e]	972.86 ± 23.27[a]	858.81 ± 16.54[b]	745.45 ± 21.30[d]	783.84 ± 17.89[c]	741.41 ± 18.08[d]
Cu	0.29 ± 0.01[e]	0.46 ± 0.01[b]	0.34 ± 0.01[c]	0.47 ± 0.01[b]	0.67 ± 0.01[a]	0.31 ± 0.01[d]
Fe	8.15 ± 0.18[a]	3.70 ± 0.08[d]	3.16 ± 0.04[f]	6.20 ± 0.14[b]	3.39 ± 0.08[e]	4.46 ± 0.09[c]
K	3325.83 ± 83.43[c]	3516.51 ± 111.07[b]	4158.54 ± 99.29[a]	3554.55 ± 80.98[b]	3578.28 ± 95.21[b]	3165.47 ± 67.11[d]
Mg	630.63 ± 17.51[c]	592.81 ± 11.93[d]	606.55 ± 17.32[cd]	887.89 ± 20.04[a]	821.58 ± 17.91[b]	794.79 ± 12.45[b]
Mn	0.44 ± 0.01[d]	0.70 ± 0.02[b]	0.69 ± 0.02[b]	0.68 ± 0.02[b]	0.86 ± 0.02[a]	0.55 ± 0.01[c]
Na	1209.21 ± 32.09[d]	1440.73 ± 39.94[a]	1361.45 ± 30.09[b]	1285.19 ± 26.22[c]	1253.65 ± 28.78[cd]	1204.93 ± 29.30[d]
Zn	1.65 ± 0.04[e]	2.34 ± 0.05[c]	2.19 ± 0.05[d]	2.76 ± 0.07[b]	0.37 ± 0.01[f]	3.04 ± 0.04[a]
Total	5,743.31 ± 120.58[c]	6,530.11 ± 156.09[b]	6,991.73 ± 211.17[a]	6,483.19 ± 154.05[b]	6,442.64 ± 151.48[b]	5,914.96 ± 126.98[c]

[a]Values represent means ± standard error (n=3). Mean values in the same row followed by different letters differ significantly (p<0.05).

Table 8: Seasonal variation of mineral contents in the dried sea tangle (*S. japonica*) cultured at Kijang area[a].

Mineral	Culture period					
	Feb	Mar	Apr	May	Jun	Jul
Ca	900.91 ± 22.15[a]	913.15 ± 17.93[a]	913.58 ± 23.49[a]	789.90 ± 18.30[b]	730.30 ± 13.90[c]	771.72 ± 14.14[b]
Cu	0.56 ± 0.01[c]	0.40 ± 0.01[d]	0.94 ± 0.02[a]	0.90 ± 0.02[b]	0.22 ± 0.00[e]	0.12 ± 0.00[f]
Fe	2.89 ± 0.06[e]	2.48 ± 0.06[f]	3.55 ± 0.10[d]	5.52 ± 0.15[a]	4.92 ± 0.05[c]	5.24 ± 0.15[b]
K	3683.32 ± 84.26[c]	4020.03 ± 114.39[a]	3847.32 ± 93.92[b]	3643.64 ± 47.59[c]	3260.86 ± 69.51[d]	3182.01 ± 41.93[d]
Mg	644.72 ± 17.15[d]	723.22 ± 15.96[c]	593.00 ± 18.77[e]	836.84 ± 25.04[ab]	852.96 ± 19.57[a]	820.82 ± 17.93[b]
Mn	0.77 ± 0.02[a]	0.62 ± 0.01[d]	0.73 ± 0.02[b]	0.47 ± 0.01[e]	0.65 ± 0.02[c]	0.31 ± 0.00[f]
Na	1378.52 ± 32.66[b]	1435.11 ± 43.64[b]	1613.96 ± 36.48[a]	1173.17 ± 25.60[d]	1141.14 ± 33.21[d]	1286.39 ± 25.75[c]
Zn	1.44 ± 0.03[f]	2.13 ± 0.06[d]	2.35 ± 0.05[c]	2.53 ± 0.08[b]	3.04 ± 0.06[a]	1.69 ± 0.05[e]
Total	6613.13 ± 165.73[b]	7097.14 ± 138.26[a]	6975.43 ± 183.16[a]	6452.97 ± 132.95[b]	5,994.09 ± 84.62[c]	6068.3 ± 96.61[c]

[a]Values represent means ± standard error (n=3). Mean values in the same row followed by different letters differ significantly (p<0.05).

Table 9: Seasonal variation of mineral contents in the dried sea tangle (*S. japonica*) cultured at Wando area[a].

Amino acid	Culture period					
	Feb	Mar	Apr	May	Jun	Jul
Aspartic acid	1574.47 ± 46.38[a]	1422.98 ± 34.04[b]	1415.36 ± 27.26[b]	1250.12 ± 35.73[c]	1087.68 ± 21.32[d]	1032.76 ± 25.39[d]
Threonine*	693.70 ± 15.40[a]	721.08 ± 14.38[a]	718.71 ± 15.30[a]	636.35 ± 20.0[b]	526.62 ± 14.11[c]	446.02 ± 9.84[d]
Serine	773.96 ± 17.51[a]	677.69 ± 14.53[b]	674.86 ± 8.57[b]	604.14 ± 13.78[c]	515.94 ± 3.45[d]	487.02 ± 9.61[d]
Glutamic acid	1718.74 ± 43.11[a]	1623.86 ± 51.29[b]	1615.31 ± 38.57[b]	1502.92 ± 34.24[c]	1223.20 ± 9.64[d]	1089.29 ± 23.09[e]
Proline	676.85 ± 18.80[a]	565.19 ± 11.37[b]	562.15 ± 16.05[b]	542.35 ± 12.24[b]	497.45 ± 9.05[c]	400.25 ± 6.27[d]
Glycine	895.55 ± 18.66[a]	821.67 ± 20.32[b]	817.98 ± 21.71[b]	765.58 ± 22.93[c]	622.11 ± 8.24[d]	536.65 ± 14.07[e]
Alanine	1105.96 ± 29.35[a]	995.41 ± 27.59[b]	990.79 ± 21.90[b]	724.24 ± 14.78[c]	602.70 ± 10.81[e]	666.54 ± 16.21[d]
Cystine	N.D.	N.D.	N.D.	N.D.	N.D.	N.D.
Valine*	592.06 ± 12.43[c]	773.56 ± 18.45[a]	769.75 ± 24.25[a]	694.82 ± 15.83[b]	557.45 ± 0.26[d]	384.18 ± 10.56[e]
Methionine	345.06 ± 10.97[a]	321.79 ± 7.45[b]	319.66 ± 7.33[b]	275.21 ± 5.95[c]	210.27 ± 2.02[d]	188.26 ± 3.41[e]
Isoleucine*	435.32 ± 9.14[d]	623.87 ± 14.91[a]	620.55 ± 18.74[a]	586.15 ± 13.93[b]	496.92 ± 4.37[c]	424.00 ± 9.11[d]
Leucine*	1094.16 ± 26.91[b]	1174.31 ± 23.05[a]	1168.03 ± 30.03[a]	997.35 ± 23.10[c]	842.37 ± 5.94[d]	656.88 ± 12.04[e]
Tyrosine*	366.60 ± 8.37[a]	299.66 ± 8.00[b]	297.48 ± 6.15[b]	285.19 ± 6.73[c]	272.69 ± 2.88[d]	222.90 ± 5.42[e]
Phenylalanine*	636.80 ± 12.54[b]	697.82 ± 15.93[a]	693.92 ± 19.42[a]	585.45 ± 15.41[c]	512.50 ± 4.42[d]	393.85 ± 11.04[e]
Histidine	325.11 ± 7.44[c]	394.50 ± 11.23[a]	394.62 ± 9.63[a]	342.34 ± 4.47[b]	304.46 ± 3.28[d]	279.06 ± 3.68[e]
Lysine*	653.32 ± 17.38[b]	700.03 ± 15.45[a]	696.74 ± 22.06[a]	513.45 ± 15.36[c]	480.60 ± 6.28[d]	430.39 ± 9.40[e]
Arginine	511.63 ± 11.16[c]	660.01 ± 15.48[a]	658.02 ± 14.11[a]	641.75 ± 14.39[a]	544.90 ± 3.66[b]	482.10 ± 5.36[d]
Total	12399.29 ± 293.74[a]	12473.43 ± 379.31[a]	12413.91 ± 280.61[a]	9986.94 ± 217.90[b]	9217.87 ± 129.32[c]	7783.17 ± 155.80[d]

[a]Values represent means ± standard error (n=3). Mean values in the same row followed by different letters differ significantly (p<0.05).

*Essential amino acid

Table 10: Seasonal variation of total amino acid contents in the dried sea tangle (*S. japonica*) cultured at Kijang area[a].

in this study was similar to the content observed in other seaweeds, comprising from 1% to 3% of dry matter [20,21]. The ash content varied from 14.3% to 18.4% in our samples. It has been reported that the ash content fluctuates depending on the species, geographical location, and season investigated [22,23].

Component sugar compositions of Kijang and Wando samples were high in the following order: fucose, galactose, glucose, mannose, and so on. Polysaccharide of seaweed generally classified into cytoskeleton, intercellular mucoid, and storage polysaccharides, most of 2M HCl hydrolyzed polysaccharide from *S. japonica* in this study originated from storage polysaccharide [24]. In the Kijang and Wando samples, there was a variation in the sugar content depending on culture period (P<0.05). Galactose content of Kijang samples were higher than that of Wando in the all culture period (p<0.05).

Major fatty acid of Kijang and Wando samples is myristic acid (14:0), palmitic acid (16:0), oleic acid (18:1), linoleic acid (18:2),

Amino acid	Culture period					
	Feb	Mar	Apr	May	Jun	Jul
Aspatic acid	1311.08 ± 30.09[a]	1370.98 ± 38.01[a]	1350.66 ± 28.74[a]	1248.37 ± 37.12[b]	1197.13 ± 32.41[c]	987.30 ± 29.07[d]
Threonine*	662.58 ± 15.35[a]	686.76 ± 15.65[a]	685.58 ± 15.73[a]	593.84 ± 11.69[b]	519.70 ± 12.70[c]	403.39 ± 3.13[d]
Serine	622.40 ± 14.98[b]	648.74 ± 7.75[a]	643.88 ± 12.84[ab]	584.90 ± 15.87[c]	539.02 ± 13.64[d]	460.76 ± 11.59[e]
Glutamic acid	1490.24 ± 37.35[b]	1563.56 ± 30.46[a]	1541.39 ± 40.47[ab]	1490.61 ± 30.71[b]	1257.52 ± 8.65[c]	1170.02 ± 18.63[d]
Proline	518.72 ± 12.20[bc]	544.67 ± 13.83[a]	536.49 ± 10.88[ab]	511.15 ± 8.81[c]	485.48 ± 13.65[d]	384.76 ± 7.06[e]
Glycine	754.48 ± 14.36[b]	787.88 ± 16.64[a]	780.47 ± 12.97[a]	686.67 ± 14.51[c]	544.26 ± 9.64[d]	526.76 ± 12.53[d]
Alanine	913.89 ± 10.07[b]	955.28 ± 19.29[a]	945.36 ± 19.00[ab]	867.07 ± 24.20[c]	726.10 ± 12.95[d]	718.06 ± 19.23[d]
Cystine	N.D.	N.D.	N.D.	N.D.	N.D.	N.D.
Valine*	710.09 ± 15.14[b]	743.50 ± 16.17[a]	734.50 ± 17.42[ab]	581.42 ± 15.16[c]	515.00 ± 13.11[d]	458.48 ± 11.91[e]
Methionine	295.23 ± 6.77[b]	312.42 ± 7.13[a]	305.21 ± 6.73[ab]	227.70 ± 6.28[c]	211.71 ± 6.35[d]	194.27 ± 3.56[e]
Isoleucine*	572.64 ± 15.91[b]	601.11 ± 12.59[a]	592.24 ± 14.98[ab]	524.88 ± 6.90[c]	510.59 ± 6.78[c]	443.58 ± 5.14[d]
Leucine*	1077.85 ± 31.36[b]	1131.59 ± 21.26[a]	1114.74 ± 32.66[ab]	941.52 ± 28.54[c]	828.06 ± 12.26[d]	626.19 ± 19.45[e]
Tyrosine*	274.85 ± 5.60[b]	292.06 ± 6.82[a]	284.09 ± 6.35[ab]	224.07 ± 4.65[c]	193.09 ± 2.31[e]	205.54 ± 5.27[d]
Phenylalanine*	640.38 ± 14.90[b]	673.29 ± 21.10[a]	662.28 ± 15.18[ab]	551.35 ± 13.22[c]	490.96 ± 9.25[d]	368.61 ± 7.66[e]
Histidine	363.34 ± 10.38[b]	368.28 ± 8.41[ab]	376.21 ± 9.25[a]	313.23 ± 8.08[c]	283.56 ± 5.28[d]	241.80 ± 4.95[e]
Lysine*	642.57 ± 20.25[b]	671.81 ± 13.22[a]	664.74 ± 14.67[a]	542.88 ± 11.71[c]	493.55 ± 12.33[d]	431.80 ± 9.64[e]
Arginine	606.52 ± 13.83[a]	627.61 ± 14.36[a]	627.62 ± 12.39[a]	529.72 ± 14.13[b]	515.74 ± 14.81[b]	458.52 ± 13.99[c]
Total	11456.88 ± 261.03[b]	13776.53 ± 366.56[a]	11845.47 ± 251.11[b]	10498.01 ± 363.89[c]	9862.18 ± 7.70[d]	7609.83 ± 206.81[e]

[a]Values represent means ± standard error (n=3). Mean values in the same row followed by different letters differ significantly (p<0.05).
*Essential amino acid

Table 11: Seasonal variation of total amino acid contents in the dried sea tangle (*S. japonica*) cultured at Wando area[a]

α-linolenic acid (18:3), arachidonic acid (20:4), and lignoceric acid (24:0). Many researches on seaweed fatty acid composition have been reported [16,25-34], but there have been various fatty acid contents since which one is chosen for analysis among about 50 selling fatty acid standards. Moreover, fatty acid compositions of the seaweed are generally varied by analyzing its sampled part. In all the data, most fatty acid composition showed a variation with harvest area and culture period. Low fatty acids, such as lauric acid (12:0), stearic acid (18:0), and arachidic acid (20:0), showed different compositions on harvest area and culture period without tendency. These results are similar with previous report [16]. Linoleic acid, γ-linolenic acid, and arachidonic acid in Kijang samples and γ-linolenic acid, arachidonic acid, and lignoceric acid in Wando samples increased with culture period, whereas α-linolenic acid in Kijang samples and stearic acid in Wando samples decreased. Both Kijang and Wando samples decreased with culture period in saturates, while those of polyenes increased.

In the mineral contents of Kijang and Wando *S. japonica* samples, the results show that *S. japonica* is rich in K and Na with moderate amounts of Ca and Mg whereas Cu, Fe, Mn, and Zn are present in small quantities.

Major amino acid of Kijang and Wando *S. japonica* samples are glutamic acid, aspartic acid, leucine, alanine, glycine, valine, phenylalanine, but cystine was not detected. Glutamic acid and aspartic acid occupied over 20% in the total amino acid. It is known that amino acid of seaweed is generally composed of high contents in neutral and acidic amino acids such as alanine, aspartic acid, glycine, and proline [24], but *S. japonica* contained low glycine and proline contents. Sulfur amino acid, cysteine and cysteine, was not detected, methionine, histidine, and tyrosine were included in small amount. The average percentages of essential amino acids (EAA) in Kijang and Wando *S. japonica* samples were 39.5%, 37.8%, which is higher than the EAA requirement (32.3%) suggested by the Food and Agriculture Organization [35]. The amino acid composition observed in this study was similar to previous studies [36], where the sum of the average percentage of three amino acids, glutamic acid (13.5%), aspartic acid

(11.9%), and alanine (7.9%), comprised the greatest proportion (33.3%) of TAA composition. Noda [37] suggested that the former three amino acids (glutamic acid, aspartic acid, and alanine) might produce the flavors specific to Nori (Porphyra). TAA content decreased at the end of the culture period. This phenomenon has also been observed in other seaweeds such as *Enteromorpha prolifera*, *C. fulvescens*, and *Codium fragile* [38].

In conclusion, we have ascertained that the monthly nutritional composition of *S. japonica* affected by harvest area and culture period from February to July 2011. *S. japonica* in Kijang and Wando showed the highest crude protein content in February and the highest carbohydrate content in July. Fucose was the most abundant and galactose the second most abundant in the monosaccharide composition profiles. Significant increases of the major fatty acids in Kijang (C18:2 n-6 and C20:4 n-6) and Wando (C18:3 n-6) were observed as the culture period progressed. The highest mineral content of both Kijang and Wando samples is potassium and followed by sodium, calcium, magnesium, and so on. In the total amino acid contents, Kijang samples increased from February to April but decreased from May to July, while Wando samples increased on March but decreased from April to July.

Acknowledgment

This study received financial support from the Ministry of Oceans and Fisheries, Republic of Korea.

References

1. Zemke-White WL, Ohno M (1999) World seaweed utilization: an endof- century summary. J Appl Phycol 11: 369-376.

2. Mohamed S, Hashim SN, Abdul Rahman H (2012) Seaweeds: A sustainable functional food for complementary and alternative therapy, Trends Food Sci Tech 23: 83-96

3. FAO (2009) FAO Yearbook of Fishery and Aquaculture statistics.

4. Park JI, Woo HC, Lee JH (2008) Production of bio-energy from marine algae: status and perspectives. Korean Chem Eng Res 46: 833-844.

5. Shin TS, Z Xue, YW Do, SI Jeong, HC Woo, et al. (2011) Chemical Properties of Sea Tangle (*Saccharina. japonica*) Cultured in the Different Depths of Seawater. Clean Technology 17: 395-405.

6. Patarra RF, Paiva L, Neto AI, Lima E, Baptista J (2011) Nutritional value of selected macroalgae. J Appl Phycol 23: 205-208.

7. Sun SM, Cho SY, Shin TS, Chung GH, Ahn CB, et al. (2012) Variation in the chemical composition of *Capsosiphon fulvescens* with area and during the harvest period. J Appl Phycol 24: 459-465.

8. Wheeler WN, Weindner W (1983) The effects of external inorganic nitrogen concentration on the metabolism of growth and activities of key carbon and nitrogen assimilatory enzymes of *Laminaria saccharina*(Phaeophyta) in culture. J Phycol 19: 91-96.

9. Ohno M, Crichley AT (1997) Seaweed cultivation and marine ranching.

10. Cosson J (1973) Action de la temperature et de la lumiere sur le de velopment du gametophyte de *la Laminaria digitata*(L.) Lam.(Pheophycee, Laminariales).

11. Kang RS, Koh CH (1999) Germination and Growth of *Laminaria japonica* (Phaeophyta) Microscopic Stages under Different Temperatures and Photon Irradiances. J. Korean Fish Soc 32: 438-443.

12. AOAC (1995) Official Methods of Analysis.

13. Amza T, Amadou I, Kamara MT, Zhu K, Zhou H (2010) Chemical and nutrient analysis of gingerbread plum (*Neocarya macrophylla*) seeds. Adv J Food Sci Technol 2: 191-195.

14. Chaplin MF, Kennedy JF (1994) Carbohydrate Analysis: A Practical Approach. Oxford University Press, Oxford, New York.

15. Iverson SJ, Lang SLC, Cooper MH (2011) Comparison of the Bligh and Dyer and Folch methods for total lipid determination in a broad range of marine tissue. Lipids 36: 1283-1287.

16. Khotimcheno SV, Kulikova IV (2000) Lipids of different parts of the Lamina of *Laminaria japonica* Aresch. Botanica Marina 43: 87-91.

17. Rosemberg G, Ramus J (1982) Ecological growth strategies in the seaweeds Gracilaria follifera and Ulva sp.: soluble nitrogen and reserve carbohydrates. Mar Biol 66: 251-259.

18. Shin DM, SR An, SK In, JG Koo (2013) Seasonal Variation in the Dietary Fiber, Amino Acid and Fatty Acid Contents of Porphyra yezoensis. Kor J Fish Aquat Sci 46: 337-342.

19. Rotem A, Roth-Bejeranu N, Arad SM (1986) Effect of controlled environmental conditions on starch and agar contents of Gracilaria sp. J Phycol 22: 117-121.

20. Hwang EK, Amano H, Park CS (2008) Assessment of the nutritional value of Capsosiphon fulvescens (Chlorophyta): developing a new species of marine macroalgae for cultivation in Korea. J Appl Phycol 20: 147-151.

21. Manivannan K, Thirumaran G, Devi GK, Anantharaman P, Balasubramanian T (2009) Proximate composition of different group of seaweeds from Vedalai coast waters (Gulf of Mannar): southeast coast of India. Middle-East J Sci Res 4: 72-77.

22. Ruperez P (2002) Mineral content of edible marine seaweeds. Food Chem 78: 23-26.

23. Kaehler S, Kennish R (1996) Summer and winter comparisons in the nutritional value of marine macroalgae from Hong Kong. Bot Mar 39: 11-17.

24. Park YH, Jang DS, Kim SB (1997) Processing of the sea food. Hyungsul Press, Seoul, Korea.

25. Choe SN, Choi KJ (2000) Fatty Acid Compositions of Sea Algaes in The Southern Sea Coast of Korea. Kor J Food Nutr 15: 58-63.

26. Dembitsky VM, Rosentsvet OA, Pechenkina EE (1990) Glycolipids, Phospholipids and Fatty Acids of Brown Algae Species. Phytochem 29: 3417-3421.

27. Fleurence J, Gutbier G, Mabeau S, Leray C (1994) Fatty Acids from 11 Marine Macroalgae of The French Brittany Coast. J Appl Phycol 6: 527-532.

28. Harwood JL (1980) Plant Acyl Lipids: Structure, Distribution and Analysis. Academic Press, New York.

29. Hayashi K, Kida S, Kato K, Yamada M (1974) Component Fatty Acids of Acetone-Soluble Lipids of 17 Species of Marine Benthic Algae. Nipp Suis Gak 40: 609-617

30. Jamieson JM, Reid EH (1972) The Component Fatty Acids of Some Marine Algal Lipids. Phytochem 11: 1423-1432.

31. Kaneniwa M, Itabashi J, Takagi T (1987) Unusual 5-olefinic Acids in the Lipids of Algae from Japanese Waters. Nipp Suis Gak 53: 861-866.

32. Kato M, Ariga N (1983) Studies on Lipids of Marine Algae: Sterol and Fatty Acid Composition of Marine Algae. GifuDaigaku 18: 53-55.

33. Kim M, Dubacq JP, Thomas JK, Giraud G (1996) Seasonal Variations of Triacylglycerols and Fatty Acids in Fucus serratus. Phytochem 43: 49-55.

34. Pohl P, Zurheide F (1979) Fatty Acids and Lipids of Marine Algae and the Control of Their Biosynthesis by Environmental Factor. Pharma Science, Walter de Gruyter, New York.

35. FAO/WHO (1973) Energy and protein requirements.

36. Shin TS, Xue Z, Do YW, Jeong SI, Woo HC, et al. (2011) Chemical Properties of Sea Tangle (*Saccharina. japonica*) Cultured in the Different Depths of Seawater. Clean Technol 17: 395-405.

37. Noda N (1993) Health benefits and nutritional properties of nori. J Appl Phycol 5: 255-258

38. Jung KJ, Jung CH, Pyeun JH, Choi YJ (2005) Changes of food components in Mesangi (*Capsosiphon fulvescens*), Gashiparae (Enteromorpha prolifera), and Cheonggak (*Codium fragile*) depending on harvest times. J Korean Soc Food Sci Nutr 34: 687- 693.

PERMISSIONS

LIST OF CONTRIBUTORS

Tessema A and Mohammed A
Department of Biology, Wollo University, Dessie, Ethiopia

Birhanu T
Tehulederie Wereda Office of Water Resource Development, Hayq, Ethiopia

Negu T
Kemissie Zonal Agriculture Office, Ehiopia

Soundarapandian P, Dinakaran GK and Varadharajan D
Centre of Advanced Study in Marine Biology, Faculty of Marine Sciences, Annamalai University, Parangipettai–608 502, Tamil Nadu, India

Carmona-Osalde Claudia and Miguel Rodriguez-Serna
National Autonomous University of Mexico (UNAM), Faculty of Science, Multidisciplinary Teaching and Research Unit, Sisal, Aquaculture Biotechnology Area, Mexico

Puerto-Novelo Enrique
Center for Research and Advanced Studies of the IPN (CINVESTAV-IPN) Unit MÉRIDA, Mexico

Chih-Chiu Yang and Kam-Chiu Lai
College of Life Science, National Taiwan University, Taipei, Taiwan

Shiu-Nan Chen, Chung-Lun Lu, Sherwin Chen and Wen-Liang Liao
Institute of Fisheries Science, National Taiwan University, Taipei, Taiwan

Bartholomew W Green
US Department of Agriculture, Agriculture Research Service, Harry K. Dupree Stuttgart National Aquaculture Research Center, Stuttgart, Arkansas USA

Kevin K Schrader
US Department of Agriculture, Agricultural Research Service, Natural Products Utilization Research Unit, Thad Cochran National Center for Natural Products Research, University, Mississippi, USA

Ogunlela AO and Adebayo AA
Department of Agricultural and Biosystems Engineering, University of Ilorin, Ilorin Nigeria

Patrick Saoud I, Ghanawi J and Nasser N
Department of Biology, American University of Beirut, Lebanon

Naamani S
Department of Biology, Beirut Arab University, Beirut, Lebanon

Haniffa MA
Centre for Aquaculture Research and Extension, St. Xavier's (Autonomous) College, Palayamkottai, 627002, Tamil Nadu, India

Meeran Mohideen
Centre for Aquaculture Research and Extension, St. Xavier's (Autonomous) College, Palayamkottai, 627002, Tamil Nadu, India
Institute for Research in Molecular Medicine, University Sains Malaysia, Pulau Penang, 11800, Malaysia

Quaiyum MA, Jahan R, Jahan N, Akhter T and Islam M Sadiqul
Department of Fisheries Biology & Genetics, Bangladesh Agricultural University, Mymensingh-2202, Bangladesh

Zainuddin, Haryati and Siti Aslamyah
Department of Fisheries, Faculty of Marine Science and Fisheries, Hasanuddin University, Sulawesi Selatan 90245, Indonesia

Islam MA, Biswas S, Rahman M, Uddin AMM, Asaduzzaman M, Rahman MS and Munira S
Department of Biochemistry and Molecular Biology, University of Rajshahi, Rajshahi-6205, Bangladesh

Asadujjaman M and Hossain MA
Department of Fisheries, University of Rajshahi, Rajshahi-6205, Bangladesh

Manirujjaman M
Department of Biochemistry, Gonoshasthaya Samaj Vittik Medical College and Hospital, Gono University, Savar, Dhaka-1344, Bangladesh

Mahsa Javadi Moosavi
M.Sc Graduated of Aquaculture, Gorgan University of Agricultural Science and Natural Resources, Faculty of Fishery and Environmental Science, Golestan, I.R. Iran

Vali-Allah Jafari Shamushaki
Assistance Professor, Department of Fisheries, Gorgan University of Agricultural Sciences and Natural Resources, Gorgan, Iran

El-Sayed G Khater and Samir A Ali
Agricultural Engineering Department–Faculty of Agriculture–Benha University 13736, Egypt

Adeniyi Bashir Tunde
Aquaculture and Fisheries Research Programme, Institute of Food Security, Environmental Resources and Agricultural Research (IFSERAR), Federal University of Agriculture, Abeokuta, Nigeria

Kuton MP
Department of Marine Sciences, University of Lagos, Akoka, Lagos, Nigeria

Ayegbokiki Adedayo Oladipo
Food Security and Socio-Economic Research Programme, Institute of Food Security, Environmental Resources and Agricultural Research (IFSERAR), Federal University of Agriculture, Abeokuta, Nigeria

Lawal Hakeem Olasunkanmi
Department of Agricultural Economics and Farm Management, Federal University of Agriculture, Abeokuta, Nigeria

Georgios Bellos and and Helen Miliou
Department of Applied Hydrobiology, Faculty of Animal Science and Aquaculture, Agricultural University of Athens, GR-11855 Athens, Greece

Panagiotis Angelidis
Laboratory of Ichthyology, Faculty of Veterinary Medicine, Aristotle University of Thessaloniki, GR 54124 Thessaloniki, Greece

El-Sayed G Khater and Samir A Ali
Agricultural Engineering Department–Faculty of Agriculture–Benha University 13736, Egypt

Orapint Jintasataporn and Ruangvit Yoonpundh
Faculty of Fisheries, Kasetsart University, 10900, Bangkok, Thailand

Siti-Ariza Aripin
Faculty of Fisheries, Kasetsart University, 10900, Bangkok, Thailand

School of Fisheries and Aquaculture Sciences, Universiti Malaysia Terengganu, 21030, Terengganu, Malaysia

Innifa Hasan
Department of Zoology, Handique Girls college Guwahati, Assam, India

Mrigendra Mohan Goswami
Department of Zoology, Gauhati University, Assam, India

Ferosekhan S, Sahoo SK, Giri SS, Saha A and Paramanik M
Central Institute of Freshwater Aquaculture, Kausalyaganga, Bhubaneswar-751 002, India

Malik M Khalafalla
Department of Aquaculture, Faculty of Aquatic and Fisheries Sciences, Kafrelsheikh University, 33516–Kafr El-sheikh, Egypt

Abd-elaziz M A El-Hais
Department of Animal Production, Faculty of Agriculture, Tanta University, Egypt

Natalia Ballesteros, Sara I Pérez-Prieto and Sylvia Rodríguez Saint-Jean
Centro de Investigaciones Biológicas (CSIC), C/ Ramiro de Maeztu 9, 28040 Madrid, Spain

Néstor Aguirre
Instituto de Física Fundamental (CSIC), C/ Serrano 123, 28006 Madrid (Spain)

Julio Coll
Instituto Nacional de Investigaciones Agrarias (INIA), Crta La Coruña km7, Madrid 28040, Spain

Kazi Belal Uddin, Sanjib Basak, Yahia Mahmud and Muhammad Zaher
Riverine Sub-Station (Lake Fisheries), Bangladesh Fisheries Research Institute (BFRI), Rangamati Hill District, Bangladesh

Mohammad Moniruzzaman
Riverine Sub-Station (Lake Fisheries), Bangladesh Fisheries Research Institute (BFRI), Rangamati Hill District, Bangladesh

Department of Marine Bio-Materials and Aquaculture, Feeds & Foods Nutrition Research Center, Pukyong National University, Busan, 608-737, Republic of Korea

Sungchul C Bai
Department of Marine Bio-Materials and Aquaculture, Feeds & Foods Nutrition Research Center, Pukyong National University, Busan, 608-737, Republic of Korea

Gabriel Marcos Domingues de Souza, Gislayne Trindade Vilas-Bôas and Laurival Antônio Vilas-Boas
Departamento de Biologia Geral, Universidade Estadual de Londrina, CP 10.011, CEP 86057.970, Londrina/PR, Brazil

Lucienne Garcia Pretto-Giordano and Ronaldos Tamanini
Departamento de Medicina Veterinária Preventiva, Universidade Estadual de Londrina, CP 10.011, CEP 86057.970, Londrina/PR, Brazil

Túlio Oliveira de Carvalho
Departamento de Matemática, Universidade Estadual de Londrina, CP 10.011, CEP 86057.970, Londrina/PR, Brazil

Ângela Teresa Silva-Souza
Programa de Pós-graduação em Ciências Biológicas, Universidade Estadual de Londrina, CP 10.011, CEP 86057.970, Londrina/PR, Brazil
Departamento de Biologia Animal e Vegetal, Universidade Estadual de Londrina, CP 10.011, CEP 86057.970, Londrina/PR, Brazil

Mauro Caetano Filho
Departamento de Biologia Animal e Vegetal, Universidade Estadual de Londrina, CP 10.011, CEP 86057.970, Londrina/PR, Brazil

Cipriano A, Burnell G, Culloty S and Long S
Aquaculture and Fisheries Development Centre School of Biological, Earth and Environmental Sciences University College Cork Distillery Fields North Mall Campus Cork, Ireland

Titik Budiati
Food Technology Department, State Polytechnic of Jember, 68121, Jember, Indonesia

Gulam Rusul
Food Technology Division, School of Industrial Technology, University of Science, Malaysia, 11800, Penang, Malaysia

Wan Nadiah Wan-Abdullah and Rosma Ahmad
Bioprocess Technology, School of Industrial Technology, University of Science, 11800, Penang, Malaysia

Yahya Mat Arip
School of Biological Science, University of Science, Malaysia, 11800, Penang Malaysia

Belal IEH
Ibrahim E H Belal Department of Aridland Agriculture, Faculty of Food and Agriculture, United Arab Emirates University, P. O Box 15551, Al-Ain, UAE

El-Tarabily KA
Department of Biology, Faculty of Science, United Arab Emirates University, PO Box 15551, Al-Ain, UAE

Kassab AA
Department of Anatomy and Embryology, Faculty of Veterinary Medicine (Moshtohor), Benha University, Egypt

El-Sayed AFM
Oceanography Department, Faculty of Science, Alexandria University, Alexandria, Egypt

Rasheed NM
Aquaculturist, MAHY Khoory, Aquaculture Centre, P. O. Box 11944, Dubai, UAE

Olaleye Ibukun Grace
Fisheries and Aquaculture unit, Institute of Oceanography, University of Calabar, Cross River State, Nigeria

Om AD
Fisheries Research Institute (FRI), Tanjong Demong, 22200 Besut, Terengganu, Malaysia

Sharif S and Jasmani S
Institute of Tropical Aquaculture (AQUATROP), University Malaysia Terengganu, 21030 Kuala Terengganu, Malaysia

Sung YY and Bolong AA
School of Fisheries and Aquaculture Science, University Malaysia Terengganu, 21030 Kuala Terengganu, Terengganu, Malaysia

Alexandra Grasteau, Thomas Guiraud and Michel Le Hénaff
Bordeaux University, CNRS UMR EPOC, Talence, France

Patrick Daniel
Laboratoire des Pyrénées et des Landes, Mont de Marsan, France

Ségolène Calvez
LUNAM University, Oniris, UMR INRA BioEpAR, Nantes, France

Valérie Chesneau
Groupement de Défense Sanitaire Aquacole d'Aquitaine, Mont de Marsan, France

Jyotirmayee Pradhan
Government College (Autonomous), Angul-759143, Odisha, India

Basanta Kumar Das
Central Institute of Freshwater Aquaculture (CIFA), Kausalyaganga, Bhubaneswar-751 002, Odisha, India

Christelle Leung and Bernard Angers
Group for Interuniversity Research in Limnology and
Aquatic Environment (GRIL)
Department of biological sciences, Université de
Montréal, C.P. 6128, Succursale Centre-Ville, Montreal,
Quebec, Canada H3C 3J7

Pierre Magnan
Group for Interuniversity Research in Limnology and
Aquatic Environment (GRIL)
Department of Chemistry-Biology, Université du
Québec à Trois-Rivières, 3351 Boulevard des Forges,
C.P. 500, Trois-Rivières, QC, Canada

Besay Ramírez, Fernando Tuya and Ricardo Haroun
Research Group on Biodiversity and Conservation,
Center for Biodiversity and Environmental
Management, University of Las Palmas de Gran
Canaria, Las Palmas 35017, Spain

Leonor Ortega
Research Group on Biodiversity and Conservation,
Center for Biodiversity and Environmental
Management, University of Las Palmas de Gran
Canaria, Las Palmas 35017, Spain
Philippe Cousteau "Union of the Ocean" Foundation,
C/ General Oraá 26, 28006 Madrid, Spain

Daniel Montero
Aquaculture Research Group, University of Las Palmas
de Gran Canaria, P.O. Box 56, 35200 Telde, Spain

**Fedorovykh JV, Ponomarev SV, Bakaneva JM,
Bakanev NM, Sergeeva JV, Bakhareva AA, Grozesku
JN and Egorova VI**
Astrakhan State Technical University (ASTU),
Department of Aquaculture and Water Bioresources,
414056, Rus-sian Federation, Astrakhan, Tatishev st,
16, Russia

El-Drawany
Department of Zoology, Faculty of Science, Zagazig
University, Egypt

Chatterjee NR, Sahoo D and Chetri C
Department of Aquaculture, West Bengal University
of Animal and Fishery Sciences, Belgachia,
Kolkata-700037, India

Jae-Ho Hwang and Sung-Ju Rha
College of Fisheries and Ocean Science, Chonnam
National University, Yosu 550-749, Korea

Nam-Gil Kim
Department Marine Biology and Aquaculture,
Gyeongsang National University, 445 Inpyeong-dong,
Tongyeong-si, Gyeongsangnam, 650-160, Korea

Hee-Chul Woo
Department of Chemical Engineering, Pukyong
National University, 365 Sinseon-ro, Yongdang-dong,
Nam-gu, Busan, 608-739, Korea

Seon-Jae Kim
Department of Marine Bio Food, Chonnam National
University, Yeosu 550-749, Korea

Tai-Sun Shin
Division of Food Nutrition Science, Chonnam National
University, Gwangju 500-757, Korea

Index